D1351333

RTC Limerick
3 9002 00003789 6

# SYMMETRY RULES FOR CHEMICAL REACTIONS

# SYMMETRY RULES FOR CHEMICAL REACTIONS

## Orbital Topology and Elementary Processes

RALPH G. PEARSON

**Northwestern University**
**Evanston, Illinois**

A WILEY-INTERSCIENCE PUBLICATION

JOHN WILEY AND SONS, New York • London • Sydney • Toronto

**Library of Congress Cataloging in Publication Data:**

Pearson, Ralph G.
Symmetry rules for chemical reactions.

"A Wiley-Interscience Publication."
Includes bibliographical references and index.
1. Chemical reaction, Conditions and laws of.
2. Molecular orbitals. 3. Symmetry (Physics)

I. Title.
QD501.P337 541'.39 76-10314
ISBN 0-471-01495-8

Printed in the United States of America

10 9 8 7 6 5 4 3 2 1

# PREFACE

This book is intended to be a detailed account of methods for deciding whether a given reaction path is likely to have a large activation energy (to be forbidden) or a small one (to be allowed). The methods used usually depend on the presence of one or more symmetry elements that persist in going from reactants to products by the specified reaction path. However rules are also given based on more general topological properties.

The results are similar to the symmetry-based selection rules found for spectroscopic transitions. When symmetry is present, the use of group theory allows quite rigorous statements to be made. The absence of symmetry weakens these statements but does not destroy them.

It is hoped that the book will be useful to chemists (and perhaps others) who are concerned with the mechanisms of chemical reactions. A molecular orbital (MO) approach is used throughout, since it seems that at present this is the most powerful way to discuss molecular structures, chemical bonding, changes in chemical bonds, and the excited electronic states of molecules. Except for brief reviews, it is assumed that the reader has had a previous introduction to MO theory, group theory, and spectroscopy.

In addition to MO symmetry properties, the orbital energies are also important. For this reason some space is devoted to both theoretical and experimental methods for determining these quantities.

The largest portion of the text presents a series of exercises in the application of these various topics. At the least it should serve the purpose of familiarizing readers with the shapes and properties of the valence-shell MOs of molecules more complex than the simple diatomics. One can learn how the molecular geometry, the kinds of chemical reactions, and the reaction mechanisms used derive from these MOs. The key (but not exclusive) role played by the frontier orbitals (the highest occupied MO, HOMO and the lowest unoccupied MO, LUMO) is stressed.

I apologize for using such units as kilocalories per mole and angströms rather than alternate units recommended by the IUPAC. This was at first done quite unconsciously, but since I, together with many other chemists, prefer these units, I have not changed them. The units for entropy, which occur in only two or three places, are calories per mole-degree.

I wish to express my thanks to many colleagues who gave generously of their expertise in diverse areas. Without this help I could not have attempted to cover the wide range of chemical reactions that exists. The errors and misconceptions that persist are entirely my own.

Thanks are due to the Chemistry Departments of the University of Hawaii and the California Institute of Technology for hospitality and refuge during the writing of this book.

RALPH G. PEARSON

*Evanston, Illinois*
*March 1976*

# CONTENTS

# SYMMETRY RULES FOR CHEMICAL REACTIONS

# CHAPTER 1

# SELECTION RULES FOR CHEMICAL REACTIONS

The word "symmetry" comes from the Greek *syn-metron*, to measure together. This implies that we consider two or more things to see symmetry. This can mean two or more things that are the same in a certain sense, such as the left- and right-hand sides of our bodies, or entities whose relative proportions are of interest, such as the size of the head related to that of the torso. As human beings we have a great interest in symmetry because our concepts of beauty are closely related to it. Unsymmetric things are considered ugly, unless we can see a deeper lying pattern of regularity, which again creates symmetry.

In physical science we are deeply dependent on symmetry. In the sense of order, pattern, and regularity, it is clear that it would be hopeless to try to understand Nature, unless such order existed. In fact, whenever we come to understand some phenomenon of Nature it is because we have perceived in some way the symmetry that is involved.

According to Heisenberg, "physicists learned from mathematicians that the symmetry of a problem as a rule produces a conservation law. All the conservation laws that we know in physics—the conservation of energy, momentum, angular momentum, etc.—rest upon fundamental symmetries in the underlying natural law."

For example, the symmetry of space, implied by the statement that all coordinate systems are equivalent, leads to conservation of momentum. Symmetry in time leads to conservation of energy. Conservation of angular momentum depends on the rotational homogeneity of space. In quantum-mechanical terms we find that any quantity represented by $A$ and obeying the equation $HA = AH$ will be a constant of the motion and be conserved. $A$ may be an operator, such as $(\partial/\partial x)$, or the result of an operation, such as the interchange of two particles. $H$ is the Hamiltonian operator. If $HA = AH$, we say that $A$ and $H$ commute, which is a type of symmetry.

The general experience in science has been that early recognition of as much symmetry as possible in any problem will lead to the quickest solution to the problem. Indeed in many cases, if we can recognize the full symmetry, an immediate but incomplete solution is possible. One of the great virtues of symmetry arguments is that the conclusions drawn from them are usually exact, that is, firm "yes" or "no" answers. This means, of course, a corresponding lack of detail in the answers.

## USE OF GROUP THEORY

Fortunately a mathematical tool has been developed that allows use of symmetry properties to be made in an exact and complete manner. This tool is group theory. The modern chemist usually receives a certain amount of training in this field. We will assume that the reader has had some exposure to group theory. Similarly we will assume a knowledge of elementary quantum mechanics and the fundamentals of MO theory. A list of useful references is appended to the end of the chapter, for those who need review.

In chemistry we might be concerned with the symmetry of packing of units in crystalline lattices. This is conveniently done by the use of space group theory. The development of x-ray crystallography as a routine method of obtaining the structures of solid substances would not have been possible without the full use of symmetry properties.

Ordinarily the chemist is more concerned with the symmetry of molecular shape, for example, the tetrahedral structure of the methane molecule. The mathematical tool to use here is point-group theory. The nuclei of the molecule are represented by a set of points, which are interchanged by the various symmetry operations. Each molecule belongs to one of the various point groups $C_{3v}$, $T_d$, $O_h$, and so on. Each of these is characterized by the number and kind of symmetry elements it possesses.

For each symmetry element (plane, axis, improper axis, and center), there is a corresponding operation. These operations form the group. Each group is further characterized by the number and kind of symmetry species associated with it. If a thing or function has any symmetry at all, it must be resolvable into these various symmetry species. Things or functions are called bases for representations of the group. The symmetry species are defined by certain irreducible representations.

There are several important, general conclusions that can be drawn by the application of point group theory to the properties of a molecule. One has to do with the symmetry properties of the wavefunction, $\Psi_i$, which is a solution of the Schrödinger equation,

$$H\Psi_i = E_i\Psi_i \tag{1}$$

where $H$ is the molecular Hamiltonian. A series of wavefunctions, or eigenfunctions, $\Psi_i$, exist. Correspondingly there are a series of energy values, or eigenvalues, $E_i$. $\Psi_0$ and $E_0$ would refer to the ground-state values.

Now it can be shown that the results of operating on the molecular wavefunction with symmetry operators of the group produce irreducible representations of the point group. The converse of this is Wigner's theorem, which states that all eigenfunctions of a molecular system belong to one of the symmetry species of the group

This important result applies to the exact wavefunctions of the molecule. However it immediately implies the restriction that satisfactory approximate wavefunctions should be constructed so that they also belong to the various symmetry species. In particular, linear combination of atomic orbitals (LCAO) MO theory is so restricted. Each MO, $\varphi_i$, satisfies the equation

$$h\varphi_i = \varepsilon_i \varphi_i \tag{2}$$

where $h$ is a one-electron operator, including an average interelectronic potential. Then each $\varphi_i$ must be a proper symmetry orbital.

Since each MO is a LCAO, the coefficients of the various AOs are strongly or even completely determined by symmetry. If one MO is known, others can often be projected out by the use of the symmetry operators. Of course, the larger the basis set of AOs used, the less completely symmetry alone defines the MOs. Also, the less symmetry the molecule has, the less completely the MOs are prescribed.

Certain approximate theories of chemical bonding are at the same time amazingly simple and amazingly successful. The best known are the Hückel theory of $\pi$ electrons in organic chemistry, and the crystal-field theory of $d$ electrons in inorganic chemistry. In both cases only a very few AOs are taken into consideration, and only molecules with some elements of symmetry are considered. It is clear that these theories are successful largely because of the extensive use of symmetry to define the properties of the MOs formed.

The total wavefunction of the molecule, $\Psi_i$, depends both on nuclear motions and electronic motions. According to the Born–Oppenheimer approximation, it is usually possible to separate the two kinds of motion. Then the wavefunction is a product of an electronic wavefunction $\psi(r, R)$ and a vibrational wavefunction, $\chi_{(R)}$.

$$\Psi_{in} = \psi_{i(r, R)}\chi_{n(R)} \tag{3}$$

The capital letter $R$ refers to nuclear positions and the lower case letter $r$, to electronic positions.

The symmetry of $\Psi$ is now the product of the symmetries of $\psi$ and $\chi$. The latter is always totally symmetric (invariant to all symmetry operations of the molecule) in the lowest vibrational state. Hence the appropriate symmetry is that of the electronic state given by $\psi_{(r, R)}$. This state is described by a wavefunction in which the nuclear positions are assumed fixed in space, while the electronic coordinates are rapidly varying. However $\psi$ depends parametrically on $R$.

The intensity of a spectroscopic transition depends on the transition moment

$$\langle P \rangle_{if} = \int \Psi_i P \Psi_f \, d\tau \tag{4}$$

where $P$ is the perturbation causing the transition between states $i$ and $f$. Usually $P$ is the dipole operator, whose symmetry is the same as that of the Cartesian coordinates $x$, $y$, and $z$. In Raman spectroscopy the quadrupole operator is important, having the symmetries of $R_x$, $R_y$, and $R_z$, where $R_x$ stands for rotation about the $x$ axis, and so on.

The integral in (4) is over the electronic and nuclear coordinates covering all space. This enables us to use a general result from group theory. In order to be different from zero, the integrand in (4) must be totally symmetric. If it were not, there would be a negative contribution from some portion of space canceling each positive contribution and the integral would be zero.

This means that the direct product of $\Psi_i$, $P$, and $\Psi_f$ must contain the totally symmetric representation. This is written as

$$\Gamma_{\Psi_i} \times \Gamma_{\Psi_f} \subset \Gamma_P \tag{5}$$

where $\Gamma$ stands for the symmetry species of each function and $\subset$ is read as "must contain." For the 0–0 transition (ground-vibrational states) it is seen that

$$\Gamma_{\psi_i} \times \Gamma_{\psi_f} \subset \Gamma_P \tag{6}$$

This is the usual selection rule for transitions from one electronic state to another. It should be mentioned that the two states must also have the same spin multiplicity for the transition to be allowed.

In MO theory, electronic transitions almost always involve one electron being promoted from an originally occupied to an empty (or half-empty) orbital. In that case (6) becomes

$$\Gamma_{\varphi_i} \times \Gamma_{\varphi_f} \subset \Gamma_P \tag{7}$$

Next we consider excitations between different vibrational levels of the same electronic state, that is, infrared and Raman spectroscopy. The selection rule is like (6) or (7) except that $\chi_i$ and $\chi_f$ appear. As indicated earlier, each vibrational wavefunction must belong to one of the symmetry species of the point group of the molecule. In particular the first excited vibrational level ($v = 1$) has the symmetry of the nuclear motions corresponding to the vibration itself.

The properties of the various vibrational modes of a molecule are important for our purposes. If we consider all possible motions of all the nuclei of a molecule, they can always be resolved into translation of the entire molecule (3 degrees of freedom), rotation of the molecule about its center of mass (3 or 2 degrees of freedom), and internal movements of the nuclei with respect to each other ($3n - 5$ or $-6$ degrees of freedom, where $n$ is the number of atoms in the molecule). In the case of a molecule in its equilibrium configuration, these internal motions correspond to the vibrational modes of the molecule. Each type of motion can be rigorously classified as belonging to one of the symmetry species of the point group to which the molecule belongs.

The translational and rotational motions of the molecules are of little further interest. They correspond to fixed relative nuclear positions, and do not correspond to any change in the potential energy $E$ or wavefunction $\Psi$ of the system. The vibrational motions of the molecule, however, correspond to changes in these quantities. If continued, furthermore, the vibrations give changing nuclear positions that correspond to the possible chemical reactions of the molecule.

Given the composition of the molecule and its geometry, the number of vibrations of each symmetry type is completely fixed. However it is a difficult task to determine exactly what nuclear motions correspond to each vibration. These nuclear motions are along normal coordinates, and they depend on the (unknown) potential energy function of the molecule as well as on its makeup. What are defined, however, are symmetry coordinates. These are simply symmetry-adapted linear combinations (SALC) of various nuclear displacements. They can be selected in more than one way. A common procedure is to take SALCs of bond lengths of equivalent bonds, or bond angles of equivalent bonds.

Symmetry coordinates of the same species can then be mixed together by the potential field of the molecule to form the normal coordinates. However this can only happen if the potential corresponds, at least approximately, to that for a harmonic oscillator for each normal coordinate. This means that the energy must be at a minimum for a zero value of the coordinate. In the case of an arbitrary arrangement of nuclei, this condition will not always be met. Then the symmetry coordinates exist as a complete set, but the normal coordinates do not.

Whether we consider motions of the nuclei corresponding to changes in symmetry coordinates or to the actual vibrations of the molecule is of minor consequence. The useful conclusion is that all possible nuclear motions can be resolved into sets of motions, each with a definite symmetry label. If we define a reaction coordinate as a set of changing nuclear positions leading to an isomerization or decomposition of the molecule, then the reaction coordinate will also be of a definite symmetry. In the usual case it will be of a single species. In rare cases it may be a mixture of two species of different symmetry.

The above arguments are also valid if two or more molecules in interaction with each other are considered. A supermolecule is formed that belongs to a definite point group and whose nuclear motions are completely symmetry prescribed. Of course the point group of any molecule or pseudomolecule is only defined at certain values of the internuclear coordinates. A slight motion can change one point group to another. However we can follow these changes point by point. Also quite large changes in bond angles and bond distances can sometimes occur without changing the point group. A totally symmetric vibrational mode has the property of not changing the point group, except at well defined limits.

## POTENTIAL ENERGY SURFACES

The total energy $E$ of a molecule consists of the potential and kinetic energies of the electrons and nuclei. In addition there are small magnetic energies. The electronic energy and the coulombic energy of the nuclei represent a "potential" energy under whose influence the nuclei carry out their vibrations. This potential energy must be represented by a $3n - 6$ (or $3n - 5$ for a linear molecule) dimensional hypersurface in a $3n - 5$ or $(3n - 4)$ dimensional space. Each electronic state of a molecule is characterized by its own hypersurface.

Usually we are concerned only with the lowest surface corresponding to the ground state. If we now move about the nuclei of a molecule (or several molecules) we encounter a number of energy minima. These correspond to the several sets of isomers or reaction products that the original molecule can form. Sometimes such an energy minimum may correspond to a hole in the hypersurface, with all internuclear coordinates having fixed values. More commonly the minima exist in long valleys stretching out to infinity. This is because some atoms are already so far away that increasing their separation has no effect on the energy.

The subject of reaction mechanisms has to do primarily with discovering how the nuclei can pass from one valley into another most rapidly. Two adjacent valleys are separated from each other by a mountain pass. The top of the pass is called a "saddle point" or "col." A collection of nuclei at this point, or very near it, is called an "activated complex." The activated complex is said to be in the transition state (TS), in the state of being converted from reactants to products.

The saddle point usually has an energy well above the energy of the valley floors. This difference in energy is the barrier that must be overcome for the forward and reverse reactions. The activated complex must possess at least this amount of extra energy, with due allowance for the zero-point vibrational energy of both reactants and activated complex.

The concept of a chemical reaction occurring by passage over a barrier in the potential energy hypersurface goes back to 1915 and work by Marcelin. Other workers, including Evans and Polanyi, Pelzer and Wigner, and London, made important early contributions. However the development of the TS theory, as it is called, is usually attributed to Eyring. Chapter III of *The Theory of Rate Processes* is still the most complete discussion of potential energy surfaces and rates available.[1]

A process in which the reactant molecule or supermolecule passes over a single saddle point is called an "elementary reaction." A reaction mechanism may consist of a single elementary process, or a sequence of such processes. A single-step mechanism is also called a "concerted reaction."

Now three factors determine the rate of an elementary reaction

(a) the height of the energy barrier,
(b) the probability or entropy of the activated complex,
(c) dynamic effects.

Experimentally the rate is characterized by the rate constant, $k$, which may be of any order, and which usually obeys the Arrhenius relation

$$k = Ae^{-E_a/RT} \tag{8}$$

The frequency factor $A$ is determined by (b) and (c) above, and the Arrhenius activation energy $E_a$, is closely related (but not equal) to the height of the energy barrier.

The actual path or trajectory of a system in undergoing an elementary process is difficult to visualize in the multidimensional space. However we can simplify the problem greatly by putting restrictions on the reaction process. For example, Fig. 1 shows the potential energy plotted as contour lines for the reaction

$$\mathrm{H + ClH \longrightarrow HCl + H} \tag{9}$$

where the further restraint is made that the three atoms are colinear. In this case there are only two internuclear coordinates, $R_{\mathrm{HCl}}$ and $R_{\mathrm{ClH}}$.

A special reaction trajectory is the dashed line shown in Fig. 1, going from the floor of one valley to the other and passing over the saddle point. This is called the "reaction coordinate" and is defined as the minimum energy path going from the ground-state reactants to the activated complex. The activated complex is the linear symmetric molecule Cl—H—Cl. Such a molecule, when stable, has four vibrational modes. There are two bending modes, which are degenerate and of $\Pi_u$ species, a symmetric stretch of $\Sigma_g$ symmetry, and an asymmetric stretch of $\Sigma_u$ species. The point group is $D_{\infty h}$.

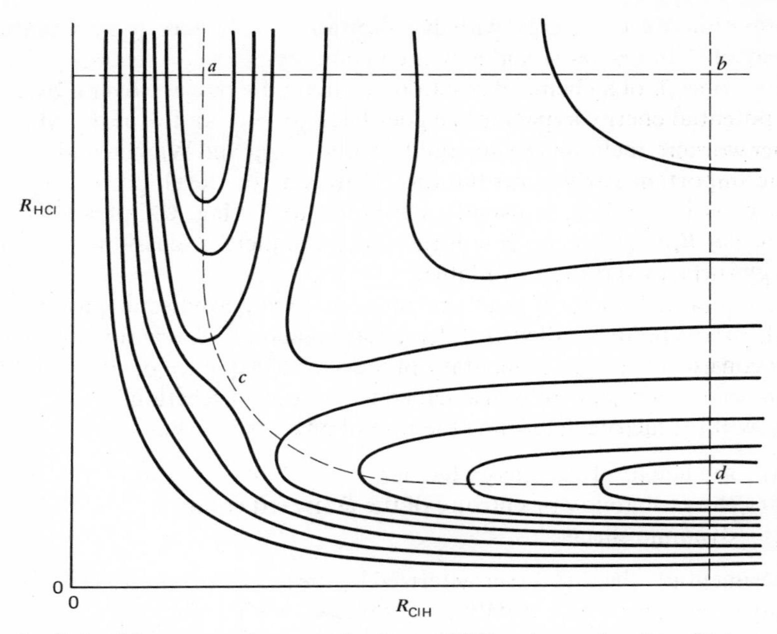


**Figure 1.** Potential energy contour map for linear HClH system as function of internuclear distances; $a$ and $d$ are potential energy minima; $b$ is a maximum and $c$, a saddle point.

It can be seen in Fig. 1 that the reaction coordinate corresponds to the asymmetric stretch of a stable molecule but it differs in that the motion would not be harmonic, because we are on a potential energy maximum rather than a minimum. The symmetric stretch is seen as a motion orthogonal to the reaction coordinate. This is a true normal mode since it exists at an energy minimum. The two bending modes also exist at minima, by virtue of our initial assumption. They cannot be shown in our three-dimensional figure. However Fig. 2 shows a potential energy surface for a linear symmetric $XY_2$ molecule as a function of both the bending mode and the symmetric stretch. The reaction coordinate at all points other than the saddle point also corresponds to a vibration of the unsymmetrical molecule H—ClH. The point group is now $C_{\infty v}$, and the reaction coordinate is of $\Sigma$ symmetry.

Figure 3 shows a further simplification that can be made even for complicated elementary reactions. The potential energy is plotted as a function of the reaction coordinate. Usually the reaction coordinate is a rather complex mixture of changing nuclear positions. It starts with the reactants, at their equilibrium geometries and finishes with the products, also in their equilibrium positions. Intervening is the activated complex and the corresponding

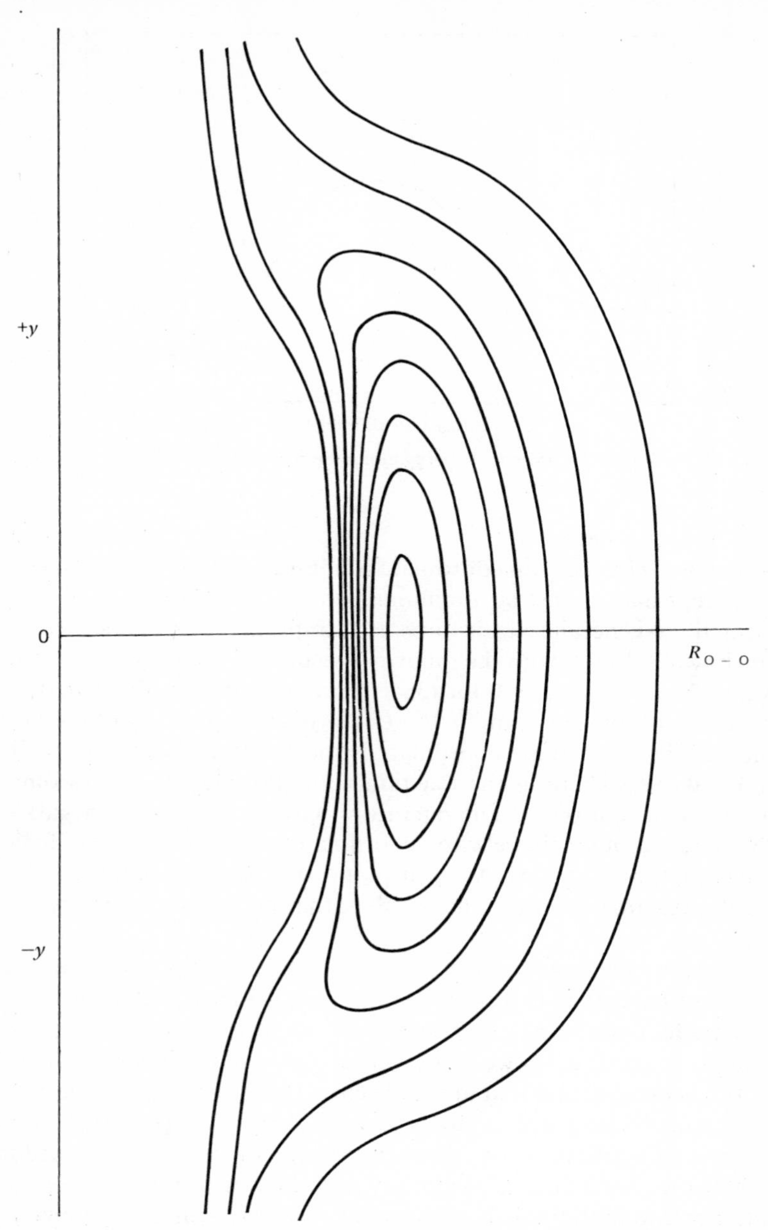


**Figure 2.** Potential surface for linear, symmetric molecule such as $CO_2$; abscissa is O—O distance, and ordinate, $y$, is perpendicular distance of C from the O—O line. (From G. Herzberg, *Molecular Spectra and Molecular Structure*, Vol. 3 Litton Educational Publishing, 1966. Reprinted by permission of Van Nostrand Reinhold Company.)

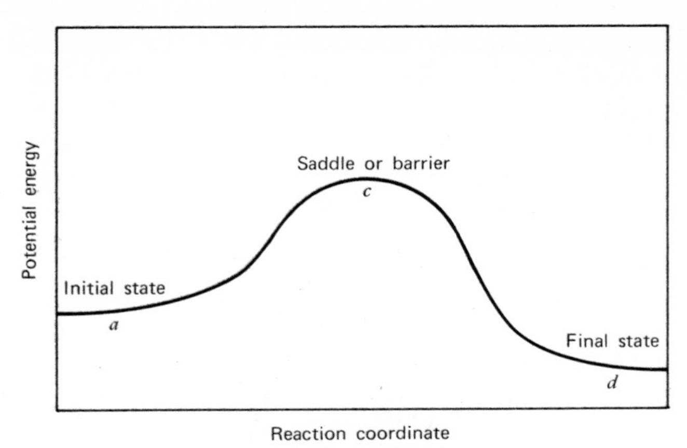


**Figure 3.** Potential energy along the reaction coordinate.

energy barrier. The reaction coordinate is not well defined in the intervening region, except that it must be a minimum energy path.

We can try to generalize the features of a simple and well-defined case, such as that shown in Fig. 1, with the following results.[2] The points at the bottom of a valley have $(\partial E/\partial Q_i) = 0$ for $(3n - 7)$ coordinates (in the general non-linear case), with the curvature $(\partial^2 E/\partial Q_i^2) > 0$. Here $Q_i$ is any of the normal coordinates. The remaining coordinate is the reaction coordinate $Q$. It has $(\partial E/\partial Q) > 0$ everywhere in the reaction zone (far ends of valleys omitted) except at the saddle point. The curvature may be positive or negative. In other words, except for the reaction coordinate, all other degrees of freedom have their optimum values. At each point on Fig. 3 all bond angles and bond distances are assumed to have values that minimize the energy, except along $Q$.

One difficulty emerges from the foregoing detailed description of the reaction coordinate. It is highly restrictive, and we may well wish to probe reaction paths that do not correspond to all of these special conditions. Fortunately in these cases we can use Fig. 3 as a much less restricted definition. It has two features that are sufficient: (1) it is a minimum energy path from reactants to activated complex, and (2) it has $(\partial E/\partial Q) \neq 0$. As we will show, this is sufficient to fix the symmetry species of the reaction coordinate, which will then be a sum of symmetry coordinates of that species. Other possible nuclear motions will correspond to the remaining symmetry coordinates. Some will be true vibrational modes and some will not, depending on whether $(\partial E/\partial Q_i)$ is zero. Any arbitrary reaction path will contain components of all the symmetry coordinates. However only some components will contribute to the reaction coordinate. The other motions will be orthogonal to the reaction coordinate and contribute nothing to it.

The activated complex is usually defined in the same way for the restricted case,[2] and the general case: $(\partial E/\partial Q_i) = 0$ for all normal coordinates, including the reaction coordinate. However, $(\partial^2 E/\partial Q_i^2)$ will be positive for $(3n - 7)$ coordinates and $(\partial^2 E/\partial Q^2)$ will be negative for the reaction coordinate. Thus the reaction coordinate is always a normal mode for the activated complex, according to this definition. Again this is more restrictive than necessary, and we will only assume that the reaction coordinate is a sum of symmetry coordinates, usually of a single species. The same remarks apply to the reaction coordinates at the bottom of a potential well, except that here the normal coordinates all exist.

Note that there can be only one reaction coordinate for each activated complex. A saddle point can connect only two valleys. Figure 4 clarifies this point. We imagine an unstable species such as equilateral $H_3$, which can decompose in three equivalent ways into $(H_2 + H)$. Figure 4 shows that symmetric $H_3$ cannot be an activated complex since a lower energy path

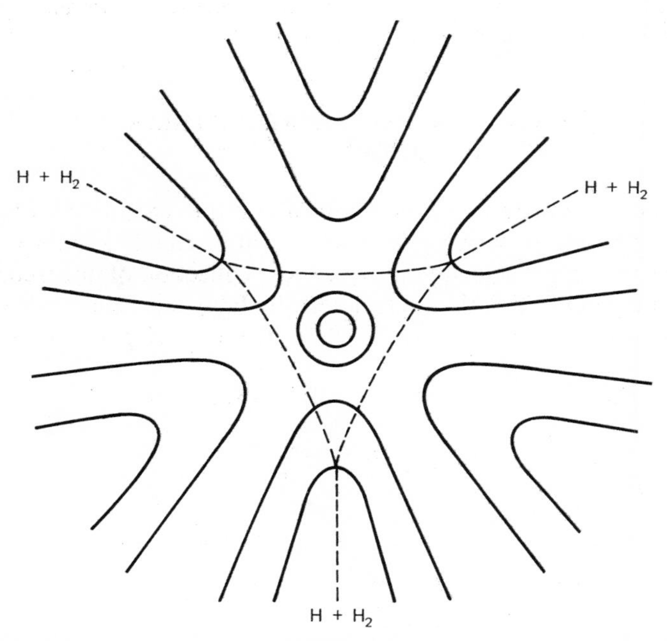


**Figure 4.** Schematic plot of potential surface for exchange reaction $H + H_2 \rightarrow H_2 + H$; the central point is symmetric $H_3$, which is at a maximum in energy; the dashed lines are the reaction coordinates actually followed.

always exists for passage from one valley into another. However the true activated complex may be only slightly distorted $H_3$, which can then serve as a model for it.

It may also be seen that the vibration of an equilateral $H_3$ molecule that leads to ($H_2$ + H) is a degenerate mode of $E'$ symmetry.

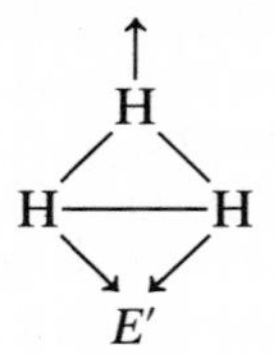


It is generally true that degenerate vibrational modes correspond to more than two equivalent ways of dissociating or rearranging a molecule. Degenerate modes exist only for highly symmetric molecules for which such possibilities exist. Therefore we conclude that, for the activated complex at least, the reaction coordinate cannot be degenerate. Two other symmetry restrictions on the reaction coordinate may be mentioned:[2b] $Q$ must be symmetric with respect to a symmetry operation of the TS that leaves reactants and products unchanged. $Q$ must be antisymmetric with respect to a symmetry operation that interchanges products and reactants.

## APPLICATION OF PERTURBATION THEORY: THE JAHN–TELLER EFFECT

We have not yet drawn any conclusions about the reaction coordinate that are very restrictive, or which give much useful mechanistic information. To do so we will apply perturbation theory to a collection of interacting nuclei as represented in Fig. 5. This was first done by Jahn and Teller,[3] who deduced the first-order Jahn–Teller (FOJT) effect. The next step was by Öpik and

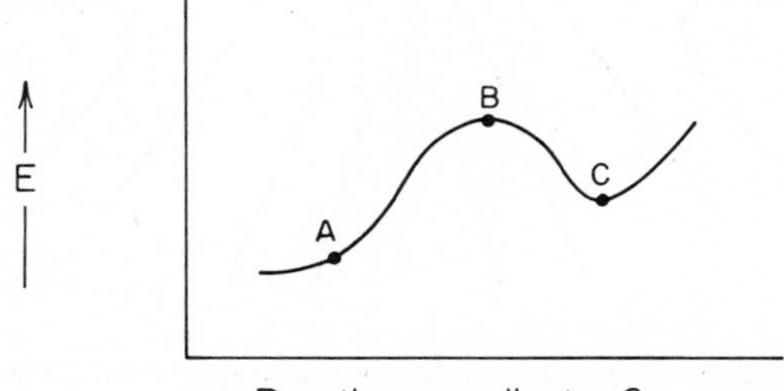


**Figure 5.** Potential energy plotted against reaction coordinate, $Q_0$. $A$, $B$, and $C$ are points at which perturbation theory (PT) might be applied.

Pryce,[4] who defined the second-order Jahn–Teller (SOJT) effect. The important step of trying to predict reaction mechanisms from such a treatment was made by Bader.[5]

Before proceeding it will be helpful to state clearly just what it is that we are trying to do. The overall goal is to make predictions about the probability or improbability of various reaction mechanisms. This will be done by trying to decide something about the height of the energy barrier in Fig. 5. While other factors can influence reaction rates, the part played by the activation energy is usually overriding, because of the exponential dependence of the rate constant on $E_a$.

If we can formulate rules for estimating that the energy barrier is large or small for various reaction paths (i.e., various reaction coordinates), this will be a great step forward in eliminating certain mechanisms and retaining others. Our procedure will not be to make any calculations of energy barriers, but to see what symmetry alone has to say about the probable shape of Fig. 5 and others like it.

We need to have an expression relating $E$ to $Q$ that involves symmetry-dependent terms. We pick an initial point $Q_0$, which may be point $A$ on Fig. 5. The wave equation is assumed to be solved giving rise to a number of eigenvalues $E_0$, $E_1 \cdots E_k$, and the corresponding electronic eigenstates $\psi_0$, $\psi_1 \cdots \psi_k$. We now move the nuclei a small distance $Q$ along the reaction coordinate. The new ground-state energy is now calculated by second-order perturbation theory (PT).

After distortion the Hamiltonian may be written as a Taylor–Maclaurin series

$$H = H_0 + \left(\frac{\partial U}{\partial Q}\right)Q + \frac{1}{2}\left(\frac{\partial^2 U}{\partial Q^2}\right)Q^2 \cdots \tag{10}$$

where $H_0$ is the original Hamiltonian, $U$ is the nuclear–nuclear and nuclear–electronic potential energy, and $Q$ is the small displacement along the reaction coordinate. We truncate the series at $Q^2$. The Hamiltonian of a system must be totally symmetric, that is, invariant to all the symmetry operations of the system. Therefore $(\partial U/\partial Q)$ must have the same symmetry as $Q$, since the product of one function with another of the same symmetry always contains the totally symmetric representation. For the same reason $(\partial^2 U/\partial Q^2)$ must be totally symmetric, since $Q^2$ is.

The energy becomes

$$E = E_0 + \left\langle \psi_0 \left| \frac{\partial U}{\partial Q} \right| \psi_0 \right\rangle Q + \left\langle \psi_0 \left| \frac{\partial^2 U}{\partial Q^2} \right| \psi_0 \right\rangle \frac{Q^2}{2} + \sum_k \frac{\left[\left\langle \psi_0 \left| \frac{\partial U}{\partial Q} \right| \psi_k \right\rangle Q\right]^2}{(E_0 - E_k)} \tag{11}$$

$E_0$ is the original energy at point $Q_0$, the next two terms are the first-order perturbation energy, and the last term is the second order perturbation energy.[6] The bracket symbol indicates integration over the electronic coordinates only.

At the same time the wavefunction changes from $\psi_0$ by mixing in the various excited states to various degrees

$$\psi = \psi_0 + \sum_k \frac{\left\langle \psi_0 \left| \frac{\partial U}{\partial Q} \right| \psi_k \right\rangle}{(E_0 - E_k)} \psi_k \tag{12}$$

This new wavefunction has not yet been normalized.

The first-order perturbation energy simply averages the effect of changing the nuclear positions over the original electronic distribution. Recall that $\psi^2$ is an electron-density function. The second-order perturbation energy is an effect due to a change in the electron distribution. Clearly this lowers the energy, as can be seen in (11). The quantity $(E_0 - E_k)$ always has a negative value. Both $E_0$ and $E_k$ are negative, for bound states, with $E_0$ having the greater magnitude.

Note that the kinetic energy operator for the electrons and the inter-electronic repulsion energy are included in $H_0$ of (10). Thus they do not change in the first order. Eventually, for larger $Q$, they will change as a result of changes in the wavefunction as given by (12).

All of the terms in (11) and (12) have symmetry labels attached to them. The rule that integration over all space can only be nonzero if the integrand is totally symmetric can now be invoked. We first consider the term in (11) linear in $Q$. There are two possibilities for $\psi_0$; it may be either degenerate or nondegenerate. If it is degenerate, then its direct product with itself (symmetric direct product) always contains at least one symmetry species besides the totally symmetric one.

Except for linear molecules, as Jahn and Teller showed,[3] there will always be at least one nonsymmetric vibrational mode of the molecule that has the same symmetry as $\psi_0^2$. Hence distortion along this mode will always lower the energy for $Q$ of one sign. The effect of this distortion will be to destroy the original symmetry and also the degeneracy in $\psi_0$. This is the FOJT effect, which states that (except for linear molecules),[27] degenerate electronic states cannot exist since they cause structural instability. Even for linear molecules a related Renner–Teller effect may remove the degeneracy.[7]

We assume then that $\psi_0$ is nondegenerate. This means that $(\partial U/\partial Q)$, and hence $Q$ also, must be totally symmetric for the integral $\langle\psi_0|\partial U/\partial Q|\psi_0\rangle$ to have a nonzero value. Now it is clear that for every point on Fig. 5, *except* at maxima or minima, that this integral does have a nonzero value. It is simply the slope of the plot of $E$ against $Q$. We conclude that except at maxima or minima, *all* reaction coordinates belong to the totally symmetric representation.[8]

Alternatively we can consider that $Q$ is an entirely arbitrary movement of the nuclei. We can now resolve it into a sum of either normal coordinates or symmetry coordinates. Only those components that are totally symmetric can contribute to a nonzero value for $(\partial E/\partial Q)$, but the one certain characteristic of the reaction coordinate is that this slope is nonzero. Therefore only symmetric nuclear movements can contribute to the reaction coordinate. This is a necessary, but not sufficient condition. Some symmetric motions will *not* contribute to the reaction coordinate. Except at potential maxima or minima, nonsymmetric modes *never* contribute.

We next consider the terms in (11) that are quadratic in $Q$. The first of these terms is always nonzero, since $(\partial^2 U/\partial Q^2)$ is totally symmetric. Furthermore it is always a term that increases the energy. It corresponds to a restoring force that tends to bring the nuclei back to $Q_0$. It stems from the circumstance that $\psi_0^2$ was the optimum electron distribution for the original nuclear configuration.

The last term in (11) is a negative, or energy-decreasing term. The integral $\langle\psi_0|\partial U/\partial Q|\psi_k\rangle$ will be nonzero only if the condition is fulfilled

$$\Gamma_{\psi_0} \times \Gamma_{\psi_k} \subset \Gamma_Q \tag{13}$$

that is only those excited states whose symmetry matches correctly with those of the ground state and the reaction coordinate can be effective. This is most critical, since from (12) we see that it is necessary for mixing of states to occur to change the wavefunction. If $\psi_0$ remains little changed, there is no way that we can go from one nuclear configuration to another that is quite different without a horrendous increase in energy.

If we are on the rising or falling parts of Fig. 5, then the reaction coordinate is totally symmetric. This means that only excited states, $\psi_k$, of the system which have the *same* symmetry as $\psi_0$ can mix in. On maximum or minimum points we still have the restriction of (13), but $Q$ can be of any symmetry, except that $Q$ cannot be degenerate for maximum points (activated complexes).

There is an interesting connection between the requirements that $Q$ can be of any symmetry for maxima and minima, and that $Q$ must be totally symmetric elsewhere. Only a minute distortion removes a system from the maximum or minimum point. The same nuclear motions that were nonsymmetric at the extremum, must now become symmetric. Obviously this can only happen because the point group changes, and there is a reduction in symmetry This is a characteristic of a nonsymmetric vibrational mode. It always lowers the symmetry and changes the point group.

We can easily discover what new point group is formed from an original one for each nonsymmetric mode. All we need do is look at tables for reduction in symmetry. The new point group is one in which the original mode becomes totally symmetric.

Figure 6 shows the reduction in symmetry in going from a linear symmetric molecule, $XY_2$, of $D_{\infty h}$ point group, to a bent molecule of $C_{2v}$ symmetry. The mode which produces the change is the $\Pi_u$ bending mode. It can be seen that this becomes an $A_1$ bending mode in the bent molecule. The $\Sigma_g$ and $\Sigma_u$ stretches become $A_1$ and $B_2$ modes, respectively. The other component of $\Pi_u$ becomes a rotation of the molecule, of $B_1$ symmetry. A linear molecule has only two moments of inertia compared to the three moments of other molecules. Hence there is an extra vibrational mode in the linear case.

Any nuclear movement which results in the creation of a new element of symmetry must always put the potential energy at either a maximum or minimum value. The existence of minimum values is, of course, the reason why we find symmetry in molecules in their normal configurations. A molecule $XY_4$ in which the nuclei are arranged in a planar form of $D_{4h}$ symmetry is either at a maximum or minimum energy in that configuration.[8]

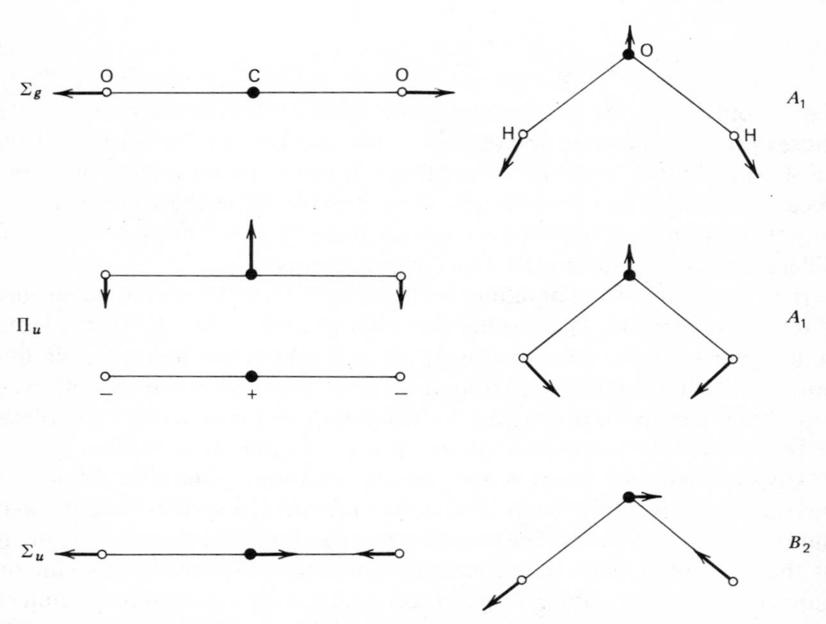


**Figure 6.** Vibrations of a linear, symmetric molecule, such as $CO_2$, in $D_{\infty h}$ point group and in $C_{2v}$ point group (bent).

Since the potential energy must always be an extremum for all the non-symmetric motions, we see that these will usually be normal modes. That is, $(\partial E/\partial Q_i) = 0$ and $(\partial^2 E/\partial Q_i^2) > 0$. Occasionally the curvature will go to zero as the nuclei move along a reaction coordinate, and then become negative. This is the mechanism whereby the point group of a reacting system can change en route to the TS. Otherwise the point group must remain the same, since the totally symmetric reaction coordinate cannot change it.

## THE SECOND-ORDER JAHN–TELLER EFFECT

In order to go further toward the goal of predicting reaction mechanisms, it is necessary to make more assumptions. Let us examine the energy-decreasing terms of (11), those due to the mixing in of excited states. Intuitively we expect the lower-lying excited states of the correct symmetry to make the greatest contribution to the sum over all excited states. This can be verified by actual calculation in some cases.[5] The important feature is not in the denominator, $(E_0 - E_k)$, which also favors the lower-lying states, but in the integral $\langle \psi_0 | \partial U/\partial Q | \psi_k \rangle$. This integral is very much like an overlap integral. Two states which correspond to quite different energies will have a small value of the integral, even if symmetry does not make the integral zero.

While it is always dangerous to ignore the sum of an infinite number of terms, each of which may be very small, there is one case where it may be permissible. That is when one excited state lies very close to the ground state. In such a case we can predict that something very similar to the Jahn–Teller effect will occur.[4] A distortion will result that pushes the two states further apart. The symmetry of the distortion is predicted by (13). This is the second-order, or pseudo, Jahn–Teller (SOJT) effect.

Early attempts to find such distortions were not very successful.[9] For that matter, attempts to find distortions due to the FOJT effect have often been unsuccessful. What does seem to happen in both instances is that structures predicted to be unstable are not in fact found. The observed structures are not slightly distorted, however, but so different that their relationship to the hypothetical structures is obscure.

It thus is quite plausible that a nuclear configuration which is truly unstable, an activated complex, may be subject to the SOJT effect. We can hypothesize that every activated complex is characterized by at least one low-lying excited state, whose symmetry, together with that of the ground state, determines the mode of decomposition. Unfortunately this hypothesis is not subject to experimental test because excited states of activated complexes cannot be directly observed.

Before resorting to further dependence on theory, we can make and test directly another assumption: namely, that the easiest mode of reaction or decomposition of a normal molecule is governed by the lowest-lying excited state, according to (14). This is the assumption made originally by Bader.[5] We now start at the bottom of a potential well, such as point $C$ in Fig. 5. The reaction coordinate can be of any symmetry.

Consider the ozone molecule, which easily decomposes into an oxygen molecule and an oxygen atom

$$O_3 \longrightarrow O_2 + O(^3P) \tag{14}$$

The activation energy is about 24 kcal. The ozone molecule is bent and of $C_{2v}$ symmetry. We will use this point group in many of our examples since it contains enough symmetry elements to be useful, and not so many as to be confusing. Figure 7 shows the four symmetry species of the $C_{2v}$ point group. There are no degeneracies.

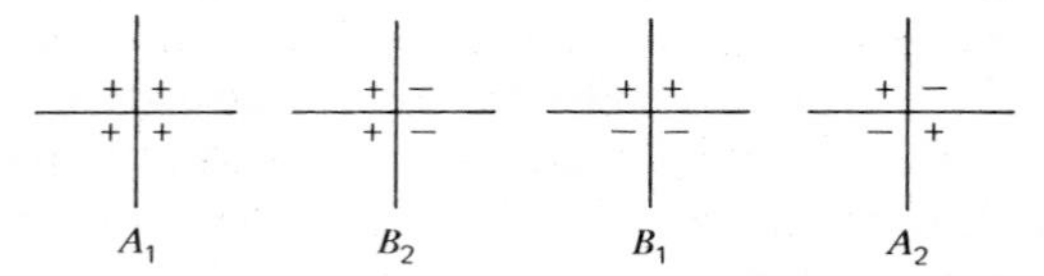


**Figure 7.** The four symmetry species of the $C_{2v}$ point group; two mirror planes are shown; the twofold $z$ axis is perpendicular to their intersection [the convention used for $B_1$ and $B_2$ is that recommended by R. S. Mulliken, *J. Chem. Phys.*, **23**, 1997 (1955)], the $x$ axis is vertical.

The kind of dissociation of a bent, triatomic molecule shown in (14) clearly corresponds to an initial reaction coordinate of $B_2$ symmetry. This is the asymmetric stretch. The ground state of $O_3$, which is a diamagnetic, closed-shell molecule, is $^1A_1$. Therefore the low-lying excited state needed must be $^1B_2$. Unfortunately the lowest excited singlet state of $O_3$ is $^1B_1$ or $^1A_2$. The $^1B_2$ state lies higher.

There is no vibration of $O_3$ of $B_1$ or $A_2$ type. These symmetries correspond to rotations of the molecule. Therefore the first excited states of $O_3$ are irrelevant to its decomposition and the theory may still be applicable. It can be tested by extension to a reaction, very closely

$$SO_2 \longrightarrow SO + O(^3P) \tag{15}$$

related, which requires very much more energy than reaction (14), about 130 kcal. Again the first excited states are probably $^1B_1$ and $^1A_2$.

More important is the observation that the relevant excited state of $O_3$ lies lower (2.2 eV for $^3B_2$ and 4.7 eV for $^1B_2$) than that of $SO_2$ (3.7 eV for $^3B_2$

and 5.3 eV for $^1B_2$).[10] This is consistent with the greater reactivity of the ozone molecule. Since all excited states of $O_3$ seem to lie lower than their counterparts in $SO_2$, we can correlate the blue color of the ozone molecule with its reactivity, whereas $SO_2$ is colorless.

Indeed we usually find that colored molecules are more reactive, or less stable, than their colorless counterparts. The usual chromophoric groups in organic chemistry are unsaturated, such as C=C, C=O, C≡N, and $NO_2$. They are also centers of reactivity. Free radicals such as $NO_2$, which is brown, and $NO_3$, which is blue, are usually colored and reactive. Their stable counterparts, such as $NO_2^-$ and $NO_3^-$, are colorless and inert. The methyl radical has its absorption bands strongly red-shifted compared to $CH_4$. Such colors and shifts are evidence for low-lying excited states.

Table 1 has some data for a series of diatomic molecules. The maximum of the first absorption band, the dissociation energy, and the stretching force constant are all given. For the alkali molecules and their hydrides, and for $H_2$, the first excited state is of $^1\Sigma_u$ type, which is the wrong symmetry to match with the ground state, $^1\Sigma_g$, and the reaction coordinate, which must be $\Sigma_g$. In other cases the excited states have not been identified. Nevertheless, the table shows considerable, if not perfect, correspondence between the energy of the first absorption band and both the dissociation energy and the force constant.

### Force Constants

The force constant is included because (11) is much more closely related to the force constant, $f$, than to the dissociation energy. We start with the molecule at its equilibrium position. Hence the term linear in $Q$ is zero. Equation (11) may be rewritten as

$$E = E_0 + \tfrac{1}{2}f_{00}Q^2 + \tfrac{1}{2}f_{0k}Q^2 \tag{16}$$

The sum of $f_{00}$ and $f_{0k}$ is then the experimental force constant. The first term is the classical force constant. It gives the restoring force due to the original electron distribution and is always positive. The second term is always negative and gives a reduction in the restoring force due to the rearrangement of the electron density. It has been defined by Salem as the "relaxability" of the molecule along the coordinate $Q$.[11]

We see that the theory is in fact predicting force constants rather than dissociation energies, when applied to stable molecules. Now it has often been supposed that the force constant for a bond and its dissociation energy are closely related. Table 2 shows a number of approximate values for these two properties for a number of single bonds. There is indeed a strong correlation; as $f$ increases, so does $D_0$. Also for double and triple bonds, we know that

**Table 1 Some Properties of Diatomic Molecules**[a]

| Molecule | $\nu_{max}$, $cm^{-1}$ | $D_0$, eV | $f$, md/Å |
|---|---|---|---|
| $H_2$ | 91,690 | 4.48 | 5.76 |
| $Li_2$ | 14,070 | 1.03 | 0.25 |
| $Na_2$ | 14,680 | 0.73 | 0.17 |
| $K_2$ | 11,680 | 0.51 | 0.10 |
| $Na_2$ | 11,000 | 0.89 | 0.08 |
| $Cs_2$ | 10,000 | 0.45 | 0.07 |
| LiH | 23,870 | 2.52 | 1.03 |
| NaH | 23,640 | 2.04 | 0.78 |
| RbF | 46,510 | 5.17 | 1.36 |
| RbCl | 40,320 | 4.39 | 0.77 |
| RbBr | 35,710 | 3.91 | 0.67 |
| RbI | 30,860 | 3.34 | 0.49 |
| RbH | 18,870 | 1.69 | 0.51 |
| NaBr[b] | 32,370 | 3.82 | 1.04 |
| NaI[b] | 25,120 | 3.14 | 0.94 |
| $Cu_2$ | 20,410 | 2.04 | 1.32 |
| $Ag_2$ | 22,980 | 1.70 | 1.18 |
| $Au_2$ | 19,650 | 2.26 | 2.11 |
| AgCl | 31,600 | 3.2 | 1.83 |
| AgBr | 31,300 | 3.0 | 1.68 |
| AgI | 31,200 | 2.7 | 1.45 |

[a] Data from G. Herzberg, *Spectra of Diatomic Molecules*, 2nd ed., Van Nostrand, Princeton, N. J., 1950; R. W. B. Pearse and A. G. Gaydon, *Identification of Molecular Spectra*, 2nd ed., Wiley, New York, 1963.
[b] Matrix isolation.[14]

both the force constants and dissociation energies are roughly twice and thrice those of similar single bonds.

In spite of this, there is no theoretical reason why force constants and dissociation energies should be related. One is the curvature of a potential energy curve, the other is the depth of the potential well. If the harmonic oscillator approximation remained valid for vibrational energies up to the dissociation limit, then some relationship between the two would exist. As it is, different degrees of anharmonicity make it impossible to correlate the two properties. Tables 1 and 2 show that $f$ and $D_0$ are really related only for small groups of similar molecules.

**Table 2 Some Average Properties of Chemical Bonds**[a]

| Bond | $f$, md/Å | $D_0$, kcal |
|---|---|---|
| F—H | 9.7 | 135 |
| C—H | 7.9 | 99 |
| O—H | 7.8 | 110 |
| N—H | 6.2 | 93 |
| C—O | 5.1 | 84 |
| C—C | 4.9 | 83 |
| Cl—H | 4.8 | 103 |
| C—N | 4.8 | 70 |
| C—Cl | 3.2 | 79 |
| S—H | 4.3 | 81 |
| Br—H | 4.1 | 88 |
| I—H | 3.2 | 71 |
| P—H | 3.1 | 76 |
| As—H | 2.9 | 59 |
| Si—H | 2.8 | 70 |
| C—I | 2.1 | 57 |

[a] These values are not constants, but depend on the molecule in which the bond occurs. Furthermore, in polyatomic molecules such single-bond properties are only approximately defined.

Accordingly it is quite dubious whether the symmetry of the lowest excited state can select the reaction coordinate, as originally proposed. Let us take a simple example. The easiest reaction of the water molecule is dissociation into H and OH.

$$H_2O \longrightarrow H + OH \tag{17}$$

This reaction requires a reaction coordinate of $B_2$ symmetry, as noted for the similar (15) and (16). The symmetric dissociation of the water molecule,

$$H_2O \longrightarrow 2H + O \tag{18}$$

requires about twice as much energy as (17) and needs a reaction coordinate of $A_1$ symmetry.

Unfortunately the lowest excited state of water is $^1A_1$, just as is the ground state.[12] The symmetric, high-energy dissociation (18) is predicted to be favored. Examination of force constants shows that the symmetric stretch constant $f_1$ is indeed smaller than the asymmetric stretch constant $f_3$, Reaction (18) does *begin* to occur more easily than (17). As the bonds lengthen, however, there are changes in the relative positions of the first two excited states, the $^1B_2$ state becoming lower. This then favors the asymmetric dissociation (17).[13]

If we look at a series of similar molecules, such as $H_2O$, $H_2S$, $H_2Se$, and $H_2Te$, we find that dissociation either according to (17) and (18), requires progressively less energy. At the same time the energy of the $A_1$ and $B_2$ excited states approaches the ground-state energy more closely. That is, the uv absorption shifts toward the red as the molecular stability decreases. This is very consistent with (11), of course.

Also in Table 1, it is found that the alkali halide or hydride molecules have a similar relationship between their dissociation energies and the excited-state energy gap. Indeed, since 8066 $cm^{-1}$ = 1 eV, the two numbers are nearly equal in most cases. This is a result of the nature of the excited state which is either weakly repulsive or weakly attractive,[14] that is, nonbonding.

The lowest energy absorptions of the alkali halide molecules are charge-transfer bands. The ground state of NaCl is naturally very ionic. The first excited state is very nonpolar, which accounts for its low bonding power

$$Na^+, Cl^- \xrightarrow{h\nu} Na^0, Cl^0 \longrightarrow Na + Cl \qquad (19)$$

Another class of complexes showing charge-transfer bands are the weak complexes between halogen molecules and Lewis bases, such as the amines. Here we find exactly the opposite behavior. The stronger the complex, the lower in energy is the charge-transfer band.[15] For example,

$$I_2 + NH_3 = I_2, NH_3 \qquad (20)$$

$$\Delta H = -4.8 \text{ kcal} \qquad h\nu = 5.3 \text{ eV}$$

$$I_2 + (CH_3)_3N = I_2, N(CH_3)_3 \qquad (21)$$

$$\Delta H = -12.1 \text{ kcal} \qquad h\nu = 4{,}7 \text{ eV}$$

This behavior seems inexplicable in terms of PT. Nevertheless it is completely expected according to the theory of such charge-transfer complexes.[15,16]

The contrasting behavior of the alkali halides and the iodine complexes can be reconciled to PT. The ground state of the latter complexes is nonionic, and the excited state is ionic.

$$I_2, NH_3 \xrightarrow{h\nu} I_2^-, NH_3^+ \qquad (22)$$

This is just the opposite of the alkali halide case. Since dissociation of both NaCl and of $I_2$, $NH_3$ is into neutral species, the former requires extensive charge rearrangement upon dissociation (19), whereas the latter does not. Consequently mixing of excited states is very important in the dissociation of alkali halides, whereas it is not important for the dissociation of weak complexes of the halogens. This is a consequence of (12).

While it is clear that it is dangerous to predict the lowest energy process of a molecule from its lowest excited state, the situation with regards to relating force constants to excited states is more favorable (see p. 83). Even here considerable caution is required. First of all, bending force constants almost always are smaller than stretching-force constants. This is because $f_{00}$ in (16) is much smaller for a bending motion than for a stretching motion. This situation usually cannot be turned about even by a favorable value of $f_{0k}$.

Secondly, a given excited state may not be effective at promoting a given type of nuclear motion, even if the symmetries are correctly matched.[11] For example, Fig. 8 shows the various nonsymmetric normal vibrations of the ethylene molecule. The first $B_{1g}$, $B_{2u}$, and $B_{3u}$ modes are primarily C—H stretches. In addition there is an $A_g$ breathing mode, which is also a C—H stretch.

The lowest excited state of ethylene is the $\pi$–$\pi^*$ transition, which gives an excited state of $B_{3u}$ symmetry. Since the ground state is $A_g$, this predicts that the $B_{3u}$ stretching mode will have the lowest force constant. The additional assumption is that all C—H modes have similar values of $f_{00}$, which seems

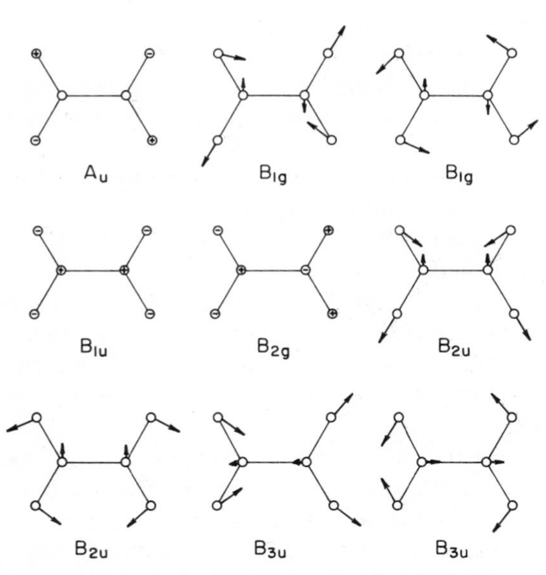


**Figure 8.** The nontotally symmetric vibrational modes of ethylene, point group $D_{2h}$.

reasonable. The actual force constants are $A_g = 6.20 \times 10^5$ dynes cm$^{-1}$; $B_{1g} = 6.15 \times 10^5$; $B_{2u} = 6.01 \times 10^5$, and $B_{3u} = 6.14 \times 10^5$. There is no preference for the $B_{3u}$ mode.

### The Transition Density

To understand this, it is necessary to examine in more detail what happens when an excited state mixes into the ground state according to (12), and to see the physical significance of the relationship (13). Suppose only a single excited state $\psi_k$ is mixed with $\psi_0$. The mixing coefficient is $\lambda$, which by definition is small. We then have

$$\psi = \psi_0 + \lambda\psi_k \tag{23}$$

$$\psi^2 \approx \psi_0^2 + 2\lambda\psi_k\psi_0 \tag{24}$$

Since the square of the wavefunction is an electron-density function, we can write (24) as

$$\rho \approx \rho_{00} + 2\lambda\rho_{0k} \tag{25}$$

where $\rho_{0k}$ is the change in electron density, called the "transition density."[17] Where $\rho_{0k}$ is positive, the electron density increases, and where it is negative, the electron density decreases. Averaged over all space, of course, the transition density must be zero.

The symmetry of $\rho_{0k}$ is simply that of $\psi_0\psi_k$. The meaning of symmetry in this case is simply the way in which $\rho_{0k}$ changes its sign from plus to minus in different parts of the molecule. The positive nuclei will move in the direction of increased electron density and away from regions of decreased electron density. Thus the nuclear motions, $Q$, will simply follow the pattern established by $\rho_{0k}$. (More logically, $\rho_{0k}$ will follow the pattern established by $Q$.) This is the physical significance of (13).

Now in the example of ethylene, the $B_{3u}$ excited state is a $\pi$–$\pi^*$ transition. The transition density is concentrated in regions above and below the plane, since both the $\pi$ and $\pi^*$ orbitals have a node in the molecular plane. The C—H vibrations, which lie in the plane, obviously cannot be much affected by such a transition density.

It seems inevitable that we cannot rely on the symmetry rule (13) unless we have a fairly good idea of the nature of the excited state involved. We are next faced with the dilemma that the excited states of most molecules are not yet well assigned, even if experimentally measurable. Also if we consider bimolecular and termolecular processes, the relevant excited states during the course of a concerted reaction, are excited states of the supermolecule. As such they are not experimentally accessible.

Even if we have the case of a molecule whose excited states are well understood, there is another difficulty remaining. The excited states of perturbation theory are not exactly the real excited states of a molecule. They are instead excited states of the molecule with frozen nuclear positions. Most excited electronic states have equilibrium geometries rather different from those of the ground state (see Chapter 6). Hence the required states correspond to excited vibrational levels of the true states. These have not only different nuclear positions, but also different electron density patterns. A saving feature is that, as a result of the Franck–Condon effect, the excitations observed in nature are to the vibrationally excited states with the same nuclear positions as the ground state.

## MOLECULAR ORBITAL THEORY

At this point it is necessary to abandon (11) and (12), which are exact, and use a more approximate theory that can give more detail. Instead of using the exact wavefunctions $\psi_0$ and $\psi_k$, which are unknown, we must use approximate wave functions, which are known, or at least better known. The logical procedure is to use MO theory. There are several great advantages:

1. The theory is highly developed and familiar to chemists.
2. Excited states are easily described.
3. Molecular orbitals are correctly labeled by symmetry, just as are the exact wavefunctions.

The last point is most significant since it is the symmetry properties that we wish to use. If these are correct, then no error is made in using MO theory in place of more exact theory. Note that we are not eventually concerned with the evaluation of integrals, but only in deciding whether a certain integral is zero or nonzero. If it is nonzero, we would also like to know if it is large or small.

We start by examining the relationship between the exact electronic wavefunction, which we will now refer to as $\psi_e$, and the approximate many-electron wavefunction, $\psi$, of MO theory. The latter is written as an antisymmetrized product of one-electron functions, $\varphi_m$, called "molecular orbitals," such as the Slater determinant,

$$\psi = \frac{1}{\sqrt{n!}} |\varphi_1(1)\alpha(1)\varphi_1(2)\beta(2) \cdots \varphi_m(n)\beta(n)| \tag{26}$$

where $\alpha$ and $\beta$ are spin functions, and $n$ is the number of electrons. The various $\varphi$ values are almost always written as a LCAO, $\sigma_l$, the LCAO MO method.

$$\varphi_m = \sum_l a_{ml}\sigma_l \tag{27}$$

The values of the coefficients, $a_{ml}$, are found by solving the secular equation

$$|H_{lj} - S_{lj}\varepsilon| = 0 \tag{28}$$

where the $H_{lj}$ are the integrals of $H = h_1 + h_2 + h_3$ over the atomic orbitals $\sigma_l$ and $\sigma_j$. The overlap integral is $S_{lj}$. The $h_1$, $h_2 + h_3$ are one-electron Hamiltonians, as in (2), and $\varepsilon$ is a one-electron energy.

The best possible solution to (26) is called the Hartree–Fock solution. The set of AOs, $\sigma_l$, taken initially is called the basis set. To get the Hartree–Fock solution, it would be necessary to take an infinite-basis set, but good results can still be obtained with a limited-basis set. Often a minimal basis set is taken consisting only of the valence-shell AOs of the atoms involved.

At this point we are assuming that all integrals are evaluated exactly, including the difficult interelectronic repulsions. Thus we have an ab initio calculation. Even so the Hartree–Fock solution is still not the exact solution, $\psi_e$. It is in error by the correlation energy. This is the reduction in energy that results because the electronic motions are correlated. That is the electrons, at any instant, tend to stay as far apart as possible. Equation (26) does not allow for this possibility. To get the exact solution requires that $\psi$ be written as an infinite sum of Slater determinantal wavefunctions, instead of a single one as in (26).

While this is not practical, it is still possible to write the wavefunction as a sum of perhaps two or three determinantal wavefunctions. This procedure is called inclusion of configuration interaction and is a means of reducing the interelectronic repulsion energy. It should be mentioned that for open-shell molecules it is often necessary to replace (26) by a sum of two or more determinants in any case.

### Configuration Interaction

The solution of the secular equation (28) gives a number of eigenfunctions, $\varphi_m$, and a number of corresponding eigenvalues, $\varepsilon_m$. The solution of (28) is greatly simplified by symmetry, if symmetry exists. Each MO, $\varphi_m$, is a symmetry-adapted LCAO and belongs to one of the symmetry species of the molecule. We assume that the lowest energy MOs are doubly occupied giving rise to a *configuration:*

$$(\varphi_1)^2(\varphi_2)^2(\varphi_3)^2 \cdots (\varphi_m)^2 \tag{29}$$

Sometimes the highest energy orbitals of the configuration are singly occupied, as must always be true for paramagnetic species. This may give rise to several alternate choices, very similar in energy, and which then require configuration interaction between the alternatives.

Each configuration such as (29) has a definite symmetry, which is simply the symmetry of the product of the individual MOs. A filled subshell (all MOs of the same energy and symmetry doubly occupied) is always totally symmetric. Each configuration also corresponds to a single-determinantal wavefunction of the form of (26). In configuration interaction, only configurations of the same symmetry can interact. This follows because the interaction consists of evaluating integrals of the form $\langle\psi_1|H|\psi_2\rangle$. Since $H$ is totally symmetric, then $\psi_1$ and $\psi_2$ must have the same symmetry for the integral to be nonzero. Also $\psi_1$ and $\psi_2$ must have the same total electron spin.

In addition there is a more powerful theorem, Brillouin's theorem,[18] which governs configuration interaction in closed-shell cases. This says that interaction is possible only between configurations in which at least two electrons are in different orbitals. This is important because different configurations are approximations to excited states.

Normally there is a large energy gap between the occupied MOs and the remaining empty MOs. This is the case for which a single configuration is a good approximation. The empty MOs are called "virtual orbitals." Now an approximation to the various excited states of the system can be obtained by writing the wavefunctions as the Slater determinants of the excited configurations.

$$(\varphi_1)^2(\varphi_2)^2 \cdots (\varphi_i)^1 \cdots (\varphi_f)^1 \tag{30}$$

$\varphi_i$ is the MO that is occupied in the ground state, and $\varphi_f$ is the MO that is occupied in its place in the excited state. The lowest excited state should be formed by excitation of an electron from the highest occupied molecular orbital (HOMO) into the lowest unoccupied on (LUMO). Figure 9 shows several possible excited states, $\psi_1$, $\psi_2 \cdots$ as well as the ground state, $\psi_0$, according to MO theory. In this picture the orbital energy difference, $(\varepsilon_f - \varepsilon_i)$, should be equal to the experimental excitation energy, $h\nu$. Also doubly excited configurations, which are important in configuration interaction normally are not experimentally observable states.

Unfortunately for this simple model, there is a fundamental error. The virtual orbitals are always calculated much too high in energy, even in an accurate ab initio calculation. The calculation of all orbitals are of the self-consistent field (SCF) type. That is, $h$ in (2) is always made to be consistent with the orbital occupancy that results from solving the secular equation (28). The energy of each orbital, $\varepsilon_m$, is calculated as the energy of an electron in the field of the $n - 1$ remaining electrons. However a virtual orbital contains no electrons. Its energy is calculated as the energy of a virtual electron in the field of $n$ electrons. Hence the energy is much too high.

The correct way to calculate the energies of excited states is to solve the SCF equations for the configurations corresponding to the various excited states. Configuration interaction becomes more important, because there are usually a number of excited states close in energy. The lowest excited states are singly excited, which usually means-open shell configurations.

### Ionization Potentials

If we calculate excited-state energies explicitly, we find that orbital energy differences are not equal to the excitation energy, but that changes in interelectronic repulsion between ground and excited states must be explicitly included. This means that excitation from the HOMO to the LUMO need not be the lowest excited-state process.Thus we cannot pick out even the order of excited states from the ground-state properties alone. Nor can we tell differences in orbital energies from visible and ultraviolet (UV) spectroscopy, except as an approximation.

Partly compensating for these difficulties is the fact that individual orbital energies can be directly found by the measurement of ionization potentials (IP). It is a good approximation, as Koopmans originally showed,[19] to equate the IP of an electron in the $m$th orbital to the negative of the energy of that orbital.

$$\mathrm{IP}_m = -\varepsilon_m \tag{31}$$

This is not an exact relationship, since the proof of equality requires that the wavefunction of the remaining electrons does not change on ionization.

This is not true, of course, and the electron density does rearrange after loss of one electron. This lowers the energy requirement for ionization. However there is also a decrease in correlation energy. There two errors tend to cancel and the result is that Hartree–Fock orbital energies tend to be calculated about 10% too high compared to experimental IPs.[20]

The new technique of photoelectron spectroscopy[21] allows the determination of as many as 10–12 orbital energies in the neutral molecule. The problem of identifying each is still formidable (see p. 242). In simple molecules the IP pattern observed is very powerful evidence for the validity of MO theory. For example, methane shows two ionization potentials at 12.75 eV and at 22.4 eV. Molecular orbital theory predicts two kinds of occupied orbitals involving the valence-shell AOs. There is a lower energy orbital of $a_1$ symmetry, and a higher energy, triply degenerate set of orbitals of $t_2$ symmetry. These have Hartree–Fock calculated energies in the neighborhood of 26 and 14 eV, respectively.

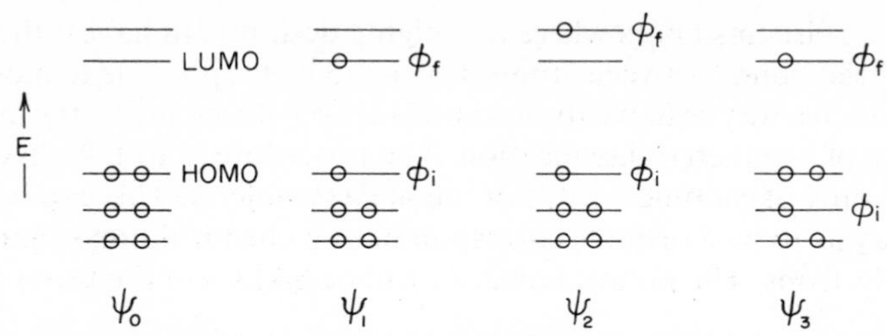


**Figure 9.** Molecular orbital configurations for the ground state, $\Psi_0$, and several excited states, $\Psi_1$, $\Psi_2$, and $\Psi_3$, of a molecule, with highest occupied (HOMO) and lowest unoccupied (LUMO) orbitals generating the first excited state in simple theory.

There is no evidence for an IP corresponding to removal of an electron from an $sp^3$-hybridized orbital. While MOs are to a considerable degree figments of man's imagination, they seem to correspond better to physical observables than do the orbitals of valence-bond theory. Also visible and UV spectra are interpretable by means of MO theory (Fig. 9) if corrections are made for changes in interelectronic repulsion. Valence-bond theory is not suitable for a simple interpretation of excited states.

The molecular orbitals that we have reference to are those that are the solutions to the secular determinant (28). These solutions are called the "canonical orbitals." The occupied orbitals generate a wavefunction such as (26), whose square is an electron-density function. Now it turns out that it is possible to make linear combinations of the occupied canonical MOs and to obtain new orbitals. If properly orthogonal and normalized, these new orbitals give a Slater determinant identical with the original one, (26). They obviously give the same electron density, and identical values of the total energy and other properties whose values depend only on $\psi^2$.

There can be advantages to forming new linear combinations of the canonical MOs. For example, in the case of $CH_4$, linear combinations of the $a_1$ and $t_2$ orbitals generate four MOs that are *equivalent* and *localized*.[22] These correspond to our concept of four equal C—H bonds much better than do the delocalized canonical orbitals. Other highly localized MOs have been discussed by England, Ruedenberg, and Salmon.[23]

We pay a price for such simplification in that the relation to IPs and spectra has been lost. More important, the precise symmetry classification of the orbitals has been lost, by mixing together MOs of different symmetries. The symmetry of the state is still conserved, however, since the value of $\psi$ is unchanged. In the same way, valence-bond theory can give a wavefunction that eventually corresponds to the same electron density and symmetry as the Hartree–Fock wavefunction. The state is again well described, but at the cost of inner detail.

There are circumstances where it is highly desirable to have a theory that uses localized bonds between atoms. Fortunately it is possible to modify MO theory in such a way as to partly accomplish this without losing the important advantage of symmetry classification. The procedure is to take linear combinations only of canonical MOs of the *same* symmetry. This can be done in such a way as to give orbitals corresponding to chemical bonds and to lone pairs of electrons. The bonds, however, will be SALCs of the usual chemical bonds.

To see how this can be done, we follow the procedure of Thompson, who showed that chemical bonds and lone pairs in a molecule can be used as a basis for symmetry classification.[24] A Lewis diagram of the molecule is drawn, showing bonds and lone pairs. The rules of Gillespie and Nyholm[25] are used to locate the lone pairs in space. Now the symmetry operations of the molecular point group are performed. A set of characters is written down for all of the bonds and lone pairs, according to their transformation properties.

For an example, we take the water molecule, as shown in Fig. 10. The characters for the bonds and lone pairs are

| $C_{2v}$ | $E$ | $C_2$ | $\sigma(xy)$ | $\sigma(yz)$ |
|---|---|---|---|---|
| $\Gamma_b$ | 2 | 0 | 2 | 0 |
| $\Gamma_{lp}$ | 2 | 0 | 0 | 2 |

These characters reduce into those of $a_1$ and $b_2$ for the two bonds, and $a_1$ and $b_1$ for the lone pairs. Figure 10 shows schematically the corresponding *bond orbitals* and *lone-pair orbitals*. Note that it is the sum and difference of two O—H bonds that have symmetry, not each bond by itself. Also it is the sum and difference of two lone pairs, originally put in tetrahedral hybrid orbitals, which have symmetry.

To appreciate Fig. 10 more fully, let us examine the canonical MOs of $H_2O$. Table 3 shows the results of an ab initio calculation by Shull and Ellison.[25] The $1s$ orbital of oxygen is much lower in energy than the other basis set atomic orbitals and forms the $1a_1$ MO almost uncontaminated. The valence-shell MOs give the configuration

$$(2a_1)^2(b_2)^2(3a_1)^2(b_1)^2$$

exactly the same symmetries as our bonds and lone pairs.

We must identify our $b_2$ and $b_1$ bond and lone-pair orbitals with the canonical MOs. There are no other occupied orbitals of the same symmetry to mix with. We cannot use the empty MO of $b_2$ symmetry, since that would change the electron distribution. The $b_2$ and $b_1$ orbitals are indeed highly localized, the latter being simply the $p$ orbital on oxygen. The $b_2$ MO is indeed like the difference of two O—H bond orbitals as shown in Fig. 10.

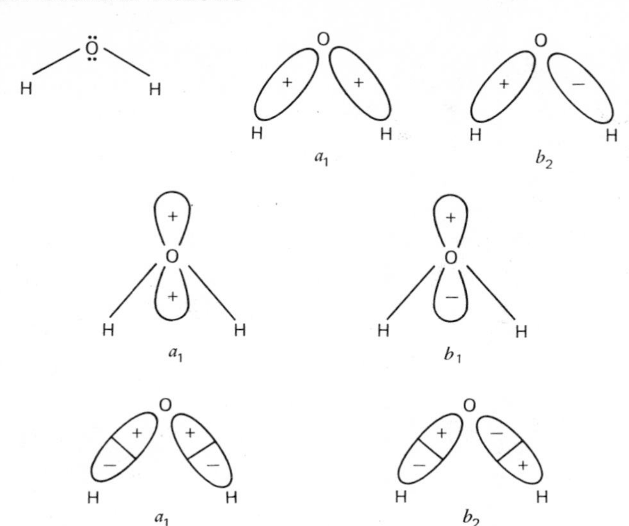


**Figure 10.** The valence bonds and lone pairs of the water molecule used as a basis for symmetry classification; the sum and difference of the two OH bond orbitals are $a_1$ and $b_2$, and the sum and difference of the lone-pair orbitals are $a_1$ and $b_1$.

We have two occupied MOs of $a_1$ symmetry. By subtracting approximately twice the $2a_1$ orbital from the $3a_1$ orbital, we obtain an orbital with no bonding characteristics and localized on the oxygen atom. It is chiefly the $2s$ orbital of O. The remaining orbital of $a_1$ symmetry is constructed to be orthogonal to the previous one. It contains all of the O—H bonding power of both $2a_1$ and $3a_1$, and resembles the $a_1$ bond orbital of Fig. 10.

Table 3 also shows two empty antibonding orbitals of $a_1$ and $b_2$ symmetry. These are also shown schematically in Fig. 10. They are the sum and difference of two localized antibonding orbitals. For any molecule, if we represent localized bonds by orbitals without a node between the two bonded atoms, there will be an equal number of antibonding localized orbitals with a node. The correct symmetry adapted bonding and antibonding orbitals will always be linear combinations of all the *equivalent* bonds or antibonds in the molecule. The total number and symmetry of such combinations, plus that of the lone-pair combinations, will give us the total number of canonical molecular orbitals of each symmetry type. Furthermore the relative energies will be roughly determined. Canonical MOs corresponding to bonding orbitals are lower in energy than those corresponding to lone-pair orbitals (nonbonding). These in turn are lower in energy than MOs corresponding to antibonding orbitals.

**Table 3 Coefficients of Atomic Orbitals in Molecular Orbitals of $H_2O$[a]**

| | $1s_O$ | $2s_O$ | $2p_{zO}$ | $1s_H + 1s_H$ | $2p_{yO}$ | $1s_H - 1s_H$ | $2p_{xO}$ | $-\varepsilon_{calc}$ | $-\varepsilon_{exp}$ |
|---|---|---|---|---|---|---|---|---|---|
| $1a_1$ | 1.000 | 0.016 | 0.002 | −0.003 | — | — | — | 557.3 eV | — |
| $2a_1$ | −0.029 | 0.845 | 0.133 | 0.178 | — | — | — | 36.2 | — |
| $1b_2$ | — | — | — | — | 0.543 | 0.776 | — | 18.6 | 18.5 |
| $3a_1$ | −0.026 | −0.460 | 0.828 | 0.334 | — | — | — | 13.2 | 14.7 |
| $1b_1$ | — | — | — | — | — | — | 1.000 | 11.8 | 12.6 |
| $4a_1$ | −0.086 | −0.833 | −0.642 | 1.061 | — | — | — | −13.7 | — |
| $2b_2$ | — | — | — | — | −1.013 | 1.230 | — | −15.9 | — |

[a] From F. O. Ellison and H. Stull, *J. Chem. Phys.*, **23**, 2348 (1955). Experimental orbital energies from IPs.

In addition there is always recourse to semiempirical calculations. Such methods always give the number and symmetry of the MOs derived from the valence-shell AOs. In addition they give a better estimate of the individual orbital energies. However, as we shall see, there are many cases where it is not necessary to make even crude calculations since symmetry plays the dominant role.

For many molecules, such as $SO_2$ or $C_6H_6$, it is not possible to construct a single Lewis diagram. In such cases it is necessary to use the delocalized MOs found by Hückel or extended Hückel-type calculations. The bonds cannot conveniently be used as a basis set.

## MOLECULAR ORBITALS AND REACTION PATHS

We now return to the problem of calculating favorable reaction paths by means of PT. We wish to replace the exact equations (11)–(13) by their equivalents in terms of MO theory. What we have shown is that the exact calculation of energies by MO theory is difficult. However it is easy to obtain the number, approximate energy and symmetry of all the MOs generated by the valence shells of the constituent atoms.

From this set of orbitals are obtained single configurations corresponding to the ground state and various excited states. The relative order of the excited states is not well known, but their symmetries, and that of the ground state, are well defined. Equations (11) and (12) are replaced by equivalent expressions in which the exact wavefunctions are replaced by determinantal wavefunctions for each configuration. As far as symmetries are concerned, the product of $\psi_0\psi_k$ is replaced by $\varphi_i\varphi_f$. The transition density is also proportional to $\varphi_i\varphi_f$. The selection rule (13) becomes

$$\Gamma_{\varphi_i} \times \Gamma_{\varphi_f} \subset \Gamma_Q \tag{32}$$

Only one-electron excitations are normally considered, since their contribution is usually much greater than that due to two-electron excitations.[5] However there are cases where the two-electron excited states become dominant. These are cases of strong configuration interaction, where electron repulsion must be reduced.

If we abandon the idea that only the lowest excited state is important, as most of the evidence suggests, we are in a position to consider *all* of the excited states generated by the valence-shell MOs. Since we know that we can describe the bonding in both products and reactants by means of these same MOs (or at least MOs generated by the same set of AOs), it seems likely that all of the important excited states contributing to (11) and (13) will be included.

Certainly chemical reaction usually consists of breaking the bonds that connect certain atoms in the reactants, and forming new bonds between other atoms in the product. All MOs are bonding between certain atoms (no intervening node) antibonding for other atoms (a node intervening) and nonbonding between the remaining atoms (too far apart to interact). A model in which electrons transfer from some of the originally occupied MOs into some of the originally empty ones, seems extremely plausible as a method for breaking old bonds and forming new ones.

It must be kept in mind that the indicated transfer of electrons is only virtual. What is actually happening, according to PT, is that the originally empty MOs are mixed with the originally full ones to produce new MOs appropriate to the reacting system. By the time the products have been formed, the original MOs of the reactants have been completely scrambled. However the number of occupied and empty MOs will remain the same. They will have different compositions in terms of the AOs that constitute the basis set, but each MO of the reactants will be *correlated* with an MO of the products in a well-defined way. This correlation is of great value in estimating the energy barrier for the reaction.

## STATE AND ORBITAL CORRELATIONS

We return again to the exact equations (11)–(13) and apply them to a process whereby a set of reactants become products following a well-defined reaction coordinate. The reaction coordinate must be totally symmetric at all points, $Q_0$, except at maxima and minima. At these points symmetry elements may be created or destroyed. There may be some symmetry elements that are conserved during the course of the concerted reaction. With respect to the symmetry operations of these elements, the reaction coordinate is always totally symmetric, even at maxima and minima.

It follows that only excited-state wavefunctions of the same symmetry as that of the ground state can mix during the course of the reaction, the symmetry being referred only to those elements that are conserved. Thus the state symmetry cannot change during the reaction. The assumption is made that the system stays on the same potential energy surface throughout (adiabatic assumption). Reactions in which a jump is made to another surface (nonadiabatic reaction) are known, and will be discussed later (p. 456).

In addition, electron spin is conserved, since only states of the same multiplicity can mix in an adiabatic process. If the reactants before interaction had spins $S_i$, then after interaction the spin must take up values given by the vector sum $S = \sum_i \mathbf{S}_i$. For example, two reactants of spins $S_1$ and $S_2$ will interact to give spins having the possible values ranging from $|S_1 + S_2|$,

$|S_1 + S_2 - 1|$, $|S_1 + S_2 - 2| \ldots$, $|S_1 - S_2|$. This is just a statement that spin must change by integral values as pairs of electrons neutralize each other's spin.

It may be that a potential energy surface that constitutes the ground state for one set of internuclear coordinates, becomes an excited-state surface for another set of internuclear distances. That is, two potential energy surfaces may cross each other. It can be shown that this will not occur if the two states have the same symmetry. This is the noncrossing rule.[26] The explanation lies in the mixing of states of the same symmetry.

If two states with wavefunctions $\psi_1$ and $\psi_2$, have energies $E_1$ and $E_2$ very nearly equal at some point, then second-order PT is no longer useful, and instead PT for degenerate levels must be used. Only two states need be considered and only the first order perturbation, $H' = (\partial U/\partial Q)Q$, needs to be used.

The perturbed energies are the roots of the secular equation

$$\begin{vmatrix} H'_{11} - E & H'_{12} \\ H'_{12} & H'_{22} - E \end{vmatrix} = 0 \tag{33}$$

These roots are

$$E = \frac{E_1 + E_2}{2} \pm \left(\frac{(E_1 - E_2)^2}{4} + H'^2_{12}\right)^{1/2} \tag{34}$$

Now crossing of the two surfaces corresponding to $\psi_1$ and $\psi_2$ requires that $E_1 = E_2$. However this can only be true if the interaction $H'_{12}$ is equal to zero. Since $H'_{12} = \langle\psi_1|\partial U/\partial Q|\psi_2\rangle Q$ is zero when $\psi_1$ and $\psi_2$ have different symmetries, two states of different symmetries are allowed to cross. However $H'_{12}$ is *not* zero for states of the same symmetry, and the energies cannot be the same.†

The resulting situation is shown in Fig. 11*a*, with $\psi_1$ originally being the ground-state wavefunction, and $\psi_2$ being an excited state. The two surfaces approach each other with opposite slopes in this case. Since these slopes are strongly perturbed only at the intended crossover point, they continue on as shown. The original ground-state surface now has become an excited-state surface, and vice versa.

A reaction occurring so that the system stays on the lowest-energy surface is still classified as adiabatic. This is because the crossing has been avoided. While $\psi_1$ and $\psi_2$ have been thoroughly mixed, the lower surface has been changed from mainly $\psi_1$ to mainly $\psi_2$. In MO terms, $\psi_1$ and $\psi_2$ correspond to different configurations.

† This is rigorously true only for diatomic molecules. For polyatomic molecules $H'_{12}$ may be zero, in special cases, even for states of the same symmetry. H. C. Longuet-Higgins, *Proc. Roy. Soc.*, **A344**, 147 (1975).

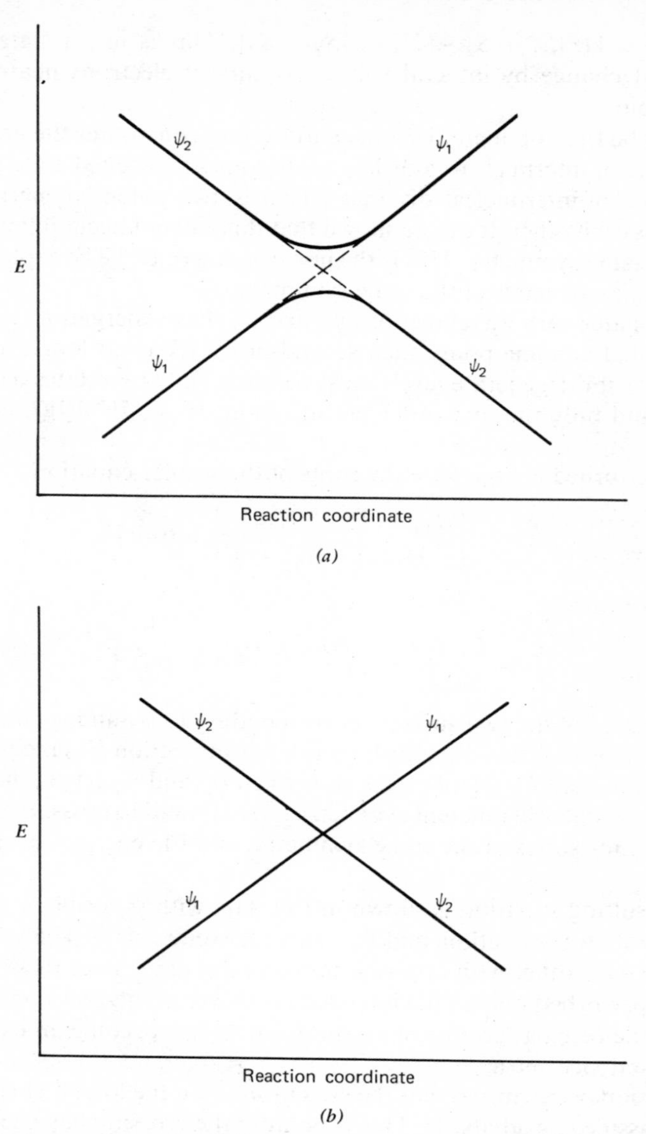


**Figure 11.** (*a*) Noncrossing of two states of same symmetry; (*b*) crossing of two states of different symmetry.

In an adiabatic reaction in which crossing of surfaces does occur, Fig 11*b*, could lead to reactants in their ground state being converted into products in an excited state. This is an energetically unfavorable situation. Depending on the excitation energy and on the exothermicity of the ground-state reaction, it still might occur. We have come, however, to our first example of a reaction that is forbidden by symmetry arguments. A reaction in which the reactants and products have different ground-state symmetries cannot occur as a ground-state reaction. It can occur either with an excited state of the reactants, or with an excited state of the products, as can be seen in Fig. 11*b*.

We can have a similar situation in which surfaces of different multiplicities cross, even though they have the same symmetry. Reactions with reactants and products that do not have ground states obeying the vector sum rule for spins are forbidden.

Unfortunately these rules are not very useful for most reactions. The great majority of molecules have totally symmetric ground states with $S = 0$. They would also be totally symmetric for the lower-symmetry point group that is conserved. Interconversions of these molecules would all be freely allowed by the above rules. There is one exceptional class of molecules, however, where the rules are significant. These are atoms and free radicals, especially the simpler ones.

### The Wigner–Witmer Rules

The earliest application of symmetry and spin rules to chemical reactions is due to Wigner and Witmer,[27] who considered the possible states of a diatomic molecule formed by combining two atoms, for example,

$$\mathrm{O + H \longrightarrow OH} \tag{35}$$

$$\mathrm{Cl + Cl \longrightarrow Cl_2} \tag{36}$$

The first Wigner–Witmer rule for such reactions was just the spin-conservation rule, already mentioned. The second rule was a disguised symmetry rule. The principle of conservation of orbital angular momentum was used instead. This was in analogy to the spin rule, which conserved spin-angular momentum.

The rule is that the orbital angular momentum quantum number, $\Lambda$, of the diatomic molecule takes up the values given by all possible values of $(M_{L_1} + M_{L_2})$. The $M_L$ values for each atom range from L, L − 1, down to −L. The L values are 0 for an $S$ state, 1 for a $P$ state, and so on. When $\Lambda = 0$, the diatomic molecule is in a $\Sigma$ state, $\Lambda = 1$ is a $\Pi$ state, and so on. That is, for a diatomic molecule the symmetry type is determined by the orbital angular momentum quantum number $\Lambda$. For homonuclear molecules it is also necessary to determine the $g$ and $u$ character.

The correlation of states can be continued all the way to the states of the united atom. For example, in (35) OH becomes F. This correlation is useful because it tells a great deal about the way in which the energies of the states of separated atoms change as the atoms approach each other. Ignoring nuclear repulsion, some states go up in energy and others go down. If we know something about the atomic states and their energies, we can tell which states of the diatomic molecule are the more stable. In (36), the united atom is Se. The *g* and *u* properties of the atomic states of Se are the same as those of the $Cl_2$ molecular states with which they correlate.

According to the Wigner–Witmer rules, O, $^3P_g$, and H, $^2S_g$, give $^4\Sigma$, $^2\Sigma$, $^4\Pi$ and $^2\Pi$ states of OH. Two Cl atoms in their $^2P_g$ ground states give $^3\Sigma_g$, $^1\Sigma_g$, $^3\Pi_g$, $^1\Pi_g$, $^3\Delta_g$ and $^1\Delta_g$ states of $Cl_2$. These results are readily visualized by considering which atomic *p* orbital of O, for example, is oriented toward the H atom. This *p* orbital contains either one or two electrons initially. In some cases the correlation of free-atom states and diatomic molecule states can be quite complex.[28]

Extension of the Wigner–Witmer rules to reactions of an atom with a diatomic molecule, or of two diatomic molecules with each other, can be made provided the products are linear molecules.[29] The resultant $\Lambda = M_L \pm \Lambda_2$, or $(\Lambda_1 \pm \Lambda_2)$. The spin rules are the same as for all reactions, $S = \sum_i \mathbf{S}_i$.

The generalized Wigner–Witmer rules form a set of selection rules based on symmetry and spin. Certain reactions are allowed, and others are forbidden. The reactions covered by the rules are unfortunately very restricted, although certainly many of them are important in flames, explosions and in photochemistry.

An attempt to extend these rules to polyatomic, nonlinear molecules proved very discouraging.[29] The orbital angular momentum is quenched in nonlinear molecules, so that $\Lambda$ is not a good quantum number. Its place must be taken by the symmetry of the state. As we have already shown, the use of state symmetry is not very restrictive for chemical reactions. If we include the possibility that mechanisms exist (vibronic coupling) for allowing states of different symmetry to mix, so that nonadiabatic processes can occur, the situation looks even bleaker.

## Orbital Correlation

A more useful procedure was initiated as early as 1927 by Hund and greatly extended by Mulliken.[30] This was again a correlation of the energy levels of diatomic molecules with those of the separated atoms on the one hand, and those of the united atom on the other hand. The important step was made of correlating the individual MOs and AOs, rather than the states. Starting with the separated atoms, we allow them to approach each other. Atomic orbitals that have a positive overlap will interact and form MOs. These MOs must eventually become AOs of the united atom.

The correlation rules are the same as for states:

1. The symmetry of each MO remains unchanged.
2. MOs of the same symmetry cannot cross.
3. No orbitals are gained or lost. Starting with the united atom, it is convenient to use tables for reduction in symmetry. Indeed these tables were first formed for just such purposes.[27,31] An $s$ orbital of an atom has $S_g$ symmetry in the $K_h$ point group that characterizes a free atom. A molecule containing a single atom has the highest possible symmetry. Formation of a diatomic molecule lowers the symmetry to $D_{\infty h}$ for like nuclei, and $C_{\infty v}$ for unlike nuclei.

In $D_{\infty h}$ an $S_g$ species becomes $\Sigma_g$. A $P_u$ species, which is that of an atomic $p$ orbital, is split into $\Sigma_u$ and $\Pi_u$, and so on. This gives us all the orbitals derived from the united atom. For the separated atoms, an individual AO has no symmetry. It is necessary to take the sum and difference of each identical pair of orbitals in the homonuclear case. These can be immediately classified by inspection. Two $s$ orbitals can give only $\Sigma$ symmetry, but the sum will be $g$ and the difference will be $u$, and so on. Note that the superscripts in $\Sigma^+$ and $\Sigma^-$ are not necessary for MOs since only $\Sigma^+$ can occur.†

Figure 12 shows the familiar correlation diagram for like nuclei. A similar one can be constructed for unlike nuclei, except there will be no $u$ and $g$ subscripts. Not only does Fig. 12 show the symmetries of all MOs formed by two like nuclei, it also gives a rough estimate of the energy of each. This varies, of course, with the internuclear separation. But we know if the energy is going up or down as the separation changes. Due allowance must be made for a change in the scale of energy on the left- and right-hand sides of Fig. 12.

The theoretical basis for the application of rules (1)–(3) is that of the one-electron approximation. If electrons did not repel each other, then an orbital, which is simply a one-electron wavefunction, would obey the same rules that an exact state wavefunction would. Correlation diagrams such as Fig. 12 are exact only to the extent that MO theory (or the central-field approximation for atoms) is exact. This restriction does not destroy the validity of these diagrams since we know that MO theory has equal symmetry validity for both products and reactants in a chemical change. It will be necessary to make allowances for the effect of interelectronic repulsion in the use of the diagrams, however.

† The symmetries of MOs will be shown henceforward by lower case symbols, state and general symmetries by upper case.

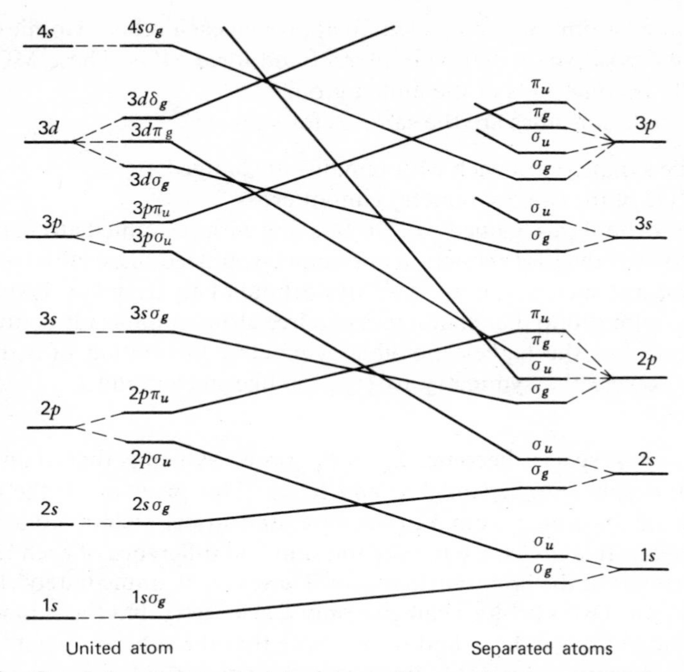


**Figure 12.** Correlation diagram for bringing together two like atoms; note that the energy scales are quite different on the left- and right-hand sides of the diagram.

Surprisingly, little extension of orbital correlation diagrams was made in the 30 years following the work of Hund and Mulliken. This may be an indication of how far they were ahead of their times, or an indication of how far theoretical chemistry lagged behind theoretical physics.

### Reactions of Hydrogen

In 1955 Griffing applied MO correlation methods to some reactions of hydrogen atoms and hydrogen molecules.[32] The conclusions drawn were that

1. Two saturated molecules react with high activation energy.
2. A radical (atom) and a molecule react with low activation energy.
3. Two radicals react with zero activation energy.

These conclusions agreed with known experimental facts. No conclusions based on symmetry properties were reached. It will be informative to re-examine these reactions, and to apply the full power of symmetry arguments.

It is helpful that for collections of hydrogen atoms, exact ab initio calculations of energy can readily be made. This simply serves as a check on any conclusions drawn. We start only with symmetry properties and with rough estimates of energy.

We start with the known reaction

$$H + H_2 \longrightarrow H_2 + H \tag{37}$$

for which the activation energy is 7.5 kcal. Accurate calculation of the ground-state potential energy surface has been made, and a linear symmetric activated complex is indicated.[33] Figure 13 shows the MO correlation diagram, including the activated complex. In this diagram, and in others to come, there is an ambiguity between the orbital energies shown and the total energy. The changes in total energy determine energy barriers to reaction and experimental activation energies.

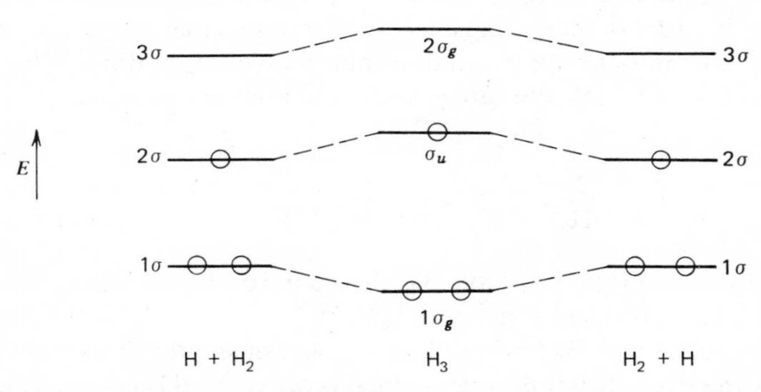


**Figure 13.** Molecular orbital correlation diagram for reaction $H_2 + H \rightarrow H + H_2$ with a linear $H_3$ molecule as the activated complex; $C_{\infty v}$ point group maintained.

The relationship between total energy $E$ and orbital energies, $\varepsilon_i$ is as follows:

$$E = \sum_{\text{occ}} \varepsilon_i + \text{nuclear repulsion–electron repulsion} \tag{38}$$

The summation is over all occupied MOs with a weighting factor of 2, if doubly occupied. The correction for nuclear repulsion potential energy is obvious, but the reduction in total energy by the sum of the interelectronic repulsions is not. It comes in because in MO theory each pairwise interelectronic repulsion is counted twice. The potential energy between electrons 1 and 2 is counted once in evaluating the average field in which electron 1 moves, and then again in evaluating the average field for electron 2. In a neutral nonpolar molecular, the two corrections of (38) tend to cancel each other, and $E$ is often approximated as a sum of orbital energies. In polar molecules and ions, the corrections cannot be ignored.

It will be useful to make a rough correction for nuclear repulsion and electron repulsion in displaying orbital energies. The intent will be to show more clearly whether total energies go up or down in passing from left to right on the diagram.

In the $H + H_2$ reaction all orbitals are of $\sigma$ symmetry, since only the $1s$ valence orbitals are used. The point group is $C_{\infty v}$ for a linear approach. A higher point group, $D_{\infty h}$, is found only for the activated complex. The reaction coordinate is also of $\Sigma$ symmetry, so that no difficulty exists in the selection rule (32). The activated complex has two valence-shell excited states, $(1\sigma_g) \rightarrow (\sigma_u)$ or $(\sigma_u) \rightarrow (2\sigma_g)$. In either case the transition density is $\Sigma_u$, and the reaction coordinate is the asymmetric stretch, leading to $H_2 + H$. The total state symmetry remains the same, being $\Sigma$.

The levels of reactants, products and activated complex all correlate with each other in order of increasing energy. No appreciable activation energy is suggested. Our present approach does not enable estimation of the experimental value of 7.5 kcal. We can consider the related reactions

$$H^- + H_2 \longrightarrow H_2 + H^- \tag{39}$$

$$H^+ + H_2 \longrightarrow H_3{}^+ \tag{40}$$

using the same diagram. Reaction (39) should have a higher activation energy than (37). The theoretical value is 10 kcal.[34]

We can predict that $H_3{}^+$ will not dissociate but should be a stable adduct, since only the strongly bonding $1\sigma_g$ orbital is occupied. This is experimentally correct, the only discrepancy being that $H_3{}^+$ has an equilateral triangular structure.[35] We should be able to predict the instability of linear $H_3{}^+$ toward bending, but unfortunately we cannot. A reaction coordinate of $\Pi_u$ symmetry is needed, but we have only $\sigma$ orbitals from our valence shell. Of course higher energy $\pi$ orbitals do exist (Fig. 12) formed from the $2p$ orbitals of H. The reason for the bending of linear $H_3{}^+$ must be attributed to the low force constant for bending compared to dissociation. That is, $\langle\psi_0|\partial^2 U/\partial Q^2|\psi_0\rangle$ is much smaller for the $\Pi_u$ mode than the $\Sigma_u$ mode.

The atomic composition of the MOs as we go from left to right in Fig. 13 are of interest. Calling the three H atoms in order A, B, and C, the initial unnormalized MOs are

$$1\sigma \quad (s_B + s_C)$$

$$2\sigma \quad s_A$$

$$3\sigma \quad (s_B - s_C)$$

The TS MOs are approximately

$$\begin{array}{ll} 1\sigma_g & (\sqrt{2}s_B + s_A + s_C) \\ \sigma_u & (s_A - s_C) \\ 2\sigma_g & (\sqrt{2}s_B - s_A - s_C) \end{array}$$

and the final MOs are

$$\begin{array}{ll} 1\sigma & (s_A + s_B) \\ 2\sigma & s_C \\ 3\sigma & (s_A - s_C) \end{array}$$

These show clearly how the original three $\sigma$ states mix together to form the MOs of the activated complex, and how these mix to form the MOs of the products.

For three hydrogen atoms in an equilateral triangular structure, the MOs in the $D_{3h}$ point group are

$$\begin{array}{ll} a'_1 & (2s_B + s_A + s_C) \\ e' & (s_A - s_C) \quad \text{and} \quad (2s_B - s_A - s_C) \end{array}$$

with the $a'_1$ orbital being very stable and the $e'$ orbitals quite unstable. This is consistent with the stability of $H_3{}^+$. However $H_3$ would have a configuration $(a'_1)^2(e')$. This is an $E'$ state, which is Jahn–Teller unstable, being doubly degenerate. The predicted distortion is given by the symmetric direct product, written as $(E' \times E')^+ = A'_1 + E'$, so that only an $E'$ vibration can remove the degeneracy. This vibration corresponds to distorting to $C_{2v}$, so that dissociation into $H_2 + H$ occurs (see p. 12).

The next example is the molecular process between two hydrogen molecules

$$H_2 + D_2 \longrightarrow 2HD \tag{41}$$

where the atomic partners interchange, as would be shown best by isotope labeling. The occurrence of (41) is very doubtful, though some controversy exists.[36] Actually isotope exchange between $H_2$ and $D_2$ usually occurs by dissociation into atoms, followed by (37). The overall activation energy is 52 kcal (for the dissociation into two atoms) plus 7.5 kcal for (37), or 60 kcal. It would appear that the direct molecular reaction (41) has a very much larger energy barrier.

Figures 14*a* and *b* show the MO correlation diagram. A broadside collision mechanism is assumed. The point group is $D_{2h}$, although the TS has a higher symmetry, $D_{4h}$. Figure 14 also shows the possible combinations of 1*s* orbitals on H in both point groups. It can be seen that there are difficulties. No vacant MOs exist of the same symmetry as that of the filled MOs in $D_{2h}$. Therefore

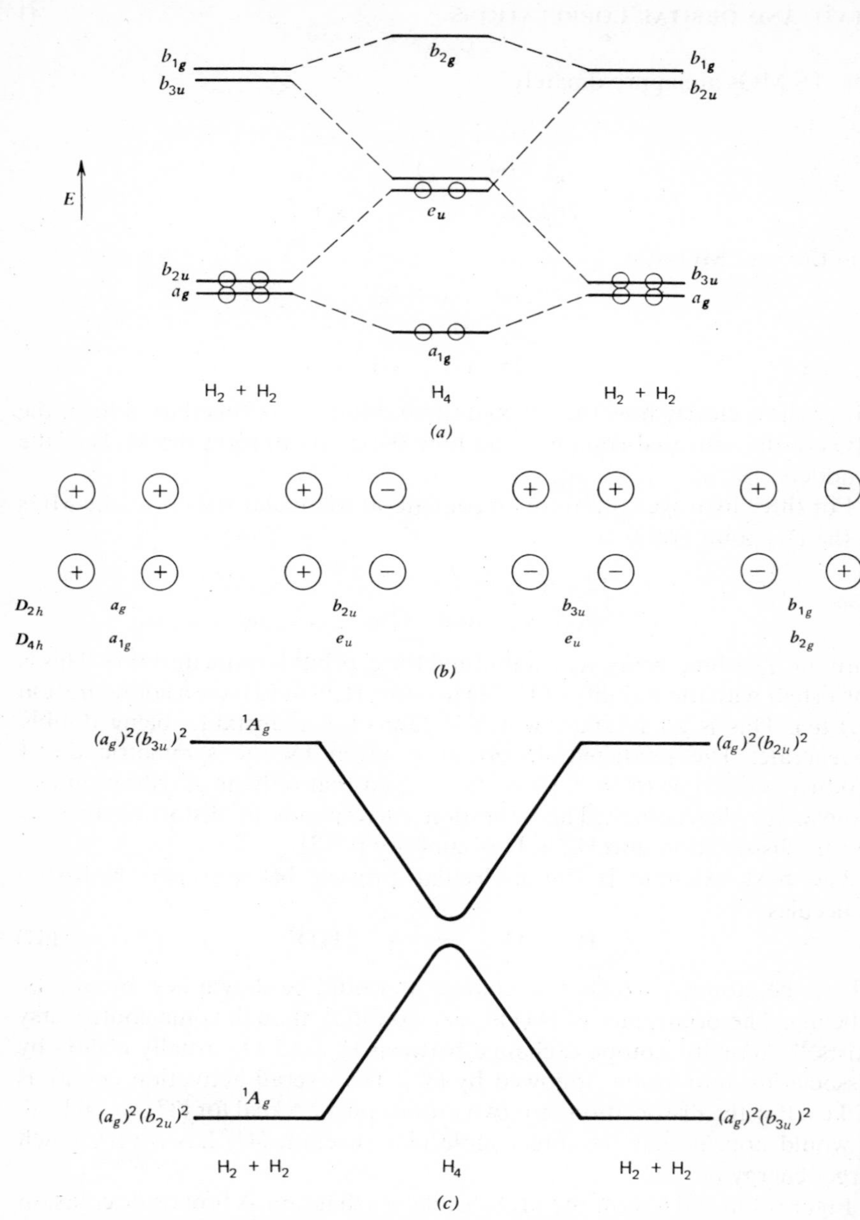


**Figure 14.** (*a*) Molecular orbital correlation diagram for reaction $H_2 + H_2 \rightarrow H_2 + H_2$ with square $H_4$ as the activated complex and $D_{2h}$ point group maintained; (*b*) linear combinations of four *s* orbitals and their species in $D_{2h}$ and $D_{4h}$ (plane of molecule is *xy* plane); (*c*) state correlation diagram for $H_2 + H_2$ exchange reaction.

no low-lying excited states exist of matching symmetry. Also on following the correlation through to the products, a crossing of levels occurs. The $b_{2u}$ and $b_{3u}$ levels become degenerate in the TS, also becoming of the same symmetry, both $e_u$.

At this point one might expect the two orbitals to mix by invoking the noncrossing rule. Indeed it is a characteristic of degenerate orbitals in MO theory that any linear combination of them is also a canonical solution. Hence the $b_{3u}$ and $b_{2u}$ orbitals are thoroughly scrambled in $D_{4h}$ symmetry. However they reemerge after the TS as if they had been uninterrupted. The question is whether two electrons originally in a $b_{2u}$ orbital can emerge in a $b_{3u}$ orbital after crossing.

It turns out that they can, but by virtue of state mixing, rather than orbital mixing. It is states to which the noncrossing rule applies rigorously, and not orbitals. Configuration interaction mixes the states. This starts to happen even before the crossing point, since the states are of the same symmetry. It may be mentioned, however, that there are circumstances where states of the same symmetry can intersect at a point, or along a line in a multi-dimensional space.[37]

Configuration interaction mixes the $(a_g)^2(b_{3u})^2$ configuration with the lower energy $(a_g)^2(b_{2u})^2$ configuration continuously during the reaction. At the TS both configurations are mixed with equal weight. After the TS, the lower energy $(a_g)^2(b_{3u})^2$ configuration predominates.

Now configuration interaction, as described previously is solely a method for reducing interelectronic repulsion. The basic equation for the variation of potential energy with reaction coordinate is given by (11), which may be written as

$$E = E_0 + aQ + bQ^2 \tag{42}$$

The coefficients $a$ and $b$ contain only changes in nuclear–nuclear and nuclear–electronic potential energy. Changes in electron–electron repulsion and in electron kinetic energy are much smaller, coming in first as coefficients of $Q^3$, $Q$ being small by definition. Therefore configuration interaction is an inefficient way of mixing states. It only becomes large when two states have nearly the same energy.

The net result is shown in Fig. 14*c*, as a state correlation diagram. The reaction occurs on the ground-state surface throughout. Both configurations which are mixed are of $^1A_g$ state symmetry, and the noncrossing rule prevails. While the crossing implied at the orbital level is avoided, the fact that it was *intended* leads to a large potential energy barrier. We still must go to the high energy TS before near degeneracy of the $b_{2u}$ and $b_{3u}$ orbitals causes effective configuration interaction. The existence of the large barrier is shown by ab initio calculations. The energy of square $H_4$ is 142 kcal above the energy of two $H_2$ molecules.[38] Because of the large energy barriers usually created by crossing situations such as are shown in Fig. 14, it is customary to refer to such reactions as being *forbidden by orbital symmetry*.

The first such use of diagrams showing crossing was by Longuet-Higgins and Abrahamson,[39] who showed both orbital and state correlation. However their work was inspired by an earlier paper of Woodward and Hoffmann, who also first used the terms "forbidden," or "allowed," by orbital symmetry.[40] The latter authors also exemplified the use of correlation diagrams in a series of remarkable papers.[41] Certainly Woodward and Hoffmann deserve the major credit for the recent realization by chemists of the dominant role played by MO symmetry in chemical reactions.

Even if a square planar TS is ruled out for (41), it is still possible that some other arrangement of the four H atoms would be lower in energy and could serve as the activated complex. Possible structures are linear, rectangular, rhombic, tetrahedral, triangular, and so on. Detailed calculations show that all of these structures are also much too high in energy.[36,37,42] The lowest energy structure is linear $H_4$, but this can only lead to isotope exchange if free atoms are also produced.

Approach of two $H_2$ molecules in a linear fashion turns out to be allowed by orbital symmetry (no crossover of orbitals). Also certain other modes of approach are allowed.[42,43] This in itself does not guarantee a low-energy TS, since the intermediates thus generated are still high in energy because of poor bonding, or else cannot lead directly to the stable products.

The important result obtained is that an activated complex, which a priori was very promising, is ruled out. A square planar TS seemed ideal for making new bonds as the old ones were breaking. Another possibility is the perpendicular approach of two $H_2$ molecules to form an $H_4$ complex with tetrahedral symmetry. This can then dissociate in three equivalent ways, two of which lead to exchange.

The correlation diagram is given in Fig. 15. It is important to note that although the perpendicular approach of two $H_2$ molecules generates a $D_{2d}$ point group, only the symmetry elements of $D_2$ are conserved throughout. Figure 15 shows that three $C_2$ axes remain the same for the $D_{2d}$ configuration of both the approach and departure. The $S_4$ axes and the dihedral mirror planes, $\sigma_1$ and $\sigma_2$, are different for approach and departure, and so they must not be used to classify the levels.

The same problems of Fig. 14 are found in Fig. 15. There are no empty orbitals of the same symmetry as the filled ones, therefore, mixing cannot occur until near the TS. A crossing of the $b_2$ and $b_1$ levels occurs. The TS has two electrons in high energy $t_2$ orbitals. The reaction is again forbidden by orbital symmetry. There are two ways in which this is shown: one is the crossover of the originally occupied $b_1$ level and the originally empty $b_2$ level.

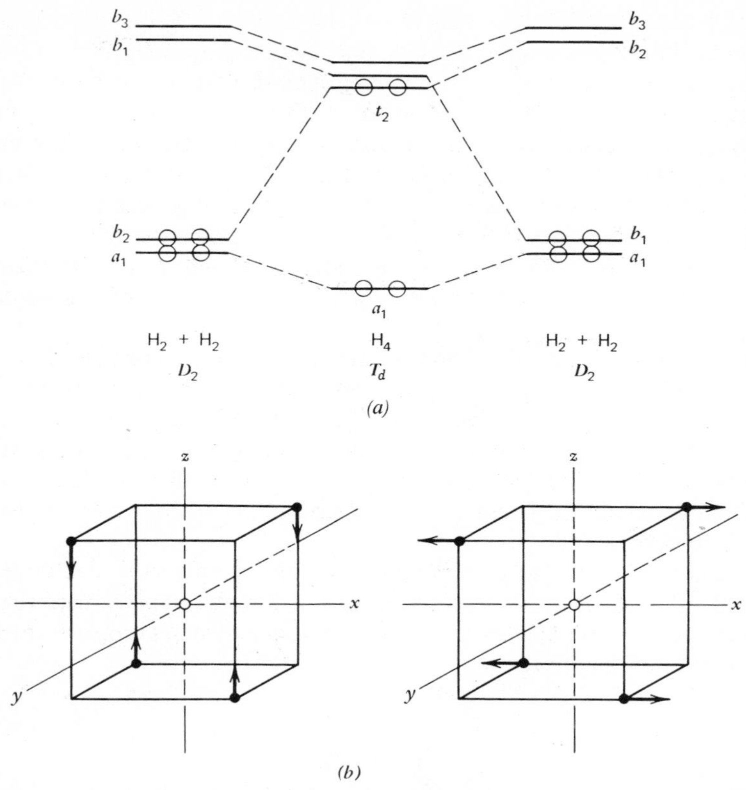


**Figure 15.** (*a*) Molecular orbital correlation diagram for $H_2 + H_2 \rightarrow H_2 + H_2$ with tetrahedral $H_4$ as the activated complex and $D_2$ point group maintained; (*b*) only $C_2$, $C_2'$, and $C_2''$ are preserved (the *x*, *y*, and *z* axes).

The other is the absence of filled and empty orbitals of the same symmetry.

The second criterion, which comes directly from perturbation theory, is often easier to use.[44] It requires only the MOs of the reactants and not those of the products. Also, the mathematical statement that two orbitals have the same symmetry is accompanied by a simple physical picture. Two orbitals of the same symmetry may have a net positive overlap, two orbitals of different symmetry always have zero overlap. Use will be made of this property in later examples.

The second general result from perturbation theory lies in the prediction of reaction coordinates at maxima and minima of potential energy by way of the symmetry requirement of (14) or (32). This can be applied to the transition states of both Figs. 14 and 15. In both cases the reaction coordinate has already been defined by the drawing of the figures. For Fig. 14 it is of $B_{1g}$ symmetry. This is the nuclear motion that takes $D_{4h}$ into $D_{2h}$.

We must show that two electrons in $e_u$ orbitals give rise to states that are consistent with this reaction coordinate. Since $2e_u$ orbitals can hold 4 electrons, we already know that several states of nearly the same energy must exist. The direct product $(E_u \times E_u)^+ = A_{1g} + B_{1g} + B_{2g}$. Of these the $B_{1g}$ state is lowest[38] (we consider only singlet states because of the spin change rule). The excited state $A_{1g}$ is of the right symmetry to give a $B_{1g}$ reaction coordinate. The state $B_{2g}$ on mixing would give a reaction coordinate of $A_{2g}$ symmetry, but there is no vibrational mode of this symmetry for $H_4$.

The $H_4$ molecule of $T_d$ symmetry has an $(a_1)^2(t_2)^2$ configuration. The possible singlet states are $^1E$, $^1T_2$ and $^1A_1$. Of these the $^1E$ is the lowest.[42] This state, being orbitally degenerate, is Jahn–Teller unstable. It will decompose by an $E$ mode (see p. 78). This is exactly the reaction coordinate needed in Fig. 15 to convert $T_d$ into $D_{2d}$. Note that tetrahedral $H_4$ is unstable by a first-order Jahn–Teller effect. Square $H_4$ is unstable by a pseudo-Jahn–Teller effect.

We next need an example of a reaction involving only saturated molecules which is allowed by orbital symmetry. Consider the hypothetical reaction in which three $H_2$ molecules collide to form an activated complex which is a regular hexagon of H atoms

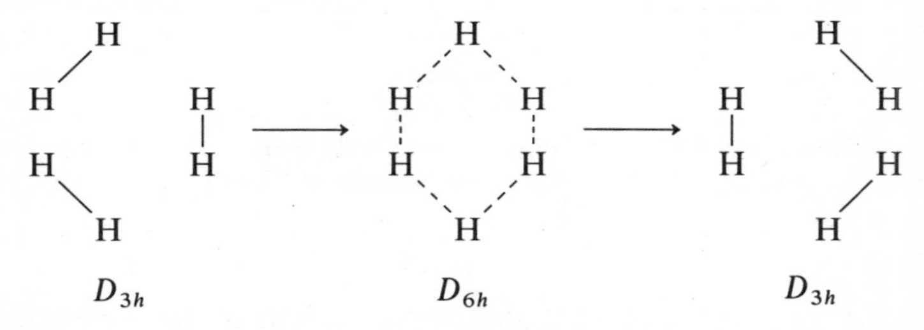


This can clearly lead to exchange. It is not a very plausible mechanism, of course, because entropy considerations make the formation of the $H_6$ ring very unlikely. Still, we are interested in the potential energy surface and in showing that no large barrier exists for the reaction.

We assume a symmetric approach of $3H_2$ molecules, so that the point group is $D_{3h}$, becoming $D_{6h}$ at the transition state. Carrying out the symmetry operation of $D_{3h}$ on six $1s$ orbitals we get the reducible representation

| | $E$ | $C_3$ | $C_2$ | $\sigma_h$ | $S_3$ | $\sigma_v$ |
|---|---|---|---|---|---|---|
| $\Gamma$ | 6 | 0 | 2 | 6 | 0 | 2 |

$$\Gamma = 2A_1' + 2E'$$

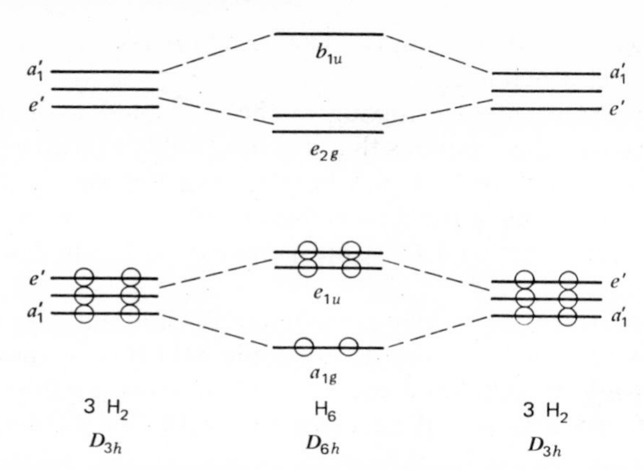


**Figure 16.** Molecular orbital correlation diagram for $3H_2 \rightarrow 3H_2$ with an activated complex, $H_6$, of $D_{6h}$ symmetry; the elements of $D_{3h}$ are preserved.

which gives us two MOs of $a_1'$ and two of $e'$ symmetry. These are shown in Fig. 16, along with the corresponding MOs in $D_{6h}$ symmetry. Initially, one $a_1'$ and one $e'$ pair are of the same energy, since they are simply the occupied bonding orbitals of the three $H_2$ molecules. The other $a_1$–$e'$ set are the empty antibonding partners, also equal in energy.

There are filled and empty MOs of the same symmetry. These can mix smoothly to give the MOs of the TS. There is no crossing over of filled and empty levels, and no large energy barrier is indicated. This does not mean that the activation energy is zero, since $H_6$ is certainly not stable with respect to $3H_2$. However we can confidently expect the activation energy to be much less than the large values found for the bimolecular reactions of $H_2$.[45] The relative stability of symmetric $H_6$ is also shown by the moderately large energy gap between the filled $e_{1u}$ levels and the empty $e_{2g}$ levels. The reaction coordinate needed at the TS is of $B_{1u}$ type, which is consistent with an $(e_{1_u}) \rightarrow (e_{2_g})$ excitation.

### Dissociation Reactions

The last example of reactions of $H_2$ molecules is the unimolecular dissociation into hydrogen atoms

$$H_2 \longrightarrow 2H \qquad (43)$$

This reaction is allowed as a ground-state process by the Wigner–Witmer rules, and hence it must be allowed by orbital symmetry as well. This is a little confusing because the only MOs that the valence shell of $H_2$ provides are $\sigma_g$, occupied, and $\sigma_u$, empty. The reaction coordinate must be the stretching mode, which is $\Sigma_g$.

Dissociation must involve mixing of the $\sigma_u$ orbital with the $\sigma_g$ orbital. The symmetry rule then requires that it is the doubly excited $(\sigma_u)^2$ configuration which mixes with $(\sigma_g)^2$. It has been known for some time that this is precisely what happens in the dissociation of $H_2$.[46] In the limit of complete dissociation, both $(\sigma_g)^2$ and $(\sigma_u)^2$ must appear with equal weights in the wavefunction.

The reason, of course, is interelectronic repulsion and the need for configuration interaction to reduce it. In simple MO theory, dissociation into two atoms would be predicted equally with dissociation into $H^-$ and $H^+$. Both sets of products would be consistent with the wavefunction corresponding to $(\sigma_g)^2$,

$$\psi = \tfrac{1}{2}[(A + B)(1)(A + B)(2)](\alpha(1)\beta(2) - \beta(1)\alpha(2)) \tag{44}$$

at infinite separation of the two nuclei. In (44) A and B are meant to be $1s$ orbitals on nuclei A and B, respectively. This gives an electron density

$$\psi^2 = \tfrac{1}{4}[A^2(1)A^2(2) + A^2(1)B^2(2) + A^2(2)B^2(1) + B^2(1)B^2(2)] \tag{45}$$

Instead let us write the wavefunction as the difference of the two configurations, $(\sigma_g)^2$ and $(\sigma_u)^2$:

$$\psi = \tfrac{1}{2}[(A + B)(1)(A + B)(2) - (A - B)(1)(A - B)(2)](\alpha(1)\beta(2) - \beta(1)\alpha(2)) \tag{46}$$

The electron density now becomes

$$\psi^2 = \tfrac{1}{2}[A^2(1)B^2(2) + A^2(2)B^2(1)] \tag{47}$$

which correctly prevents both electrons from being on the same nucleus at the same time. By means of configuration interaction the MO wavefunction becomes identical with the valence-bond wavefunction, which does not have the error of dissociation into the wrong products.

The independence of (43) on one-electron excited states can be rationalized by realizing that the electrons do not move from one atomic orbital to another, as is characteristic of most chemical reactions. They remain in the same two AOs, A and B. The situation is somewhat different for dissociation of a heteronuclear molecule

$$HCl \longrightarrow H + Cl \tag{48}$$

Here both the bonding and antibonding MOs are of $\sigma$ symmetry. This matches with the reaction coordinate, also of $\Sigma$ symmetry.

In reactions such as (48), dissociation is accompanied by single-electron excitations, $(\sigma) \rightarrow (\sigma^*)$, as well as by two-electron excitations. The role of the singly excited states is to shift the valence electron density from being concentrated on chlorine in the molecule, to being equally divided between H and Cl in the products.[47]

The dissociation of a polyatomic molecule into two radicals may also be considered. The reaction coordinate may initially be nontotally symmetric. For example, the dissociation of methane

$$CH_4 \longrightarrow CH_3 + H \tag{49}$$

requires a $T_2$ vibration to break the $T_d$ point group. This can occur with the aid of $(t_2) \rightarrow (a_1^*)$ excitation. The point group becomes $C_{3v}$, and the reaction coordinate, $A_1$. The remaining process is very much the same as for a heteronuclear diatomic molecule. We can conclude that there are few orbital symmetry barriers to reactions in which bonds are broken, but no new bonds are formed. There will still be restrictions of the Wigner–Witmer type on spin multiplicity and the states of the reactants and products. The same remarks apply to the reverse processes in which new bonds are made, but no bonds are broken.

Of course reactions (47)–(49) require large amounts of energy, as do bond-breaking reactions in general. We are not making any attempt to estimate how much energy is required. Nor in more complex molecules, do we usually attempt to predict which of several possible bonds may break. There is a rough correlation between the facts that $CH_4$, HCl, and $H_2$ have strong bonds, and that they do not absorb until the UV spectrum, but this correlation does not enable us to predict bond strengths with any accuracy.

What we are looking for is the presence of an unexpected barrier to reaction besides the thermodynamic energy requirement. This barrier would be due to symmetry restraints. Figure 17 shows potential energy plots for unimolecular reactions allowed (*a*), and forbidden (*b*), by orbital symmetry.

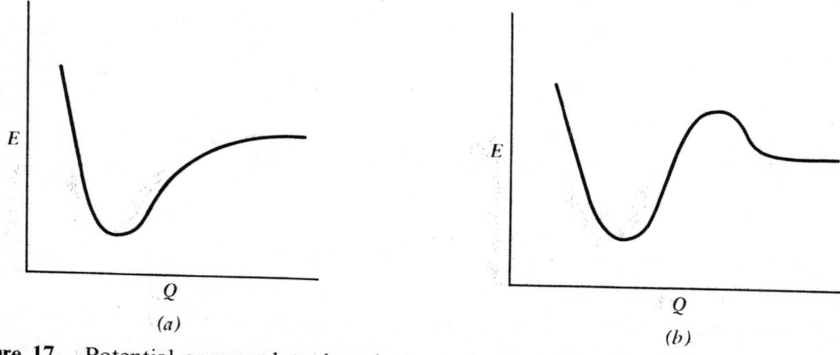


**Figure 17.** Potential energy plotted against reaction coordinate for dissociation reactions: (*a*) allowed and (*b*) forbidden, by orbital symmetry.

Energy profiles of type (*b*) are not expected for simple dissociations of molecules. We expect to find them instead for reactions where bonds are made as well as broken. In such cases electrons must be transferred from one part of the molecule to another. This process is strongly subject to orbital symmetry restraints.

## REACTIONS OF DIATOMIC MOLECULES

Instead of taking up the unimolecular reactions of complex molecules, it is easier to consider further the bimolecular reactions of diatomic molecules. Especially homonuclear molecules have a high degree of symmetry, and symmetry effects should be very pronounced. The usual sequence of MO stability for $X_2$ molecules is (Fig. 12)

$$(\sigma_g)(\sigma_u)\underline{(\pi_u)(2\sigma_g)}(\pi_g)(2\sigma_u)$$

with the underlining of $\pi_u$ and $\sigma_g$ meant to imply that they are close in energy. Only MOs from the valence shell are shown. Figure 18 shows schematically the appearance of these orbitals. Note that symmetry means simply the way in which the sign of the wavefunction changes in different parts of the molecule.

We can obtain values of the orbital energies directly by photoelectron spectroscopy. Table 4 shows the IPs of the three HOMOs and the electron affinity (EA) of the LUMO of the halogens. The configuration is seen to be

$$(1\sigma_g)^2(1\sigma_u)^2(2\sigma_g)^2(\pi_u)^4(\pi_g)^4(2\sigma_u)^0$$

so that the EA is the negative orbital energy of the antibonding $2\sigma_u$ orbital. Note that the energy is negative, meaning that the orbital is binding for an electron.

It is still antibonding for the molecule. This can be seen because the EAs of the free halogen atoms are *greater* than for the molecules. For Cl it is 3.86 eV, for example. Thus the process

$$Cl_2 \longrightarrow 2Cl \tag{50}$$

requires some 35 kcal more energy than the reaction

$$Cl_2^- \longrightarrow Cl + Cl^- \tag{51}$$

Bonding in the halogens is a result of the $2\sigma_u$ orbital being empty. Bond breaking in the halogen molecule results when electrons are added to this orbital. Because of the electronegativity of these elements, this is what normally happens in chemical reactions. Hence the $2\sigma_u$ orbital and its symmetry are critical for the reactions of the halogens.

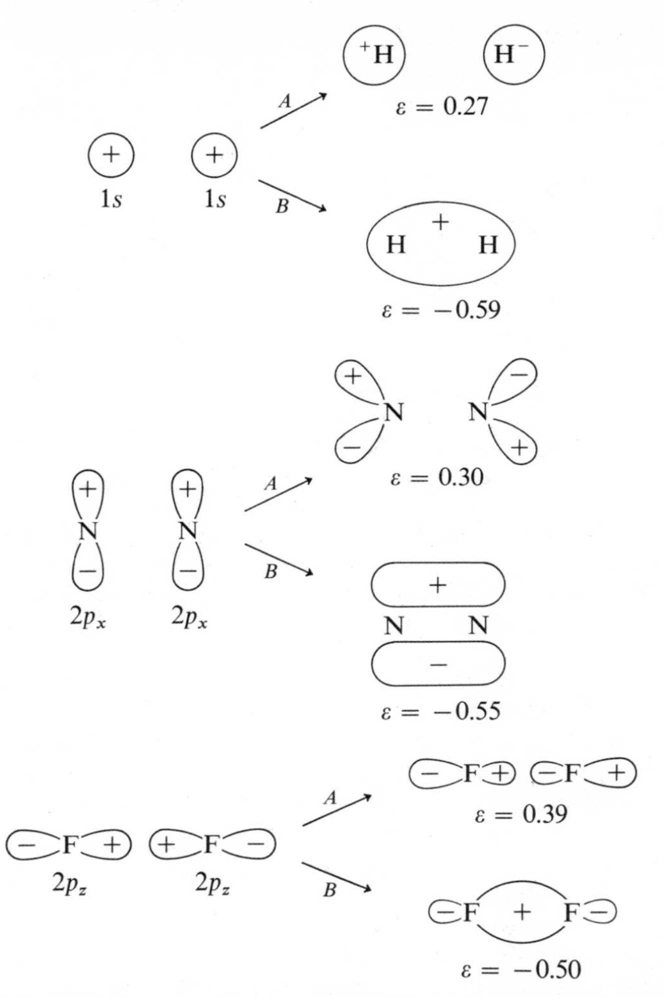


**Figure 18.** Molecular orbitals of homonuclear diatomic molecules formed from sums and differences of atomic orbitals. Critical MOs of $H_2$, $F_2$, and $N_2$ are shown; B = bonding and A = antibonding (energies in atomic units; 1 AU = 27.21 eV). [After B. J. Ransil, *Rev. Mod. Phys.*, **32**, 245 (1960).]

**Table 4 Ionization Potentials and Electron Affinities of the Halogen Molecules**

| | Adiabatic Ionization Potential, eV[a] | Electron Affinity, eV[b] |
|---|---|---|
| $F_2$ | 15.83 $\pi_g$ | 3.08 |
| $Cl_2$ | 11.30 $\pi_g$ | 2.38 |
| $Cl_2$ | 13.85 $\pi_u$ | |
| $Cl_2$ | 15.65 $2\sigma_g$ | |
| $Br_2$ | 10.33 $\pi_g$ | 2.51 |
| $Br_2$ | 12.30 $\pi_u$ | |
| $Br_2$ | 14.20 $2\sigma_g$ | |
| $I_2$ | 9.18 $\pi_g$ $J = \frac{1}{2}$ | 2.58 |
| $I_2$ | 9.83 $\pi_g$ $J = \frac{3}{2}$ | |
| $I_2$ | 10.71 $\pi_u$ $J = \frac{1}{2}$ | |
| $I_2$ | 11.69 $\pi_u$ $J = \frac{3}{2}$ | |
| $I_2$ | 12.58 $2\sigma_g$ | |

[a] Values from S. Evans and A. F. Orchard, *Inorg. Chim. Acta*, **5**, 81 (1971).
[b] Values from W. A. Chupka, J. Berkowitz, and D. Gutman, *J. Chem. Phys.*, **55**, 2724, 2733 (1971).

Identification of observed IPs with the right orbital is not trivial. A variety of criteria can be used.[21] One of the most reliable is the analysis of the vibrational fine structure that appears in the photoelectron spectrum. This fine structure results from the formation of the product cation in a vibrationally excited state, for instance,

$$N_2 + h\nu \longrightarrow N_2{}^{+*} + e \text{ (electron)} \tag{52}$$

The cation has a different internuclear separation at equilibrium from that of the parent molecule. Because of the Franck–Condon principle, removal of an electron ($10^{-15}$ s) is faster than nuclear vibration ($10^{-13}$ s). Hence ionization removes an electron at an internuclear separation which is accessible only as a vibrationally excited state for the ion.

Figure 19 shows the fine structure in the photoelectron spectrum of the $N_2$ molecule. From the spacing, it is easy to calculate the frequencies for the stretching mode of the ion in its various electronic states. Comparison of these frequencies with those of the parent molecule enables a decision to be made as to the bonding, nonbonding, or antibonding nature of the orbital from which the electron has been removed. In this case the conclusion is that the first IP (at 15.6 eV) is weakly bonding, the second (at 16.9 eV) is strongly bonding, and the third (at 18.8 eV) is weakly antibonding. The

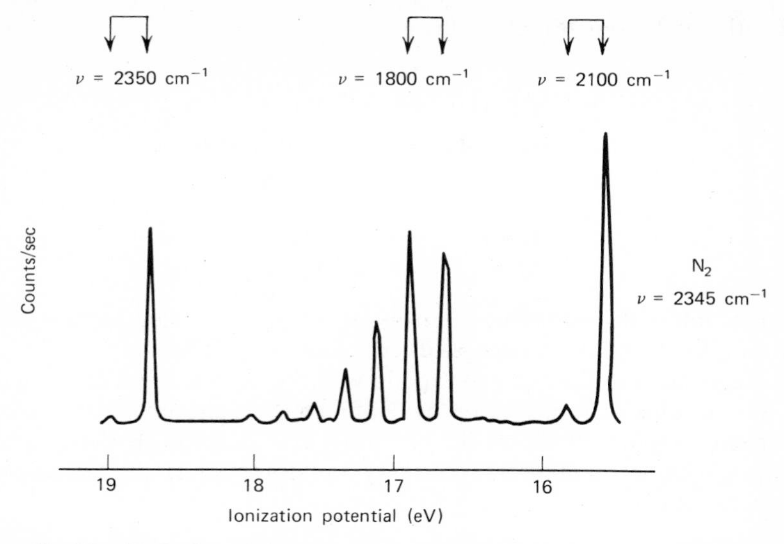


**Figure 19.** The He I photoelectron spectrum of molecular nitrogen; the vibrational fine structure is shown and also the vibrational frequency of $N_2$ is given. [Reproduced with permission from D. A. Sweigart and J. Daintith, *Sci. Prog. Oxf.*, **59**, 325 (1971).]

assignments are to $2\sigma_g$, $\pi_u$, and $1\sigma_u$, in that order.[48] The configuration for $N_2$ is accordingly

$$(1\sigma_g)^2(1\sigma_u)^2(\pi_u)^4(2\sigma_g)^2$$

Note that $(1\sigma_u)$ is actually antibonding, even though it is very stable in energy. It is essentially the difference of the $2s$ orbitals on the two nitrogen atoms.

After this brief review, we can consider the reactions of two homonuclear diatomic molecules with each other. The logical assumption is that a broadside collision with a four-center TS is energetically the most favorable reaction path. The hydrogen–iodine reaction is a classical example of this postulated mechanism.

$$\begin{matrix} \text{H—H} \\ \text{I——I} \end{matrix} \longrightarrow \begin{matrix} \text{H}\cdots\text{H} \\ / \quad \backslash \\ \text{I}\cdots\cdots\text{I} \end{matrix} \longrightarrow \begin{matrix} \text{H} & \text{H} \\ | & | \\ \text{I} & \text{I} \end{matrix} \qquad (53)$$

It was quite a surprise when, in 1965, Sullivan showed that this was not what happened at all, but that $H_2$ reacted with two I atoms at low temperature, and with one I atom at high temperature.[49]

$$\begin{aligned} I_2 &\rightleftharpoons 2I \\ I + H_2 &\rightleftharpoons IH_2 \\ IH_2 + I &\longrightarrow 2HI \end{aligned} \tag{54}$$

or

$$\begin{aligned} I + H_2 &\longrightarrow HI + H \\ H + I_2 &\longrightarrow HI + I \end{aligned} \tag{55}$$

The reaction with two iodine atoms is shown as a two-step process (sticky collision) but it may be a concerted termolecular reaction.

At any rate, it is necessary to explain why the expected reaction (53) does not occur. This was first done by Hoffmann[50] using orbital correlation arguments. Figure 20 shows the resulting MO diagram in the $C_{2v}$ point group. To understand it more readily we show below the reduction in symmetry scheme for $D_{\infty h}$ going to $C_{2v}$.

$$\begin{aligned} \sigma_g &\longrightarrow a_1 \qquad & \pi_u &\longrightarrow a_1 + b_1 \\ \sigma_u &\longrightarrow b_2 \qquad & \pi_g &\longrightarrow a_2 + b_2 \end{aligned}$$

Also for simplicity not all the MOs of the TS $H_2I_2$ are shown, since they give no new information.

The critical feature is the crossing of the $a_1$ and $b_2$ orbitals. This is characteristic of a forbidden reaction. Of course configuration interaction will prevent an actual crossing. Nevertheless a large energy barrier is indicated since in the TS the occupied $a_1$ orbital (originally the $\sigma_g$ orbital of $H_2$) will

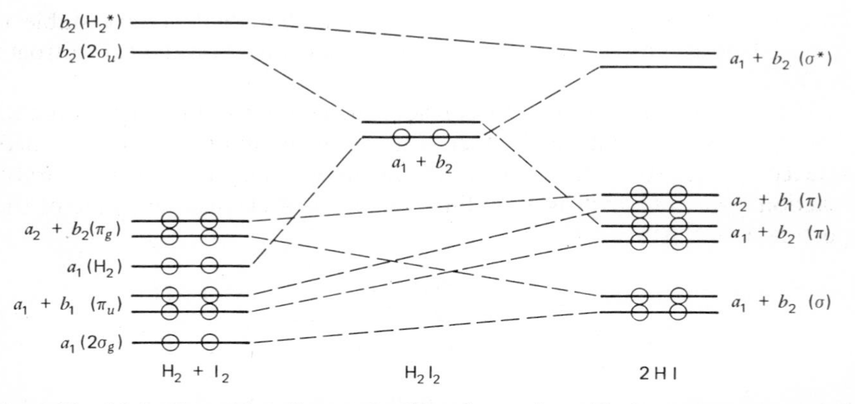


**Figure 20.** Molecular orbital diagram for $H_2 + I_2$ reaction with $C_{2v}$ symmetry preserved throughout; only the crossing MOs of the TS are shown ($\sigma_g$ and $\sigma_u$ orbitals of $I_2$ not shown).

have risen in energy, while the empty $b_2$ orbital (originally the $2\sigma_u$ orbital of $I_2$) has dropped in energy.

In perturbation theory terms, the problem is the absence of empty orbitals of $a_1$ symmetry to mix in. At this point one should ask about the absence of empty orbitals of $a_2$ and $b_1$ symmetry, since filled orbitals of these symmetries exist. This creates no problem, obviously, since smooth correlation is found with filled ground-state orbitals of the same symmetry. The question now is whether or not this could have been anticipated without drawing the MOs of the products as well.

The answer is affirmative, since these orbitals are not directly concerned with the chemical reaction itself. The $\pi_u$ and $\pi_g$ electrons essentially are lone pair electrons in $5p$ orbitals of the iodine atoms both in $I_2$ and in HI. Their dependence on excited states to change their wavefunctions, while not zero, is minimal. The same is true for the $\sigma_g$ and $\sigma_u$ electrons of the $I_2$ molecule which correlate with $\sigma$ orbitals of HI. These are, in both cases, essentially the $5s$ electrons of iodine.

The problem then reduces to the fact that the bonding $2\sigma_g$ orbital of $I_2$ and the $\sigma_g$ orbital of $H_2$ are both of $A_1$ symmetry, while their antibonding partners are of $B_2$ symmetry. There is no way that the filled orbitals and the empty orbitals can mix in order to break the original bonds and create new bonds. This can be shown quite vividly by diagrams such as Fig. 21. The different symmetries, as mentioned earlier, mean zero overlap. With zero overlap, there is no way the orbitals can mix, or in equivalent terms, no

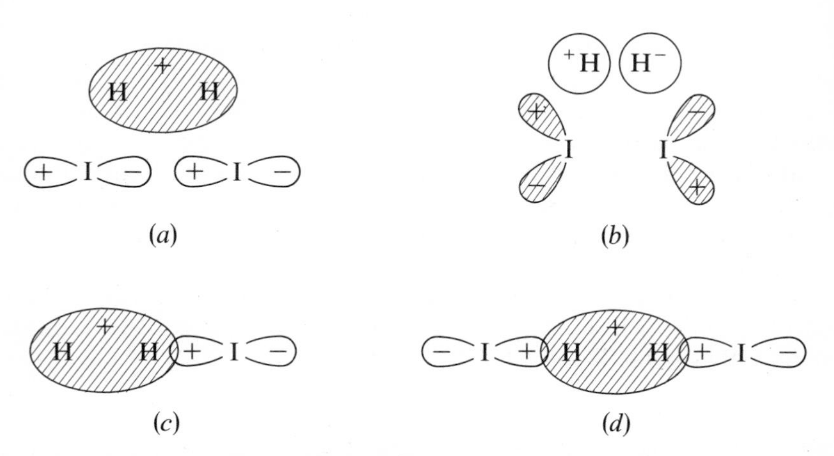


**Figure 21.** Schematic overlaps of filled (shaded) and empty (unshaded) MOs of $H_2$, $I_2$, and I: (*a*) zero overlap, electron flow not allowed; (*b*) nonzero overlap, but electron flow will create double bond, and cannot lead to reaction; (*c*) nonzero overlap for $H_2$ molecule with iodine atom; (*d*) nonzero overlap for termolecular reaction of $H_2$ with two I atoms.

way that electrons can flow from the filled orbitals to the empty ones to break the required bonds of the reactants.

The use of diagrams such as Fig. 21 is a very simple, but powerful, way of describing bimolecular reactions. We focus our attention on both the filled and empty MOs that relate to the bonds we need to break. Positive overlap between them corresponds to an allowed reaction. The larger the overlap, the more allowed is the reaction, that is, the lower will be the energy barrier, which can exist even for an allowed reaction.

For example, Fig. 21 shows that the mechanisms found for the $H_2$–$I_2$ reaction are indeed allowed. The half-filled $p$ orbital on the iodine atom can either accept an electron from the filled $\sigma_g$ orbital of $H_2$, or donate an electron to the empty $\sigma_u$ orbital. Presumably the former process predominates because of the electronegativity difference. Note that in either process, the bond between the two H atoms is broken and that new bonds between H and I are formed as a result of the positive overlap in the region between the nuclei.

The halogens react with each other to form interhalogen compounds.

$$Br_2 + I_2 \longrightarrow 2IBr \tag{56}$$

The complete MO correlation diagram for even this simple reaction gets quite cluttered, since there are 16 valence-shell MOs. We need look only at the $2\sigma_g$ and $2\sigma_u$ orbitals for each molecule. Just as for the $H_2$–$I_2$ reaction, the empty orbitals in $C_{2v}$ are $B_2$ and the filled orbitals are $A_1$. The reaction is forbidden.

Experimentally, reactions such as (56) are very rapid in some solvents. In polar solvents mechanisms in which ions, such as $I_2Br^+$ and $Br^-$, are formed may be operating. Our analysis does not include this possibility. What is also known is that very careful studies show that in dry, pure solvents reaction (56) and related ones are very slow. Also peculiar rate laws are found, not simple second-order ones.[51] A mechanism involving three molecules of halogen seems implicated. This would correspond to a reaction allowed by orbital symmetry, if a six-membered ring were formed. In the gas phase, the corresponding reactions occur either on the walls of the reaction vessel, or by slow free-atom chain processes.[52]

If we examine all planar, four-center reactions of homonuclear diatomic molecules with each other, orbital-symmetry forbiddenness will be found. An important example is the $N_2$–$O_2$ reaction

$$N_2 + O_2 \longrightarrow 2NO \tag{57}$$

This reaction is multiply forbidden since neither the $\sigma$ nor $\pi$ bonds can convert smoothly. Thermodynamically NO is 43 kcal less stable per mole than its elements. Yet once formed it is indefinitely stable. The activation energy for its decomposition is 50 kcal, which is remarkable, considering that the reaction is so much downhill thermodynamically. This is a good example of a symmetry-imposed barrier. It may be worthwhile making the reminder that if a symmetry barrier exists for a forward reaction, then it also exists equally for the reverse reaction.

A reaction that has often been proposed in gas-phase kinetics is of the following type:

$$2ClO \longrightarrow Cl_2 + O_2 \tag{58}$$

Since ClO is a free radical, it is particularly attractive to consider a two-step mechanism.

$$2ClO \longrightarrow \begin{array}{c} O{-}O \\ / \qquad \backslash \\ Cl \qquad\quad Cl \end{array} \tag{59}$$

$$\begin{array}{c} O{-}O \\ / \qquad \backslash \\ Cl \qquad\quad Cl \end{array} \longrightarrow O_2 + Cl_2 \tag{60}$$

The dimerization (59) is allowed, but the concerted decomposition (60) is not, even if the oxygen molecule is formed in the $^1\Delta_g$ state to preserve electron spin. We argue here from microscopic reversibility, since $Cl_2$ and $O_2$ cannot react to form $Cl_2O_2$, because of symmetry restrictions.

What actually happens in reaction (58) is the allowed series of reactions[53]

$$2ClO \longrightarrow Cl + ClO_2 \tag{61}$$

$$Cl + ClO_2 \longrightarrow Cl_2 + O_2 \tag{62}$$

The relatively stable analog of $Cl_2O_2$, $F_2O_2$ can be prepared. Its mode of decomposition is complex, F atoms being produced.[54]

$$F_2O_2 \longrightarrow F + O_2F \tag{63}$$

The reaction of $K_2$ and $Br_2$ is an interesting extreme of a very good reducing agent coupled with a very good oxidizing agent. Nevertheless the four-center mechanism

$$K_2 + Br_2 \longrightarrow 2KBr \tag{64}$$

does not occur. What actually happens, as shown by molecular beam experiments, is quite complicated.[55]

$$\begin{array}{l} K_2 + Br_2 \longrightarrow K_2^+ + Br_2^- \longrightarrow K_2Br + Br \\ K_2Br \longrightarrow KBr + K \end{array} \tag{65}$$

An electron jump from $K_2$ to $Br_2$ dominates the process.

The reverse of reaction (60) is an addition reaction to an unsaturated linkage. These are forbidden as planar four-center reactions. Consider the *cis* addition of $H_2$ to $N_2$.

$$\begin{array}{c} N\equiv N \\ + \\ H{-}H \end{array} \longrightarrow \begin{array}{c} N{=}N \\ |\quad | \\ H\quad H \end{array} \qquad (66)$$

In $C_{2v}$ the MOs are correlated as shown in Fig. 22. The same problem of crossing of an $a_1$ and a $b_2$ orbital occurs as in the $H_2$–$I_2$ reaction. However in this case the $b_2$ orbital is a $\pi_g$ orbital on nitrogen. The $a_1$ is, as before, the filled $\sigma_g$ orbital of $H_2$. The absence of an empty orbital of $a_1$ symmetry is the feature that makes (66) a forbidden reaction in PT.

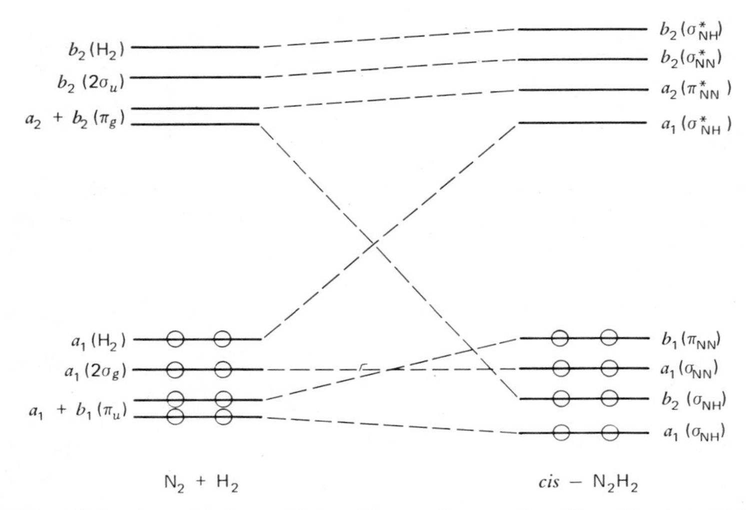


**Figure 22.** Molecular orbital correlation diagram for reaction $H_2 + N_2 \rightarrow$ *cis*-$N_2H_2$; the $C_{2v}$ point group is maintained ($\sigma_g$ and $\sigma_u$ orbitals of $N_2$ not shown).

So far we have only examined planar four-center TS. What about the feasibility of other directions of approach of two diatomic molecules and other geometries for the TS? Figure 23 shows that a perpendicular approach of $N_2$ and $H_2$ still leads to zero overlap of the critical orbitals, and hence to a forbidden reaction. However a skew approach leads to a net positive overlap, and to an allowed reaction. This is what Woodward and Hoffmann call "antarafacial" addition.[41] The corresponding *cis* addition process is called "suprafacial" addition. Figure 23 also shows that neither a perpendicular nor skew approach will make the $H_2$–$I_2$ reaction allowed.

In the skew approach, the point group becomes $C_2$, only a twofold axis passing through the centers of both molecules remaining. In this point group

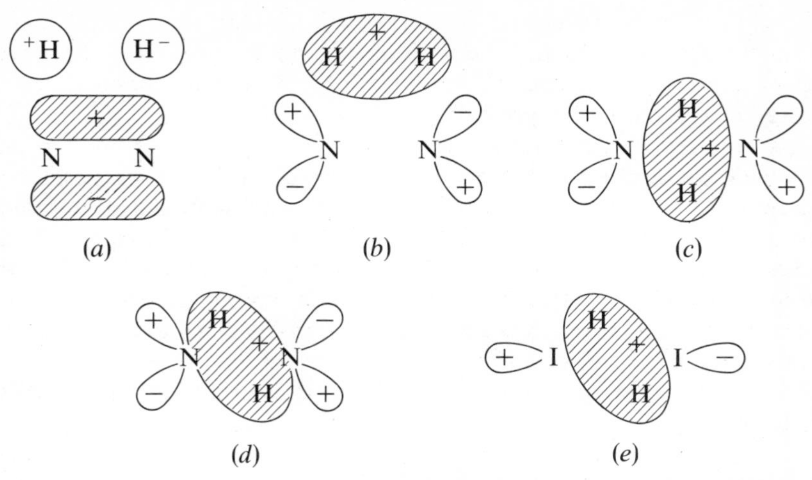


**Figure 23.** (*a*) Electron flow from $N_2$ to $H_2$ is forbidden in $C_{2v}$; (*b*) electron flow from $H_2$ to $N_2$ also forbidden; (*c*) perpendicular approach also leads to zero overlap; (*d*) skew approach leads to positive overlap and allowed (antarafacial) reaction; (*e*) skew approach in $H_2 + I_2$ still gives zero overlap.

both the $\pi_g$ orbital of $N_2$ and the $\sigma_g$ orbital of $H_2$ are of $A$ symmetry. (The second $\pi_g$ orbital is of $B$ symmetry). However the $\sigma_u$ orbital of $I_2$ is of $B$ symmetry, changing sign on rotation. A $\pi_u$ orbital would become of $A$ symmetry if pointing toward the second molecule, but would be of $B$ symmetry if oriented 90° from that direction.

The stable structure of diimide $N_2H_2$ is *cis* or *trans* coplanar, with the *trans* form somewhat more stable. The diimide that is formed by antarafacial addition is of neither of these geometries. It has very distorted bond angles and distances. This is a rather general result. If a symmetrical reaction coordinate forbidden by orbital symmetry is changed slightly to one of lower symmetry, it may become allowed. Obviously by even small distortions, all elements of symmetry can be lost, all orbitals become of the same symmetry, and all reactions are formally allowed. However there is still an energetic price to pay. An overlap which was zero in one point group will still be small, even if nonzero, in a lower symmetry point group produced by a small distortion. This means very weak interaction and mixing. Large distortions are needed to get large overlaps. Products are formed with serious departures from normal bond distances and angles. The strain energy is part of the energy barrier in this allowed, but still unfavorable, reaction path.

Isotope-exchange reactions are all forbidden by planar four-center paths. There are a series of isotope-exchange reactions that have been reported to occur by four-center mechanisms.[56] These include $^{14}N_2$–$^{15}N_2$, $^{16}O_2$–$^{18}O_2$,

$^{13}C^{16}O$–$^{12}C^{18}O$, and even $H_2$–$D_2$. The kinetic data are usually accumulated by shock tube methods and the rate laws found are abnormal. The interpretations are certainly not straightforward,[36,57] but the possibilities for allowed nonplanar paths should be examined.

Two features are prominent in the experimental results. One is that only molecules which are highly excited vibrationally can undergo exchange. The second is that $^{16}O_2$–$^{18}O_2$ exchange occurs with a low activation energy (about 40 kcal) and with a simple bimolecular rate law. In the case of the $O_2$ exchange, a reaction coordinate with a skew approach and with a $C_2$ point group is allowed. The two $\pi_g$ orbitals of each oxygen molecule split into A + B in $C_2$ point group. Since each is only half-filled, the two orbitals of $A$ symmetry, and the two of $B$ symmetry can interact.

A dimer of composition $O_4$ would be formed. It would have a twisted $C_2$ structure,

$$^{16}O_2 + {}^{18}O_2 \longrightarrow [O_4] \longrightarrow 2\,{}^{16}O^{18}O \qquad (67)$$

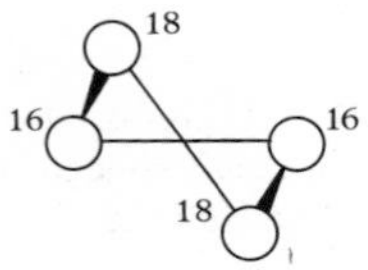

and would lead to exchange since all four oxygens become equivalent. In this case the skew structure is not a serious disadvantage because the normal bond angles of peroxides, such as $H_2O_2$ are very nearly the skew angles indicated in (67). This results because the bonds are formed to the hydrogen atoms using the same $\pi_g$ orbitals employed in forming $O_4$.

Of course $O_4$ is not a stable dimer, very little evidence for its existence being found. For example, the van der Waals heat of condensation of $O_2$, and of $N_2$, is only $-1.6$ kcal. This is about the same as for the inert gases of similar molecular weight. But we need postulate only that it be formed without any large energy barrier. If $O_4$, with all oxygen–oxygen bond distances the same, lies 40 kcal above $2O_2$, this will be consistent with the experimental results.

Next consider the $^{14}N_2$–$^{15}N_2$ system. In the $C_2$ point group the filled $\pi_u$ and empty $\pi_g$ orbitals both become A and B. We could have dimer formation with the same skew structure shown in (67). However in this case the structure is very distorted compared to the expected one. The molecule $N_4$, if it existed, would have one of two structures. It could be a planar molecule, derived from diimide, or it could be tetrahedral, just as is the related molecule $P_4$.

$$2N_2 \longrightarrow \begin{matrix} N{=}N \\ | \quad | \\ N{=}N \end{matrix} \quad \text{or} \quad N_4 \text{ (tetrahedral)} \qquad (68)$$

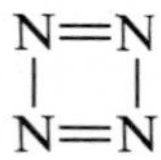

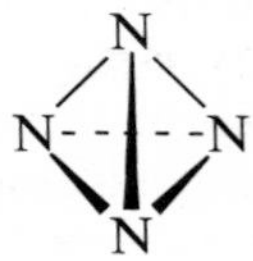

The tetrahedral structure should be strain free, and appears to be readily formed by a perpendicular approach of the two nitrogen molecules. If we attempt to form it in this way, we find that its formation is forbidden by orbital symmetry. The MO configuration for tetrahedral $N_4$ would be[58]

$$(a_1)^2(t_2)^6(2a_1)^2(e)^4(2t_2)^6(t_1)^0(3t_2)^0.$$

If we drop the symmetry to $D_{2d}$, for the point group during perpendicular approach, we have the following correlations:

| $T_d$ | $A_1$ | $E$ | $T_1$ | $T_2$ |
|---|---|---|---|---|
| $D_{2d}$ | $A_1$ | $A_1 + B_1$ | $A_2 + E$ | $B_2 + E$ |

For the two nitrogen molecules we must take the sum and difference of all equivalent orbitals, since these are the functions which have $D_{2d}$ symmetry. We obtain the following correlations by carrying out the symmetry operations of the group, and seeing how each combination transforms.

$$(\sigma_g + \sigma_g) = A_1 \qquad (\pi_u \pm \pi_u) = A_1 + B_2$$
$$(\sigma_g - \sigma_g) = B_2 \qquad (\pi_g \pm \pi_g) = A_2 + B_1$$
$$(\sigma_u \pm \sigma_u) = E \qquad (\pi_g' \pm \pi_g') = E$$
$$(\pi_u' \pm \pi_u') = E$$

We can draw the orbital correlation diagram, as shown in Fig. 24. There is a crossing of $b_1$ and $b_2$ levels, the latter arising from the $(2\sigma_g)$–$(2\sigma_g)$ combination. This is filled in the starting molecules but is empty in tetrahedral $N_4$. Hence we cannot form the latter by the direct approach of two nitrogen molecules. The formation is also forbidden in a skew approach of $D_2$ symmetry.

This result seems paradoxical in view of the existence of $P_4$, which uses the same valence shell AOs as $N_4$. However it does not imply that $N_4$ is remarkably unstable, but simply that a large energy barrier exists for its formation from two $N_2$ molecules. In fact $N_4$ might very well exist in a potential well of some depth, though less deep than that for $2N_2$.[58]

Since the experimental activation energy for the exchange reaction is 79 kcal, this might well be the barrier for formation of tetrahedral $N_4$. More likely is the possibility that skew $N_4$ is formed. Vibrational excitation of $N_2$ could aid in its formation since skew $N_4$ is simply vibrationally excited planar $N_4$, as in (68). It is well known from unimolecular reaction rate theory that rapid exchange of vibrational energy between various modes occurs at high levels of excitation. Similar arguments could be applied to isotope exchange in carbon monoxide.

We have seen in the preceding sections that reactions of the simple diatomic molecules are very sensitive to orbital symmetry restrictions. Chemists have known for a long time that the reactions of these molecules,

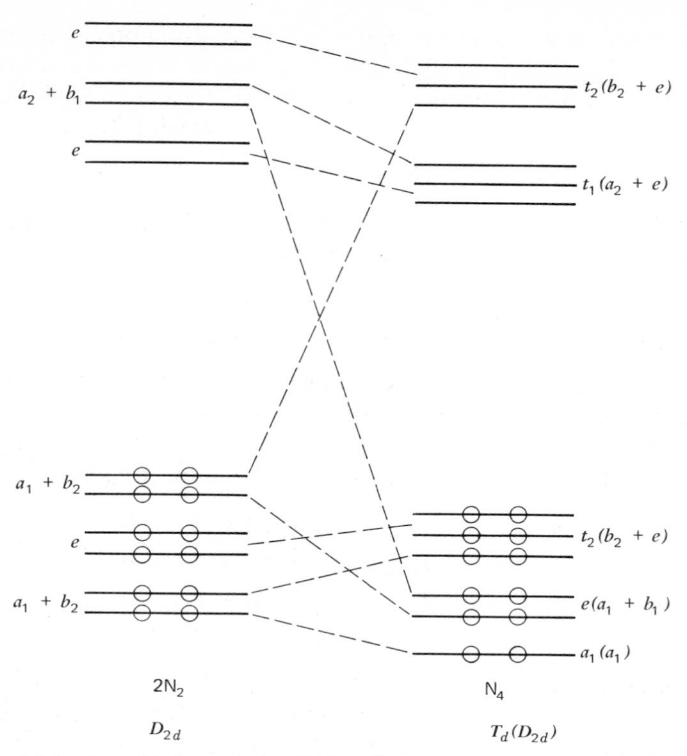


**Figure 24.** Molecular orbital correlation diagram for $2N_2 \rightarrow N_4$ (tetrahedral); the elements of $D_{2d}$ are preserved (MOs coming from linear combinations of the 2*s* orbitals on each nitrogen are not shown).

especially with each other, occur by extremely complex mechanisms. Free-radical chains are common, with the accompanying phenomena of explosions, induction periods, inhibition, and so on. It is a sobering thought to realize that today, nearly 200 years after Lavoisier, we still do not know the detailed mechanisms of reaction of molecular oxygen.

Sensitivity to catalysts is another characteristic of reactions of diatomic molecules, especially $H_2$, $O_2$, and $N_2$. Reactions that are blocked by orbital symmetry are prime candidates for catalytic assistance. One of the ways in which catalysts function is to circumvent symmetry-forbidden processes by creating symmetry allowed paths. Examples will be given in later chapters.

## Some Reactions of Free Atoms

Free-radical chains are resorted to in the reactions of diatomic molecules because the elementary reactions of free atoms and radicals are remarkably free from symmetry restrictions. Of course the Wigner–Witmer rules dictate the electronic state of the products, but an end-on attack of an atom on a molecule seems always allowed. These can be addition reactions, for example,

$$\begin{aligned} CO + O &\longrightarrow CO_2 \\ SO_2 + O &\longrightarrow SO_3 \end{aligned} \qquad (69)$$

or atom abstraction reactions, such as (36) or (55).

Broadside collisions of atoms and radicals with diatomic molecules may also occur. These can give reactions only in certain cases since symmetry barriers exist. Consider the general reaction

$$X + \begin{matrix} H \\ | \\ H \end{matrix} \longrightarrow X\begin{matrix} \diagup H \\ \\ \diagdown H \end{matrix} \qquad (70)$$

where X is an atom other than hydrogen. $H_2X$ may go on to form a linear molecule or may decompose into HX + H. A general correlation diagram can be drawn, using $s$ and $p$ orbitals on X.[59a] The point group is $C_{2v}$, as shown in Fig. 25.

The occupancy of the $p$ orbitals on X is critical. We obviously want two electrons in the $b_2$ orbital ($p_y$) and zero electrons in the $a_1$ orbital ($p_z$). The latter is the $p$ orbital pointing at the hydrogen molecule, the former is at

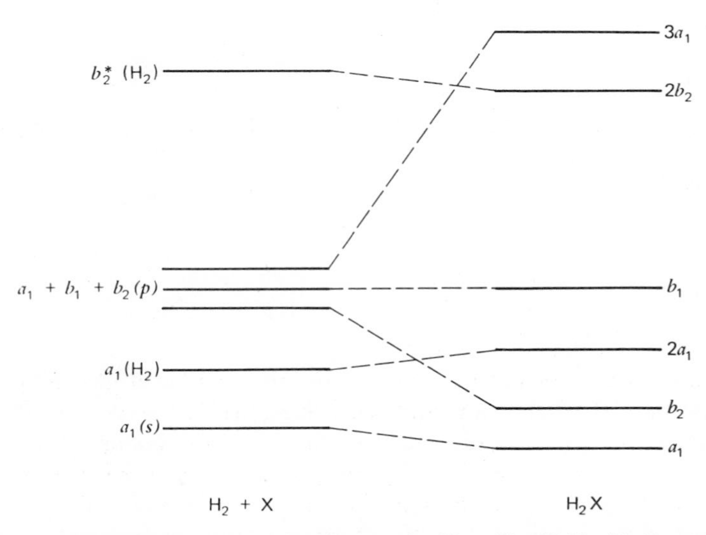


**Figure 25.** Molecular orbital correlation diagram for $H_2 + X \rightarrow H_2X$ with $C_{2v}$ point group preserved.

right angles, but also in the molecular plane. The prediction is that no ground-state atoms can react. This prediction is supported by data on the reactions of monatomic ions with $H_2$.[59b] Also an ab initio calculation of the Cl—$H_2$ potential energy surface shows that the broadside approach of Cl gives an energy barrier 20 kcal higher than for an end-on approach.[60]

Spin-paired excited states of atoms could react, however. For example, the $^1D$ oxygen atom has the $(p_x)^2(p_y)^2(p_z)^0$ configuration available. Indeed $^1D$ oxygen atom does appear to react by insertion into $H_2$, but $^3P$ oxygen atom reacts by H-atom abstraction.[13]

From the viewpoint of symmetry, the reaction of free atoms and radicals with halogen molecules is essentially the same as reaction with $H_2$. End-on attack is predicted, rather than broadside. This is supported by more detailed theoretical arguments for the reactions of Li and H.[61] There is obviously zero overlap between the half-filled *s* orbital of these atoms and the empty $(2\sigma_u)$ orbital of a halogen molecule in a broadside collision. Hence electron transfer cannot occur.

More important, the reaction of free atoms and radicals with an ethylenic double bond is essentially the same as (70) from symmetry. The initial products of broadside attack would be cyclic three-membered rings, which might react further.

$$X + C{=}C \longrightarrow X\langle^{C}_{C} \qquad (71)$$

End-on approach is sterically difficult, but addition to one carbon atom would correspond.

$$X + C{=}C \longrightarrow X{-}C{-}C \qquad (72)$$

The latter process is the normal reaction for ground-state atoms and radicals.

As expected, $^3P$ oxygen atom adds, according to (72), the resulting diradical then dissociating to $CH_3$ and HCO in the case of ethylene. However the $^1D$ oxygen atom can add to give ethylene oxide directly as in (71).[62] Since the addition reaction is very exothermic, the initial product may still decompose, unless quenched. Similarly $^1S$ and $^1D$ carbon atoms will add to olefins to give cyclic carbenes, which immediately rearrange to allenes.[63]

$$\mathrm{C} + \mathrm{>C{=}C<} \longrightarrow \text{cyclic carbene} \longrightarrow \mathrm{>C{=}C{=}C<} \qquad (73)$$

Ground-state $^3P$ carbon atoms do not react to give allenes, in accordance with symmetry prediction.

## THE BOND-SYMMETRY RULE

As we go on to consider the reactions of polyatomic molecules, the complete MO correlation schemes become hopelessly complex. Even the MOs of the reactants alone are too numerous to be handled conveniently. Fortunately it is not necessary to look at all this detail. Only a few critical orbitals dominate all the correlation diagrams, and it is usually easy to pick out which ones they are. They are the only ones we need to consider.

As we have seen, many of the MOs of the products will differ but little from those of the reactants. They are orbitals which keep their electrons localized or delocalized on the same atoms. The greatest changes will occur in those MOs that correspond to the changes in bonding. Both originally empty and filled MOs of the correct symmetry will mix to form the MOs giving the new bonding situation.

Often it is easy to identify these critical MOs. It is even easier if we take advantage of the fact (p. 30) that the bonds themselves can be assigned symmetry labels. We then need to mix filled bond orbitals with empty antibond orbitals to produce the new bond orbitals. With respect to the elements of symmetry that are conserved, only bond orbitals and antibond orbitals of the same symmetry can mix. More important, the new bond that is formed must have the same symmetry as the bond that was broken. If this is not consistent with the actual bonds made and broken in a reaction, the reaction is a forbidden one.

We can reach the same conclusion by using orbital correlation arguments. If the symmetries of the bonds that are made and the bonds that are broken always match up in pairs, then the noncrossing rule guarantees that the corresponding orbitals will not cross. If they do not cross, leading to a hypothetical excited state product, the reaction is allowed. The situation is illustrated in Fig. 26. This shows the canonical MOs of reactants being converted to bond orbitals of products, and these in turn to the bond orbitals

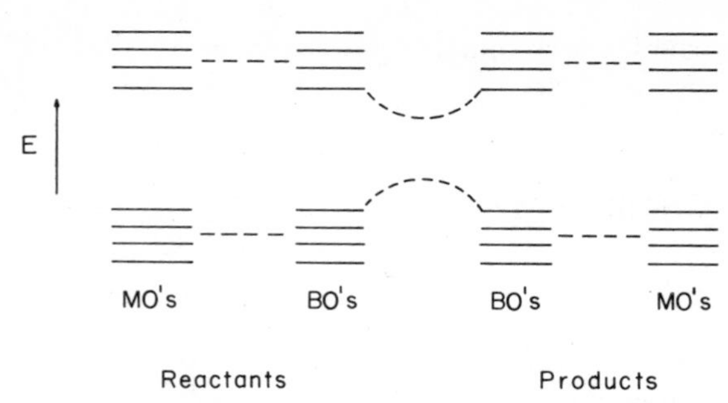


**Figure 26.** Scheme for an allowed reaction; the canonical MOs of reactants and products can be converted into an equivalent number of bond orbitals, and mixing of filled and empty bond orbitals can only occur if they are of the same symmetry.

of the reactants. The final correlation is to canonical MOs of the products. Every filled canonical MO corresponds in symmetry to a filled orbital which is usually a bond orbital, sometimes a lone-pair orbital (nonbonding), and very rarely, an antibond orbital. Figure 26 shows a filled bond orbital and an empty antibond orbital of the same symmetry combining to form a new bond orbital.

We have accordingly derived a very simple and powerful symmetry rule:[47] *a reaction is allowed if the symmetry of the bonds that are made is the same as the symmetry of the bonds that are broken.* The symmetry is related only to those elements that are conserved in going from reactants to products. Lone pairs of electrons must be considered in addition to bonds, if lone pairs become bonding, or if bonding electrons become lone pairs.

We apply the rule to see if the direct molecular formation of phosgene from CO and $Cl_2$ is allowed. Assume a broadside collision with a $C_{2v}$ point group, and draw Lewis diagrams for reactants and products.

$$\mathrm{Cl{-}Cl} + \mathrm{:C{=}O} \longrightarrow \mathrm{Cl_2C{=}O} \tag{74}$$

$$a_1 \qquad\qquad a_1 \qquad\qquad\qquad a_1 + b_2$$

The lone pairs on Cl and O can be ignored. The essential electrons are in the Cl—Cl $\sigma$ bonding orbital of $a_1$ symmetry, and in a lone-pair orbital on carbon, mainly $2s$ and also of $a_1$ symmetry. These four electrons must finish in the two C—Cl bonds. For two identical bonds, we must take the sum and difference (cf. Fig. 10), which gives bond orbitals of $a_1$ and $b_2$ symmetries. The bonds do not match up, and the reaction is forbidden.

Experimentally phosgene synthesis from the gaseous elements is a complex chain reaction involving chlorine atoms.[64] A similar forbidden reaction would be the molecular decomposition of formaldehyde.

$$\begin{array}{c} H \\ \diagdown \\ C{=}O \\ \diagup \\ H \end{array} \longrightarrow \begin{array}{c} H \\ | \\ H \end{array} + :C{=}O \qquad (75)$$

$$a_1 + b_2 \qquad\qquad a_1 \qquad a_1$$

Decomposition instead involves a free-radical process.

Another forbidden reaction is the molecular dissociation of ethane.

$$C_2H_6 \longrightarrow C_2H_4 + H_2 \qquad (76)$$

The reverse process, the suprafacial addition of hydrogen to ethylene is forbidden, just as is reaction (66). Assume a $D_{3h}$ eclipsed conformation for ethane which should facilitate concerted loss of two H atoms. The $C_{2v}$ point group is generated as the atoms start to leave. Figure 27 shows the bonds as orbitals. The orientation of the $\pi$ bond of ethylene is toward the $H_2$ molecule. Both bonds formed are $a_1$, but the sum and difference of the two C—H bonds is $a_1 + b_2$.

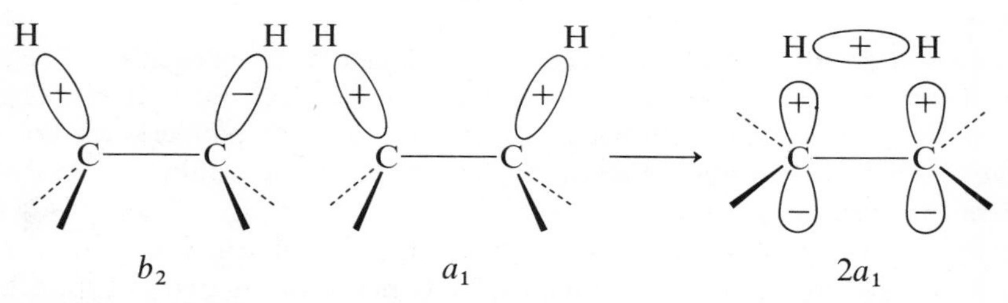


**Figure 27.** Suprafacial elimination of $H_2$ from $C_2H_6$ is forbidden by bond symmetries.

The concerted suprafacial cycloaddition of two ethylene molecules to form cyclobutane is a classical example of a forbidden reaction. It was first discussed by Woodward and Hoffmann in symmetry terms.[65] The point group is $D_{2h}$, but the classification is made only with respect to the two mirror planes that bisect the bonds of cyclobutane. These are sufficient; if a reaction is forbidden by some of the symmetry elements, it will also be forbidden by the complete number. The converse statement is not true, since a reaction which is forbidden by the full symmetry of the group may be allowed by all elements but one.

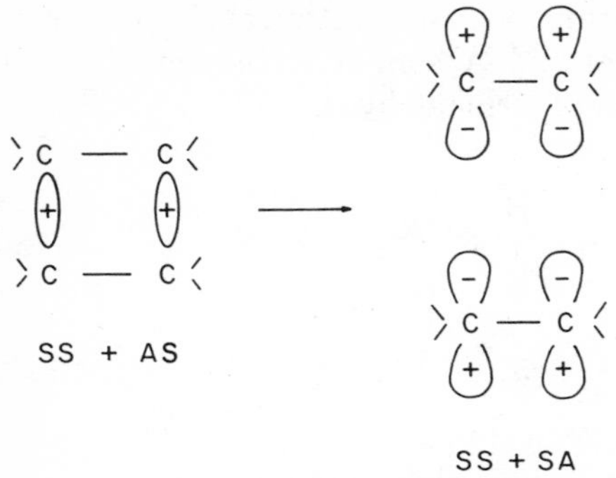

SS + AS

SS + SA

Figure 28. Symmetries of the two $\sigma$ bonds of $C_4H_8$ and the two $\pi$ bonds of $2C_2H_4$, with respect to two mirror planes; only the sum of the two orbitals is shown in each case.

Figure 28 shows that localized bond orbitals are used and that the key factor is that the symmetries of the bonds made and the bonds broken are not the same. In $D_{2h}$ the bonds made (the $\sigma$ bonds) are $A_{1g}$ and $B_{3u}$. The bonds broken (the $\pi$ bonds) are $A_{1g}$ and $B_{2u}$. There are two plausible allowed mechanisms. One is addition at one end of each ethylene molecule only, leading to a diradical intermediate

$$2C_2H_4 \longrightarrow \dot{C}H_2\text{—}CH_2\text{—}CH_2\text{—}\dot{C}H_2 \tag{77}$$

The point group would be $C_{2v}$ and the bonds made and broken would each be $A_1$ and $B_2$. To make this work out the two odd electrons of the diradical must be put in the antisymmetric combination of lone-pair orbitals on the two end atoms. Closing of the diradical to cyclobutane would be a separate elementary reaction (p. 378).

The second allowed mechanism is antarafacial addition. The point group becomes $C_2$, in which low symmetry the bonds again match in symmetry, being A and B. The product would be cyclobutane badly distorted by the twisting mode. The bulk of the experimental evidence supports the diradical mechanism.[66] The decomposition of cyclobutane from the PT-viewpoint has been discussed in detail by Wright and Salem.[67]

Unlike ethylene dimerization, the 1,4 addition of ethylene to butadiene is an allowed process, the well known Diels–Alder reaction. Figure 29 shows the reverse reaction, the decomposition of cyclohexene into ethylene and butadiene. The bonds that are broken (the 3–4 $\pi$ bond and the 1–2 and 5–6 $\sigma$ bonds) are of $2A' + A''$ symmetry. The bonds to be made (the 1–6, 4–5, and 2–3 $\pi$ bonds) are also $2A' + A''$. A disrotatory (see p. 72) twist of the hydrogen atoms on carbon atoms 2 and 5 is needed to preserve the $C_s$ point group.

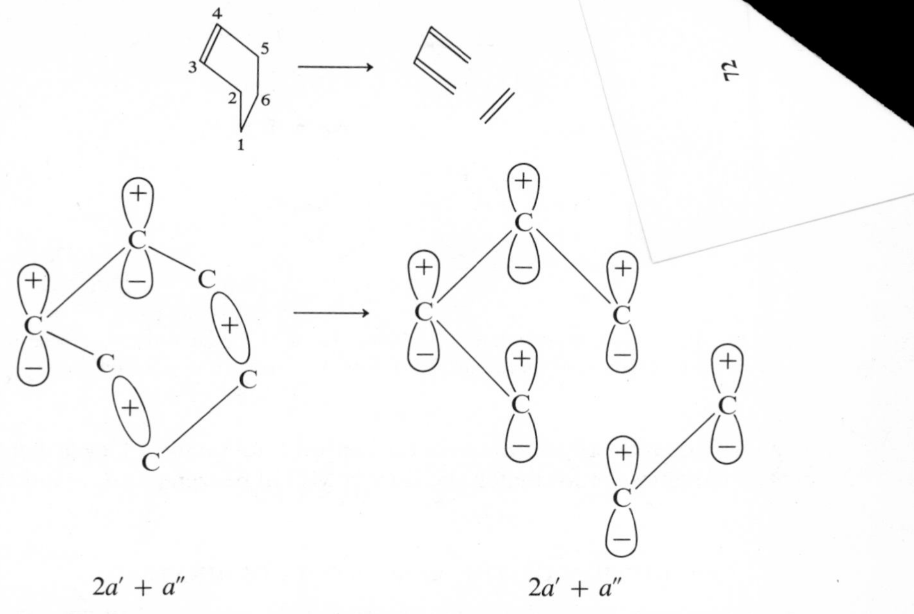


**Figure 29.** Bond symmetries for reverse of Diels–Alder reaction; only the sum of the two carbon—carbon $\sigma$ bonds is shown, and the lowest MO of butadiene; a plane of symmetry bisects the 3,4 and 1,6 bonds.

The examples of cycloaddition given can be extended to any number of molecules containing any number of $\pi$ orbitals. Both suprafacial and antarafacial additions of various parts can be considered. Woodward and Hoffmann[41] have shown that the general rule is that the reaction is allowed if the total number of elements $(4q + 2)_s$ and $(4r)_a$ is odd. Here $q$ and $r$ are the number of electrons involved in suprafacial and antarafacial addition, respectively. This rule can be extended to other types of reaction with cyclic TS, called, as a class, "pericyclic reactions."[41]

Our last example of the bond symmetry rule is the thermal isomerization of Dewar benzene into benzene. We can anticipate that the reaction is forbidden since Dewar benzene is 60 kcal less stable than benzene. It would not exist at all if it were not for a symmetry barrier. We select it as an example because the $\pi$ bonds of benzene cannot be represented as localized bonds. Hence we must use the occupied canonical MOs to represent them as delocalized bonds.

Figure 30 shows the bond symmetries. In $D_{6h}$ the occupied MOs of benzene are $(a_2u)^2(e_1g)^4$. But only the symmetry elements possessed in common with Dewar benzene are useful. These are for the $C_{2v}$ point group.

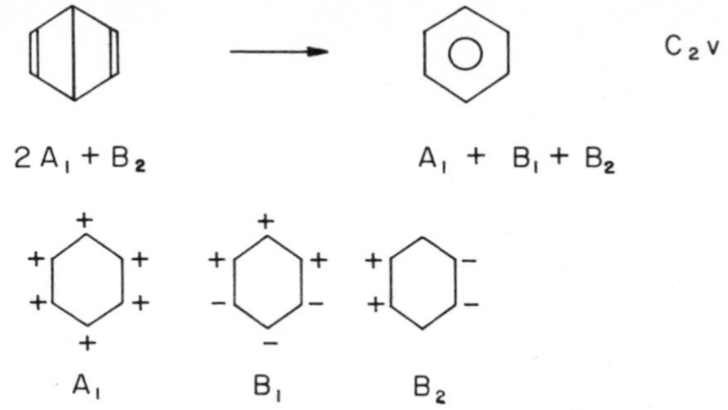

Figure 30. Bond symmetries in forbidden isomerization of Dewar benzene to benzene; the three occupied $\pi$ orbitals of benzene are shown with their symmetries given for $C_{2v}$.

The forbiddenness results because the central long bond of Dewar benzene of $A_1$ symmetry cannot match the third $\pi$ MO of benzene of $B_1$ symmetry.

## PREDICTION OF THE REACTION COORDINATE

Up till now the examples have all been of the kind where the reaction coordinate is first selected, and then to see if the bond symmetries or the orbital correlation diagram are consistent with the selection. The fundamental symmetry rule $\Gamma_{\varphi_i} \times \Gamma_{\varphi_f} = \Gamma_Q$ has been used by first selecting $Q$. It should be possible to reverse this procedure. That is, we pick $\varphi_i$ and $\varphi_f$ as being necessary for breaking or making certain bonds and then let their product select $Q$.[8]

This procedure is most easily applied to a unimolecular reaction, since in other cases we must still select an initial configuration of the interacting molecules. The unimolecular reaction starts with the defined equilibrium configuration. We use another classical example, the ring-opening reaction of cyclobutene to form butadiene. As Woodward and Hoffmann showed,[40] this can happen in two ways that are physically distinct and experimentally differentiable (by isotope labeling). Both terminal methylene groups can twist in the same direction (conrotation), or in opposite directions (disrotation). In the $C_{2v}$ point group of the cyclobutene molecule, the two motions are of $A_2$ and $B_1$ symmetries, as shown in Fig. 31.

Figure 31 also shows the four bond orbitals of the reactant that are pertinent. These are the filled and empty $\sigma$ and $\sigma^*$ orbitals of the 1–4 carbon—carbon bond, and the filled and empty $\pi$ and $\pi^*$ orbitals of the 2–3 double bond. Both of the bonds selected must be destroyed in the ring-opening process. A bond can only be broken by removing electrons from a

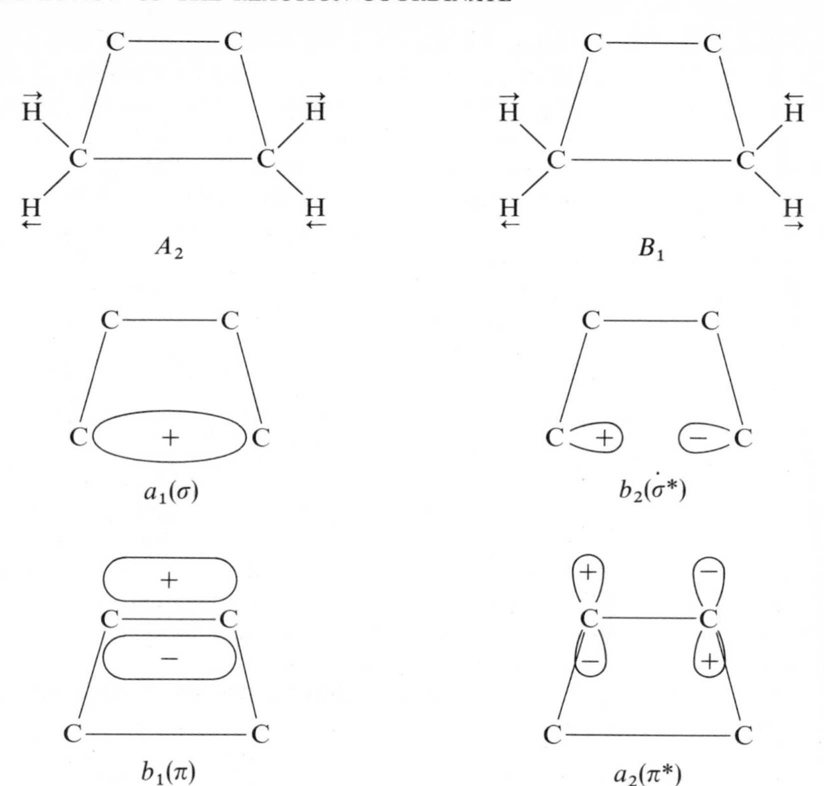


**Figure 31.** (*a*) Ring opening in cyclobutene, conrotatory opening is $A_2$ and disrotatory opening is $B_1$; (*b*) critical orbitals for cyclobutene.

bonding orbital, or by adding electrons to an anti-bonding orbital. This explains our choice of orbitals.

First consider that the bonds are made and broken by transferring electrons from the $\sigma$ orbital of $A_1$ species to the $\pi^*$ orbital of $A_2$ species. At the same time, electrons are transferred from the $\pi$ orbital ($B_1$) to the $\sigma^*$ orbital ($B_2$). In both cases the reaction coordinate is $(A_1 \times A_2) = A_2$, and $(B_1 \times B_2) = A_2$, a conrotatory ring opening. This agrees with the experimental result. The conrotatory twist destroys the two planes of symmetry of $C_{2v}$, and leaves only the $C_2$ axis. In the lower point group both the $a_1$ and $a_2$ orbitals become $a$ and both $b_1$ and $b_2$ become $b$. This is necessary for allowed mixing of the orbitals. The four selected orbitals are smoothly converted into the four $\pi$ orbitals of butadiene by the methylene twisting.

A disrotatory twist would leave only a single mirror plane. The point group would be $C_s$. Both filled orbitals become $a'$, and both empty orbitals become $a''$. No mixing of filled and empty orbitals would be possible. Mixing of the two filled orbitals would not change the bonding. We might also consider the effect of transferring electrons from $\sigma$ to $\sigma^*$, and from $\pi$ to $\pi^*$. These mixings both predict a reaction coordinate, $A_1 \times B_2 = B_2$, $A_2 \times B_1 = B_2$, which is an asymmetric distortion of the molecule in the plane. This cannot lead to ring opening.

We can, of course, also look at the reverse process of ring closing with exactly the same results. The ring-closing reaction of the allyl cation

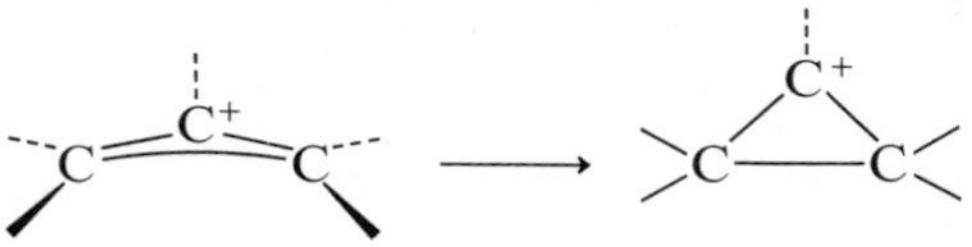

(78)

would require electrons to be moved from the filled $\pi$ orbital of $B_1$ symmetry to an empty $\sigma^*$ orbital of $A_1$ symmetry. The reaction coordinate is $A_1 \times B_1 = B_1$, or disrotatory. The empty $\sigma^*$ orbital of $A_1$ symmetry is selected because in the cyclopropyl cation we have formed a new carbon—carbon $\sigma$ bond of that symmetry. The electrons needed for this bond can only come from the delocalized $\pi$ electrons of the allyl system.

The cyclopropyl anion would differ in having two more electrons in a $\pi$ orbital of $A_2$ symmetry.

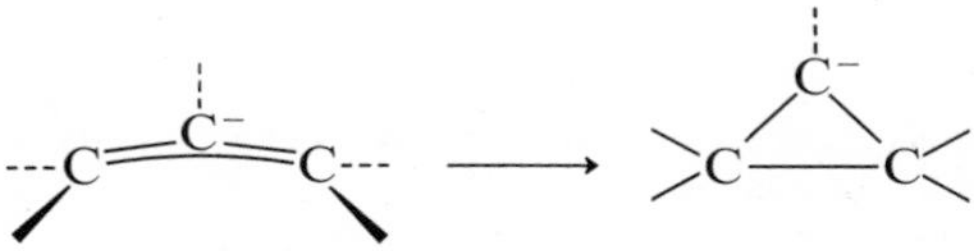

(79)

In the product the two extra electrons appear as a lone pair on the central carbon. This is of $B_1$ symmetry, if we consider the molecule to remain planar. Consequently we must transfer the electrons from the filled $a_2$ orbital into the empty $\sigma^*$ orbital of $A_1$ symmetry. This makes the reaction coordinate $A_2$, and the ring closure conrotatory. The $\pi$ electrons in the $b_1$ orbital of the reactant must be left there to become the $b_1$ lone pair of the product.

Woodward and Hoffmann developed a general rule for these ring closures which are called "electrocyclic reactions."[41] If the number of $\pi$ electrons is $4q$, the process is conrotatory; if it is $(4q + 2)$, the process is disrotatory. Here $q$ is an integer. The behavior of free radicals is not predicted by the rule (see p. 385).

Predicting the reaction coordinate for bimolecular reactions is usually not so straightforward. If we do not commit ourselves to an initial reaction coordinate, the question that must be answered is: In what point group do

$\varphi_i$ and $\varphi_f$ become of the same symmetry? This is more readily answered by inspection, and trial and error, than by group theory. One looks for the best positive overlap of $\varphi_i$ and $\varphi_f$.

If we select an initial reaction coordinate, and then find it forbidden, the symmetries of the mismatched orbitals will give us a signal for a better reaction coordinate. But if the original choice was very far from the best one, the signal may be inadequate. For example in the $H_2$–$I_2$ reaction with a four-center collision mechanism, the critical orbitals are $a_1$ and $b_2$ (Fig. 20). The direct product $(A_1 \times B_2) = B_2$, tells us that a correction in the original reaction coordinate should be made of $B_2$ symmetry. This is an asymmetric distortion in the plane, for example,

```
 H---H                 H---H
 /     \      B2        \     \
/       \   ------->     \     \          (80)
I---------I               I-----I
```

Comparison with Fig. 21 shows that this distortion indeed leads to a small positive overlap, where before the overlap was zero. We have come to a better reaction coordinate, but still would be quite ignorant of the complex mechanism that actually operates. In the same way in the forbidden $H_2$–$D_2$ four-center mechanism, the symmetries, $B_{2u} \times B_{3u} = B_{1g}$, predict a new reaction coordinate which takes the $D_{2h}$ point into $C_{2h}$. This is an in plane distortion similar to (80).

When two nitrogen molecules attempt to form tetrahedral $N_4$, the mismatched orbitals (corresponding to $\varphi_i$ and $\varphi_f$) are of $B_2$ and $B_1$ symmetry. Their direct product in $D_{2d}$ is $A_2$. A four-atom molecule in the $D_{2d}$ point group has no vibrational mode of $A_2$ symmetry. This can only be interpreted as meaning that the reaction coordinate continues to be totally symmetric. However instead of proceeding along the original direction ($Q$ positive), the path followed should be back to the reactants ($Q$ negative).

## THE JAHN–TELLER PHENOMENON

The classical example of a reaction coordinate which is selected by the symmetry of electronic energy levels is the first order Jahn–Teller effect (FOJT). This has proved to be a frustrating phenomenon from the experimental point of view, since its unambiguous detection is not easy. The theory is also much more difficult than appears at first glance.[68] The fundamental selection rule for the FOJT effect is

$$\Gamma_{\psi_0} \times \Gamma_{\psi_0} \subset \Gamma_Q \tag{81}$$

The proof for (81) was first given by Jahn and Teller using brute-force methods, testing all possible point groups. A more sophisticated proof has been given by Ruch and Schönhofer.[69]

The latter proof shows that a degenerate wavefunction $\psi_0$ gives rise to a nonsymmetric electron distribution, which is incompatible with the presence of certain symmetry elements. There can be no proper or improper axes of rotation, $C_n$ or $S_n$, with $n$ greater than 2. But these symmetry elements are necessary for degeneracies to exist. The important question is still left unanswered: how much distortion is needed to remove the degeneracy? Even the direction of the distortion is not very sharply defined, in spite of (81).

This can be made clearer by examining a famous case of the FOJT; the tetragonal distortion that is found for six-coordinated complexes of copper(II), a $d^9$ system. The electronic state of a regular octahedral structure would be $E_g$. The predicted mode of distortion is also $E_g$, since $(E_g \times E_g)^+ = A_{1g} + E_g$, and a symmetric mode cannot change the point group. However the $E_g$ mode is degenerate, and the two components are not equivalent. Figure 32 shows the appropriate symmetry coordinates, $Q_2$ and $Q_3$.

Displacement from the octahedral configuration generates a potential energy function

$$E = E_0 + aQ_2 + bQ_3 + cQ_2^2 + dQ_3^2 \tag{82}$$

where $a$ and $b$ are negative and $c$ and $d$ are positive. The final distortion could be any mixture of $Q_2$ and $Q_3$, giving rise to rhombic as well as tetragonal distortion. In principle, $a$ and $b$ can be found from first-order PT for degenerate levels, but this requires the evaluation of integrals. Both $c$ and $d$ can be identified with ordinary bond-stretching force constants. An additional complication is that (82) is inadequate for large displacements, and

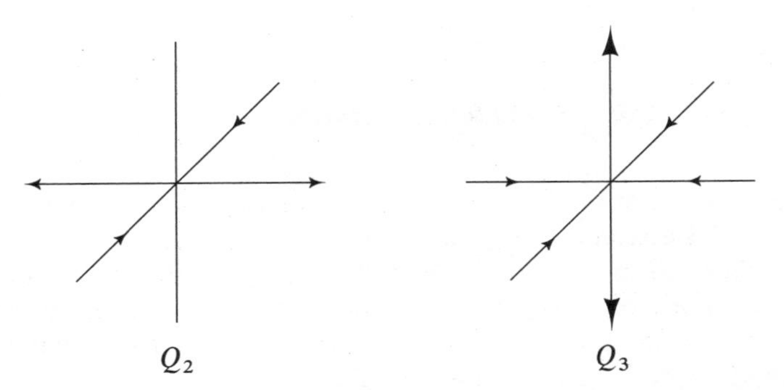


**Figure 32.** The two components, $Q_2$ and $Q_3$ of the $E_g$ vibrational mode in an octahedral molecule; in $Q_3$ the $z$ displacement is twice as large as $x$ and $y$.

terms in $Q_2^3$ and $Q_3^3$ must be added.[70] Finally spin-orbit coupling must be taken into account since this is another mechanism for removing orbital degeneracy.[71]

Jotham and Kettle have given a detailed analysis of the possible structural consequences of the Jahn–Teller effect.[72] In addition to the major rule of (81), two additional rules are necessary: (1) a new point group is accessible only if the Jahn–Teller active mode becomes totally symmetric in the new point group and (2) there can be no intermediate point group (such that there is an ordered reduction in symmetry in passing from original to intermediate to final point group) in which rule (1) is obeyed. Rule (1) follows because an infinitesimal displacement destroys the original point group and $Q$ must become symmetric to preserve continuity. Table 5 shows a listing of the final geometries for various starting geometries and electronic states.

Distortion of a symmetric complex usually leads to more than one equivalent structure. If the energy barrier between them is small, they will interconvert rapidly as a result of vibrational motion. This constitutes, in fact, a breakdown of the Born–Oppenheimer approximation, since electronic and vibrational wavefunctions cannot be separated. The resulting situation is called the "dynamic Jahn–Teller effect." Spectroscopic evidence for this dynamic situation is abundant.

The evidence for a static Jahn–Teller effect, where the barrier is large enough to freeze one structure for at least $10^{-9}$ s, is less certain. There is some question as to whether it has ever been observed.[72] Even the copper(II) examples are suspect, since only the $Q_3$ distortion is normally observed. The rhombic $Q_2$ distortion is small, if it exists at all. Other factors may well be responsible for the tetragonal structures commonly observed.[73] For example, it is possible that they result from mixing of the 4$s$ orbital of copper with the $3d_{z^2}$. In this case there would be a SOJT effect instead of a first-order one. This would account for the absence of the $Q_2$ distortion.

However if the FOJT effect is not large, it is logical to wonder if the SOJT effect should be of any importance. The FOJT effect relates to states equal in energy and SOJT, to levels that are only close. In both cases a distortion occurs that pushes the states apart. It seems that this should be much more efficient for equal, or nearly equal, states. As it turns out, this is not necessarily so, symmetry again playing a vital role.

## Can SOJT Effects be Large?

The physical bases for FOJT and SOJT are really quite different. This can be seen best by an MO interpretation. The FOJT effect arises from incompletely filled shells. These shells are doubly and triply degenerate MOs that by nature are not only of the same symmetry, but of the same kind.

**Table 5 Geometric Consequences of the Jahn–Teller Effect**[b]

| Parent Point Group | Jahn–Teller Active Vibrations | Electronic States Split | Ground-state Symmetries Consistent with the Operation of a Jahn–Teller Effect |
|---|---|---|---|
| $O_h$ | $E_g$ | $E_g, E_u, T_{1g}, T_{1u}, T_{2g}, T_{2u}, G_{3/2g}, G_{3/2u}$ | $D_{4h}, D_{2h}$ (rhombus) |
| | $T_{2g}$ | $T_{1g}, T_{1u}, T_{2g}, T_{2u}, G_{3/2g}, G_{3/2u}$ | $D_{3d}, D_{2h}$ (rectangle), $C_{2h}, C_i$ |
| $T_d$ | $E$ | $E, T_1, T_2, G_{3/2}$ | $D_{2d}, D_2$ |
| | $T_2$ | $T_1, T_2, G_{3/2}$ | $C_{3v}, C_{2v}, C_s, C_i$ |
| $T_h$ | $E_g$ | $E_g, E_u, T_g, T_u, G_{3/2g}, G_{3/2u}$ | $D_{2h}$ |
| | $T_g$ | $T_g, T_u, G_{3/2g}, G_{3/2u}$ | $C_{2h}, S_6, C_i$ |
| $D_{6h}$ | $E_{2g}$ | $E_{1g}, E_{2g}, E_{1u}, E_{2u}$ | $D_{2h}, C_{2h}$ |
| $D_{4h}$ | $B_{1g}$ | $E_g, E_u$ | $D_{2h}$ (rhombus) |
| | $B_{2g}$ | $E_g, E_u$ | $D_{2h}$ (rectangle) |
| $D_{3h}$ | $E'$ | $E', E''$ | $C_{2v}, C_s$ |
| $C_{6h}$ | $E_{2g}$ | $E_{1g}, E_{1u}, E_{2g}, E_{2u}$ | $C_{2h}$ |
| $C_{4h}$ | $2B_g$ | $E_g, E_u$ | $C_{2h}$ |
| $C_{3h}$ | $E'$ | $E', E''$ | $C_s$ |
| $C_{6v}$ | $E_2$ | $E_1, E_2$ | $C_{2v}, C_2$ |
| $C_{4v}$ | $B_1$ | $E$ | $C_{2v}$ |
| | $B_2$ | $E$ | $C_{2v}$ |
| $C_{3v}$ | $E$ | $E$ | $C_s, C_1$ |
| $D_{3d}$ | $E_g$ | $E_g, E_u$ | $C_{2h}, C_i$ |
| $D_{2d}$ | $B_1$ | $E$ | $D_2$ |
| | $B_2$ | $E$ | $C_{2v}$ |
| $S_4$ | $2B$ | $E$ | $C_2$ |

| | | | |
|---|---|---|---|
| $T_h$ | $G_g$ | $G_g, G_u, H_g, H_u, I_{5/2g}, I_{5/2u}$ | $T_h, D_{3d}, C_{2h}, S_6, C_i$ |
| | $2H_g$ | $T_{1g}, T_{1u}, T_{2g}, G_g, G_u, H_g, H_u$<br>$G_{3/2g}, G_{3/2u}, I_{5/2g}, I_{5/2u}$ | $D_{5d}, D_{3d}, D_{2h}, C_{2h}, C_i$ |
| $D_{\infty h}$ | none[a] | | |
| $D_{5h}$ | $E'_1$ | $E'_2, E''_2$ | $C_{2v}, C_s$ |
| | $E'_2$ | $E'_1, E''_1$ | $C_{2v}, C_s$ |
| $C_{5h}$ | $E'_1$ | $E'_2, E''_2$ | $C_s$ |
| | $E'_2$ | $E'_1, E''_1$ | $C_s$ |
| $C_{\infty v}$ | none[a] | | |
| $C_{5v}$ | $E_1$ | $E_2$ | $C_s, C_1$ |
| | $E_2$ | $E_1$ | $C_s, C_1$ |
| $D_{6d}$ | $B_1$ | $E_3$ | $D_6$ |
| | $B_2$ | $E_3$ | $C_{6v}$ |
| | $E_2$ | $E_1, E_5$ | $D_2, C_{2v}, C_2$ |
| | $E_4$ | $E_2, E_4$ | $D_{2d}, S_4$ |
| $D_{5d}$ | $E_{1g}$ | $E_{2g}, E_{2u}$ | $C_{2h}, C_i$ |
| | $E_{2g}$ | $E_{1g}, E_{1u}$ | $C_{2h}, C_i$ |
| $D_{4d}$ | $B_1$ | $E_2$ | $D_4$ |
| | $B_2$ | $E_2$ | $C_{6v}$ |
| | $E$ | $E_1, E_3$ | $D_2, C_{2v}, C_2$ |

[a] Linear molecules may be distorted by the Renner–Teller effect.
[b] From reference 72.

Examples are $d$ orbitals in inorganic chemistry and $\pi$ orbitals in organic chemistry.

An incompletely filled shell can give rise to degenerate states; for example, the configuration $(e)^1$ or $(e)^3$ gives an $E$ state. A nondegenerate state can also result, for example for $(e)^2$, but then there will be another state of similar energy for at least one multiplicity. These states differ only in the amounts of electron repulsion, which means that, at the one-electron level of approximation, they also are degenerate states. An example would be the two lowest singlet states of cyclobutadiene. These cases were first called "pseudo-Jahn–Teller systems" by Longuet-Higgins and Coulson.

It seems eminently reasonable to use this nomenclature. Both effects arise from the same cause: an incompletely filled shell of MOs. The true Jahn–Teller effect and the pseudo-Jahn–Teller effect may reasonably be lumped together, even though the energy lowering in the latter case is governed by the quadratic terms of (11). The magnitudes of the effects will be similar.

The other possibility for a quadratic energy lowering is when a molecule has filled shells, but where there is only a small energy difference between the highest filled orbital and the lowest empty one. This will again give rise to at least one low-lying excited state. Let us agree that only these cases be called "second-order Jahn–Teller." (Depending on the nature of the critical, or frontier, orbitals, this is the case where large deformations of the molecule may result.

The FOJT distortion will always split the degenerate states, and also the orbitals. Furthermore both will normally become of different symmetries in the new point group. Rare exceptions may occur if the point group is of very low symmetry. Since the states are of different species, they can no longer interact under the influence of $(\partial U/\partial Q)$, which like $Q$ itself, is totally symmetric. Furthermore, the orbitals, being of different species, can no longer interact at the monoexcited level. They can only do so at the much less efficient diexcited level, in short, to reduce interelectronic repulsion by configuration interaction.

However in the second-order case $\varphi_i$ and $\varphi_f$ must become the same species in the new point group since their direct product must be totally symmetric to match $Q$. This means that they can continue to interact under the influence of the perturbation $(\partial U/\partial Q)$. Large distortions become possible in some cases because there is a favorable mechanism whereby the wavefunction can adjust to changing nuclear positions. Mixing of $\varphi_i$ and $\varphi_f$ can produce a new orbital consistent with changing chemical bonds, as we have shown.

This favorable mechanism is absent in the FOJT and pseudo systems. Since the original symmetric shape of the molecule is usually selected because it would be stable in the absence of the JT effect, there will be strong forces resisting any distortion. In addition to the elastic forces indicated in

(11), interelectronic repulsions resist distortion. The reason is that in the degenerate case the electrons are delocalized to the greatest possible extent. Distortion forces the electrons to localize more in the MOs that are stabilized.

The existence of symmetry in a molecule creates degeneracies not only in the MOs of the molecule, but also in the electronic states. The role of the FOJT effect is to destroy the degeneracy by destroying the symmetry. The role of the SOJT effect is to push the levels even further apart. This raises the interesting question as to why any symmetry at all should persist in molecules. It is a matter of experience that we find the two O—H bond distances in water equal to each other, but why should this be so? Of course in the $C_{2v}$ point group we cannot have any degenerate orbitals. However a $B_2$ distortion in principle could lower the energy, rather than raise it.

Any element of symmetry is associated with an energy maximum or minimum, with respect to a coordinate $Q$ that destroys it. The potential energy function becomes

$$E = E_0 + \frac{1}{2}\left\langle \psi_0 \left| \frac{\partial^2 U}{\partial Q^2} \right| \psi_0 \right\rangle Q^2 + \sum_k \frac{\left\langle \psi_0 \left| \frac{\partial U}{\partial Q} \right| \psi_k \right\rangle^2 Q^2}{(E_0 - E_k)} \tag{83}$$

If the second term in $Q^2$ is larger than the first, the energy *must* be a maximum. Distortion along $Q$ will occur spontaneously. If $Q$ is nonsymmetric, all elements of symmetry created at $Q_0$ will be destroyed. If the first term in $Q^2$ is larger than the second, the energy is a minimum and the elements of symmetry due to $Q_0$ will be conserved. From this point of view, symmetry exists in a molecule whenever it is consistent with a large gap between the occupied and the empty MOs.

However the nature of the orbitals must also be taken into account. Two orbitals close in energy may still not be efficient in changing the shape of a molecule. For example, in octahedral complexes of the transition metal ions, the $e_g$ and $t_{2g}$ orbitals lie close to one another (2-3 eV) and are not completely filled. Since $(E_g \times T_{2g}) = (T_{1g} + T_{2g})$, the allowed vibrations are $T_{1g}$ and $T_{2g}$. The former does not exist for $XY_6$ molecules, but $T_{2g}$ gives rise to a rectangular $D_{2h}$ structure, which is not observed experimentally.

The reason is that the $e_g$ and $t_{2g}$ orbitals are of the same kind, largely $d$ orbitals on the central metal. Mixing two $d$ orbitals together produces another $d$ orbital, rotated about some axis. Such an effect does not improve the main bonding in an octahedral complex, and instead hinders it.

Suppose there was a bent molecule, $H_2X$, which had a small gap between an occupied orbital of $a_1$ symmetry and a filled orbital of $b_2$ symmetry. The $B_2$ mode would distort spontaneously making the bond lengths unequal, but there is no reason why this would stop with the molecule $H_2X$ intact. Instead complete dissociation into HX and H could be the result. Thus small distortions from symmetric structures are not demanded. If a molecule is marked as being unstable by either a FOJT or SOJT effect, it is likely to rearrange very extensively, either dissociating, or taking up quite a different structure. Notice that (83) is valid only for small $Q$. If there are large distortions, we must return to (11), applying it pointwise along the reaction coordinate.

## MORE ON FORCE CONSTANTS

It would be advantageous to have some independent evidence as to the magnitude of integrals such as $\langle\psi_0|\partial U/\partial Q|\psi_k\rangle Q = H_{0k}$, since they play a critical role in PT. They also appear in the theory of vibronic coupling, which makes forbidden absorption bands become partly allowed. According to (12), a molecular vibration mixes in to the ground-state wavefunction some excited-state wavefunctions. The same mixing also happens to each of the excited states. An electronic transition due to light absorption may be forbidden, according to (5), for pure states, but it will be partly allowed for the contaminated states.

The ratio of intensities of forbidden to allowed states by this mechanism becomes

$$\frac{I_f}{I_a} = \frac{H_{0k}^2}{(E_0 - E_k)^2} \tag{84}$$

since the intensity depends on the square of the transition moment [see (4)]. For simplicity we are assuming that only a single excited state, $\psi_k$, is vibronically coupled. Experimentally $(I_f/I_a)$ has a range from $10^{-4}$ to $10^{-1}$. Taking $|E_0 - E_k| = 25{,}000\ \text{cm}^{-1}$, we find that $H_{0k} = 250 - 8330\ \text{cm}^{-1}$, or $H_{0k}^2/|E_0 - E_k| = 2.5 - 2500\ \text{cm}^{-1}$. This includes the range of vibrational energies and justifies our contention that the second term in $Q^2$ in (83) can sometimes exceed the first term in $Q^2$.

The second method of evaluating $H_{0k}$ is to use accurate wavefunctions and carry out direct integration. Summation over all the excited states is not possible, of course, but at least in one case it can be done. Take the case of a single hydrogen atom and displace the nucleus from the center of the charge cloud of the electron by an infinitesimal amount.[74] The potential energy is given by (83) or (16). Furthermore we know that $f_{00} = f_{0k}$ in this case since the potential energy change for a simple translation of the hydrogen atom is zero. The electron cloud perfectly follows the nucleus. The evaluation of $f_{00}$ is easy and so is the summation over all the excited states of the hydrogen atom. As it turns out, 93% of $f_{0k}$ comes from the continuum of excited states corresponding to an unbound electron.

This is a rather alarming result, since it suggests that relaxation of the electron cloud on perturbation has little to do with the lowest-lying states, contrary to our assumptions. A series of ab initio calculations on diatomic molecules reveals the difficulty.[75] Use is made of the relationship

$$\left(\frac{\partial^2 E}{\partial Q^2}\right)_{Q_0} = f_{00} + f_{0k} \tag{85}$$

The second derivative of $E$ with respect to $Q$ can be found by pointwise calculation of the energy. The value of $f_{00}$ can be calculated from the wavefunction $\psi_0$, at $Q_0$ (the equilibrium internuclear distance), and $f_{0k}$ is calculated by difference.

Now the results obtained depend on whether $\psi_0$ is locked on fixed points in space, or whether it is locked on the nuclei. For $N_2$ the following was found:

$$\left.\begin{aligned} f_{00} &= 6018.68 \text{ AU} \\ f_{0k} &= -6017.21 \text{ AU} \end{aligned}\right\} \text{ locked on space}$$

$$\left.\begin{aligned} f_{00} &= 2.885 \text{ AU} \\ f_{0k} &= -1.935 \text{ AU} \end{aligned}\right\} \text{ locked on nuclei}$$

One atomic unit $= 15.57 \times 10^5$ dynes $\text{cm}^{-1}$. The enormous values in the locked on space result are largely due to the work needed to move the $1s$ electrons of the two nitrogen nuclei. It seems clear that chemistry is little concerned with hypothetical energies of this kind. Accordingly we assume that our wavefunctions consist of linear sums of atomic orbitals which continue to be centered on the nuclei even if the nuclei move. This gives us the much more reasonable results shown for "locked-on nuclei." These are due almost entirely to changes in energy experienced by the valence electrons. Inner-shell electrons automatically follow the nuclei by the use of the continuum excited states, as was previously indicated. The valence-shell electrons follow the nuclei by the same continuum states, but far more importantly, they change their relative positions by using the low-lying states arising from the valence-shell orbitals.[76] Thus force constants are indeed determined by these low-lying states, as we previously assumed.[77]

The normal coordinates and their accompanying potential functions define a unique set of force constants, given by $(f_{00} + f_{0k})$, for each vibrational mode. However these constants are of little value for complex molecules. There is no easy way to determine them even if we have all the fundamental frequencies. Furthermore, if determined for one molecule, they would be of little use in other molecules. Instead it is much more useful to use force constants such as those of the generalized valence force field. These consist of bond-stretching and bond-bending constants, together with various interaction constants.

For example, in a linear symmetric molecule such as $CO_2$ we would use the potential function,

$$2U = f_R(R_1^2 + R_2^2) + f_\alpha(\alpha)^2 + 2f_{RR}(R_1R_2) \tag{86}$$

where $R_1$ and $R_2$ are the displacements of the first and second C—O bonds, and $\alpha$ is the bending angle from linearity. This potential function can be fitted to the experimental frequencies to give

$$\begin{aligned} \nu_1 &= \frac{g(\mu)}{2\pi}[f_R + f_{RR}] = 1340 \text{ cm}^{-1} \qquad \Sigma_g^+ \\ \nu_2 &= \frac{h(\mu)}{2\pi} f_\alpha = 667 \text{ cm}^{-1} \qquad \Pi_u \\ \nu_3 &= \frac{j(\mu)}{2\pi}[f_R - f_{RR}] = 2349 \text{ cm}^{-1} \qquad \Sigma_u^+ \end{aligned} \tag{87}$$

where $g$, $h$, and $j$ are known functions of the atomic masses. The question now is whether PT tells us anything about these force constants. The symmetries of $\nu_1$, $\nu_2$, and $\nu_3$ are shown.

The procedure to be used is due to Bader.[78] We had earlier showed that, while a rough correlation existed between force constants and UV absorption spectra (Tables 2 and 3), it was not good enough to enable us to predict with any certainty. Thus we give up hope of predicting constants such as $f_R$ and $f_\alpha$. However we can tell something about the interaction constant $f_{RR}$.

The MO configuration of $CO_2$ is $(1\sigma_g)^2(1\sigma_u)^2(2\sigma_u)^2(\pi_u)^4(\pi_g)^4(2\pi_u)^0(3\sigma_g)^0$. The ground state is $\Sigma_g^+$. The lowest-energy electronic transition is $(\pi_g) \rightarrow (2\pi_u)$. The states generated are $\Sigma_u^+$, $\Sigma_u^-$, and $\Delta_u$. Of these the $\Sigma_u^+$ state is lowest in energy. The PT prediction is that the $\Sigma_u^+$ vibration is favored over $\Sigma_g^+$ and $\Pi_u$ by electron relaxation effects. We must still take into account the classical contribution $f_{00}$, however, which is smaller for bending than for stretching modes.

This explains why $\pi_u$ is the lowest-energy vibration, being the bending mode. However (87) shows us that the force constants for $\nu_1$ and $\nu_3$ are the same except for the effect of the interaction constant. In order to favor the asymmetric stretch over the symmetric, it is only necessary that $f_{RR}$ have a positive value. This is indeed the case, with $f_R = 15.5 \times 10^5$ and $f_{RR} = +1.3 \times 10^5$ dynes cm$^{-1}$, from the experimental values of $\nu_1$ and $\nu_3$, and the values of $g$ and $j$.

We have shown that apparently the sign of bond-interaction constants can be predicted from the symmetries of the ground and first excited states. We examine the vibrational mode which has the same symmetry as $\rho_{0k}$. If all bond lengths increase, then $f_{RR}$ is negative. If some increase and some decrease, then $f_{RR}$ is positive. Alternatively, $f_{RR}$ has the sign that favors the vibrational mode.

Burdett has greatly extended Bader's early work.[79] In suitable circumstances PT can be very useful in understanding infrared vibrational frequencies and intensities. For example, in the metal hexacarbonyls $Cr(CO)_6$, $Mo(CO)_6$, and $W(CO)_6$ there are very characteristic strong bands at about 2000 $cm^{-1}$, which are due mainly to C—O stretching. By symmetry these divide into three different bands of $A_{1g}$, $E_g$, and $T_{1u}$ species. The MO scheme for these complexes, and the UV absorption spectra, predict electronic transitions of increasing energy $T_{1u} < E_g < A_{1g}$.[80] Since it is very likely that the classical force constant $f_{00}$ is the same for all three C—O stretches, the electron-relaxation term, $f_{0k}$, should dominate the differences. This seems to be the case since $\nu_{A_{1g}} > \nu_{E_g} > \nu_{T_{1u}}$.

The intensities of the infrared absorption depend on the change in dipole moment with changing coordinate $Q$. Vibrational motion always mixes in excited states of matching symmetry. The resulting $\rho_{0k}$ for the vibrations will usually create a corresponding dipole moment. The equation for $\mu$, the total dipole moment, becomes

$$\mu = \mu_{00} + 2Q \sum_k \frac{\left\langle \psi_0 \left| \frac{\partial U}{\partial Q} \right| \psi_k \right\rangle}{(E_0 - E_k)} \mu_{0k} \tag{88}$$

Consequently any vibration which has a strong relaxation, as a result of a low-lying excited state, will have a large value of $(\partial\mu/\partial Q)$ and hence a correspondingly large intensity.[81]

## REFERENCES


1. S. Glasstone, K. J. Laidler, and H. Eyring, *The Theory of Rate Processes*, McGraw-Hill, New York, 1941, Chapter 3.
2. (a) J. N. Murrell and K. J. Laidler, *Trans. Faraday Soc.*, **64**, 371 (1968) and J. N. Murrell and G. L. Pratt, ibid., **66**, 1680 (1970); (b) J. W. McIver, Jr., and R. S. Stanton, *J. Am. Chem. Soc.*, **94**, 8618 (1972); (c) K. Fukui, *J. Phys. Chem.*, **74**, 4161 (1970).
3. H. A. Jahn and E. Teller, *Proc. Roy. Soc.*, **A161**, 220 (1937).
4. U. Öpik and M. H. L. Pryce, *Proc. Roy. Soc.*, **A238**, 425 (1957).
5. R. F. W. Bader, *Can. J. Chem.*, **40**, 1164 (1962).
6. R. McWeeney and B. T. Sutcliffe, *Methods of Molecular Quantum Mechanics*, Academic, New York, 1969.
7. R. Renner, *Z. Phys.*, **92**, 172 (1934).
8. R. G. Pearson, *Theor. Chim. Acta*, **16**, 107 (1970).

9. D. H. W. den Boer and H. C. Longuet-Higgins, *Mol. Phys.*, **5**, 387 (1962); B. J. Nicholson and H. C. Longuet-Higgins, ibid., **9**, 461 (1965).
10. (a) H. J. Maria, P. Larson, M. E. McCarville, and S. P. McGlynn, *Acc. Chem. Res.*, **3**, 368 (1970); (b) W. A. Goddard, III, T. H. Dunning, Jr., W. J. Hunt, and P. J. Hay, ibid., **6**, 368 (1973).
11. L. Salem, *Chem. Phys. Lett.*, **3**, 99 (1969).
12. G. Herzberg, *Electronic Spectra of Polyatomic Molecules*, Van Nostrand, Princeton, 1966. This is the prime reference for excited electronic states of molecules.
13. R. F. W. Bader and R. A. Gangi, *Chem. Phys. Lett.*, **6**, 312 (1970).
14. M. Oppenheimer and R. S. Berry, *J. Chem. Phys.*, **59**, 5058 (1971).
15. R. S. Mulliken and W. B. Person, *Ann. Rev. Phys. Chem.*, **13**, 107 (1962).
16. R. L. Flurry, Jr., *J. Phys. Chem.*, **69**, 1927 (1965).
17. H. C. Longuet-Higgins, *Proc. Roy. Soc.*, **A235**, 537 (1956).
18. L. Brillouin, *Actual. Sci. Ind.*, **71**, (1933); **159**, (1934).
19. T. Koopmans, *Physica*, **1**, 104 (1934).
20. C. R. Brundle, M. B. Robin, and H. Basch, *J. Chem. Phys.*, **53**, 2196 (1970).
21. D. W. Turner, A. P. Baker, C. Baker, and C. R. Brundle, *Molecular Photoelectron Spectroscopy*, Interscience, New York, 1970.
22. J. S. Lennard-Jones, *Proc. Roy. Soc.*, **A198**, 1, 14 (1949).
23. W. England, K. Ruedenberg, and L. S. Salmon, *Fortschr. Chem. Forsch.*, **16**, 221 (1971).
24. H. B. Thompson, *Inorg. Chem.*, **7**, 604 (1968).
25. F. O. Ellison and H. Shull, *J. Chem. Phys.*, **23**, 2348 (1955).
26. E. Teller, *J. Phys. Chem.*, **41**, 109 (1937); K. R. Naqvi and W. Byers-Brown, *Int. J. Quant. Chem.*, **6**, 271 (1972).
27. E. Wigner and E. E. Witmer, *Z. Phys.*, **51**, 859 (1928).
28. Y. N. Chiu, *J. Chem. Phys.*, **58**, 722 (1973).
29. K. E. Shuler, *J. Chem. Phys.*, **21**, 624 (1953).
30. F. Hund, *Z. Phys.*, **40**, 742 (1927); **42**, 93 (1927); R. S. Mulliken, *Rev. Mod. Phys.*, **4**, 1 (1932).
31. H. Bethe, *Ann. Phys.*, **3**, 133 (1929); R. S. Mulliken, *Phys. Rev.*, **43**, 279 (1933).
32. V. Griffing, *J. Chem. Phys.*, **23**, 1015 (1955); V. Griffing and J. T. Vanderslice, *J. Chem. Phys.*, ibid., 1035, 1039.
33. B. Liu, *J. Chem. Phys.*, **58**, 1925 (1973).
34. A. Macias, *J. Chem. Phys.*, **49**, 2198 (1968).
35. R. E. Christoffersen, S. Hagstrom, and F. Prosser, *J. Chem. Phys.*, **40**, 236 (1964).
36. D. M. Silver, *Chem. Phys. Lett.*, **14**, 105 (1972).
37. G. Herzberg and H. C. Longuet-Higgins, *Disc. Faraday Soc.*, **35**, 77 (1963).
38. M. Rubinstein and I. Shavitt, *J. Chem. Phys.*, **51**, 2014 (1969).
39. (a) H. C. Longuet-Higgins and E. W. Abrahamson, *J. Am. Chem. Soc.*, **87**, 2045 (1965); (b) D. M. Silver, *J. Am. Chem. Soc.*, **96**, 5959 (1974).
40. R. B. Woodward and R. Hoffmann, *J. Am. Chem. Soc.*, **87**, 395 (1945).
41. R. B. Woodward and R. Hoffmann, *The Conservation of Orbital Symmetry*, Verlag Chemie, Gmbh., Weinbeim/Bergstrasse, 1970.
42. C. W. Wilson, Jr., and W. A. Goddard, III, *J. Chem. Phys.*, **51**, 716 (1969).

43. B. M. Gimarc, *J. Chem. Phys.*, **53**, 1623 (1970).
44. R. G. Pearson, *Theor. Chim. Acta*, **16**, 107 (1970).
45. J. S. Wright, *Chem. Phys. Lett.*, **6**, 476 (1970).
46. C. A. Coulson and I. Fischer, *Phil. Mag.*, **40**, 386 (1949).
47. R. G. Pearson, *J. Am. Chem. Soc.*, **94**, 8287 (1972).
48. T. G. Edwards, *Theor. Chim. Acta*, **27**, 1 (1972).
49. J. H. Sullivan, *J. Chem. Phys.*, **46**, 73 (1967).
50. R. Hoffmann, *J. Chem. Phys.*, **49**, 3739 (1968).
51. J. H. Hildebrand, *J. Am. Chem. Soc.*, **68**, 915 (1946); P. Schweitzer and R. M. Noyes, ibid., **93**, 3561 (1971).
52. E. A. Fletcher and B. E. Dahneke, *J. Am. Chem. Soc.*, **91**, 1603 (1969); D. L. King, D. A. Dixon, and D. R. Herschbach, ibid., **96**, 3330 (1974).
53. M. A. A. Clyne and J. A. Coxon, *Proc. Roy. Soc.*, **303A**, 207 (1968).
54. M. C. Lin and S. H. Bauer, *J. Am. Chem. Soc.*, **91**, 7737 (1969).
55. J. C. Whitehead, D. R. Hardin, and R. Grice, *Chem. Phys. Lett.*, **13**, 319 (1972).
56. H. F. Carroll and S. H. Bauer, *J. Am. Chem. Soc.*, **91**, 7727 (1969); A. Bar-Nun and A. Lifschitz, *J. Chem. Phys.*, **51**, 1826 (1969).
57. L. L. Paulson, *J. Chem. Phys.*, **53**, 1987 (1970).
58. S. A. Kettle, *Theor. Chem. Acta*, **4**, 150 (1966).
59. (a) B. H. Mahan, *J. Chem. Phys.*, **55**, 1436 (1971); (b) J. S. Wright, *J. Am. Chem. Soc.*, **96**, 4753 (1974).
60. S. Rothenberg and H. F. Schaefer, III, *Chem. Phys.*, *Lett.* **10**, 565 (1971).
61. C. Maltz, *Chem. Phys. Lett.*, **9**, 251 (1971); G. G. Balint-Kurti and M. Karplus, ibid., **11**, 203 (1971).
62. R. J. Cvetanovic, *Adv. Photochem.*, **1**, 115 (1963).
63. J. Dubrin, C. MacKay, and R. Wolfgang, *J. Am. Chem. Soc.*, **86**, 959, 4741, 4747 (1964); P. S. Skell, J. E. Villaume, J. H. Plonka, and F. A. Fagone, ibid., **93**, 2699 (1971).
64. C. H. Bamford and C. F. H. Tipper, *Comprehensive Chemical Kinetics*, Vol. 4, p. 176, Elsevier, Amersterdam, 1972.
65. R. Hoffmann and R. B. Woodward, *J. Am. Chem. Soc.*, **87**, 2046 (1965).
66. P. D. Bartlett and G. E. H. Wallbillich, *J. Am. Chem. Soc.*, **91**, 409 (1969); H. E. O'Neal and S. W. Benson, *J. Phys. Chem.*, **72**, 1866 (1968).
67. J. S. Wright and L. Salem, *J. Am. Chem. Soc.*, **94**, 322 (1972).
68. A. D. Liehr, *Progr. Inorg. Chem.*, **3**, 281 (1962); **4**, 455 (1963); H. C. Longuet-Higgins, *Adv. Spectr.*, **2**, 429 (1961); R. Englman, *The Jahn Teller Effect in Molecules and Crystals*, Wiley-Interscience, New York, 1972; I. Bersuker, *Coord. Chem. Rev.*, **14**, 357 (1975).
69. E. Ruch and A. Schönhofer, *Theoret. Chim. Acta*, **3**, 291 (1965).
70. L. L. Lohr, Jr., *Inorg. Chem.*, **6**, 1890 (1967).
71. C. J. Ballhausen, *Theoret. Chim. Acta*, **3**, 368 (1965).
72. R. W. Jotham and S. F. A. Kettle, *Inorg. Chim. Acta*, **5**, 183 (1971).
73. P. T. Miller, P. G. Lenhert, and M. N. Joesten, *Inorg. Chem.*, **12**, 218 (1973); J. P. Fackler, Jr. and A. Avdeef, *Inorg. Chem.*, **13**, 1869 (1974); R. G. Pearson, *Proc. Natl. Acad. Sci.*, **72**, 2104 (1975).
74. W. Byers Brown and E. Steiner, *J. Chem. Phys.*, **37**, 461 (1962).

75. R. F. W. Bader and A. D. Bandrauk, *J. Chem. Phys.*, **49**, 1666 (1968).
76. R. G. Pearson, *J. Chem. Phys.*, **52**, 2167 (1970).
77. P. Empedocles, *J. Chem. Phys.*, **46**, 4474 (1967).
78. R. F. W. Bader, *Mol. Phys.*, **3**, 137 (1960).
79. J. K. Burdett, *J. Chem. Phys.*, **52**, 2983 (1970); *J. Chem. Soc.* **A**, 1195 (1971).
80. H. B. Gray and N. A. Beach, *J. Am. Chem. Soc.*, **85**, 2922 (1963).
81. J. K. Burdett, *Chem. Phys. Lett.*, **5**, 10 (1970).

## Some Useful General References for Quantum Chemistry, Group Theory, and Spectroscopy

1. P. W. Atkins, *Molecular Quantum Mechanics*, Parts I, II, and III, Clarendon, Oxford, 1970.
2. F. A. Cotton, *Chemical Applications of Group Theory*, 2nd ed., Wiley, New York, 1971.
3. R. L. Flurry, Jr., *Molecular Orbital Theories of Bonding in Organic Molecules*, Dekker, New York, 1968.
4. R. M. Hochstrasser, *Molecular Aspects of Symmetry*, Benjamin, New York, 1966.
5. W. L. Jorgensen and L. Salem, *The Organic Chemists' Book of Orbitals*, Academic, New York, 1973.
6. I. N. Levine, *Quantum Chemistry*, Vols. I and II, Allyn and Bacon, Boston, 1970.
7. S. P. McGlynn, L. G. Vanquickenborne. M. Kinoshita, and D. G. Carroll, *Introduction to Applied Quantum Chemistry*, Holt, New York, 1972.
8. K. Nakamoto, *Infrared Spectra of Inorganic and Coordination Compounds*, 2nd ed., Wiley, New York, 1971.
9. M. Orchin and H. H. Jaffe, *Symmetry, Orbitals, and Spectra*, Wiley-Interscience, New York, 1971.
10. J. A. Pople and D. L. Beveridge, *Approximate Molecular Orbital Theory*, McGraw-Hill, New York, 1970.
11. H. L. Shläfer and G. Gliemann, *Basic Principles of Ligand Field Theory*, Wiley-Interscience, New York, 1969.
12. E. G. Wilson, J. C. Decius, and P. C. Cross, *Molecular Vibrations*, McGraw-Hill, New York, 1955, Chapters 5 and 6.
13. J. R. Ferraro and J. S. Ziomek, Introductory Group Theory, Plenum, New York, 1975.

# CHAPTER 2

# OTHER SYMMETRY-BASED RULES

In the first chapter, PT was applied to the case of a collection of nuclei and electrons undergoing chemical reaction. A number of useful rules were derived, based on this theoretical analysis. It was also shown that the Woodward–Hoffmann procedure based on orbital-correlation diagrams was completely consistent with, and could be derived from, PT.

The purpose of this chapter is to present other methods for determining whether a given reaction path is an allowed, or feasible, one. Many methods have been advocated, more or less rigorously justified. Some are useful only for restricted classes of reaction. In any case, many of the procedures are simple to use and give results quite consistent with experiment.

In Chapter 1 the selection criteria for chemical reactions all depended on the continued presence, from reactants to products, of at least one useful symmetry element. The word useful is necessary since some symmetry elements are of no use in making predictions. For example, in a typical four-center forbidden reaction of two diatomic molecules, all atoms can lie in a plane of symmetry, such as $H_2 + I_2$. All bonds made and broken have the same symmetry with respect to this plane, and using this plane for predictive purposes, the reaction would appear allowed as concerted reaction. In general, planes that do not differentiate between any of the bonds changed in the reaction are not useful for predicting allowedness.

Unfortunately the great majority of all molecules belong to the $C_1$ point group. They possess no symmetry elements, except the trivial identity element. Pairs of interacting molecules are even more likely to generate a supermolecule that is $C_1$. How, then, can one make decisions about the forbiddenness or allowedness of the great majority of chemical reactions? Since all orbitals, bonds, and reaction paths are of the same symmetry type, namely $A$, it would appear that all reactions are allowed. This is obviously an unreasonable conclusion, since we expect the reactions of propylene, for example, not to differ greatly from those of ethylene, which are subject to strong symmetry restrictions.

Actually the situation is very much like that of forbidden transitions in optical spectroscopy. Reductions in symmetry do make such transitions partly allowed. There will still be an inherent forbiddenness that shows up as a low intensity of absorption. In the same way a reduction in symmetry can make a forbidden reaction partly allowed, but still one that has an abnormally large activation energy, and therefore the reaction is slow, or non-observable.

Table 1 shows some rate data for the isomerization of substituted Dewar benzenes to the corresponding benzene derivatives. The parent reaction is strongly forbidden (p. 71). This shows up as a slow rate of reaction and a large activation energy for a process that is strongly favored thermodymically

$$\longrightarrow \qquad \Delta H = -60 \text{ kcal} \qquad (1)$$

A single halogen substituent greatly increases the rate of the reaction and lowers the activation energy. A pair of halogen atoms placed 1, 4 drops the rate down even below the original value and raises the activation energy.

**Table 1 Rate Constants for Isomerization of Dewar Benzenes at 24.3°C[a]**

| | $k \times 10^6$, $sec^{-1}$ | $\Delta H^{\ddagger}$, kcal | $\Delta S^{\ddagger}$, eu |
|---|---|---|---|
| 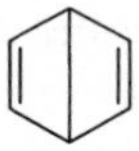 | 5.18 | 23.0 | −5.0 |
| 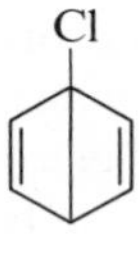 Cl | 464 | 19.1 | −9.4 |
| 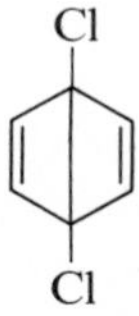 Cl, Cl | 0.0084 | 30.5 | +12.0 |
| 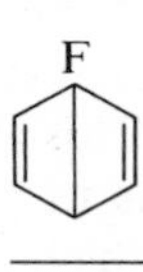 F | 1860 | — | — |

[a] Data from reference 2.

A single substituent lowers the symmetry from $C_{2v}$ to $C_s$. In this point group the bonds to be broken are $2A'$ and $A''$, the bonds to be made are also $2A'$ and $A''$. From previous discussion one can see that the troublesome $b_2$ orbital which causes the forbiddenness in Dewar benzene becomes $a'$, just as the $a_1$ orbital becomes $a'$. The reaction is formally allowed. Adding two chlorine atoms raises the symmetry to $C_{2v}$ again, and the rate falls off. Both a symmetry factor and an electronic factor are manifest in the rate data.[2] Nevertheless, the "allowed" reactions are still abnormally slow and have very large activation energies for such endothermic processes. There is no obvious reason why an activation energy should exist at all, if it were not for the lingering effects of the original symmetry barrier.

## TOPOLOGICAL CONSIDERATIONS

In order to understand what is happening it is necessary to examine more closely the way in which orbital symmetry can create a large activation energy. We emphasize again that symmetry barriers are not mysterious obstructions that appear only because we are discussing dynamic systems in a state of reaction, or because we have selected a particular reaction path. A TS is energetically unfavorable if symmetry says so, but it will be so no matter how we attain it. Furthermore its energy can be calculated by the usual ab initio methods, and it will obey the usual laws of dynamics or quasi-thermodynamics applicable to reacting systems.[3]

Activation energies even for allowed reactions exist primarily because of the Pauli exclusion principle. Each pair of electrons added to a collection of nuclei must occupy an orbital of successively higher energy. This usually means an orbital with one more nodal surface, or region where the wave-function has a zero value. In the case of LCAO MOs, a node exists wherever there is a change in sign of the coefficients of two AOs.

A nodal surface raises the electronic energy by a kinetic energy effect. However if the node exists in a region where the orbital has a small value in any case, the effect will be minimal. An example would be that of an anti-bonding MO for two atoms very far apart. As the two atoms approach each other, the effect of a nodal surface between them becomes very great. Any two polyelectronic molecules, if forced very close together, will have a large positive electronic energy because of this phenomenon. The effect of nuclear repulsion is relatively trivial, at least as soon as one has molecules more complex than two helium atoms.

In a chemical reaction, where atoms change their relative positions, some orbitals go up in energy and some go down. This depends on whether the orbitals have or do not have nodes for the atoms which approach each other, or recede from each other. The absence of a node between two atoms (same sign for the AO coefficients) means a decrease in energy as the two atoms approach because of better overlap and bonding, but this effect is usually less than the energy increase due to a node.

Now an energy barrier due to orbital symmetry results from a situation where electrons are trapped in an orbital that is going up in energy very rapidly because of the close approach of atoms separated by a nodal surface. They are trapped because no empty orbital of lower energy exists into which they can escape. Empty orbitals exist of higher energy, and these could be useful because they can be mixed with the original orbital to create a new orbital in which the effect of the nodal surface is alleviated. If the empty orbitals are of a different symmetry, this is a signal that such orbitals cannot change the nodal dilemma. The energy cannot be lowered by mixing.

Now it is obvious that these same phenomena would exist in reactions between molecules devoid of symmetry. The greatest difference is that we would not have symmetry labels as convenient tags for telling us whether mixing of an empty orbital is useful. This must be determined in some other way. It will turn out that mixing of the empty orbitals will always lower the energy. The amount by which this happens, however, can be so small as to be negligible.

The symmetry of a wavefunction depends on the way in which the sign of the wavefunction changes from plus to minus in different parts of the molecule. Hence the symmetry species is a way of indicating the nodal structure. If we change from a symmetrical molecule to a less symmetrical one there will still be a very similar nodal structure, even though exact symmetry classification is not possible. Figure 1 shows the $\pi$ and $\pi^*$ orbitals of a homonuclear diatomic molecule, $N_2$, compared to a heteronuclear one, CO.

Figure 1 shows schematically how the bonding $\pi$ orbital in CO is more heavily concentrated on the more electronegative atom, O. Conversely, in the $\pi^*$ orbital the weight is on the C atom. The orbitals of CO can be formed from those of $N_2$ by a process of deformation. In other words the orbitals of $N_2$ and of CO are topologically identical. Two things are topologically identical if they can be interconverted by continuous twisting, stretching, and other deformation. Cutting, tearing, and piercing are not allowed.

A collection of nuclei will have a ground-state wavefunction with a certain number of nodal surfaces. As the nuclei are rearranged adiabatically to correspond to a chemical reaction, the nodal surfaces will be distorted and deformed. However except for certain limiting circumstances, their number and kind will not change.[4] The wavefunction maintains its topological identity.[5] This is the more generalized equivalent of the conservation of orbital symmetry.

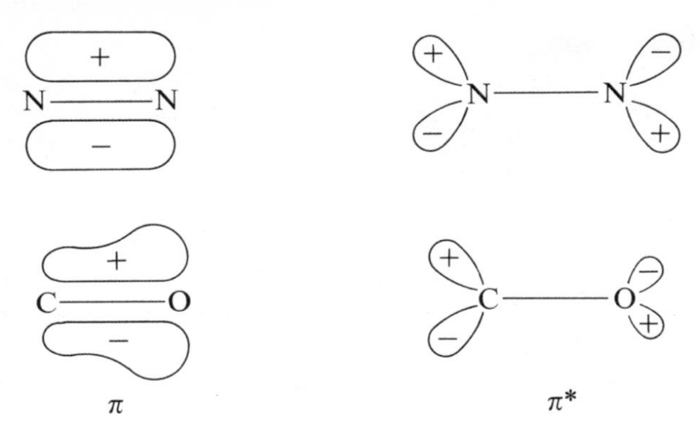


**Figure 1.** The $\pi$ and $\pi^*$ orbitals of a homonuclear ($N_2$) and heteronuclear (CO) diatomic molecule; each orbital is doubly degenerate, but only one orbital of each pair is shown.

Because of the complexity of the problem a general mathematical proof does not seem to be possible.[6] However as a result of many observations,[7] the following statement can be made: in the case of MOs if a node exists in the region between two nuclei, this node will tend to persist even though the MO changes as a result of changing nuclear positions. As a simple example, the case of three H atoms starting as ($H_2$ + H) (but in a state of interaction), going to linear $H_3$ and finally to triangular $H_3$, may be considered (p. 41). The lowest-energy MO is nodeless, the next MO has a node between atoms A and C, and the highest MO has a node between atoms A and B, as well as between B and C.

These useful properties may be used to label the MOs and to follow them throughout the course of chemical reactions.[8,9] Accordingly, correlation diagrams can be constructed and decisions about allowed and forbidden reactions made in the usual way. Decisions about the feasibility of mixing occupied and empty orbitals must now be made on the basis of overlap. Two orbitals with a large positive overlap will mix effectively. The lower-energy orbital will be lowered or raised, and the higher-energy one will be raised or lowered in energy.

An elaborate procedure has been developed for mapping the state function of the products on to that for the reactants.[5] The degree of overlap indicates whether the functions are topologically identical and accordingly, whether the reaction is allowed or forbidden. This procedure would be useful in deciding between two possible allowed paths, but is otherwise more complicated than is necessary. An easier method is illustrated by a consideration of 1,2 shifts of hydrogen or alkyl groups.[10] The method is due to Zimmerman.[9]

## Sigmatropic Reactions

This kind of reaction belongs to a family termed as "sigmatropic reactions" by Woodward and Hoffmann.[11] A $\sigma$ bond moves from one position to another in a molecule, the bond shifting over one or two $\pi$-electron systems. Sigmatropic reactions are classified by two numbers set in brackets, $[i,j]$. The numbers indicate the particular atoms along the conjugated chains to which the migrating $\sigma$ bond becomes attached. The migration of the bond across one chain only is designated as a $[1,j]$ shift. Migration over two chains is designated as $[i,j]$ with $i,j \neq 1$. Thus a 1,2 shift may also be called a [1,2] sigmatropic rearrangement. These sigmatropic reactions are usually characterized by a lack of symmetry, except perhaps in the TS.

Figure 2 shows the relevant bond orbitals of the initial state, the transition state, and the final product for (2),

$$CH_3{-}CH_2{}^{\pm} \longrightarrow H_2C\overset{H^{\pm}}{\cdots\cdots}CH_2 \longrightarrow {}^{\pm}CH_2CH_3 \quad (2)$$

There may be two, three, or four electrons in these orbitals, depending on whether the reactant is a carbenium ion, a free radical, or a carbanion. The orbitals of the reactant can be correlated with those of the TS by virtue of the

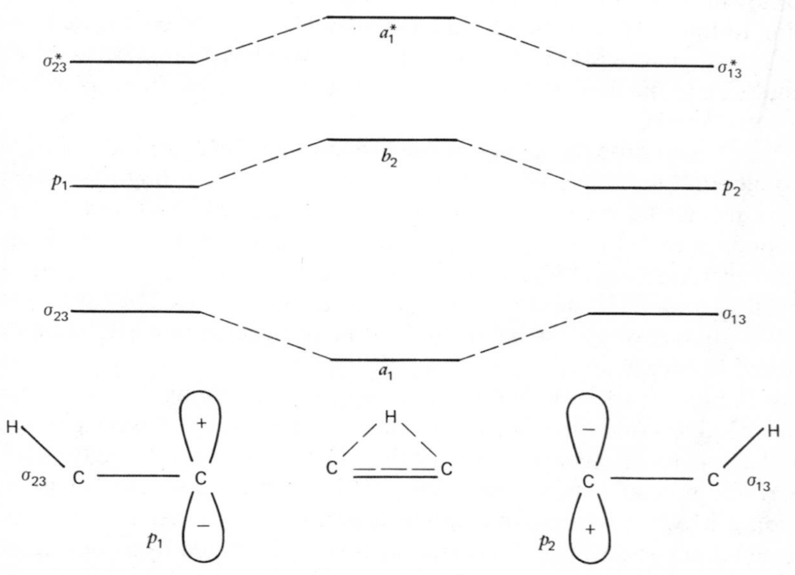


**Figure 2.** Orbital correlation for 1,2 hydrogen shift in $C_2H_5{}^+$ (two electrons), $C_2H_5$, (three electrons) and $C_2H_5{}^-$ (four electrons).

nodal properties of each. Since the TS can also be correlated with the reactant by the same criterion, a complete correlation is possible.

In Fig. 2 it is not clear that nodal properties are conserved. For example, the $p_1$ orbital becomes an antibonding orbital, so that a new node has apparently developed between AOs 1 and 2. This anomaly arises because it is convenient to use bond orbitals rather than canonical MOs. The latter are the orbitals that maintain their nodal integrity. The correlation is still possible with partial orbitals. For example, the $\sigma_{23}$ orbital of the reactant is strongly bonding between AOs 2 and 3. Only the lowest orbital of the TS has this property. In the same way only the highest orbital of the TS has a node between orbitals 2 and 3, and must therefore correlate with $\sigma_{23}^*$.

The conclusions from Fig. 2 are that a 1,2 hydrogen shift is very easy for a carbenium ion, and very difficult for a carbanion. A free radical is intermediate in difficulty. These predictions agree perfectly with experiment.[10] There is no implication that the carbanion rearrangement is forbidden, since there is no crossing of orbitals in the correlation. A high-energy barrier can still be consistent with an allowed reaction, even in cases where much more symmetry exists.

There is some symmetry in (2), because a simple example was used. The original point group is $C_s$ and that of the activated complex, $C_{2v}$. Mixing of $\sigma_{23}$ with $p_1$ and $\sigma_{23}^*$ with $p_1$ are dominant initially. All of these orbitals are $a'$. In the TS the orbitals become, in order of increasing energy, $a_1$, $b_2$, and $a_1^*$. The reaction coordinate at the TS is a motion of the proton toward one of the carbon atoms and away from the other. This nuclear motion is of $B_2$ symmetry in $C_{2v}$. The necessary condition for this to occur readily is that there be an $a_1$ orbital and a $b_2$ orbital, one filled and one empty, and close in energy. This condition is met for the carbanion reaction, where we have a true TS, that is, an unstable nuclear configuration. For the carbenium-ion reaction there is a large energy gap between the filled $a_1$ orbital and the empty $b_2$ orbital. In fact the intermediate in this case is a species of considerable stability, quite comparable to that of the carbenium ion.[12] It is a $\pi$-complex of ethylene, or protonated ethylene.

Figure 3 takes up a reaction closely related to (2). It is the [1,3] hydrogen shift, another sigmatropic reaction, such as might occur in propylene

$$CH_3{-}CH{=}CH_2 \longrightarrow H_2C\overset{H}{\cdots}CH_2\ (\text{bridged via } \underset{H}{C}) \longrightarrow CH_2{=}CH{-}CH_3 \qquad (3)$$

The relevant bond orbitals now are the bonding and antibonding pairs, $\sigma_{12}$, $\sigma_{12}^*$ and $\pi_{34}$, $\pi_{34}^*$. The MOs of the activated complex are easily found.

They have the same nodal pattern as that of four H atoms in a square array (p. 44), or of the four $\pi$ orbitals in cyclobutadiene. In the same way the TS orbitals of Fig. 2 have the same nodes as those of three H atoms (p. 43), or of the $\pi$ orbitals of an allyl group.

The orbital following can be carried out just as before. In this case a crossing occurs, the $\pi_{34}$ orbital of the reactant correlating with the $\pi_{23}^*$ orbital of the product. In propylene a 1,3-hydrogen shift is forbidden in the same sense that other reactions are forbidden when a crossing occurs.

In this case the forbiddenness results not from symmetry, but from topology. While we could call such reactions "topologically forbidden," it is probably more convenient to continue to use the less exact term "symmetry forbidden." Also since $\pi_{34}$ and $\pi_{34}^*$ do have the same symmetry label, they mix somewhat earlier in the reaction coordinate than if they relied only on configuration interaction. Hence the energy barrier produced is not so large. For this reason we might also call the [1,3] shift a reaction "partly forbidden by symmetry."

Examples (2) and (3) are cases in which the migrating hydrogen atom stays at all times on the same face of the accompanying $\pi$ system. They are called suprafacial processes.[1] We can also have antarafacial migrations, in which the migrating group passes from the top face to the bottom face.

(4)

Antarafacial migration

The MO correlation diagram for this process is also shown in Fig. 3.[9] The nodal character of the orbitals renders this an allowed process for a four-electron system. The MOs for the TS will be described later (p. 119). Antarafacial migration would be impossible for small systems, but quite possible for larger $\pi$ systems.

Figures 2 and 3 could be extended to larger systems. However this can become awkward and is not necessary. We are interested in predicting the stability of the TS. This can be done most easily by removing the migrating hydrogen atom with its electron and considering the remaining conjugated $\pi$ system. The symmetry of the lowest empty (or half-empty) orbital determines the stability of the TS formed by adding the hydrogen atom symmetrically between the two termini of migration. If these two termini have the same sign, there will be strong bonding, and the reaction has a low-energy barrier. If the

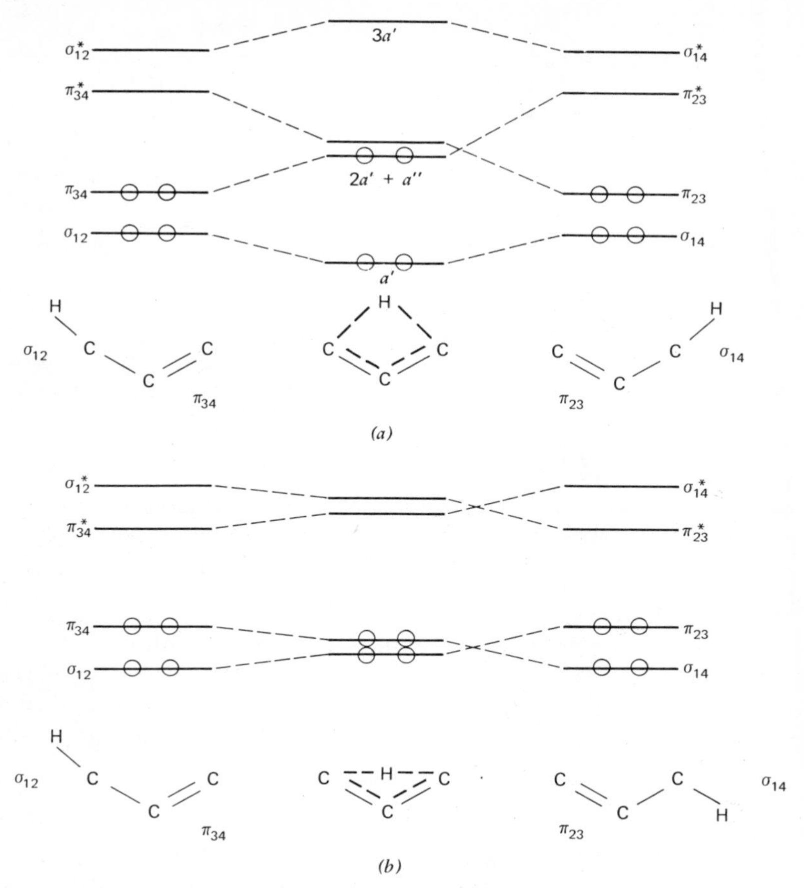


**Figure 3.** Orbital-correlation diagram for: (*a*) a forbidden 1,3 suprafacial shift of H in $C_3H_6$; (*b*) an allowed 1,3 antarafacial shift of H in $C_3H_6$.

termini have opposite signs, the binding of the H atom is weak, the barrier is high, and the reaction is forbidden (Fig. 4).

If the migrating group is not a hydrogen atom, but an alkyl or other group in which the bonding atom has *p* orbitals, the situation is again changed. A *p* orbital has two lobes of opposite sign. Each of these could be used to bond to the $\pi$ system termini, in the case where the termini have opposite signs. This is shown in Fig. 4. Clearly such a migration would occur with *inversion* of configuration at the migrating group, if it were an alkyl or silyl group, for example.[13] For such reactions, the rules for forbidden and allowed must be reversed.

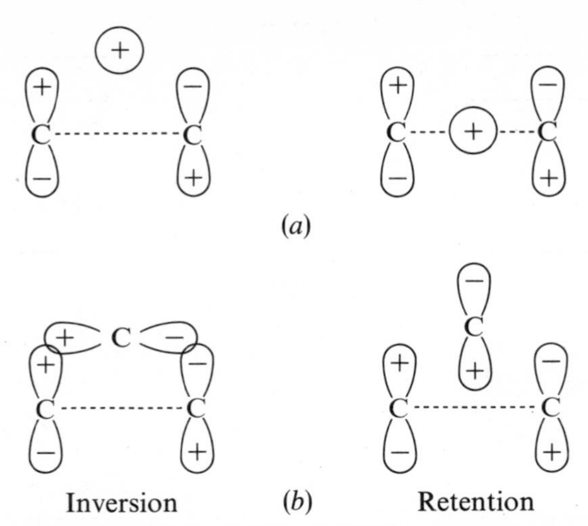


**Figure 4.** (*a*) Suprafacial migration forbidden and antarafacial migration allowed for an atom using an *s* orbital for bonding; (*b*) suprafacial migration allowed for an atom using a *p* orbital for bonding, but inversion results; suprafacial migration with retention is forbidden.

In a sigmatropic reaction a $\sigma$ bond moves from one position to another in a molecule. The bond shifts across an intervening $\pi$ system whose double bonds are reorganized in the process. If there is no intervening $\pi$ system, then there are no symmetry restrictions of the kind stated. For example, the 1,3 shift

$$CH_3{-}CH_2{-}CH_2{}^+ \longrightarrow {}^+CH_2{-}CH_2{-}CH_3 \qquad (5)$$

is fully allowed.[14]

The feasibility of [*i*,*j*] sigmatropic rearrangements where $i, j \neq 1$ may be estimated from the stability of the TS. This is done by considering the interaction of the two polyenyl radicals assumed as hypothetical intermediates.[11] For good stability it turns out that $(i + j) = (4n + 2)$, where $n$ is an integer. Thus [3,3] shifts are allowed, but [3,5] shifts are forbidden. The Cope rearrangement, shown in (34), is a [3,3] process.

### Addition Reactions

The effect of a reduction in symmetry can be seen very well, if the PT criterion of overlap of filled and empty orbitals is used. Figure 5 shows the most important overlaps in the forbidden concerted additions.

$$\begin{aligned} H_2 + C_2H_4 &\longrightarrow C_2H_6 \\ Cl_2 + C_2H_4 &\longrightarrow C_2H_4Cl_2 \end{aligned} \qquad (6)$$

Because of the high symmetry in these cases, it can be easily seen that the overlaps are exactly zero for suprafacial addition. Now we consider the addition of HCl to an olefin

$$HCl + C_2H_4 \longrightarrow C_2H_5Cl \tag{7}$$

Figure 5 shows that the overlap is no longer zero by symmetry. However it still is small, because there are positive and negative overlaps that tend to cancel. We conclude that the *cis* addition of molecular HCl to $C_2H_4$ is partly forbidden by symmetry.

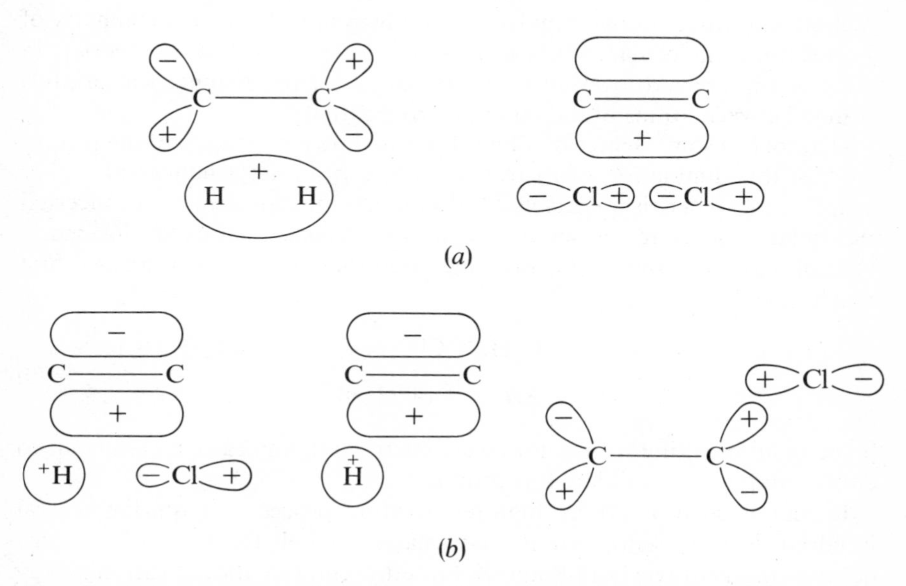


**Figure 5.** (*a*) Concerted *cis* addition of $H_2$ or $Cl_2$ to ethylene forbidden by symmetry; (*b*) concerted *cis* addition of HCl to $C_2H_4$ partly forbidden; addition of $H^+$ (or H) and of $Cl^-$ (or Cl) allowed.

Experimentally we know that the addition usually occurs stepwise, and that *trans* addition is favored. Figure 5 shows that the addition of $H^+$ (or H), and the addition of $Cl^-$ (or Cl) is allowed because very good overlap is possible. The addition could be synchronous, or it could occur in two stages. The latter is more probable from entropy considerations. Even synchronous *cis* addition could occur, if it were possible to have $H^+$ and $Cl^-$ in close proximity without bonding to each other.

Of course this is not possible, but it illustrates a point of some significance. Orbital-symmetry restrictions, as well as topological restrictions, exist because of the directed nature of covalent bonding. If the atoms consisted of spherically charged ions, plus and minus, all reaction paths would be allowed by orbital symmetry. Some would still be barred because of unfavorable electrostatic interactions. Four-center exchange reactions, such as

$$\mathrm{CsCl} + \mathrm{KI} \longrightarrow \mathrm{CsI} + \mathrm{KCl} \tag{8}$$

occur very rapidly.[15] There is no appreciable energy barrier, and in fact intermediates of some stability are formed.

This analysis implies that an important factor in the difference in forbiddenness of (6) and (7) lies in the polarity of the H—Cl bond. The difference in electronegativity of two bonded atoms always results in an asymmetry of the resulting molecular orbitals. This in turn produces some overlap in situations where zero overlap would be found for the symmetrical orbitals formed between atoms of the same electronegativity.

That (6) has some degree of allowedness is shown by studies of the reverse process, the elimination of hydrogen halides from alkyl halides at higher temperatures. There is no reasonable doubt that this can occur as a concerted molecular process, in the sense that no free radicals or ions are formed.[16] Maccoll has made the interesting suggestion that an ion pair is formed first which then decomposes,[17]

$$\begin{array}{c} \mathrm{C_2H_5Cl} \longrightarrow \mathrm{C_2H_5^+, Cl^-} \longrightarrow \mathrm{C_2H_4} + \mathrm{HCl} \\ \Delta H = +14.5 \text{ kcal} \end{array} \tag{9}$$

Since bond breaking would not be symmetrical in any case, it is difficult to either validate or invalidate this proposal.

Reactions such as (9) are high-temperature processes, requiring several hundred degrees Centigrade in many cases, though the activation energy depends markedly on both R and X. For ethyl chloride the activation energy is 56.6 kcal. This large value, in comparison with the modest thermodynamic requirement of 14.5 kcal, is evidence for a substantial barrier due to the partial forbiddenness. Substituents which increase the polarity of the C—Cl bond increase the rate and decrease $E_a$.

$$\mathrm{CH_3CHClOCH_3} \longrightarrow \mathrm{HCl} + \mathrm{CH_2} = \mathrm{CHOCH_3} \qquad E_a = 33.3 \text{ kcal} \tag{10}$$

While this result is predicted from orbital overlap arguments, it may also be explained by the ion-pair model of (9).

There is a great deal of evidence that 1,4-elimination of HX occurs much more easily than 1,2-elimination. For example, the *cis* isomer shown below reacts much more readily than the corresponding *trans* isomer.[18]

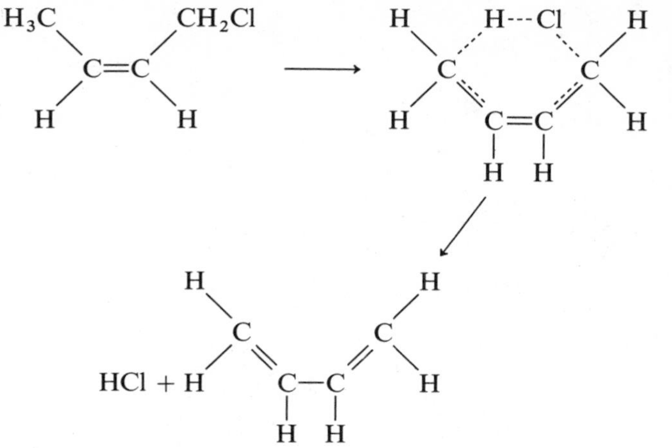

$$\mathrm{HCl} + \ldots \qquad (11)$$

The rates of elimination for 1,4-eliminations such as (11) are many fold faster (ca. $10^6$) than for similar 1,2-eliminations. This could be attributed to the extra stability of the allylic carbenium ion.

It is also consistent with the deduction that 1,4-addition (and hence elimination as well) is a symmetry allowed process for $H_2$, or $Cl_2$, and butadiene. This is shown in Fig. 6 in terms of the overlap of the highest filled $\pi$ orbital of butadiene and the $\sigma^*$ orbital for $H_2$, as well as the overlap of the lowest empty $\pi^*$ orbital of the diolefin and the $\sigma$ orbital of $H_2$. We can, of course, use the bond symmetry rule to show allowedness.

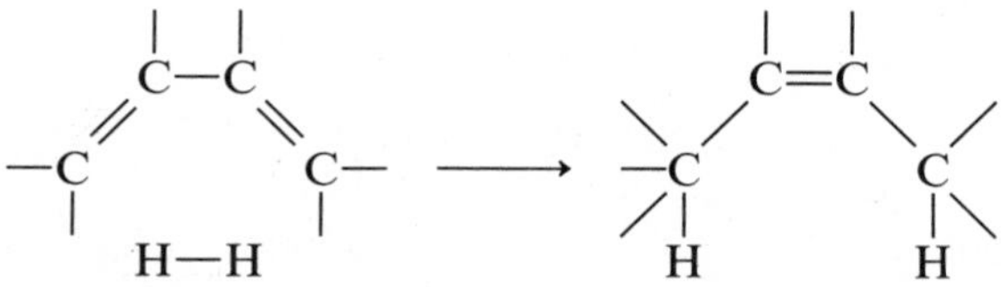

$$2a_1 + b_1 \longrightarrow 2a + b_1 \qquad (12)$$

The point group is $C_{2v}$.

The evidence for facile 1,4-elimination of halogens and hydrogen is unambiguous. For example, we have the reactions

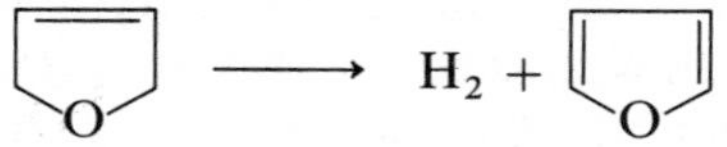

$$\longrightarrow H_2 + \qquad (13)$$

$$\longrightarrow H_2 + \quad + \text{ other products} \qquad (14)$$

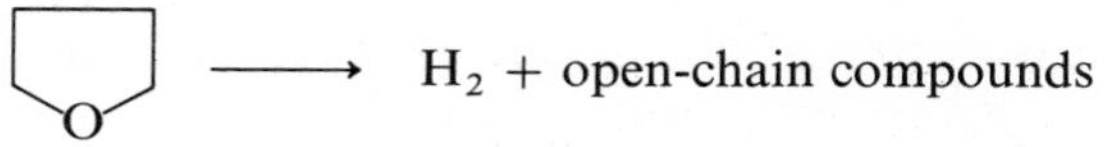

$$\longrightarrow H_2 + \text{ open-chain compounds} \qquad (15)$$

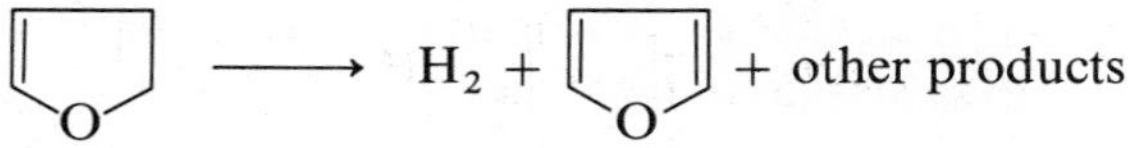

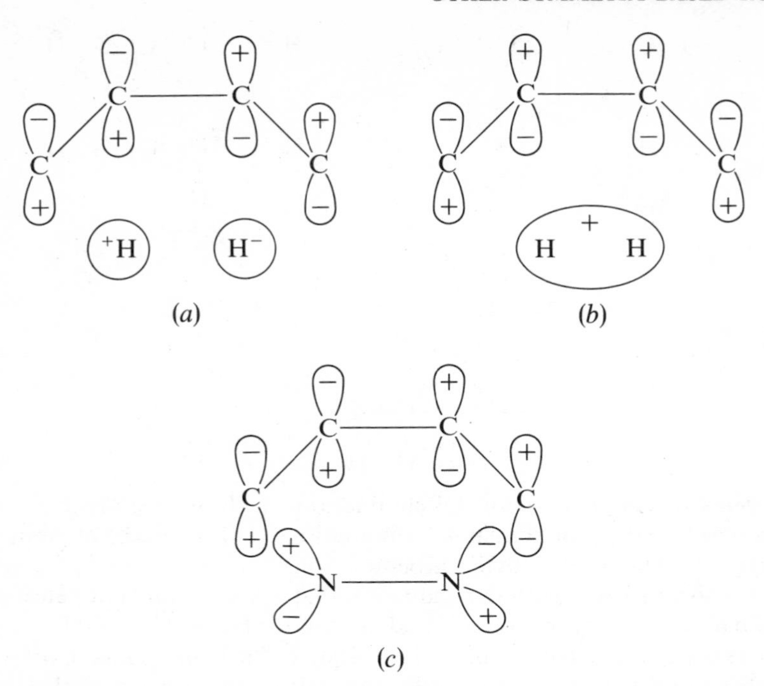


**Figure 6.** Concerted 1,4 addition of $H_2$ (or $N_2$) to butadiene allowed: (*a*) overlap of HOMO of butadiene with $\sigma^*$ orbital of $H_2$; (*b*) overlap of LUMO of butadiene with $\sigma$ orbital of $H_2$; (*c*) overlap of HOMO of butadiene with $\pi^*$ orbital of $N_2$.

Reaction (13) occurs easily and with clean kinetics.[19] No free radicals and no chain reactions are involved. Reactions (14) and (15) occur with great difficulty and by means of complex free radical chains.[20] Addition of $H_2$ 1,4 to butadiene is not found for steric reasons, but $Cl_2$ can add 1,4. Note that even at high temperatures, 1,2-elimination of molecular $H_2$ or $Cl_2$ has not been observed. Only free-radical chains are found. This is quite unlike the 1,2-elimination of HCl, which must be considered as only partially forbidden by symmetry.

While normally the effect of a reduction in symmetry is to speed up a reaction by making it more allowed, the opposite could easily occur. A reaction that is allowed in a symmetrical molecule may become less allowed in a less symmetrical molecule. Proof for this is hard to come by since naturally a substituent that destroys symmetry has effects of a steric and electronic nature as well. That is, all of the MOs are raised or lowered in such a way as to influence their degree of interaction. However overlap could also be affected, and this is a symmetry property, at least in part.

A possible example may lie in the following thermal decompositions[21]

$$\text{N{=}N ring} \longrightarrow \text{butadiene} + N{\equiv}N \qquad \text{easy} \qquad (16)$$

$$2a_1 + b_2 \qquad\qquad 2a_1 + b_2$$

$$\text{N{=}N(O) ring} \longrightarrow \text{butadiene} + N{=}N{=}O \qquad \text{difficult} \qquad (17)$$

Decomposition of the diaza ring compound into butadiene and molecular $N_2$ is fully allowed (see Fig. 6). It occurs very readily and is undoubtedly a concerted molecular process. The pyrolysis of the azoxy derivative is surprisingly much more difficult. The $C_{2v}$ point group has been destroyed by the oxygen substituent and only a plane of symmetry remains, which is of no use in classification.

An explanation lies in the asymmetric nature of the molecular orbitals of $N_2O$ compared to those of $N_2$. Figure 7 shows the form of the highest occupied (HOMO) and lowest unoccupied (LUMO) orbitals of $N_2O$. The latter is more significant, since electron transfer from butadiene to $N_2O$

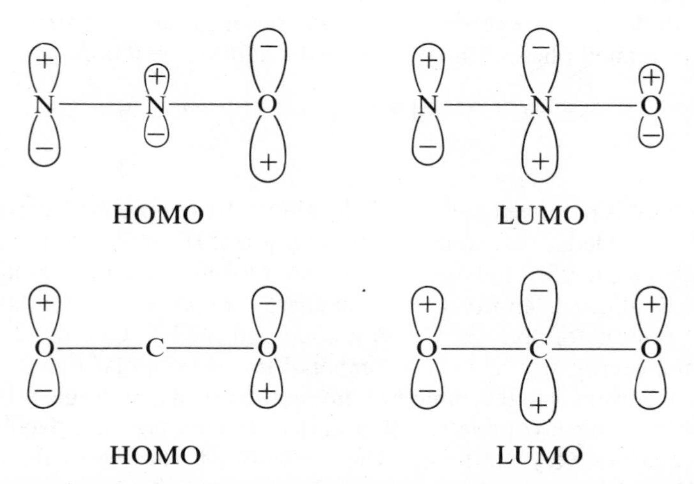


**Figure 7.** Representation of the HOMO and LUMO of nitrous oxide and carbon dioxide; the model pattern and the concentration of each orbital on the several atoms are shown.

would dominate the interaction between the two molecules. This follows simply from the greater electronegativity of N compared to C. In Fig. 6 we see a symmetrical overlap between $N_2$ and butadiene. This facilitates concerted addition, and, by microscopic reversibility, a concerted decomposition. The bonding of $N_2O$ to butadiene would be very unsymmetrical, the central N atom being bound more firmly than the terminal one. This would allow for easy breaking of one bond in the decomposition (17), but would inhibit the concerted loss of $N_2O$.

Figure 7 also shows the HOMO and LUMO of $CO_2$, a more symmetrical molecule ($D_{\infty h}$) than $N_2O$ ($C_{\infty v}$). The same nodal properties exist in both cases, but with some distortion for $N_2O$. If we take any three atoms that contribute *s* and *p* valence shell orbitals, they will generate MOs with the same nodal patterns. The MOs will be topologically identical, independent of the chemical nature of the nuclei. The same holds true for larger molecules with equal numbers of similar nuclei.

It is also true that inert atoms (not involved directly in a reaction) can be replaced by groups of atoms without affecting the nodal symmetry other than by distortion. The reason for stressing these points is to show that the topology of the MOs of an unsymmetrical molecule can be deduced from a knowledge of the symmetry of the MOs of the closest symmetrical analog. In particular, the use of bond orbitals rather than canonical MOs is very useful. Clearly the nodal character of all $\sigma$ bonds is the same, as is that of all $\pi$ bonds. The corresponding antibonding $\sigma^*$ and $\pi^*$ orbitals all have the same property of having a node between the two atoms concerned.

By concentrating on the bonds made and broken, we can often predict without difficulty the existence or nonexistence of energy barriers due to symmetry-related phenomena. Consider the addition reaction

$$\mathrm{R{-}C{\equiv}N + HCl \longrightarrow R{-}\underset{\displaystyle Cl}{\underset{|}{C}}{=}\underset{\displaystyle H}{\underset{|}{N}}} \tag{18}$$

This reaction is predicted to be partly forbidden by symmetry. The reasoning lies in the knowledge that concerted *cis* addition of $H_2$ or $Cl_2$ to an olefinic or acetylenic unsaturated linkage is forbidden. Of course *trans*, or antarafacial addition, is allowed but involves strain energy, which makes it unfavorable. Also the polarity of both the C—N $\pi$ bond and the H—Cl $\sigma$ bond work to reduce the energy barrier in (18) compared to the nonpolar bonds of (6).

Our conclusion is that symmetry is not necessary in a molecule for making deductions about favorable reaction paths. Its presence does facilitate the task of analysis very markedly. The conclusions from cases that can be analyzed exactly can be carried over to more complex molecules. This will show up very clearly in the following sections.

## VALENCE-BOND APPROACH

The first application of quantum mechanics to a chemical reaction was by London in 1928.[22] This was for a simple three-atom reaction

$$X + YZ \longrightarrow XY + Z$$

and the method used was naturally the valence-bond method developed previously by Heitler and London. Eyring and Polanyi extended this procedure by using semiempirical methods, so that potential energy surfaces for four-atom systems could be calculated.[23] For many years valence-bond theory was used to estimate the energetics of reactions. This was during the hey-day of resonance theory and the procedures used were largely based on changes in resonance energy.

During the 1950s MO theory, with configuration interaction, supplanted the more cumbersome valence-bond method for most chemical purposes. Still it should be possible to deduce chemical reaction selection rules from the latter approach to chemical binding. Indeed this has been done by van der Hart, Mulder, and Oosterhoff,[24] who showed that valence-bond theory can lead to some of the Woodward–Hoffmann rules. The method is rather mathematical and complex, and it is by no means obvious just where the difference between forbidden and allowed reactions lies.

A more useful approach has been presented by Goddard, who has developed a generalized valence-bond approach (GVB) to molecular quantum mechanics.[25] In this method every electron is in a different orbital and each orbital can be delocalized over several atoms. While the orbitals are not orthogonal, energy considerations keep this mutual overlap small.

Initially accurate ab initio calculations were made for isolated molecules and for simple molecules in course of reaction.[26] This establishes the general nature of the GVB orbitals and their changes during reaction. If the assumption is made that the general description can be transferred to other molecules, it becomes relatively easy to make predictions about reactions of even complex molecules.

The important feature is illustrated in Fig. 8 for the hydrogen-atom exchange reaction

$$D + H_2 \longrightarrow HD + H \tag{20}$$

We start with three orbitals, two of which make up the original H—H bond and the third being isolated on D. As the three atoms approach a linear symmetric activated complex, each orbital changes. In order to minimize overlap, one orbital is forced to undergo a change in phase during the exchange process. The final products contain two orbitals in phase bonding H—D,

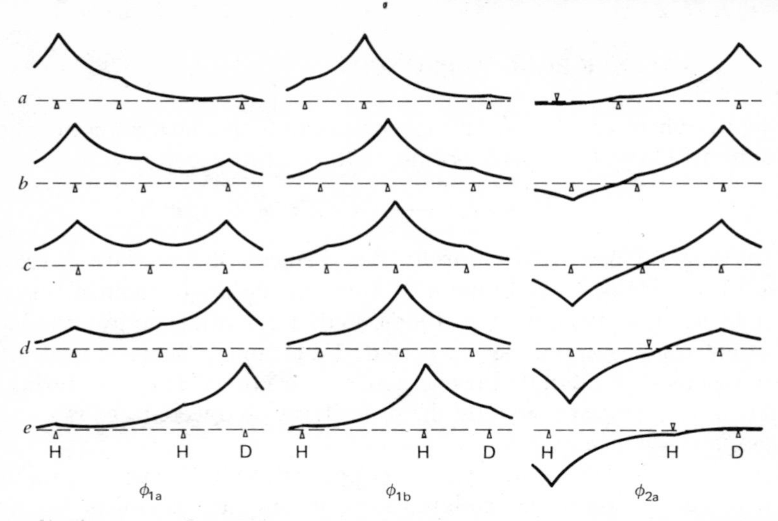


**Figure 8.** The generalized valence bond orbitals for the $H_2 + D \rightarrow H + HD$ exchange reaction; each column represents a different orbital, and each row a different nuclear configuration (row $a$ = initial stage, row $c$ = saddle point, and row $e$ = final stage). [Reprinted with permission from W. A. Goddard, III, *J. Am. Chem. Soc.*, **94**, 793 (1972). Copyright by the American Chemical Society.]

and a third orbital isolated on H, which is now out of phase with the other two.

In this case the phase change has no effect since the orbital is nonbonding in any case. In other cases the necessary phase changes lead to a bonding dilemma. For example, in the four-atom exchange reaction,

$$H_2 + D_2 \longrightarrow \begin{array}{c} {}^1\,H\text{---}D\,{}^3 \\ |\quad\;\; | \\ {}^2\,H\text{---}D\,{}^4 \end{array} \longrightarrow 2HD \tag{21}$$

suppose the 1,2 bond becomes the new 1,3 bond. The GVB orbital centered originally on atom 3 must shift to atom 2 and change its sign. As the 3,4 bond becomes the 2,4 bond, the GVB orbital on atom 2 must shift to atom 3 and change its sign. The net phase changes would be

$$\begin{array}{cc} \oplus & \oplus \\ \oplus & \oplus \end{array} \longrightarrow \begin{array}{cc} \oplus & \ominus \\ \ominus & \oplus \end{array} \tag{22}$$

which is quite inconsistent with the new bonding required. This indicates that the required phase changes cannot occur continuously and that a high-energy barrier would exist for (21).

However let us examine the allowed (p. 48) reaction between three $H_2$ molecules.

$$\begin{array}{ccc} & \overset{1}{H} & \\ {}^2H & & H^6 \\ \vdots & & | \\ {}^3H & & H^5 \\ & \underset{4}{H} & \end{array} \longrightarrow \begin{array}{ccc} & \overset{1}{H} & \\ {}^2H & & H^6 \\ | & & \\ {}^3H & & H^5 \\ & \underset{4}{H} & \end{array} \qquad (23)$$

Convert the 1,2 bond to the 2.3 bond. This changes the phase of the orbital on atom 1. Now change the 5,6 bond to the 4,5 position. This changes the phase of the orbital on atom 6. The final phase changes

$$\begin{array}{ccc} & \oplus & \\ \oplus & & \oplus \\ \oplus & & \oplus \\ & \oplus & \end{array} \longrightarrow \begin{array}{ccc} & \ominus & \\ \oplus & & \ominus \\ \oplus & & \oplus \\ & \oplus & \end{array} \qquad (24)$$

are such that a bond can form between atoms 1 and 6, as required. The old bonds can be broken and the new bond made in a smooth, continuous process.

The method is called the "orbital-phase continuity principle" (OPCP). Its use divides reaction into favored reactions (phase conditions satisfied), and unfavored reactions (phase conditions not satisfied). We could just as well employ the phrases allowed and forbidden, but symmetry plays no direct role. The analysis can be carried out for reactions quite lacking in symmetry.

Goddard has used the OPCP to analyze a number of reactions.[25] Usually the same predictions are made as for PT or orbital-correlation theory. A few differences occur, but these result from some ambiguity in the use of the method. Reactions such as (9), HX elimination from RX, are predicted to be favored, but this happens because an independent *p* orbital originally containing one of the lone pairs of electrons on X is used to bond H. Also the addition of singlet oxygen to ethylene is predicted to be favored.

$$O_2(^1\Delta_g) + C_2H_4 \longrightarrow \begin{array}{c} O{-}O \\ | \quad\;\; | \\ H_2C{-}CH_2 \end{array} \qquad (25)$$

This reaction is certainly forbidden in a planar (suprafacial) mode of addition. However it is allowed in the antarafacial mode. It has been pointed out that this mode of addition is reasonable for (25) because peroxides have skew structures in any case.[27] The OPCP presumably relates to the antarafacial mode on $O_2$.

The phase of a wavefunction refers to the mathematical signs, plus or minus, in the various parts in space. In all of our analyses of potential energy surfaces we are dealing with so-called "stationary" states. Even so, the phase is oscillating harmonically with the time according to

$$\psi_{(x,y,z,t)} = \psi^0_{(x,y,z)} \exp\left(-\frac{2\pi i E t}{h}\right) \tag{26}$$

The phase of a stationary wavefunction is not an observable quantity since $\psi^2$, which we must now write as $\psi\psi^*$, is not dependent on time. Nevertheless the relative phases for different parts of the molecule must be fixed, since they determine the symmetry and nodal characteristics. Thus it makes no difference whether we write one orbital with a plus sign or a minus sign in any part. However we must then be consistent in writing the signs in all other parts. Two noninteracting or orthogonal functions, however, have arbitrary phases with respect to each other.

## OTHER APPLICATIONS OF PERTURBATION THEORY

Perturbation theory is obviously a very logical way to approach chemical reactions.[28] The reason is that total electronic energies of molecules are usually in the range of thousands of kilocalories, whereas heats of reaction and activation energies are usually in tens of kilocalories. The effects of chemical interaction are small perturbations. In Chapter 1 PT was applied in a very special way in which only symmetry aspects were considered as major, although orbital energies played a minor role. There have been many other applications of PT to chemical reactions in which the emphasis has been on the energetics. The goal has been to calculate the energy of interaction of two or more reactants. Symmetry has not played an explicit role in these efforts. Since interaction energies depend very much on orbital overlaps, symmetry has entered in an indirect way.

There are several common features of these other PT approaches:

1. They all deal with bimolecular, or higher, reactions.
2. They all resort to numerical calculations of varying degrees of sophistication.
3. They usually use an interaction operator, $H'$, between the two (or more) reacting molecules.
4. Group-theoretic language is not used.
5. Conjugated $\pi$ systems are usually the systems examined, although some of the methods apply in principle to all molecular systems.

The earliest papers are due to Coulson and Longuet-Higgins.[29] These attempt to predict the reactivity at various positions of a $\pi$ system by calculating various properties (free-valence polarizability, charge density, and bond orders) of the ground-state molecule. These properties would then influence the reactivity toward various kinds of reagent (electrophiles, free radicals, and nucleophiles).

The first papers in which two reacting molecules were explicitly treated were by Dewar.[30] Reactants R and S were combined in an RS complex. The Hamiltonian was broken up into three parts, $H_R$, $H_S$ and $H'$, where $H_R$ and $H_S$ were the Hamiltonians of the isolated molecules and $H'$ was the interaction term. This would be the potential energy of the electrons and nuclei of molecule R acting on the electrons and nuclei of molecule S.

The calculation of the energy now depends very much on the kinds of approximation that are made, and the kinds of system considered. Dewar's examples were nonpolar $\pi$-electron molecules. Salem has recently extended the application of PT to interacting $\pi$ systems.[31] Rather than examine these special cases in detail, it is more instructive to look at more generalized treatments that can also handle polar molecules and $\sigma$-bonded systems.[32]

### Evaluation of the Energy

In a formal sense the energy can be written as

$$E = E_R^0 + E_S^0 + \langle \psi_{RS}^0 H' \psi_{RS}^0 \rangle + \sum_k \frac{\langle \psi_{RS}^0 H' \psi_{RS}^k \rangle^2}{(E_R^0 + E_S^0 - E_{RS}^k)} \tag{27}$$

$E_R^0 + E_S^0$ are the energies of the isolated molecules. The next term in (27) is the first-order PT, where $\psi_{RS}^0$ is the ground-state wavefunction for the combined system. It may be approximated as the simple product, $\psi_R^0 \psi_S^0$, of the wavefunctions for the isolated molecules. The last term in (27) is the second-order perturbation energy. It includes all the changes in energy that result from changes in the wavefunction due to the interaction between the molecules. The summation over $k$ includes all of the excited states of the combined system.

Equation (27) is valid only for small $H'$ and for closed-shell molecules. It obscures the important fact that $\psi_R^0$ and $\psi_S^0$ will also mix under the influence of the perturbation. If each is expressed as a product of MOs, then occupied MOs of the same symmetry in the two different molecules will mix. Some MOs will be lowered in energy, and some will be raised. For small perturbations, the effects just cancel and the net energy is unchanged, but if the molecules are brought very close, the net effect of the mixing is a very large increase in energy. This is the same intermolecular repulsion phenomenon discussed earlier (p. 91). Also if some of the orbitals are only half filled, there can be a net lowering of the energy due to the mixing (see Fig. 9).

The first-order perturbation energy is essentially the potential energy of interaction of two isolated distributions of charges. For ions or for polar molecules it can have a very substantial value, which might be approximated by a classical point-charge or point-dipole calculation. For hydrocarbons and other uncharged, nonpolar molecules, its value is small and may be neglected.

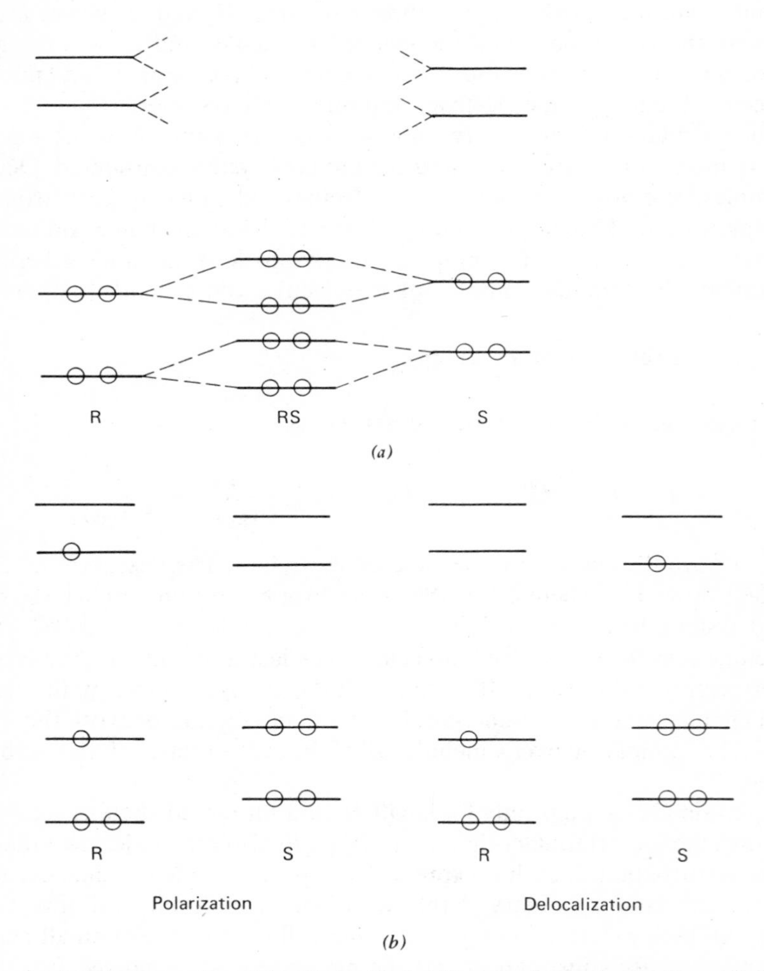


**Figure 9.** (*a*) The interaction of the filled orbitals of two reacting molecules R and S; the net energy change is small; (*b*) polarization of R molecule in field of S molecule; (*c*) charge transfer, or delocalization, between R and S molecules.

The most interesting part of (27) lies in the second-order terms. These arise from the mixing of occupied and empty MOs. There are two separate cases:[33]

1. An occupied MO of R (or S) will interact with an empty MO of R (or S).
2. An occupied MO of R (or S) will interact with an empty MO of S (or R).

Case 1 is relatively unimportant. It corresponds to the polarization of one molecule in the electrical field of a second. It is responsible for the attractive London forces that exist between molecules. Polarization of a molecule in a field always involves the mixing in of excited states in such a way as to favorably change the electron distribution. The energy is lowered as a result.

Case 2 is the chemically significant one. For a chemical reaction in which covalent bonding is important, this is the major process whereby bonds are created and destroyed. It is, of course, an energy-lowering process. Figure 9 shows the various kinds of interactions that occur between orbitals. Only one electron excitation is shown, but two-electron excitations can also become important, or even dominant.

An important example would be the case of an electron acceptor, A, reacting with an electron donor, B, to form a charge transfer complex.[34]

$$A + :B \longrightarrow A:B \tag{28}$$

Here the occupied MO is in B, and the empty MO is in A. It has been pointed out[35] that all chemical bonds can be visualized in terms of the generalized acid–base reaction (28). This emphasizes the importance of the Case 2 perturbation, which may be called the "charge-transfer process." It has also been called the "delocalization process."[33]

The interaction of *all* filled MOs with *all* empty MOs must be included in calculating the energy. It is not surprising to find that the main effect usually comes from the HOMO of one molecule and the LUMO of the second. These are the orbitals that Fukui has called the "frontier orbitals."[36] In a PT treatment of donor–acceptor reactions such as (28), Klopman has distinguished two main categories.[37] Some complexes are held together chiefly by ionic bonding. This is where the first-order perturbation of (27) dominates. Other complexes are held together mainly by covalent bonding, by the second-order perturbation between the HOMO and the LUMO. The two classes or reaction may be called charge-controlled and frontier-controlled, respectively. This helps explain the many experimental phenomena covered by the "principle of hard and soft acids and bases."[35,38]

In evaluating the second-order perturbation energy, LCAO MO theory is used, usually in a semiempirical form. The perturbation energy becomes[32,33]

$$\Delta E = E - E_{\mathrm{R}}^0 - E_{\mathrm{S}}^0 = \frac{q_{\mathrm{R}} q_{\mathrm{S}}}{\rho} + 2 \sum_m \sum_n \frac{(C_r^m C_s^n \beta_{rs})^2}{(\varepsilon_m - \varepsilon_n)} \tag{29}$$

where $q_R$ and $q_S$ are the net charges on R and S, $\rho$ is a corrected "distance" between these charges, $m$ refers to an occupied MO, and $n$ to an empty MO, and $\varepsilon_m$ and $\varepsilon_n$ are their orbital energies. The $C_r^m$ and $C_s^n$ are the coefficients of AOs $r$ and $s$ in molecules R and S (or vice versa), and $\beta$ is an exchange integral of $H'$ between these two AOs.

Of course (29) is a very rough approximation to the energy. In particular it is not good for close distances of approach when the interactions become large. Also if charge transfer is very large, because $\varepsilon_m$ and $\varepsilon_n$ approach each other, then a different approach must be used. The perturbed energies become the solution to the secular determinant

$$\begin{vmatrix} \varepsilon_m - \varepsilon & \beta_{mn} - S\varepsilon \\ \beta_{mn} - S\varepsilon & \varepsilon_n - \varepsilon \end{vmatrix} = 0 \tag{30}$$

where $\beta_{mn}$ is now the exchange integral between MOs $m$ and $n$, and S is the overlap integral between them. The lowest root of (30) is given by

$$(1 - S^2)\varepsilon = \frac{\varepsilon_m + \varepsilon_n}{2} - S\beta_{mn} + \left\{\left(\beta_{mn} - S\left(\frac{\varepsilon_m + \varepsilon_n}{2}\right)\right)^2 + \frac{(\varepsilon_m - \varepsilon_n)^2}{4}(1 - S^2)\right\}^{1/2} \tag{31}$$

The two electrons originally in orbital $m$ are now in the lowest-energy orbital. Accordingly there is a stabilization energy of roughly

$$\Delta E \simeq ((\varepsilon_m - \varepsilon_n)^2 + 4\beta_{mn}^2)^{1/2} \tag{32}$$

if we ignore S as being small. This reduces to the equivalent of the last term in (29) when $|\varepsilon_m - \varepsilon_n| \gg |\beta_{mn}|$.

The charge transfer effect in these uses of PT gives rise to what are effectively orbital symmetry selection rules. Reaction paths in which there is a good overlap between suitable filled MOs and empty MOs will be favorable paths. A forbidden reaction is one in which no such favorable overlap exists. Since the interaction Hamiltonian $H'$ is totally symmetric, as is the total Hamiltonian, nonzero overlap can only occur between orbitals of the same symmetry.

It is instructive to compare (27) with the related equation (12) of Chapter 1. For convenience this will be repeated.

$$E = E_0 + \left\langle \psi_0 \left| \frac{\partial U}{\partial Q} \right| \psi_0 \right\rangle Q + \left\langle \psi_0 \left| \frac{\partial^2 U}{\partial Q^2} \right| \psi_0 \right\rangle \frac{Q^2}{2} + \sum_k \frac{[\langle \psi_0 | \partial U/\partial Q | \psi_k \rangle Q]^2}{(E_0 - E_k)} \tag{33}$$

The first-order perturbation energy in (33) includes changes in the net electrostatic interaction between two approaching molecules. It also includes changes in potential energy if the nuclei within a molecule change their relative positions. The second-order perturbation energy in both (33) and (27) involves the same mixing of excited states, but the perturbation is quite different. In (33) the perturbation is the *change* in potential energy due to moving the nuclei, not the potential energy itself.

In fact the exact relationship between (27) and (33) is that the energy $E$ of (27) is the energy $E_0$ of (33). It is the starting point for a small displacement from $Q_0$, an initial nuclear configuration. Equation (33) is useful if we wish to evaluate, or estimate, the slope and curvature of a plot of energy against the reaction coordinate. Equation (27) is applicable if we wish to evaluate the actual energy. This suggests that the latter equation is more powerful and useful.

This is unfortunately not so. The calculation of the actual energies of polyelectronic systems is a very difficult problem in quantum mechanics. Equation (27), like (33), can only be used to estimate small changes in energy. Since the available information usually includes only very approximate MOs rather than exact wavefunctions, it is no more accurate to use (27) than (33). By the time an activated complex has been formed, (27) is useless. The MOs of the activated complex are much too different from those of the original reactants for the latter to be a good starting point.

Both (27) and (33) have their major usefulness in making decisions about the beginning of a possible reaction. Will it be a favorable one or not? Equation (27) is good only for bimolecular (or higher) reactions, whereas (33) is valid for unimolecular reactions as well. This includes the important case of the decomposition of activated complexes. Extraction of symmetry rules is easier from (33) because it was constructed for that purpose. Symmetry is only implicit in (27). In particular, whenever the reaction coordinate is not totally symmetric (at maxima or minima), (27) is of no use since $H'$ is always totally symmetric and cannot tell us anything about asymmetric nuclear movements.

One situation where (27) is superior to (33) is in evaluating several paths, all of which are allowed by symmetry. One could more easily estimate, at least for the initial stages of the reaction coordinate, whether one path was to be preferred over another. Also (27) would be more useful than (33) in deciding which of several similar reactants would be more reactive toward a given reagent.[28,39] The actual equation used for such comparative studies would be (29), or some more sophisticated variant of it.[32]

## The Method of Fukui

There is a way in which PT, in the sense of two molecules interacting, can be applied to a unimolecular process, namely, by dividing the reacting molecule into two parts, which are then considered to interact with each other. This procedure has been used successfully by Fukui. It is part of a general approach to reaction mechanisms developed by Fukui, which is based on PT, but which dispenses with most forms of calculation or group-theory analysis.[40] It is necessary in this method to know something about the valence-shell MOs of the reacting molecule or molecules.

Essentially what is needed is the topology of the orbitals. It is also important to know if an MO is concentrated on certain atoms in preference to others. The magic formula is simple: "A majority of reactions take place at the position and in the direction of maximum overlap of the HOMO and LUMO of the reactant molecules." A singly occupied orbital can take the place of either the HOMO or the LUMO. For unimolecular reactions, the molecule is divided into two parts, one of which contains the HOMO, and the other, the LUMO. An additional lemma is required for the division: "The bordering surface of the two parts of the molecule should be crossed by the newly formed bonds."

Figure 10 shows some bimolecular applications of Fukui's method. Electrophilic substitution in an aromatic molecule is exemplified in Fig. 10*a*. The form of the HOMO in napthalene is used to explain why only α nitration

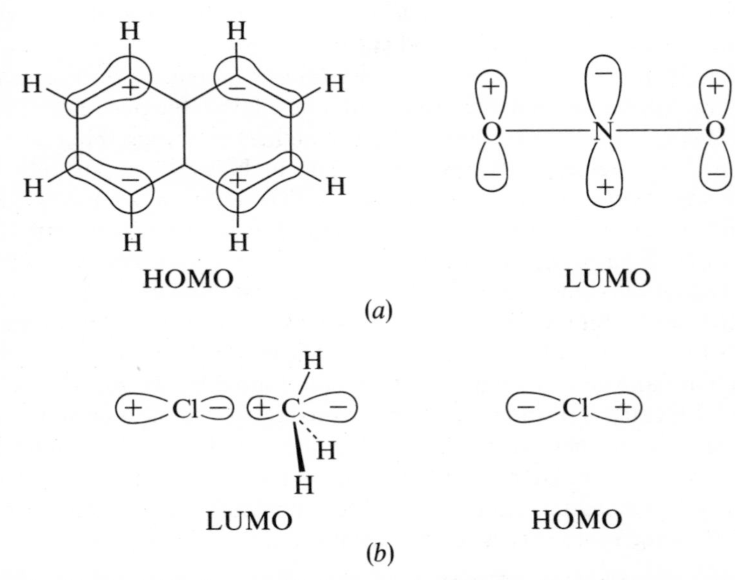


**Figure 10.** (*a*) The HOMO of naphthalene and the LUMO of $NO_2^+$; these are the critical orbitals in the nitration of $C_{10}H_8$; (*b*) the critical orbitals in the bimolecular nucleophilic substitution ($S_N2$) reaction; the HOMO is a σ orbital of the nucleophile, $Cl^-$, and the LUMO is the σ* orbital of the C—Cl bond in $CH_3Cl$.

is observed. There is a higher frontier electron density at the $\alpha$ position. Figure 10b shows the important example of $S_N2$ (bimolecular, nucleophilic) substitution at an alkyl halide.[41] This serves as a model for what may well be the most important reaction in organic chemistry.[42] The HOMO is obviously in the nucleophile, and the LUMO is the antibonding orbital that is concentrated largely between the carbon and halogen atoms. Inversion of configuration at carbon is explained by this model.

Figure 11 shows how unimolecular reactions are treated. The conrotatory ring opening of cyclobutene (11*a*) is explained by considering the $\pi$ bond as the HOMO and a $\sigma$ bond between the remaining two carbon atoms as the LUMO. The nuclear motions are always in the direction that increases the overlap between these two parts. Initially, of course, the overlap is zero because the MOs of a molecule are orthogonal to each other. More than one interaction can be considered. For example, the $\pi^*$ orbital may be taken as the LUMO and the $\sigma$ orbital as the HOMO. Figure 11 shows that this overlap also increases with a conrotatory ring opening.

More complex rearrangements are treated in stages. In the Cope rearrangement,

(34)

the first step is to form a TS consisting of two interacting allyl radicals (11*b*). These can recombine to form the product. An advantage is that secondary interactions of other orbitals can also be considered. In this way the preference for a chairform TS in (34) can be rationalized.

The Fukui method is simple and quite powerful. It works very well in most cases and has the advantages of being both pictorial and nonmathematical. It is a modern version of the old "electron-pushing" methods of mechanistic chemistry. In the case of bimolecular reactions it uses the same principles that application of (33) would suggest (see p. 57). There are some disadvantages, particularly in the case of unimolecular reactions, where it is not always clear how to select the parts of the molecule. For example, the two reactions

$$SO_2Cl_2 \longrightarrow SO_2 + Cl_2 \quad (35)$$

$$SCl_4 \longrightarrow SCl_2 + Cl_2 \quad (36)$$

appear very similar. Yet (35) is forbidden and (36) is allowed, as discussed in Chapter 4. It is by no means obvious that Fukui's method could be used to predict these results.

Also the abandonment of any reliance on exact symmetry always raises the possibility that subtle symmetry effects can be missed. The bimolecular

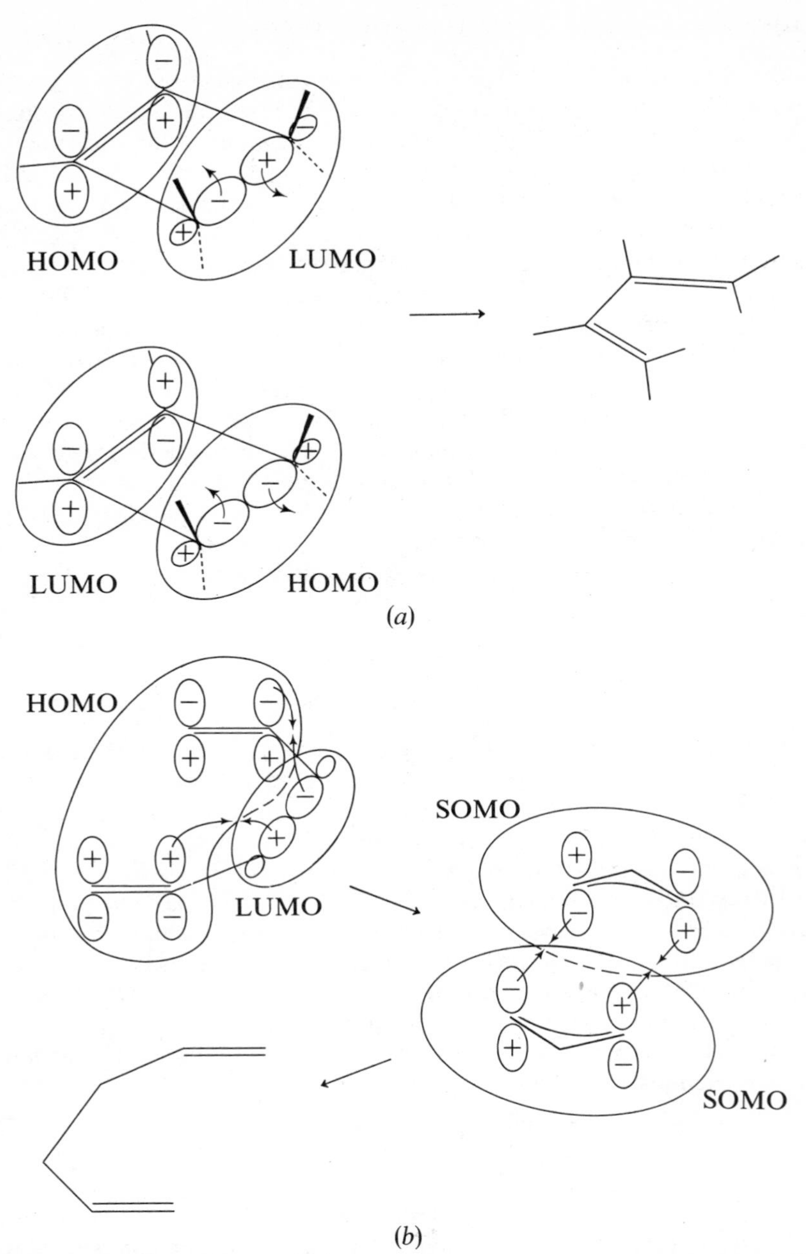


**Figure 11.** (*a*) The increasing HOMO–LUMO interaction in the ring opening of cyclobutene to butadiene (two choices of HOMO and LUMO shown); (*b*) The two stages in the Cope rearrangement of 1,5-hexadiene; the two hypothetical alkyl radicals are coupled by interaction of their singly occupied molecular orbitals (SOMO).

reaction of two acetylene molecules to form tetrahedrane is forbidde
orbital symmetry (p. 63).

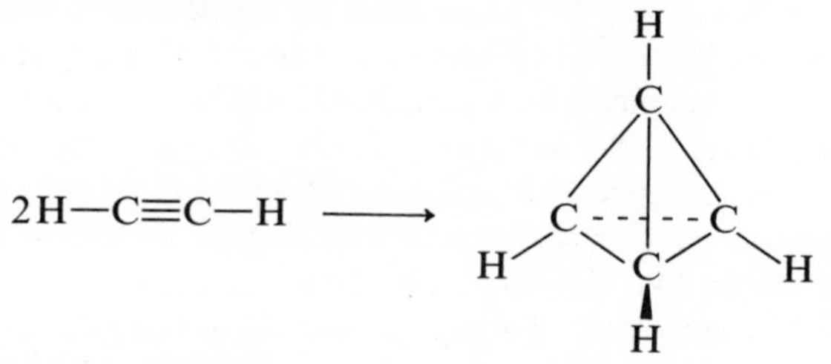

$$2\,H{-}C{\equiv}C{-}H \longrightarrow \qquad (37)$$

Simple orbital overlap considerations would suggest that it could be formed by two antarafacial–suprafacial electrocyclic additions, but this is not so since the reaction is forbidden in the skew $D_2$ reaction coordinate.

It is also dangerous to emphasize too strongly the HOMO and LUMO of a reacting system. It could well be other high energy occupied MOs or low-energy empty MOs which are the critical ones. It may be noted that in Woodward and Hoffmann's original papers it was also the HOMO that was followed to see if it remained a bonding orbital. This restriction was quickly abandoned.[1]

## METHODS BASED ON TRANSITION-STATE STABILITY

Since in the final analysis we are only interested in the magnitude of an energy barrier, why not proceed directly to a theoretical consideration of the stability of activated complexes? Any considerations that suggest extra stability for the activated complex also suggest a low-energy barrier and a favorable reaction path. Since an activated complex in many ways is like an ordinary molecule (except for its vibrational behavior along the reaction coordinate), any rules that have been developed for estimating stability of ordinary molecules may be applied to the TS.

Some caution is necessary, the most significant being the need to be sure that we are dealing with an activated complex (potential energy maximum) and not a reactive intermediate (potential energy minimum). Too much stability estimated for a particular structure may well mean that it is *not* that of the activated complex. There will be an energy barrier elsewhere en route to this structure from the starting point. This barrier may well be so high as to be prohibitive, and may exist because of symmetry effects. For example, tetrahedrane, as shown in (37), may well be a moderately stable molecule once formed. As an intermediate for carbon-isotope scrambling in acetylene, it would appear very favorable. However the orbital symmetry

en route to the formation of $C_4H_4$, not in the stability
f.

s of stability arguments for the activated complex go back
bond theory, and resonance, to estimate stabilizations.[43]
was made by Evans in 1939, who explained by resonance
els–Alder reaction readily occurred, but why two olefin
t cyclize.[44] The former reaction involved six electrons in
the interconverting $\pi$–$\sigma$ system, and hence the activated complex was aromatic and stable like benzene. The latter reaction involved only four electrons, and the activated complex would be antiaromatic and unstable like cyclobutadiene. This simple concept of aromatic and antiaromatic TSs can be extended in a most remarkable and general way, as shown by Dewar and Zimmerman.[45,46]

The requirement is that the reaction can be written with a TS containing a single cycle made up of one interacting atomic orbital from each atom in the cycle. Such reactions are called "pericyclic reactions."[1] No distinction is made between $\sigma$ bonds and $\pi$ bonds, so long as they can be considered to be interacting. Also any atom can be part of the cyclic system. Even an atom such as H, which contributes only an $s$ orbital, can take part. The AOs are assigned phases in such a way as to give positive overlaps throughout. Sometimes this will not be possible, and a sign inversion will occur between two adjacent and interacting orbitals. If no inversions occur, or if an even number occur, the system is said to be a Hückel system. If one inversion or an odd number of inversions occurs, the system is said to be a Möbius one.[46]

The stability rule is then that $(4n + 2)$ electrons are stable in a Hückel system and $4n$ electrons are stable in a Möbius system, where $n$ is an integer including zero.[47] Any system in which the number of electrons does not follow the rule is one with an unstable TS. Hence reaction through that particular activated complex is forbidden or disfavored. The stable TSs may be considered aromatic and the unstable ones, antiaromatic, according to Dewar.[28,45] This follows from the definition that aromaticity and antiaromaticity in a cyclic compound may be defined as either greater or lesser stability than for the corresponding open-chain compound.

The Hückel and Möbius stability rules derive simply from the MO energy levels in the Hückel approximation. The equations for the energies are

$$\varepsilon = q - 2\beta \cos\left(\frac{2\pi k}{n}\right) \quad \text{Hückel} \tag{38}$$

$$\varepsilon = q - 2\beta \cos\left(\frac{2\pi(k + 1)}{n}\right) \quad \text{Möbius} \tag{39}$$

where $q$ is the Coulomb integral and $\beta$ is the exchange integral. The integer $k$ runs from 0 to $n - 1$, and $n$ is the number of atoms in the ring. This is a useful mnemonic for deriving the energy levels.[46,48] A polygon with $n$ sides is inscribed inside a circle of radius $2\beta$. If a vertex is at the bottom, the Hückel levels are reproduced, both as to energy and degeneracy, as shown in Fig. 12. If a side of the polygon is at the bottom, the energy levels of the Möbius system are reproduced. The $(4n + 2)$ and $4n$ are simply the number of electrons needed to fill up the bonding MOs for each class of molecules.

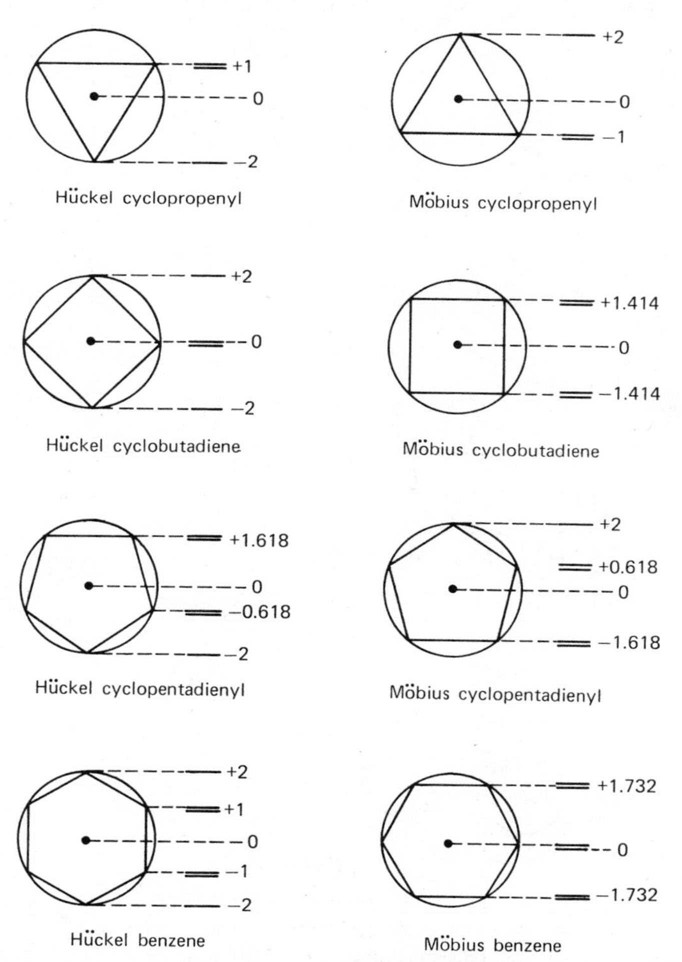


**Figure 12.** Simple device for obtaining the $\pi$-orbital energies of either Hückel or Möbius cyclic systems (after reference 46).

### Some Pericyclic Reactions

Figure 13 shows how three carbon atoms in a cycle generate a Hückel system with the three $p$ orbitals perpendicular to the ring and a Möbius system with three $p$ orbitals in the plane of the ring. In the first case two electrons can be accommodated in stable orbitals, in the second case four electrons can be placed in bonding MOs. Figure 13 also shows how the $\pi$ orbital of $C_2H_4$ and the $\sigma$ orbital of $H_2$ create a four-atom cycle that is Hückel in type. Since it also contains four electrons, it is an unstable TS. The *cis* addition of hydrogen to ethylene is forbidden because of the low stability of the TS. Note that the

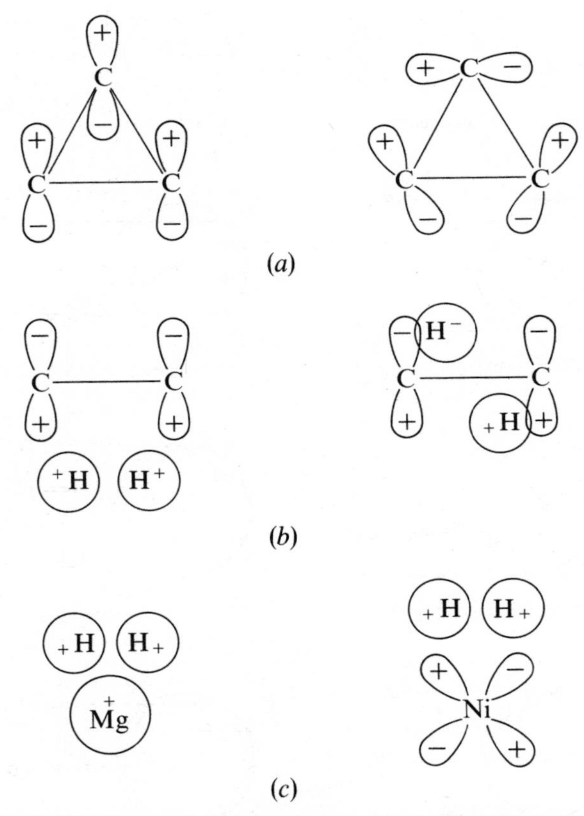


**Figure 13.** (*a*) Three $\pi$ orbitals in a cyclopropenyl ring form a Hückel system; three $p$ orbitals in the plane of the ring form a Möbius system; (*b*) suprafacial addition of $H_2$ to $C_2H_4$ creates a Hückel system, whereas antarafacial addition gives a Möbius system; (*c*) reducing electrons in an $s$ orbital on Mg create a Hückel system with $H_2$; electrons in a $d$ orbital on Ni give a Möbius system with $H_2$.

same results would be found for the reaction of two hydrogen molecules with each other, or two ethylene molecules with each other.

Figure 13 also shows that antarafacial addition of $H_2$ to $C_2H_4$ would be allowed. The activated complex is now Möbius in type and four electrons is a stable number. The use of a *d* orbital on an atom instead of an *s* or *p* orbital changes the overlap situation dramatically. A nontransition metal, such as magnesium, has its valence-shell electrons in an *s* orbital. With molecules such as $H_2$, $Cl_2$, or RCl it forms a Hückel system with four electrons. Therefore a one-step reaction such as

$$\mathrm{Mg} + \mathrm{H_2} \longrightarrow \mathrm{MgH_2} \tag{40}$$

is forbidden. However for a transition metal, where the reducing electrons may be in a *d* orbital, the system is Möbius and the one-step reaction is allowed.

$$\mathrm{Ni} + \mathrm{H_2} \longrightarrow \mathrm{NiH_2} \tag{41}$$

A very large number of important class reactions are covered by the Dewar–Zimmerman rules. Some examples are given below for Hückel TS.

[1,3] Sigmatropic shifts

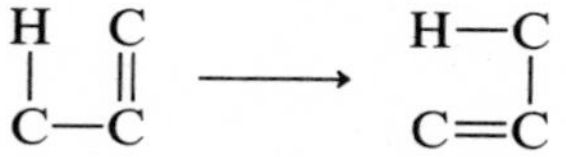

$$\begin{array}{cc} \mathrm{H} & \mathrm{C} \\ | & \| \\ \mathrm{C} & \!\!-\!\! \mathrm{C} \end{array} \longrightarrow \begin{array}{c} \mathrm{H{-}C} \\ \quad | \\ \mathrm{C{=}C} \end{array} \qquad \text{4e forbidden} \tag{42}$$

Cycloadditions

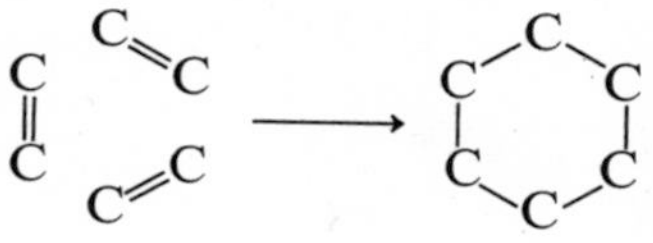

$$\begin{array}{c} \mathrm{C} \\ \| \\ \mathrm{C} \end{array} \quad \begin{array}{c} \mathrm{C{=}C} \\ \\ \mathrm{C{=}C} \end{array} \longrightarrow \text{cyclohexene (C}_6\text{ ring)} \qquad \text{6e allowed} \tag{43}$$

ene Reaction

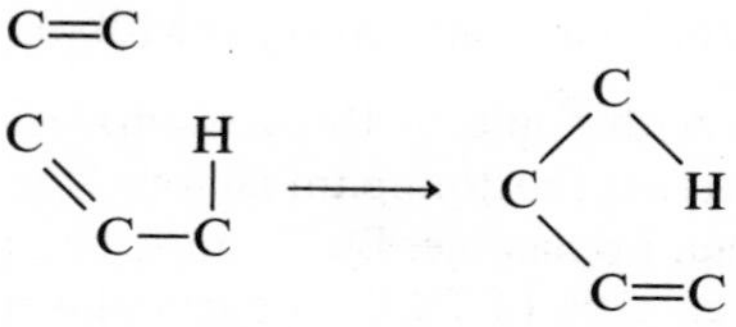

$$\begin{array}{l} \mathrm{C{=}C} \\ \\ \mathrm{C} \qquad \mathrm{H} \\ \;\; \backslash\backslash \quad\;\; | \\ \quad\;\; \mathrm{C{-}C} \end{array} \longrightarrow \begin{array}{c} \mathrm{C} \\ / \quad \backslash \\ \mathrm{C} \qquad\quad \mathrm{H} \\ \backslash \\ \mathrm{C{=}C} \end{array} \qquad \text{6e allowed} \tag{44}$$

Cope reaction

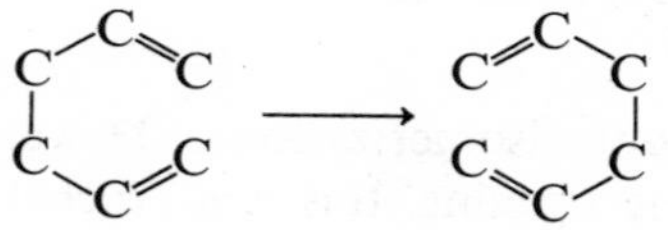

$$\text{1,5-hexadiene (C}_6\text{)} \longrightarrow \text{1,5-hexadiene (C}_6\text{)} \qquad \text{6e allowed} \tag{45}$$

Claisen reaction

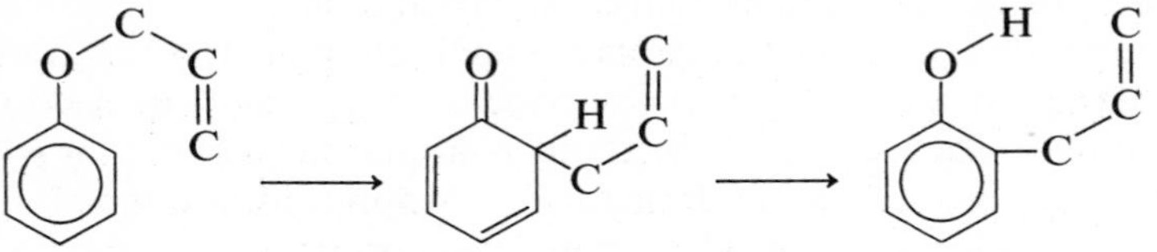

6e allowed (46)

Ester pyrolysis

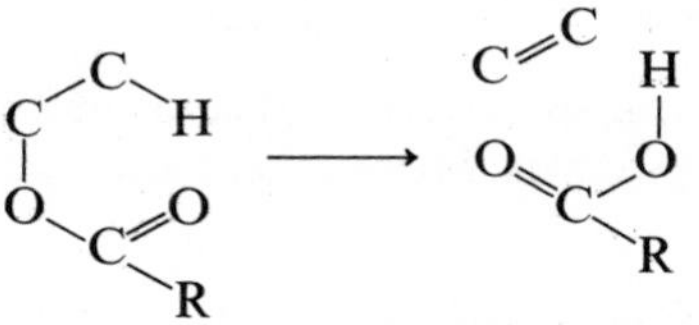

6e allowed (47)

Cope elimination (1,4 sigmatropic shift)

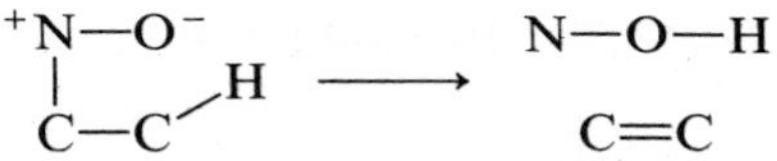

6e allowed (48)

[1,6] Sigmatropic shift

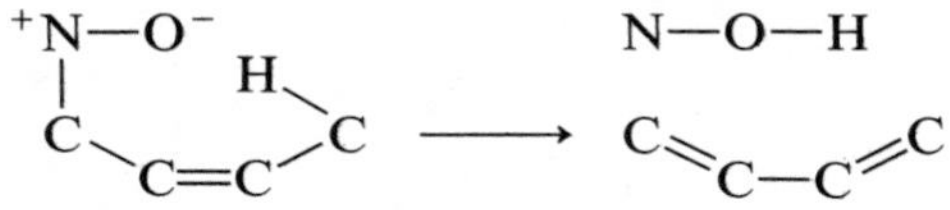

8e forbidden (49)

Electrocyclic reactions

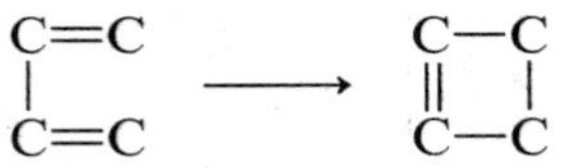

(50)

Disrotatory is Hückel and forbidden; conrotatory is Möbius and allowed.

Symmetry has little to do with some of these examples, but the nodal properties of the wave functions are the dominant factors. The method is very simple and easy to apply, when applicable. But it must be appreciated that the rules cover only a restricted class of TS.[49] In particular it is not allowed

(a) to have more than one cycle present,
(b) for an atom to contribute more than one orbital to the cycle,
(c) to have a noncyclic TS.

An illustration of (a) would be the isomerization of Dewar benzene (p. 90). Because of the disrotatory ring opening, this is a Hückel system with six electrons. Nevertheless, the reaction is forbidden. An example of (b) is found in (35) and (36). The only difference in the two reactants is an additional lone pair of electrons in $SCl_4$. That one reaction is allowed and the other is forbidden might well be deduced, but which is which?

The correct answer is reached by taking one $p$ orbital on sulfur and constructing a Möbius system. Then reaction (35) is a two-electron cycle and reaction (36), a four-electron cycle. However two orbitals on sulfur are obviously needed to form two sulfur—chlorine bonds. There are many cases where it seems reasonable to use two orbitals on a single atom. For example, it has been concluded that $C_6H_5ICl_2$ is capable of adding two chlorine atoms *cis* to an olefin in a concerted reaction.[50]

$$C_6H_5ICl_2 + C_2H_4 \longrightarrow C_6H_5I + C_2H_4Cl_2 \qquad (51)$$

Presumably this conclusion is based on counting six electrons, two from each I—Cl bond and two from the carbon—carbon $\pi$ bond, and considering the AOs to form a Hückel ring. Yet Fig. 14 shows that reaction (51) is actually

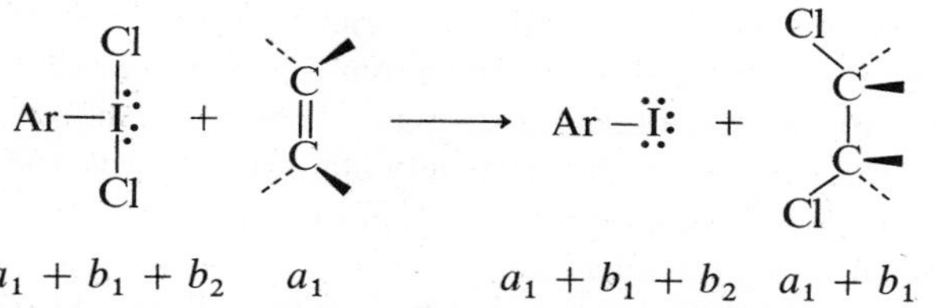

Ar—I(Cl)₂ + C=C ⟶ Ar—I: + Cl—C—C—Cl

$2a_1 + b_1 + b_2 \quad a_1 \qquad a_1 + b_1 + b_2 \quad a_1 + b_1$

**Figure 14.** Symmetries of bonds and lone pairs in reaction of $C_6H_5ICl_2$ with $C_2H_4$.

forbidden. The bond-symmetry rule is used, which is not subject to any limitations, except that all lone pairs must be properly accounted for. The two lone pairs of $C_6H_5ICl_2$ are $a_1 + b_2$ in the $C_{2v}$ point group. They lie in a plane perpendicular to the nearly linear Cl—I—Cl axis. The lone pairs of $C_6H_5I$ have the symmetry of $p_x \cdot p_y$ and $p_z$ orbitals, that is, $a_1$, $b_1$, and $b_2$. Since the two chlorine atoms are 4.90 A apart in $C_6H_5ICl_2$,[51] it is obvious that they could not add to ethylene in a concerted process in any case.

If there is a linear, or noncyclic, TS, then the concept of aromatic stabilization does not apply. An example would be the nucleophilic displacement reaction

$$I^- + CH_3Cl \longrightarrow I\text{---}CH_3\text{---}Cl^- \longrightarrow CH_3I + Cl^- \qquad (52)$$

This is a four-electron system where the orbitals can be written to give all positive overlaps, but the reaction is allowed. The concept of aromatic stabilization, or resonance stabilization, in linear systems is not clearly defined. It is also difficult in polycyclic systems. Recently rules for aromatic stabilization in complex molecules have been developed that may be combined with Evans' principle to predict favorable TS.[52]

## VALIDITY OF SYMMETRY RULES

In the previous sections we have developed a number of rules for predicting the existence of favorable or unfavorable reaction paths. These are all attempts to estimate the height of the energy barrier existing for a selected reaction coordinate. We must also take into account other factors besides the energy barrier that affect the rates of chemical reactions. Before doing this it is necessary to assess the validity or reliability of the rules themselves.

In the case of reactions lacking symmetry, the rules are not sharply defined since we cannot speak of reactions as forbidden or allowed in a strict sense. We must return to the more formal rules of Chapter 1. The question to be answered is, "if a reaction is forbidden by strict symmetry selection rules, does it, in fact, not occur?" A second question might be whether reactions allowed by symmetry do, in fact, occur. The answer here is quickly given. Since a given stoichiometric reaction has a variety of allowed paths open to it, not all of them will occur to a measurable extent. Being allowed by symmetry may be necessary, but it is not a sufficient condition.

An answer to the first question is also quickly given. All rules have exceptions, including the symmetry-selection rules. No reaction is ever completely forbidden. If a mechanism can be visualized, it is occurring to some extent.[53] However its rate may be much too small for practicality. The question still remains as to the effectiveness of symmetry rules. We start by considering the venerable Wigner–Witmer rules (p. 37).

The angular momentum aspect of the rules relates only to linear TS, but the electron-spin conservation rule is general for all reactions. Yet we find it violated in many cases, both for simple reactions and for complex ones. A few examples are

$$\mathrm{SO_2}(^1A_1) + \mathrm{O}(^3P_g) \longrightarrow \mathrm{SO_3}(^1A_1') \tag{53}$$

$$\mathrm{O^+}(^4S_u) + \mathrm{CO_2}(^1\Sigma_g^{\ +}) \longrightarrow \mathrm{O_2}^+(^2\Pi_g) + \mathrm{CO}(^1\Sigma^+) \tag{54}$$

$$\underset{\text{tetrahedral, high spin}}{\mathrm{CoCl_2[P(C_6H_5)_3]_2}} \rightleftharpoons \underset{\text{planar, low spin}}{\mathrm{CoCl_2[P(C_6H_5)_3]_2}} \tag{55}$$

All of the above are very rapid reactions. In addition we have many cases of intersystem crossing of singlet and triplet excited states. These will be discussed in Chapter 6.

## Spin–Orbit and Vibronic Coupling

The mechanism whereby the spin-conservation rule is violated is well known.[54] It occurs by spin–orbit coupling. That is, the magnetic dipole due to the spin of an electron interacts with the magnetic dipole due to the orbital motion of the electron. The energies involved are small in a chemical sense, but the addition of spin–orbit coupling to the Hamiltonian has an important consequence. No electronic state is a pure spin state. There is always some mixture of states of different spin due to $H_{SO}$. This in turn permits transitions between two electronic states, each predominantly of a different spin. Spin–orbit coupling is much greater for heavy than for light atoms. Yet we see reaction (54), involving only light atoms, occurring with one of the largest rate constants ever observed.[55]

Fortunately the same spin–orbit coupling mechanism is responsible for making partially allowed spin-forbidden transitions in electronic spectroscopy. This makes a very large body of data available for estimating the effectiveness of this mechanism. Experimentally the presence of a spin change reduces the intensity of an absorption band by a factor in the range of $10^{-3}$–$10^{-6}$. These are only average figures, since results in the range of $10^{-9}$–1 also exist. Even for atoms as heavy as iron or cobalt, a substantial lowering in probability ($10^{-3}$) exists for any event requiring a change in electronic spin.

Another mechanism that makes forbidden transition allowed in electronic spectroscopy is vibronic coupling. A molecular vibration mixes into the ground-state wavefunction some excited state wavefunctions of the proper symmetry (see p. 14). This in turn allows transitions which are otherwise forbidden by state symmetry. However this mechanism is very ineffective as a means of circumventing orbital-symmetry restrictions. The reason is very simple, energy must be conserved during a transition from one state to another.

This condition is easily met in electronic spectroscopy by the energy of the quantum of radiation absorbed. In a chemical reaction, it is only thermal energy that is available. Two electronic states must be at nearly the same energy, for the same nuclear positions, before a transition between them can be induced by vibrational coupling. If this requirement is met by exciting the lower state vibronically, then one has already risen to an energy near the top of the energy barrier and little saving in activation energy results. In addition, in most cases configuration interaction will occur to mix together two states of the same symmetry. For the usual case of totally symmetric states, this guarantees staying on the lower energy surface (adiabatic process). Vibronic coupling will only be important when two states belong to different species of the relevant point group. A nonadiabatic process can result.

It is sometimes thought that the random vibrational and other motions of a reacting system will cause the symmetry selection rules to break down. That is, a random reaction path can be of any symmetry and the instantaneous point group of a set of nuclei can be devoid of elements of symmetry. The conclusion that the symmetry rules then disappear is completely invalid, however. The symmetry rules are based on an idealized model, but the shape of the potential energy surface so deduced is unchanged by any later inclusion of dynamic effects. The way in which the nuclei move on the surface, the dynamics, can influence the rate, but in other ways (see p. 145).

Just as the spin rules sometimes break down, so the Wigner–Witmer rules for angular momentum must also break down, since it is the coupling of spin and orbital momentum that is involved. Also in reactions of linear molecules another mechanism can occur. Some examples are

$$N^+(^3P) + O_2(^3\Sigma_g^-) \longrightarrow NO_2^+(^1\Sigma^-) \tag{56}$$

$$N_2O(^1\Sigma^+) \longrightarrow N_2(^1\Sigma_g) + O(^3P_g) \tag{57}$$

The way in which these deviations occur is by coupling of the electron angular momentum with the angular momentum due to the rotation of the molecule.[54] In addition, electronic angular momentum is readily quenched in nonlinear molecules. The reaction

$$H(^2S_g) + NO(^2\Pi) \longrightarrow HNO(^1A') \tag{58}$$

proceeds directly to the ground state of the bent HNO molecule.[55] If electronic angular momentum was conserved, the correlation would be to an excited state of HNO that is repulsive.

We next look for deviations in the selection rules based on orbital correlations and PT. As pointed out in Chapter 1, these are equivalent in terms of the predictions made. We first note that a slavish adherence to mathematical symmetry can lead to incorrect predictions. This would be the case for reactions of no symmetry, or of useless elements of symmetry, as already mentioned. However it can also be true in reactions of very high symmetry. A case in point is the isomerization of cubane to cyclooctatetraene (COT).

Figure 15 shows that the $C_{2v}$ point group is preserved. The bonds that are broken are $a_1, b_1, a_1$, and $b_2$ in this point group. The bonds that are made are of the same symmetries. The reaction is predicted to be allowed, but it is in fact forbidden. This can be seen by examining a correct correlation diagram (Fig. 15).[57] The intended correlation of the $b_1$ and $b_2$ orbitals are to antibonding orbitals of those symmetries. Of course mixing of $b_1$—$b_1^*$ and

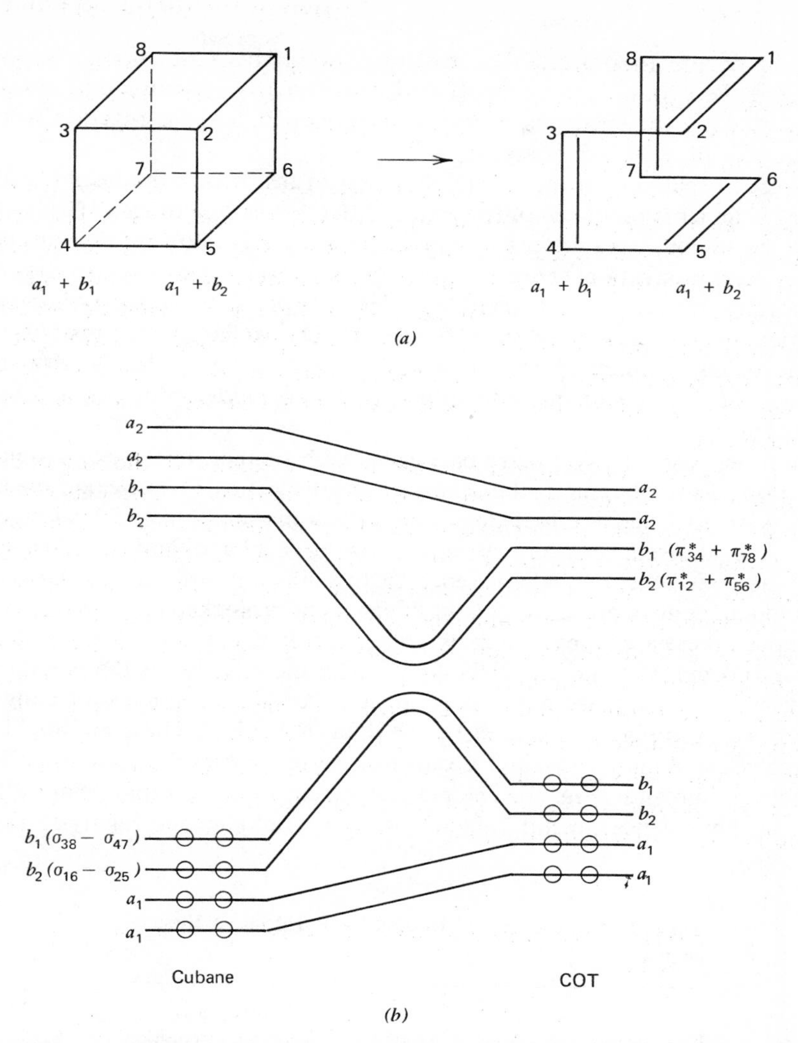


**Figure 15.** (*a*) Symmetries of bonds that are made and broken in the cubane–cyclooctatetraene (COT) isomerization; bonds are taken in pairs ($\sigma_{16} \pm \sigma_{25}$, etc.); (*b*) correlation diagram showing the intended correlation of the $b_1$ and $b_2$ $\sigma$-bond orbitals with antibonding $\pi^*$ orbitals in the product (after reference 57).

$b_2$—$b_2^*$ occurs to prevent the crossing, but in this case mixing is poor because the orbitals are far apart and are of nearly zero overlap initially. Consequently the energy barrier is well developed, and the reaction is forbidden for practical purposes.

A great weakness in the orbital symmetry rules is that the height of the energy barrier is predicted only as substantial (forbidden) or small (allowed). This leaves considerable leeway, especially since fully allowed reactions can easily have activation energies of 25–30 kcal even when exothermic. An important question must then be: "How high is a symmetry-imposed barrier?" In simple cases such as the $H_2$–$D_2$ exchange reaction, or the nitric oxide decomposition, the answer seems to be in the 50–100 kcal range. This is quite adequate to knock out the forbidden process as a useful mechanism.

In other cases the symmetry barrier is much smaller. In the ring-opening reaction of cyclobutene to butadiene, both theoretical[58] and experimental results[59] show that the forbidden disrotatory mode is only 15 kcal higher than the allowed conrotatory mode. While such a difference effectively excludes one reaction in preference to the other, it still would allow the forbidden mode to operate when the allowed one is blocked for other reasons.

Indeed there are many examples where forbidden organic rearrangements seem to occur quite facilely.[60] Unfortunately there is always the possibility that such reactions are not concerted but occur in a sequence of allowed steps. An example is illustrated by the data in Table 2. These are for [1,3] sigmatropic shifts of an alkyl group in a series of bicycloheptenes.[61] The process is shown in Fig. 16. The rearrangement must be suprafacial on the double bond since an antarafacial process would give a highly-strained

**Table 2 Stereochemical Results for Pyrolysis of Bicycloheptenes[a]**

| Reactant[b] | $E_a$, kcal | log $A$ | Rate Ratio, Allowed/Forbidden |
|---|---|---|---|
| 1 | 47 | 13.7 | 19 |
| 2 | 47 | 14.2 | 10 |
| 3 | — | — | 0.14 |
| 4 | 49 | 14.8 | — |

[a] From reference 61.
[b] Reactant 1 has R = D, R′ = H in Fig. 16, 2 has R = $CH_3$, R′ = H, 3 has R = H, R′ = D, and 4 is unsubstituted bicyclo [4.2.0] heptene.

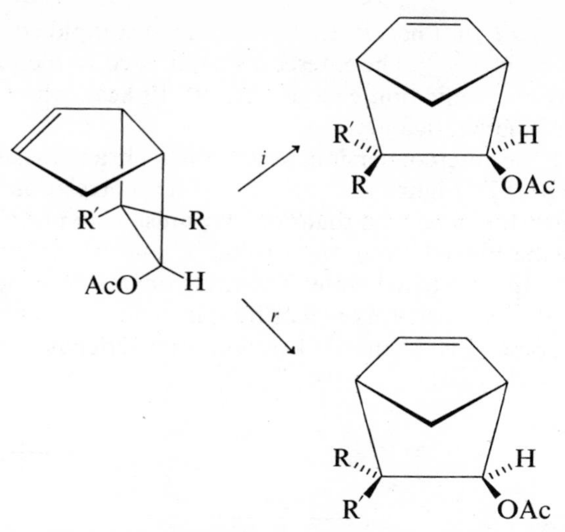


**Figure 16.** Isomerizations of bicycloheptenes: *i* refers to allowed inversion mechanism and *r*, to forbidden retention mechanism (see Table 2 for R and R′) (after reference 61).

*trans* bridged product. It will be recalled that an allowed [1,3] suprafacial shift should give inversion of configuration at the migrating alkyl group, whereas the forbidden process gives retention (p. 98).

Table 2 shows that both the allowed and forbidden processes occur to varying degrees, depending on the substituents present. The energy difference between the two processes must be small. The alternative to the forbidden reaction is biradical formation, followed by ring closure.

(59)

If a biradical is formed, it is difficult to understand the stereospecificity. Usually alkyl free radicals cannot maintain a fixed configuration. Many other forbidden reactions that occur may be cited.[61]

Reaction (55) is typical of a very large number of reactions in inorganic chemistry that violate both spin- and orbital-conservation rules. The best studied examples are four-coordinated nickel(II) complexes, which can be either paramagnetic and tetrahedral or diamagnetic and planar.

$$\underset{T_d}{\mathrm{Ni}L_4^{2+}(^3T_2)} \rightleftharpoons \underset{D_{4h}}{\mathrm{Ni}L_4^{2+}(^1A_{1g})} \tag{60}$$

For many ligands $L$ (not necessarily all the same), a rapid equilibrium exists between the two forms.[62] The interconversion is conveniently followed by nmr spectroscopy. Activation energies are 10–15 kcal, whereas $\Delta H$ for (60) is 0–5 kcal for different ligands.

Presumably the interconversion passes through an intermediate of $D_{2d}$ structure (see p. 182). Figure 17 shows the orbital-correlation scheme for the $d$ manifold only. It can be seen that not only must a spin be changed, but an electron must be moved from one orbital to another to prevent forming planar $NiL_4^{2+}$ in an excited state. The interconversion is accordingly forbidden by orbital symmetry, as well as by spin symmetry. The forbiddenness accounts for some part of the 10 kcal energy barrier not due to thermodynamics.

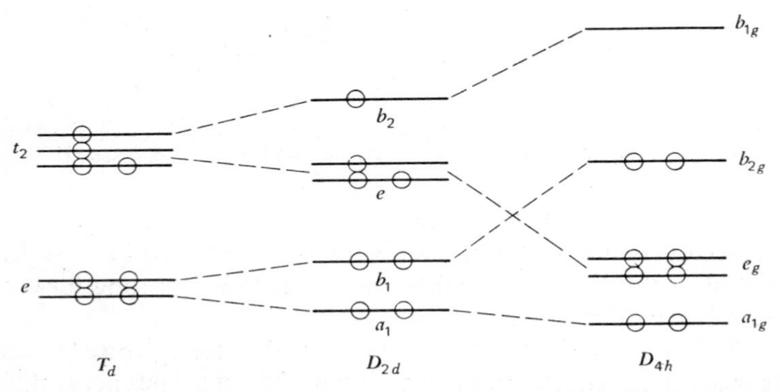


**Figure 17.** Correlation for $d$ manifold for conversion of high-spin tetrahedral complex, $NiL_4^{2+}$, to low-spin planar complex, by way of $D_{2d}$ intermediate; the four metal—ligand bond orbitals correlate without any crossover.

Actually the dual requirement of both a spin and orbital change is an advantage rather than an extra factor preventing the reaction. An examination of Fig. 17 shows that an electron must move from the $b_{1g}$ orbital to an $e_g$ orbital. The $e_g$ orbital is one of the $d_{xz}$, $d_{yz}$ pair. The $b_{1g}$ orbital is the $d_{xy}$ orbital, if we maintain the same $x$ and $y$ axes as in the $T_d$ form. These different $d$ orbitals correspond to different amounts of orbital-angular momentum about the $z$ axis. The change in orbital-angular momentum just compensates for the change in spin-angular momentum.

It may be noted that the bond symmetry rule applied to reaction (60) predicts it to be allowed. The four $\sigma$ bonds of tetrahedral $XY_4$ are $a_1$ and $t_2$. In $D_{2d}$ these become $a_1$, $b_2$, and $e$. The four $\sigma$ bonds of planar $XY_4$ are $a_{1g}$, $e_u$, and $b_{1g}$. In $D_{2d}$ these also become $a_1$, $b_2$, and $e$. The forbidden nature of (60) lies in the formation of a low-lying excited state involving the nonbonding *d* electrons only. The main bonding is not affected. This is consistent with the small value of the energy barrier, and in general explains why many "forbidden" reactions of transition-metal complexes occur with great rapidity.

Our conclusions are that the rules do work in the sense of predicting barriers that exist. In the case of spin selection rules, the restriction is one of probability, otherwise the barriers are potential energy. A major drawback is that the heights of the barriers are not readily estimated. They range from over 100 kcal to 10 kcal, or perhaps less. In the great majority of cases, reactions predicted to be forbidden do not seem to occur. This suggests that the energy barriers are quite substantial.

## ESTIMATION OF ACTIVATION ENERGIES

We could, of course, try to estimate activation energies both for forbidden and allowed reactions by the actual evaluation of integrals, or by ab initio calculations. At the semiempirical level, this has been the goal of many of the theories based on perturbation methods. Without trying to calculate energies in detail, we can see that such factors as the orbital energies of the HOMO and LUMO, or similar orbitals, the degree of overlap of the HOMO and LUMO, and net charges on the reactants are important. Particularly, it is possible to estimate the relative reactivities of a series of similar reactants.[28,32,39,63]

Even simpler procedures have been developed for estimating the energy barriers in "forbidden" four-center reactions.

$$\begin{matrix} A\text{—}B \\ \\ D\text{—}C \end{matrix} \longrightarrow \begin{matrix} A & B \\ | & | \\ D & C \end{matrix} \tag{61}$$

Earlier methods did not even take into account the orbital restrictions on such reactions, but were still fairly successful.[64] One feature that emerged from these "classical" treatments was that large polarizability in the reactant molecules lead to a lowering of the activation energy. Thus the reactions of $H_2$ and $F_2$ were generally of highest activation energy.

Jackson has presented an interesting method for estimating the barrier in reaction (61) which is based directly on its forbidden nature as portrayed by the state correlation diagram.[65] Figure 18 shows the correlation diagram for the reaction

$$\begin{matrix} A\text{—}A \\ \\ B\text{—}B \end{matrix} \longrightarrow \begin{matrix} A & A \\ | & | \\ B & B \end{matrix} \tag{62}$$

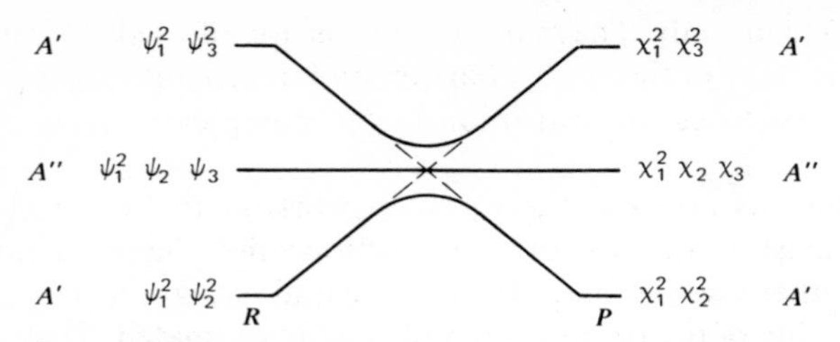


**Figure 18.** State-correlation diagram for the reaction $A_2 + B_2 \rightarrow 2AB$; the intended crossing is avoided ($R$ = reactants, $P$ = products).

The dashed line shows the plane of symmetry used to classify the states. $\psi_1$ and $\psi_2$ are molecular orbitals which are essentially the sum and difference of the bond orbitals of $A_2$ and $B_2$. $\psi_3$ and $\psi_4$ are essentially the sum and difference of the corresponding antibond orbitals. $\chi_1$, $\chi_2$, $\chi_3$, and $\chi_4$ are similar orbitals for the products, 2AB.

The total wavefunction for the reactants is given by $(\psi_1)^2(\psi_2)^2$ in the ground state. This correlates with a doubly excited state, $(\chi_1)^2(\chi_2)^2$, of the products. Because the state symmetries are the same, the intended crossing is avoided. This is an example of configuration interaction. Now several assumptions are made:

1. $\psi_1$ and $\psi_2$ have negative orbital energies equal to the bond dissociation energies of $A_2$ and $B_2$ (the interaction between the bonds is ignored). $\psi_2$ is taken as the weaker bond.
2. $\psi_3$ and $\psi_4$ have orbital energies also equal to the bond-dissociation energies but of positive sign. $\psi_3$ corresponds to the weaker bond. The same considerations apply to $\chi_1$, $\chi_2$, $\chi_3$, and $\chi_4$.
3. The energy separation at the non-crossing point is ignored.
4. The energy is a smooth function of a reaction coordinate $\lambda$, which ranges from 0 (the reactants) to 1 (the products).
5. All other energy factors are ignored.

A semiquantitative correlation diagram can now be made from the known bond energies in the system (Fig. 19). The activation energy is simply the difference in energy between the reactants and the saddle point, assumed to be at the noncrossing point. The zero of energy is that of the four separated atoms.

The same diagram can be used for the less symmetrical reaction (61), even though the plane of symmetry no longer exists. From simple geometry, the activation energy is

$$E_a = \frac{\mathrm{CD(CD + AB + AD - BC)}}{\mathrm{CD + AD}} \tag{63}$$

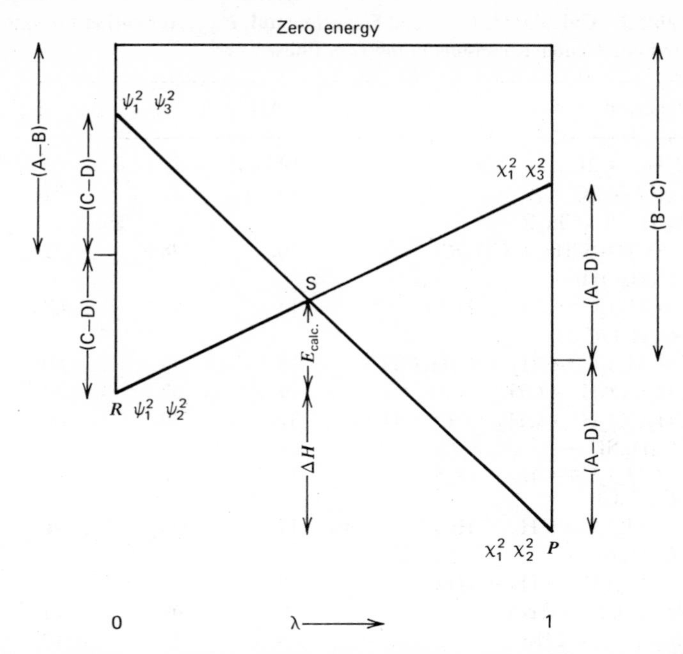


**Figure 19.** Estimation of activation energy for the reaction AB + CD → AD + BC. $\Delta H$ for the reaction is shown as well as $E_{calc}$, the calculated activation energy. A—B and so on are bond-dissociation energies; S locates the saddle point, which is the same as the avoided crossing point, and $\lambda$ measures movement along the reaction coordinate. [After reference 65.]

where AB is the strength of the stronger bond broken and CD is the weaker bond. Also AD is the weaker bond formed, and BC the stronger bond.

It can easily be seen that the activation energy decreases as the reaction becomes more exothermic, since the saddle point will move closer to $\lambda = 0$. Also, activation energies will be lowered by weakening the weakest bond in reactants or products. Table 3 shows a number of activation energies calculated from (63). The difference between $E_{calc}$ and $E_{exp}$ is not to be taken as serious. Rather it is the trend in activation energies which counts, and which is well correlated. Note that reactions involving the heavier halogen atoms have larger $E_{calc} - E_{exp}$ values. This is another example of the polarizability effect.

The forbidden cycloaddition reaction

$$2C_2H_4 \longrightarrow C_4H_8 \tag{64}$$

**Table 3 Calculated, $E_{calc}$, and Experimental, $E_{exp}$, Activation Energies for Four-Center Reactions in the Gas Phase[a]**

| Reaction | $\Delta H$ | $E_{calc}$ | $E_{exp}$ |
|---|---|---|---|
| $C_4H_8 \rightarrow 2C_2H_4$ | 19 kcal | 72 kcal | 63 kcal |
| $C_4F_8 \rightarrow 2C_2F_4$ | 50 | 82 | 74 |
| $RCH_2CH_2CH_2R' \rightarrow$ $RCH{=}CH_2 + CH_3R'$ | 20 | 78 | 72 |
| $t\text{-}C_4H_9OH \rightarrow$ $(CH_3)_2C{=}CH_2 + H_2O$ | 13 | 76 | 62 |
| $t\text{-}C_4H_9OC_2H_5 \rightarrow$ $(CH_3)_2C{=}CH_2 + C_2H_5OH$ | 18 | 74 | 60 |
| $CH_3CO_2H \rightarrow CH_4 + CO_2$ | −9 | 87 | 70 |
| $CH_3CO_2H \rightarrow CH_2{=}CO + H_2O$ | 31 | 90 | 65 |
| $t\text{-}C_4H_9SH \rightarrow$ $(CH_3)_2C{=}CH_2 + H_2S$ | 17 | 70 | 55 |
| $t\text{-}C_4H_9Cl \rightarrow$ $(CH_3)_2C{=}CH_2 + HCl$ | 17 | 74 | 46 |
| $t\text{-}C_4H_9Br \rightarrow$ $(CH_3)_2C{=}CH_2 + HBr$ | 18 | 68 | 41 |
| $Br_2 + Cl_2 \rightarrow 2BrCl$ | 0 | 49 | 17 |
| $Br_2 + I_2 \rightarrow 2IBr$ | −3 | 38 | 10 |

[a] From reference 65.

is generally considered to occur by a two step biradical mechanism (p. 70). The similar reaction of tetrafluoroethylene

$$2C_2F_4 \longrightarrow C_4F_8 \tag{65}$$

certainly occurs very much more readily than (64).[66] This is consistent with the much lower activation energy calculated in Table 3: 32 kcal for (65), compared to 54 kcal for (64). The difference is due to the low $\pi$-bond energy of $C_2F_4$, 42 kcal compared to 60 kcal for the $\pi$ bond of $C_2H_4$.

It is quite possible that reaction (65) goes by a concerted, though forbidden, mechanism. This cannot be taken as established, however, since other mechanisms are also facilitated by the high-energy $\pi$ orbital, and corresponding low-energy $\pi^*$ orbital. For example, Fig. 20 shows a plot of activation energies against the wave number for $\pi \rightarrow \pi^*$ absorption for a series of cyclic olefin molecules.[67] The activation energies are for unimolecular isomerizations of these olefins. For instance, bicyclo[2,1,0] pent-2-ene forms cyclopentadiene.

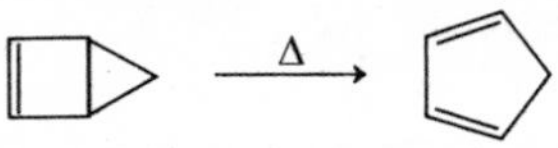

$\Delta$ (66)

This reaction probably occurs as a concerted process, even though the necessary disrotatory ring opening is forbidden.[68]

The good linearity observed in Fig. 20 suggests a common mechanism for all the reactions included, or at least a mechanism with a common rate-determining step. In fact, this is not likely to be the case since some of the reactions included are concerted processes, such as (66), and others occur by several-step mechanisms in which allylic biradicals are intermediates.[69]

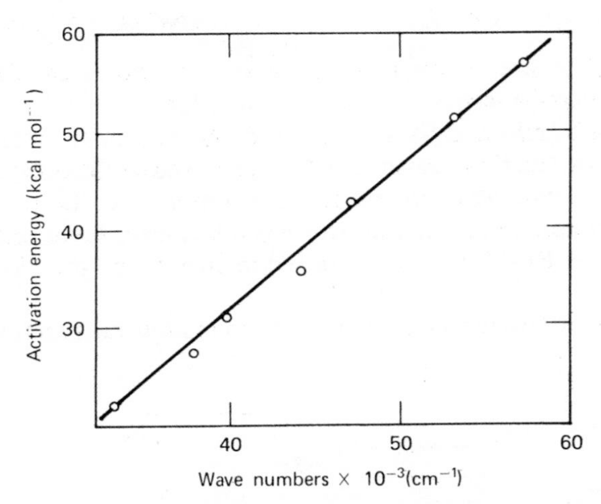


**Figure 20.** Empirical correlation between Arrhenius activation energy for isomerization of cyclic olefins and $\nu$ for the $\pi \rightarrow \pi^*$ electronic transition of the olefins. [Reprinted with permission from J. E. Baldwin and A. H. Andrist, *J. Am. Chem. Soc.*, **93**, 3289 (1971). Copyright by the American Chemical Society.]

For example, the isomerization

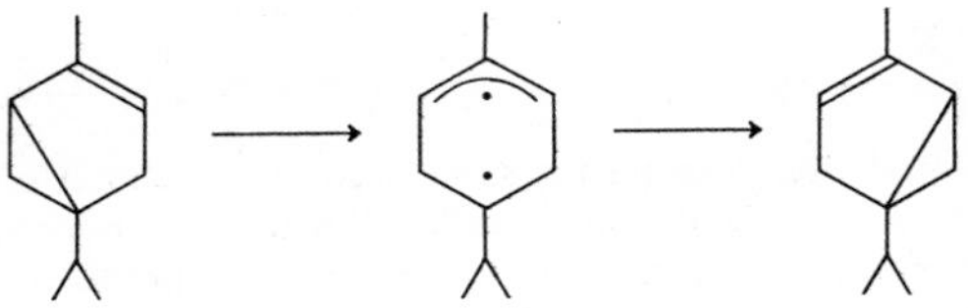

(67)

would be hard put to find a route easier than that shown.

## Substituent Effects

It has already been noted that increasing polarity of both bonds to be broken decreases the forbiddenness of four-center reactions (p. 100). This results from the highly asymmetric nature of the transition state. A similar effect is predicted if each reactant remains nonpolar overall, but with internal changes in polarity due to substituents. In this case, the reaction remains forbidden by symmetry, but changes in orbital energies occur in a favorable way.[70]

Suppose we attempt the concerted cycloaddition of tetramethoxyethylene and tetracyanoethylene.

$$\begin{matrix} (CH_3O)_2C{=}C(OCH_3)_2 \\ (NC)_2C{=}C(CN)_2 \end{matrix} \longrightarrow \begin{matrix} (CH_3O)_2C{-}C(OCH_3)_2 \\ | \quad\quad | \\ (NC)_2C{-}C(CN)_2 \end{matrix} \tag{68}$$

The substituents are selected to be electron-donating on the one hand, $OCH_3$, and electron-attracting on the other, CN. The $\pi$ and $\pi^*$ levels are raised relative to ethylene by an electron donating substituent, and lowered by an electron attracting substituent. Figure 21 shows the orbital-correlation scheme for reaction (64) compared to reaction (68). Both reactions are forbidden, but configuration interaction is much more favorable for reaction (68). That is, the HOMO has been raised in energy and the LUMO lowered in energy.

Indeed the experimental evidence on cycloaddition reactions of olefins

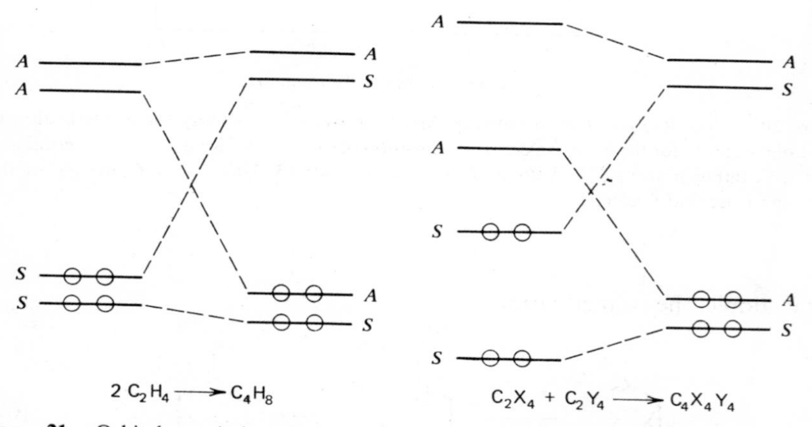


**Figure 21.** Orbital-correlation diagram for suprafacial cycloaddition of two unsubstituted olefins (on left), and two substituted olefins (on right); X is an electron-withdrawing substituent and Y, an electron-donating substituent; *A* and *S* are symmetry properties with respect to the mirror plane bisecting both double bonds. [Reprinted with permission from N. D. Epiotis, *J. Am. Chem. Soc.*, **95**, 5264 (1973). Copyright by American Chemical Society.]

suggests that concerted suprafacial (*cis*) addition occurs with strongly substituted olefins, as in (68). Two-step biradical mechanisms, or antarafacial (*trans*) additions, occur with weakly substituted olefins.[70] In valence-bond terms, the concerted nature of (68) can be explained by resonance stabilization of the antiaromatic, four-electron TS.

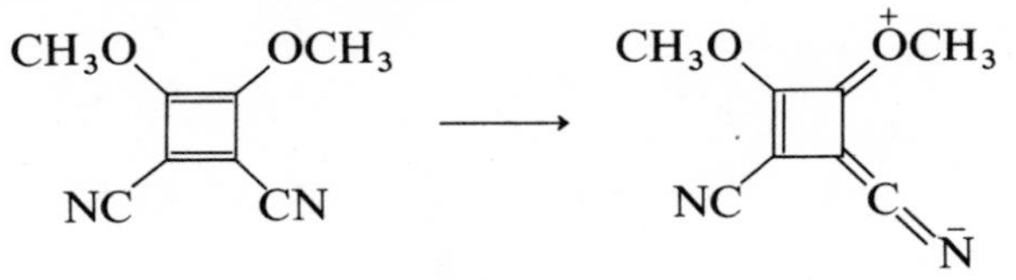

CH3O OCH3 / NC CN ⟶ CH3O $\overset{+}{O}CH_3$ / NC C=N$^-$ etc. (69)

Substituted cyclobutadiene can be taken as a model.

Figure 21 can also be used to discuss addition reactions of the type

$$C_2H_4 + X_2 \longrightarrow C_2H_4X_2 \qquad (70)$$

Raising or lowering the frontier $\sigma$ and $\sigma^*$ orbitals of $X_2$ should affect the ease of configuration interaction. It is predicted that *cis* addition will be easiest for $F_2$ and most difficult for $Br_2$ and $I_2$.[70] For the heavy halogens, two-step mechanisms will predominate, giving overall *trans* addition.

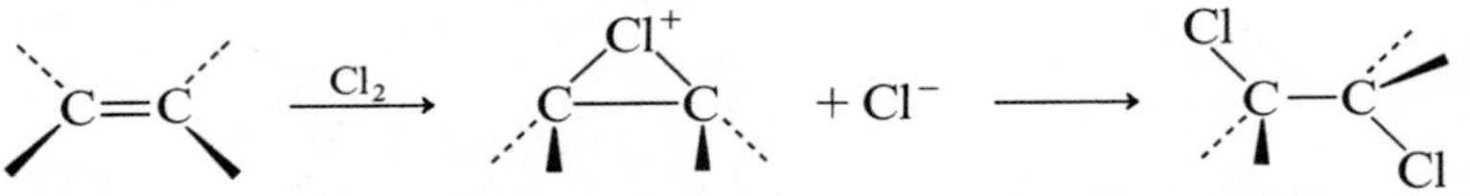

$$C{=}C \xrightarrow{Cl_2} C{-}C\ (\text{bridged } Cl^+) + Cl^- \longrightarrow Cl{-}C{-}C{-}Cl \qquad (71)$$

Each step in (71) is allowed by symmetry. Free-radical mechanisms are also possible.

Less symmetrical substitution in a reactant such as an olefin, or other molecule, can have two important effects.[71] First of all, relative reactivities can be affected, as the energies of the frontier orbitals move up or down. Secondly, the stereoselectivity, or regioselectivity, can be altered by substituents. This is a consequence of the requirement that a large positive overlap, of the HOMO and LUMO for example, is needed to produce a low activation energy.

Figure 22 shows some $\pi$- and $\pi^*$-orbital energies for several substituted ethylenes. The energies are only estimated from semiempirical calculations, but they are adequate for the purpose. Also shown are the AO coefficients for each MO, given by the relative sizes of the $p$ orbitals on each atom. Similar diagrams have been made for a number of other organic molecules.[72] The rule derivable from PT is that the HOMO and LUMO will give the greatest interaction energy (stabilization) when the atoms whose coefficients are the greatest are brought closest together. This is illustrated in Fig. 22 for an olefin undergoing concerted cycloaddition with a generalized molecule, of which only the LUMO is shown.

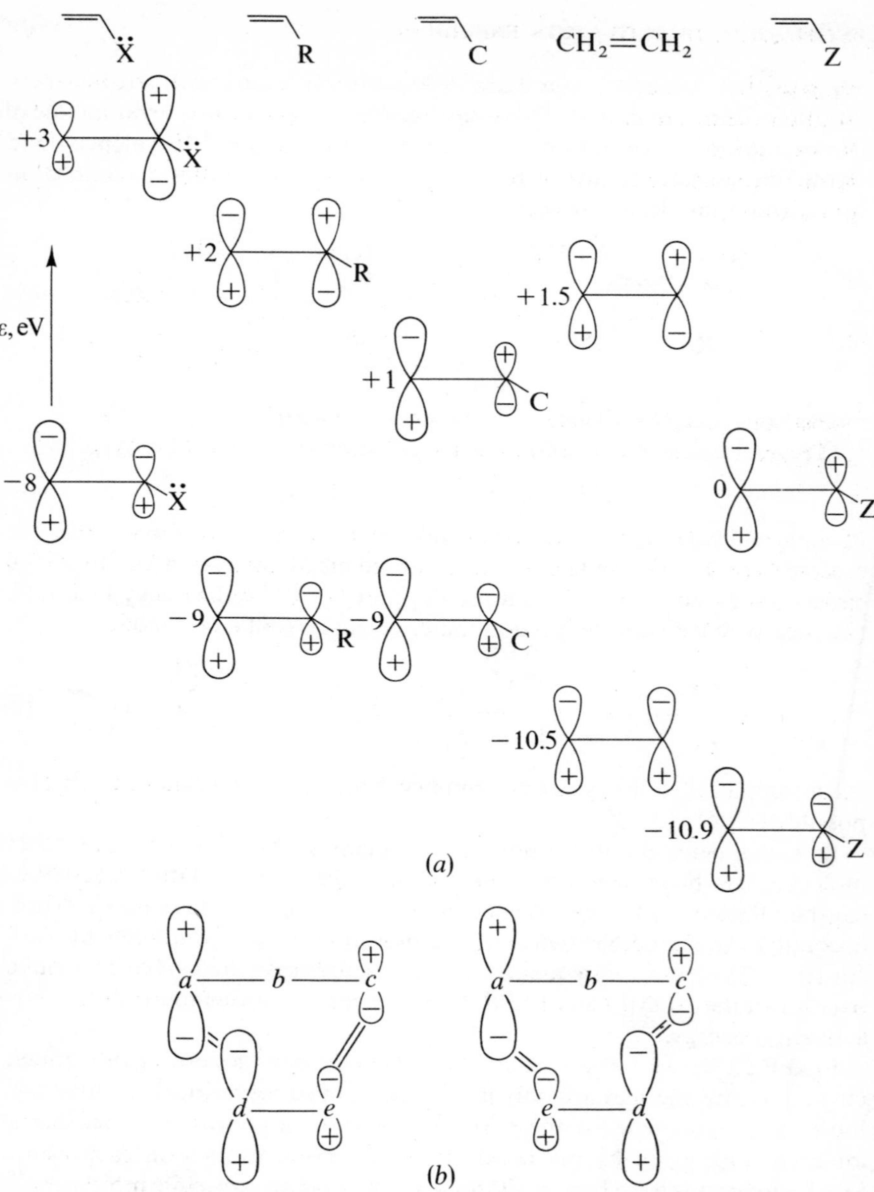


**Figure 22.** (*a*) Estimated $\pi$ and $\pi^*$ orbital energies and atomic coefficients for substituted olefins. R = alkyl; X = halogen, alkoxy, amido, and so on, C = conjugated substituent, such as vinyl; Z = CHO, CN, $CO_2R$, and so on. Energies in eV; (*b*) cycloaddition reaction showing greater stabilization on left than on right because of greater total overlap. [Reprinted with permission from K. N. Houk, J. Sims, R. E. Duke, Jr., R. W. Strozier, and J. K. George, *J. Amer. Chem. Soc.*, **95**, 7287 (1973). Copyright by the American Chemical Society.]

## OTHER FACTORS INFLUENCING THE RATE

The potential energy surface on which a reacting collection of nuclei moves is the primary factor determining the eventual rate of reaction. If this surface is known in detail, the absolute rate of reaction can be calculated, at least in principle. Strictly speaking, what is necessary is a solution to the time-dependent Schrödinger wave equation. Since this equation can only be solved approximately, there would still be great difficulties. In fact the application of scattering theory, as it is called, has not yet had a great impact on the study of chemical kinetics. It is necessary to use some aspects of classical mechanics to proceed.

The most important and common procedure is the use of the activated complex or TS theory, due largely to Eyring and his co-workers.[73] The assumptions are made that the reactants are in equilibrium with the activated complex (AC) at the top of an energy barrier, and that the rate of reaction is given by the rate of passage through the transition state in which the AC exists. The rate constant becomes

$$k = \kappa \frac{RT}{Nh} e^{-\Delta H^{\ddagger}/RT} e^{\Delta S^{\ddagger}/R} \tag{72}$$

where $(RT/Nh) = 6.0 \times 10^{12}\ \text{s}^{-1}$ at 25°C, $\Delta H^{\ddagger}$, and $\Delta S^{\ddagger}$ are the enthalpy and entropy differences between the AC and the reactants, all in their standard states. The transmission coefficient $\kappa$ takes into account dynamic effects, which will be discussed later. The choice of standard state must agree with the concentration units of the rate constant.

There are several circumstances under which the equilibrium assumption may fail. One is for very rapid reactions, another is for gas-phase reactions at low pressure, and a third is when molecules are suddenly given a very large amount of energy, as in shock tubes.[73b] In the limit of no energy barrier for the reaction itself, a reaction approaches a rate controlled by the rate of diffusion. Such rapid processes will usually not be rate limiting for a given overall reaction. Also, reactions at low pressure or in shock tubes, while of considerable theoretical interest, are not closely related to most reactions run under practical conditions. We may assume the essential validity of (72) for most applications, bearing in mind that some cases will not conform.

Again, if the potential energy surface were known in detail, everything in (72) could be calculated. This would require a knowledge not only of how the energy changes along the reaction coordinate, but also its variation along all the other internuclear coordinates of the activated complex. This would be needed to calculate the "equilibrium" geometry of the AC and the vibrational frequencies of its normal modes. This kind of information is hardly available, even for very simple systems.

We are just entering a period where it is possible to calculate energy surfaces by accurate ab initio methods.[74] These calculations are very difficult and time consuming, even on high-speed digital computers. It is necessary to make a large number of calculations at points corresponding to different internuclear positions. Also it is imperative to use a very large number of configurations to get meaningful results. That is, configuration interaction is very important, much more so than for calculations at the equilibrium configurations of stable molecules. The reason is that a single, preponderant configuration is very characteristic of stable structures.[75] Unstable structures, such as activated complexes, are not characterized by a preponderant configuration. In other words, very low-lying excited states necessarily exist.

We must fall back on empirical and semiempirical methods. Collision theory and TS theory give average, or normal, values for $\Delta S^{\ddagger}$ for various reactions. For example, the normal frequency factors, $A$, in $k = A e^{-E_a/RT}$ are

| | |
|---|---|
| $10^{13}$ sec$^{-1}$ | unimolecular |
| $10^{11}$ M$^{-1}$s$^{-1}$ | bimolecular |
| $10^{9}$ M$^{-2}$s$^{-1}$ | termolecular |

There are many methods whereby better values of $\Delta S^{\ddagger}$ can be estimated for various kinds of reaction.[76] In a similar way $\Delta H^{\ddagger}$ can be estimated, but only for very restricted classes of reaction. Our symmetry rules, of course, are also attempts to predict something about the magnitude of $\Delta H^{\ddagger}$.

### Solvent Effects

Calculations of potential energy surfaces are explicitly or implicitly for reactions in the gas phase. In solution there are an unmanageably large number of internuclear coordinates to consider. Nevertheless we must deal with chemical reactions in solution, more frequently than for the gas phase. The solvent molecules are rather arbitrarily divided into two classes. A few are considered as reactants, in the same sense that other molecules are; bonds are made and broken both to and in these molecules. The remaining solvent is usually treated as a continuous medium in which the reaction occurs. For example, the hydration energy of an ion can be broken up into two parts;[77] the first is the energy of interaction with a primary coordination layer,

$$Fe^{2+}(g) + 6H_2O(g) \longrightarrow Fe(H_2O)_6{}^{2+}(g) \quad -\Delta H = 344 \text{ kcal} \quad (73)$$

The second is the energy of interaction (Born energy) of the complex ion thus formed,

$$Fe(H_2O)_6^{2+}(g) \longrightarrow Fe(H_2O)_6^{2+}(aq) \quad -\Delta H = 186 \text{ kcal} \qquad (74)$$

The latter can be estimated by using the classical dielectric constant of the medium.

Potential energy surfaces, including any active solvent molecules, are assumed to exist in solution as well as in the gas phase. Since all molecules, charged or uncharged, polar or nonpolar, can interact with the solvent to lower the energy, a plot of potential energy against the reaction coordinate usually is affected as shown in Fig. 23. The reactants, activated complex and products all have lower energies in solution. The relative lowerings determine the change in $\Delta H^{\ddagger}$ in solution compared to the gas phase. The values of $\Delta S^{\ddagger}$ are also very sensitive to the solvent. In particular, strong solvation of ions produces large decreases in entropy. This can be explained as the "freezing" of solvent molecules about the ion.[78] The large solvation energies of ions can produce substantial activation energies for their reactions, even when fully allowed.

The way in which different solvents can affect rates of reaction is a subject too involved to pursue further here.[79] In comparison to the gas phase, the most important point is that ions are capable of existence in solutions, even when the solvent is nonpolar. In the gas phase ions exist only under high-energy conditions, such as in a discharge or in a high-temperature plasma. This often means that reactions which occur in solution do not occur in the gas phase, and vice versa. Changes in mechanism occur in the two phases. The forbidden reactions of the halogen molecules, for instance, the reverse of reaction (35), appear to take place in polar solvents. However this is because of a change in mechanism involving a sequence of allowed steps.

$$\begin{aligned} SO_2 + Cl_2 &\longrightarrow SO_2Cl^+ + Cl^- \\ SO_2Cl^+ + Cl^- &\longrightarrow SO_2Cl_2 \end{aligned} \qquad (75)$$

Just as one or more solvent molecules can lower the energy of a TS, so it is easy to imagine in many reactions that a suitably placed molecule of some kind could have a beneficial effect on the potential energy surface. This leads to the postulation of mechanisms of a highly complex, yet concerted, nature. An example would be the base-catalyzed elimination reaction of an alkyl halide.

B---H BH$^+$

C—C ⟶ C=C (76)

Cl Cl$^-$

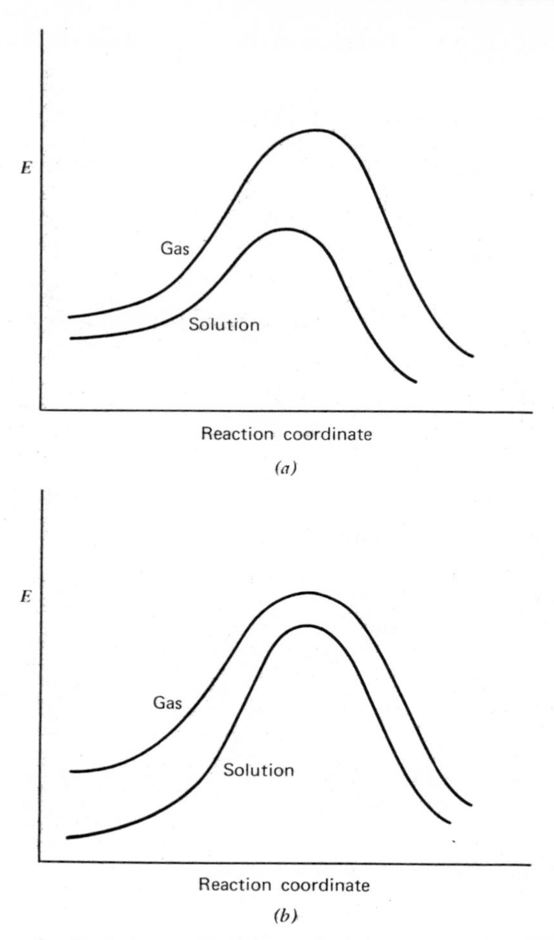


**Figure 23.** Effect of solvation on potential energies of reactants and activated complexes: (*a*) reaction speeds up in solution relative to gas phase; (*b*) reaction slows down in solution relative to gas phase.

Not shown in (76) is the simultaneous cooperation of a number of solvent molecules needed to solvate the ions produced. The idea behind these concerted processes is that all bonds are made and broken synchronously, so that high-energy intermediates such as a carbanion or a carbenium ion are avoided.

It must be true that energetically, concerted processes are favored. It is equally true that entropy factors are strongly against them. It is statistically less and less probable that more and more reactant molecules will arrange themselves in just the orientation needed for a favorable reaction path. Indeed Bordwell has shown that there is a great deal of evidence against highly concerted reactions in which the simultaneous making and breaking of anywhere from four to six bonds has been postulated.[80] Even a reaction like (76), in which only three bonds are implicated, is a relatively rare event.

A typical example of an attractive, but highly improbable, concerted reaction is the following,

$$B + HOCH_2CH_2CH_2Br \longrightarrow BH^+ + O = CH_2 + CH_2 = CH_2 + Br^- \quad (77)$$

There are six making and breaking chemical bonds in this reaction. The improbability of (77) can be seen by considering the reverse process. By microscopic reversibility, this must occur by again relating all six bonds. The reaction would be tetramolecular, not counting the solvent molecules surrounding the ions. A sequence of simple bimolecular or unimolecular processes is much more plausible.

### Termolecular Reactions

A general rule is that termolecular reactions are improbable, unless one or more of the molecules is a solvent molecule, or unless two of the molecules form an initial complex first. This complex must last for some time ($> 10^{-10}$s) and must contain the two molecules in approximately the correct orientation for reaction with the third molecule.

In the gas phase, third-order reactions are rare. Virtually the only ones known are three body recombinations of atoms, and some reactions of nitric oxide. The three-body processes, for instance,

$$2Br + M \longrightarrow Br_2 + M \quad (78)$$

have the simplicity of being nearly spherically symmetric. That is, any orientation of the three particles is likely to be suitable, provided the two bromine atoms touch. They are probably termolecular in many but not all cases. An example of the nitric oxide reactions would be the oxidation of NO to $NO_2$. The stoichiometric equation is

$$2NO + O_2 = 2NO_2 \rightleftharpoons N_2O_4 \quad (79)$$

The kinetics follow the rate equation

$$\text{rate} = k[NO]^2[O_2] \quad (80)$$

which could be termolecular.

However it is almost certain that a complex is formed first. There are two possibilities, and orbital symmetry can be used to decide between them.[81] In the first possible mechanism, a dimer $N_2O_2$ is formed reversibly

$$2NO \rightleftharpoons N_2O_2 \tag{81}$$

$$N_2O_2 + O_2 \longrightarrow 2NO_2 \tag{82}$$

The second possible mechanism has the molecule $NO_3$ forming first

$$NO + O_2 \rightleftharpoons NO_3 \tag{83}$$

$$NO_3 + NO \longrightarrow 2NO_2 \tag{84}$$

Figure 24 shows the relevant filled and empty orbitals for the reaction of the $N_2O_2$ dimer with $O_2$. Since the electronegative oxygen molecule is certainly the electron acceptor, and since its $\pi_g$ orbital is the accepting orbital, the filled MO of $N_2O_2$ must have $b_2$ symmetry as shown. There is good overlap and reaction could occur smoothly. The O—O bond is broken because electrons go into an antibonding orbital, the N—N bond is strengthened because an orbital antibonding with respect to the two nitrogen atoms is emptied. Two new N—O bonds are formed by positive overlap between the two orbitals.

All of this is very reasonable, but clearly the product being formed is $N_2O_4$, not $2NO_2$.

$$N_2O_2 + O_2 \longrightarrow N_2O_4 \tag{85}$$

This is not the same as reaction (82). In fact it is forbidden by the spin-conservation rules, if the oxygen is ground state, ${}^3\Sigma_g{}^-$. It could happen for excited state, ${}^1\Delta_g$, $O_2$. The alternative mechanism is shown in Fig. 24 also, assuming that $NO_3$ has a Y-shaped structure. While this is the expected normal structure for $NO_3$, there is also evidence for a peroxy structure, in which only one oxygen atom of $O_2$ is bonded to nitrogen.[82] In either case, both reactions (83) and (84) seem allowed by orbital and spin symmetry (however see p. 264).

The third-order reaction of nitric oxide with hydrogen,

$$2NO + 2H_2 \longrightarrow N_2 + 2H_2O \tag{86}$$

is believed to go by a mechanism in which $N_2O_2$ is formed as an intermediate

$$N_2O_2 + H_2 \longrightarrow N_2 + H_2O_2 \tag{87}$$

This is followed by a rapid reaction of $H_2O_2$ with $H_2$ to give water. Figure 24 shows that reaction (87) is allowed by symmetry and spin. The H—H bond is broken, the N—N bond is strengthened, and two new O—H bonds are formed. It is questionable whether $H_2O_2$ or two OH radicals are formed, however. The final products would be the same in either case.

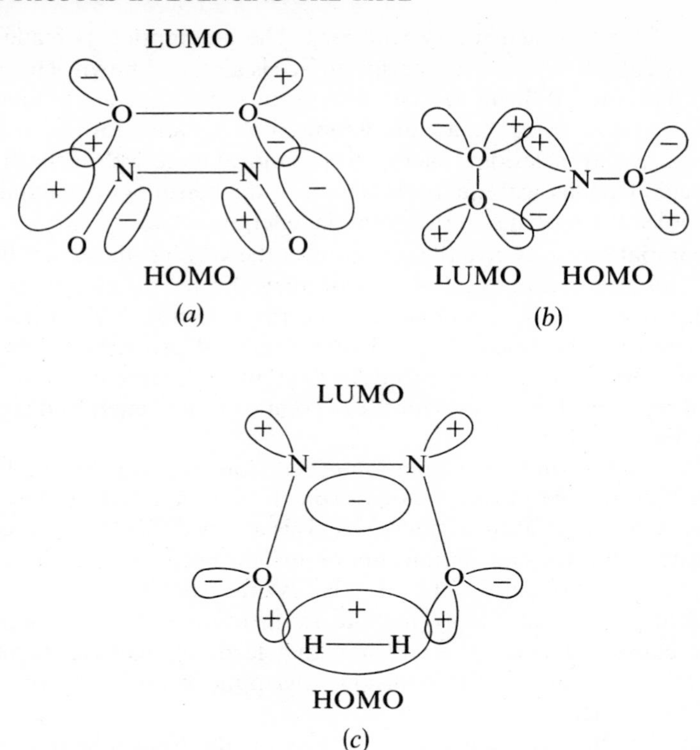


**Figure 24.** Orbital overlaps in some reactions of nitric oxide: (*a*) reaction of $N_2O_2$ and $O_2$ to form $N_2O_4$; (*b*) reaction of NO and $O_2$ to form $NO_3$; (*c*) reaction of $H_2$ and $N_2O_2$ to form $N_2$ and $H_2O_2$.

## Molecular Dynamics

The conclusions from the previous section are that nature has a preference for overall mechanisms in which a series of unimolecular or bimolecular reactions occur. These are chosen on a probability, or entropy, basis over highly concerted processes in which three or more molecules must cooperate to give a favorable potential energy barrier. The difficulty of having several molecules collaborate in a single collision is emphasized by examining the complexity of a collision between even a pair of molecules.

Molecular dynamics is the study of what happens during individual molecular processes.[83] If the consequences of a single collision with a given set of starting parameters can be calculated, then suitable averaging can give the rates of various elementary reactions. The assumption is made that a sufficiently complete potential energy surface is already known. The calculations are extremely difficult and can only be done with the use of high speed computers. It is necessary to assume an initial set of values for the rotational, vibrational, and translational energy of each molecule, as well as the direction of approach. The classical equations of motion are then integrated numerically for the molecules moving on the potential energy surface.

The computer results give the positions of the various nuclei as a function of time. This information is called a "trajectory." Some trajectories lead to the consummation of a chemical reaction, but most do not. A great number of trajectories must be calculated and then averaged properly to obtain the "reactive cross section." This property then gives the rate constant for the chemical reaction. Trajectories for the decomposition of energized molecules can also be calculated.

The first such computer calculation of reaction rates was for the $H + H_2$ reaction.[84] A number of recent calculations have been made on this, and on other three- and four-atom systems.[85] This is a very difficult way to obtain a rate constant, but the calculations are important because of the insight that they give us into the details of molecular events. They emphasize the often overlooked points that energy, angular momentum and linear momentum must be conserved, even in a collision that leads to reaction. If only two simple molecules are in the collision, this puts severe restraints on the possibility of reaction.

In fact, the information gained is a value for the transmission coefficient $\kappa$ in (72). This factor is included to take account of the possibility that not every reactant molecule that becomes an activated complex, or passes over the energy barrier, will become a product molecule. It measures the probability that some systems will be reflected backwards after passage. A familiar example is the atom recombination reaction, (78), which will not occur in the absence of a third body. The energy of recombination of the two atoms is just equal to the energy of dissociation. If it is not removed by the intervention of another molecule, it necessarily goes into vibrational energy of the diatomic molecule and causes immediate dissociation. Thus $\kappa = 0$ in this case because of energy and momentum conservation.

In general, it is very difficult to estimate $\kappa$, which is usually set equal to unity. Trajectory calculations show that $\kappa$ is very sensitive to the exact shape of the potential energy surface, as well as to the distribution of energy between translation and vibration in the reactant molecules. These kinds of energy are interconverted on crossing the barrier. Figure 25 shows two limiting cases for the nearly thermoneutral reaction

$$A + BC \longrightarrow A\text{---}B\text{---}C \longrightarrow AB + C \qquad (88)$$

In the first case the barrier comes early. That is, the B—C bond has changed little, and the A—B bond is much longer than its equilibrium value. After passage over the energy barrier, energy is released, since the path is downhill. This energy must appear as vibrational energy of the AB molecule. Thus translational energy is converted into vibrational energy.

In the second case, the barrier comes late. The molecule AB has nearly its equilibrium bond distance. The energy released must go into translational energy of the reactants. Looking now at the reverse of (88), it can be seen that for a late barrier (Fig. 25*a*) vibrational energy is effective at causing reaction, whereas translational energy is less effective. For an early barrier (Fig. 25*b*), vibrational energy is less effective than translational energy. Two molecules may have sufficient energy between them to exceed $E_0$, the height of the barrier, and still not be able to cross the barrier.

When some reactant atoms are very heavy and others are light, a special problem related to conservation of momentum exists. During the duration of a collision, it may be that heavy atoms, because of their inertia, do not have time to change their positions. This is well known in hot-atom chemistry where collisions are of very short duration. The conclusion is stated in Wolfgang's Golden Rule:[86] "Reactions requiring nuclear motions that are slow compared to collision time are forbidden."

This rule has been applied in an interesting way to the much-discussed reaction

$$H_2 + I_2 \longrightarrow 2HI \tag{89}$$

Noyes was the first to suggest that, during a four-center broadside collision of $H_2$ and $I_2$, the iodine atoms because of inertia would not move appreciably.[87] This means that the collision would necessarily be elastic, since the hydrogen molecule by itself cannot cause any change along the collision direction. Hence $\kappa$ would be effectively equal to zero. This prediction by Noyes has since been verified by extensive trajectory calculations, using an assumed potential energy surface.[88]

The conclusion is that (89) is dynamically forbidden by momentum conservation. Even though the calculations are not necessarily reliable because the potential energy surface is not known, it is quite possible that the conclusion is correct. However it must be realized that it has nothing to do with the fact that (89) is also forbidden by orbital symmetry. Since the reaction does not occur, it is not easy to discover which forbiddenness is the more effective. The experimental activation energy of 40 kcal is large enough so that the forbidden four-center process may have a similar value.

There is an experimental test that has not yet been carried out. If a reaction is run in solution, restrictions due to conservation laws are effectively wiped out. There are now many molecules that can share the necessary partitioning

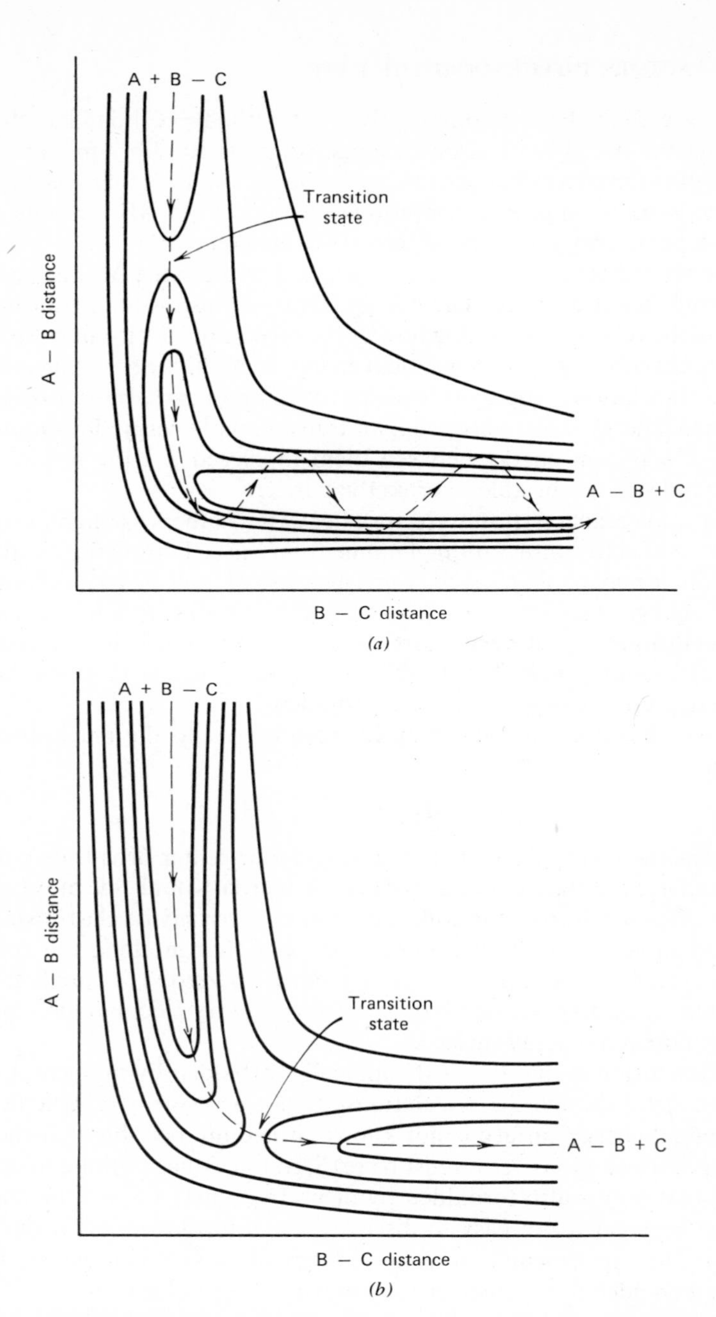


**Figure 25.** Potential energy surfaces for reaction A + BC → AB + C: (*a*) barrier comes early; translational energy is converted to vibrational; (*b*) barrier comes late; translational energy remains translational energy.

of energy and momentum. The effect is similar to that of the recombination of free radicals rather than free atoms. As soon as a radical becomes large enough, a third body is no longer needed. The energy of recombination can be distributed among the many vibrational degrees of freedom available. Accordingly, it would be of interest to study the rate and mechanism of (89) in a suitable inert solvent.

Even for the gas phase, reactions of complex molecules would be released from dynamic constraints to a considerable degree. The additional atoms would act as a reservoir, just as solvent molecules do. Also simple molecules not containing very heavy atoms such as iodine, would be less hampered by dynamic effects. For example, the decomposition of nitric oxide

$$2NO \longrightarrow N_2 + O_2 \tag{90}$$

does not occur in solution, any more than it does in the gas phase. This thermodynamically favored reaction is strongly forbidden by orbital symmetry (see p. 58). Dynamic effects cannot be responsible.

### Tunneling

In contrast to dynamic effects which are classical in nature, and which have an inhibitory effect on rates, there is another effect which is purely quantum mechanical and which increases rates of reaction. This is tunneling through a potential energy barrier.[89] That is, a reacting system need not have an energy $E_0$ in order to appear on the other side of the barrier. Tunneling depends on the De Broglie wavelength of a particle

$$\lambda = \frac{h}{mv} \tag{91}$$

It is important for electron-transfer reactions, because of the small mass of the electron.

The only chemical nucleus for which tunneling is a factor is the proton. Even for the deuteron, the efficiency falls off markedly, which provides a convenient way to study tunneling. The probability of passing through a barrier, instead of over it, also depends strongly on the height and width of the barrier. The present evidence is that some tunneling always occurs in proton transfers, hydrogen-atom transfers, and hydride-ion transfers.[90] This means that the actual energy of these hydrogen transfer systems is usually somewhat less than $E_0$ during passage. But it cannot be too much less, otherwise the probability of tunneling is small.

## THE PRINCIPLE OF MICROSCOPIC REVERSIBILITY

There is another symmetry-based factor that puts restrictions on the rates of chemical reactions in a special way. This is the principle of microscopic reversibility (PMR), or as it is often called, the principle of detailed balancing.[91] This states that any molecular process and its reverse occurs with equal rates at equilibrium. In mechanistic terms it states that, if a certain series of steps constitutes the mechanism of a forward reaction, the mechanism of the reverse reaction (under the same conditions) is given by the same steps traversed backward. Thus, it is certainly valid to exclude all other mechanisms in any case when the reverse mechanism is known.

However the PMR may be used as a basis for exclusion where neither mechanism is known beforehand. The exclusion is due to a certain lack of symmetry in the forward and reverse directions. The argument is most easily seen for the case of isotope-exchange reactions, where, if kinetic isotope effects are ignored, the mechanism must be precisely the same in both directions. However great care must be used in the correct application of the principle.[53] This can be illustrated by considering several schematic potential energy profiles with possible intermediates indicated as minima.

If the exchange reaction proceeds by but one path, then PMR does indeed require a symmetric energy profile with respect to the reaction path. If the number of intermediates formed is even, including zero, the two isotopic atoms entering and leaving become equivalent in the central TS; if odd, in the central intermediate. This is illustrated in Fig. 26*a* for an $S_N2$ reaction between $CH_3Br$ and $*Br^-$ (zero intermediates) and in Fig. 26*b* for an $S_N1$ reaction between *t*-butyl bromide and $*Br^-$ (one intermediate, $(CH_3)_3C^+ + Br^- + *Br^-$). Two or more atoms are equivalent if, within an appropriate time period, the environments seen from the nuclei of the two or more atoms are identical or are mirror images of each other.

The PMR does not permit a one-path mechanism that is not symmetric about the midpoint of the energy profile. However it does permit such a process as a two-path mechanism in which the second path is that obtained by reflecting the first in a plane at the midpoint. As shown in Fig. 26*c*, *d*, the sum of the two paths is symmetric, but the isotopic atoms never become equivalent. Since an isotope-exchange reaction occurs under equilibrium conditions, if isotope effects are not considered, the PMR also requires that exactly 50% of the exchange occurs by the solidline path and 50% of the exchange occurs by the dotted line path of Fig. 26*c*, *d*. That is, the rates are exactly the same for each mechanism since both must be in equilibrium.

Reactions of the type of Fig. 26*c* must be common and, in particular, will involve associative exchange processes in which the isotopic atoms cannot

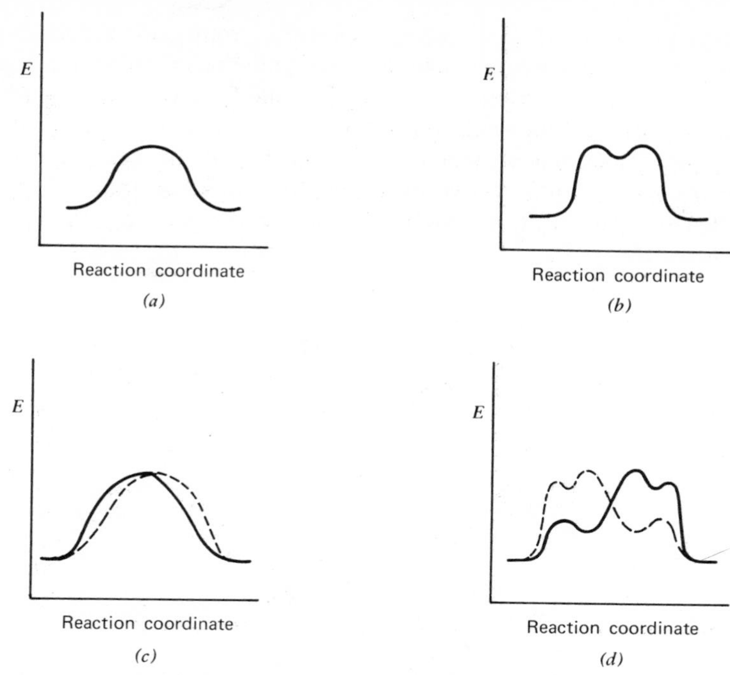


**Figure 26.** Potential energy *vs.* reaction coordinate for the exchange reactions:
(*a*) $CH_3Br + {}^*Br^-$;
(*b*) $(CH_3)_3CBr + {}^*Br^-$;
(*c*) $CH_3CH_2CH(CH_3)COCH_3 + H_2{}^*O$;
(*d*) hypothetical complex exchange reaction.

become equivalent because there are one or more asymmetric centers elsewhere in the molecule. Here the isotopic atoms are not equivalent because the two environments observed are epimeric. An example would be oxygen-isotope exchange between 3-methyl-2-pentanone and water via the ketone hydrate. Notice that there are two epimeric TSs of equal energy, each reached by its own path.

In complex mechanisms of the type in Fig. 26*d*, the positions of the isotopic atoms are structurally isomeric rather than epimeric. Because of this, there is no possibility of a system on the solid-line path crossing over to the dotted-line path at the midpoint where they apparently meet. Thus the reaction coordinates for the two paths, while otherwise identical, are located in different parts of coordinate space. This is also true in Fig. 26*c*.

Apparently no one has postulated such a multipath reaction for any homogeneous isotopic-exchange reaction. More commonly, some more symmetric reaction path will be of lower energy. However multipath processes have been proposed in heterogeneous exchange and racemization reactions.[92] The presence of a surface may often prevent the formation of symmetrical intermediates. Fig. 27 shows the details of a two-path mechanism for deuterium exchange in benzene on a metal surface in which hydrogen is added from one side of the ring and removed from the other side. The application of the above ideas to substitution reactions that are not isotopic exchanges is straightforward.

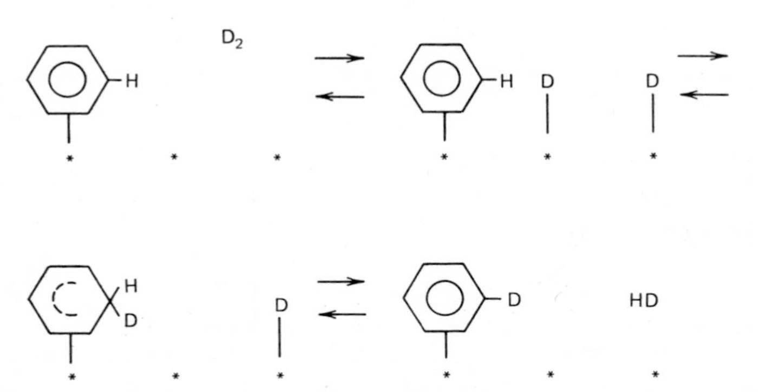


**Figure 27.** Two-path mechanism for deuterium exchange in benzene on a metal surface; Metal atoms, or adsorption sites, are represented by asterisks. [Reprinted with permission from R. L. Burwell, Jr. and R. G. Pearson, *J. Am. Chem. Soc.*, **70**, 300 (1966). Copyright by the American Chemical Society.]

## REFERENCES

1. R. B. Woodward and R. Hoffmann, *The Conservation of Orbital Symmetry*, Verlag Chemie, Gmb., Weinheim/Bergstrasse, 1970, p. 31.
2. R. Breslow, J. Napierski, and A. N. Schmidt, *J. Am. Chem. Soc.*, **94**, 5906 (1972).
3. A. A. Frost and R. G. Pearson, *Kinetics and Mechanism*, 2nd ed., Wiley, New York, 1961, p. 98.
4. K. F. Herzfeld, *Z. Naturforsch.*, **3a**, 457 (1948).
5. C. R. Trindle, *J. Am. Chem. Soc.*, **92**, 3251 (1970).
6. K. F. Herzfeld, *Rev. Mod. Phys.*, **21**, 527 (1949).
7. C. Trindle and O. Sinanoglu, *J. Am. Chem. Soc.*, **91**, 4054 (1969); H. E. Zimmerman and L. R. Sousa, ibid., **94**, 834 (1972).
8. L. C. Cusachs, M. Krieger, and C. W. McCurdy, *Int. J. Quant. Chem.*, **3**, 67 (1969).
9. H. E. Zimmerman, *Acc. Chem. Res.*, **5**, 393 (1972).
10. N. F. Phelan, H. H. Jaffé, and M. Orchin, *J. Chem. Ed.*, **44**, 626 (1967); H. E. Zimmerman and A. Zweig, *J. Am. Chem. Soc.*, **83**, 1196 (1961).

11. R. B. Woodward and R. Hoffmann, *J. Am. Chem. Soc.*, **87**, 2511 (1965).
12. L. Radom, J. A. Pople, and P. von R. Schleyer, *J. Am. Chem. Soc.*, **94**, 5935 (1972).
13. This prediction has been verified experimentally; see J. A. Berson and G. L. Nelson, *J. Am. Chem. Soc.*, **89**, 5503 (1967).
14. M. Saunders and J. J. Stofko, Jr., *J. Am. Chem. Soc.*, **95**, 253 (1973).
15. W. B. Miller, S. A. Safron, and D. R. Herschbach, *Disc. Faraday Soc.*, **44**, 292 (1967).
16. G. O. Pritchard and M. J. Perona, *Int. J. Chem. Kinet.*, **2**, 281 (1970).
17. A. Maccoll and P. J. Thomas, *Progress in Reaction Kinetics*, G. Porter (ed.), Vol. 4, Pergamon, Oxford, 1967, p. 119 ff.
18. C. J. Harding, A. Maccoll, and R. A. Ross, *J. Chem. Soc.*, **B**, 634 (1969).
19. C. A. Wellington and W. D. Walters, *J. Am. Chem. Soc.*, **83**, 4888 (1961).
20. C. H. Bamford and C. F. H. Tipper, *Comprehensive Chemical Kinetics*, Vol. 5, Elsevier, Amsterdam, 1972, p. 422.
21. J. P. Snyder, L. Lee, V. T. Bandurco, C. Y. Yu, and R. J. Boyd, *J. Am. Chem. Soc.*, **94**, 3260 (1972).
22. F. London, *Z. Elektrochem.*, **35**, 522 (1929).
23. H. Eyring and M. Polanyi, *Z. Phys. Chem.*, **B12**, 279 (1931).
24. W. J. van der Hart, J. J. C. Mulder, and L. J. Oosterhoff, *J. Am. Chem. Soc.*, **94**, 5724 (1972); D. M. Silver and M. Karplus, ibid., **97**, 2645 (1975).
25. W. A. Goddard, III, *J. Am. Chem. Soc.*, **92**, 7520 (1970); ibid., **94**, 793 (1972).
26. R. C. Ladner and W. A. Goddard, III, *J. Chem. Phys.*, **51**, 1073 (1969); W. A. Goddard, III, and R. C. Ladner, *J. Am. Chem. Soc.*, **93**, 6750 (1971).
27. P. D. Bartlett and A. P. Schaap, *J. Am. Chem. Soc.*, **92**, 3223 (1970).
28. M. J. S. Dewar, *The Molecular Orbital Theory of Organic Chemistry*, McGraw-Hill, New York, 1969, p. 192.
29. C. A. Coulson and H. C. Longuet-Higgins, *Proc. Roy. Soc.*, **A191**, 39 (1947); **A195**, 188 (1948); H. C. Longuet-Higgins, *J. Chem. Phys.*, **18**, 265, 275, 283 (1950).
30. M. J. S. Dewar, *J. Am. Chem. Soc.*, **74**, 3341, 3357 (1952).
31. L. Salem, *J. Am. Chem. Soc.*, **90**, 543, 553 (1968).
32. G. Klopman and R. F. Hudson, *Theor. Chim. Acta*, **8**, 165 (1967); A. Devaquet, *Mol. Phys.*, **18**, 233 (1969); R. Sustmann and G. Binsch, *Mol. Phys.*, **20**, 1, 9 (1971).
33. K. Fukui and H. Fujimoto, *Bull. Chem. Soc. Jap.*, **41**, 1989 (1968); ibid., **42**, 3399 (1969).
34. R. S. Mulliken, *J. Am. Chem. Soc.*, **74**, 811 (1952); *Rec. Trav. Chim.*, **75**, 845 (1956).
35. R. G. Pearson, *J. Chem. Ed.*, **45**, 581, 643 (1968).
36. K. Fukui, T. Yonezawa, and H. Shingu, *J. Chem. Phys.*, **20**, 722 (1952).
37. G. Klopman, *J. Am. Chem. Soc.*, **90**, 223 (1968).
38. R. G. Pearson, *J. Am. Chem. Soc.*, **85**, 3533 (1963).
39. W. C. Herndon and L. H. Hall, *Theor. Chim. Acta*, **7**, 4 (1967).
40. H. Fukui, *Bull. Chem. Soc. Jap.*, **39**, 498 (1966); *Fortsch. Chem. Fortsch.*, **15**, 1 (1970); *Acc. Chem. Res.*, **4**, 57 (1971).
41. H. Kato, K. Morukama, T. Yonezawa, and K. Fukui, *Bull. Chem. Soc. Jap.*, **38**, 1749 (1965).
42. L. P. Hammett, *Physical Organic Chemistry*, 1st ed., McGraw-Hill, New York, 1940, Chapter 5.

43. G. W. Wheland and L. Pauling, *J. Am. Chem. Soc.*, **57**, 2086 (1935); G. W. Wheland, ibid., **64**, 900 (1942).
44. M. G. Evans, *Trans. Faraday Soc.*, **35**, 824 (1939).
45. M. J. Dewar, *Tetrahedron Suppl.*, **8**, 75 (1966).
46. H. E. Zimmerman, *J. Am. Chem. Soc.*, **88**, 1564, 1566 (1966); *Acc. Chem. Res.*, **4**, 272 (1971).
47. E. Heilbronner, *Tetrahedon Lett.*, 1923 (1964).
48. A. A. Frost and B. Musulin, *J. Chem. Phys.*, **21**, 572 (1953).
49. N. T. Anh, *Les Regles de Woodward Hoffmann*, Ediscience, Paris, 1970, p. 145 ff.
50. See reference 28, p. 340.
51. E. M. Archer and T. G. D. von Schalckwyk, *Acta Cryst.*, **6**, 88 (1953).
52. M. J. Goldstein and R. Hoffmann, *J. Am. Chem. Soc.*, **93**, 6193 (1971); A. C. Day, ibid., **97**, 2431 (1975).
53. R. L. Burwell, Jr. and R. G. Pearson, *J. Phys. Chem.*, **70**, 300 (1966).
54. T. F. George and J. Ross, *J. Chem. Phys.*, **55**, 3851 (1971). This is an extensive analysis of the reasons for failure of symmetry-selection rules.
55. F. C. Fehsenfeld et al., *J. Chem. Phys.*, **44**, 3022 (1966); J. F. Paulson et al., ibid., 3025.
56. J. L. Bancroft, J. M. Hollas, and D. A. Ramsay, *Can. J. Phys.*, **40**, 322 (1962).
57. See reference 1, p. 32.
58. K. Hsu, R. J. Buenker, and S. D. Peyerimhoff, *J. Am. Chem. Soc.*, **93**, 2117 (1971).
59. J. J. Brauman and D. M. Golden, *J. Am. Chem. Soc.*, **90**, 1920 (1968).
60. J. E. Baldwin, A. H. Andrist, and R. K. Pinschmidt, Jr., *Acc. Chem. Res.*, **5**, 402 (1972).
61. J. A. Berson, *Acc. Chem. Res.*, **5**, 406 (1972).
62. L. H. Pignolet, W. D. Horrocks, and R. H. Holm, *J. Am. Chem. Soc.*, **92**, 1855 (1970); G. N. Lamar and E. O. Sherman, ibid., 2691.
63. P. W. Lert and C. Trindle, *J. Am. Chem. Soc.*, **93**, 6392 (1971); W. C. Herndon, *Chem. Rev.*, **72**, 157 (1972).
64. S. W. Benson and G. R. Haugen, *J. Am. Chem. Soc.*, **87**, 4036 (1965); R. M. Noyes, ibid., **88**, 4311 (1966).
65. R. A. Jackson, *J. Chem. Soc.*, **B**, 58 (1970).
66. B. Atkinson and A. B. Trenwith, *J. Chem. Soc.*, 2082 (1953); B. Atkinson and V. A. Atkinson, ibid., 2086 (1957).
67. J. E. Baldwin and A. H. Andrist, *J. Am. Chem. Soc.*, **93**, 3289 (1971).
68. M. J. Dewar and S. Kirschner, *J. Chem. Soc. Chem. Comm.*, 461 (1975).
69. J. Meinwald and D. Schmidt, *J. Am. Chem. Soc.*, **91**, 5877 (1969).
70. N. D. Epiotis, *J. Am. Chem. Soc.*, **95**, 1191 (1973); **94**, 1924, 1946 (1972).
71. K. N. Houk, *J. Am. Chem. Soc.*, **95**, 4093 (1973); K. N. Houk, J. Sims, C. R. Watts, and L. J. Luskus, ibid., 7301; N. D. Epiotis, ibid., 5624.
72. K. N. Houk, J. Sims, R. E. Duke, Jr., R. W. Strozier, and J. K. George, *J. Am. Chem. Soc.*, **95**, 7287 (1973).
73. (a) S. Glasstone, K. J. Laidler, and H. Eyring, *The Theory of Rate Processes*, McGraw-Hill, New York, 1941; (b) for recent discussions see S. J. Yao and B. J. Zwolinski, *Adv. Chem. Phys.*, **21**, 91 (1971); K. J. Laidler and A. Tweedale, ibid., 113; W. H. Miller, *J. Chem. Phys.*, **62**, 1399 (1975).

74. H. F. Schaefer, III, *The Electronic Structure of Atoms and Molecules*, Addison-Wesley, Reading, Mass., 1972.

75. C. K. Jørgensen, *Acc. Chem. Res.*, **4**, 307 (1971).

76. For example, see S. W. Benson, *Thermochemical Kinetics*, 1st ed., Wiley, New York, 1968; H. E. O'Neal and S. W. Benson, *Int. J. Chem. Kinet.*, **2**, 423 (1970); H. S. Johnston and C. Parr, *J. Am. Chem. Soc.*, **85**, 2544 (1963); N. B. Chapman and J. Shorter, *Advances in Linear Free Energy Relationships*, Plenum, London, 1972.

77. F. Basolo and R. G. Pearson, *Mechanisms of Inorganic Reactions*, 2nd ed., Wiley, New York, 1967, Chapter 2.

78. R. G. Pearson, *J. Chem. Phys.*, **20**, 1478 (1952).

79. A. A. Frost and R. G. Pearson, *Kinetics and Mechanism*, 2nd ed., Wiley, New York, 1961, Chapter 7; J. F. Coetzee and C. D. Ritchie, *Solute–Solvent Interactions*, Dekker, New York, 1969.

80. F. G. Bordwell, *Acc. Chem. Res.*, **3**, 281 (1970); **5**, 374 (1972).

81. R. G. Pearson, *Pure Appl. Chem.*, **27**, 145 (1971).

82. S. W. Benson, *J. Chem. Phys.*, **38**, 1251 (1963); W. A. Guillory, *Diss. Abst.*, **25**, 6981 (1965).

83. General references are D. L. Bunker, *Theory of Elementary Gas Reactions*, Pergamon, Oxford, 1966; H. S. Johnston, *Gas Phase Reaction Rates*, Ronald, New York, 1966; K. J. Laidler, *Theories of Chemical Reaction Rates*, McGraw-Hill, New York, 1969; R. D. Levine and R. B. Bernstein, *Molecular Reaction Dynamics*, Oxford U. P., New York, 1975.

84. F. T. Wall, L. A. Hiller, and J. Mazur, *J. Chem. Phys.*, **29**, 255 (1958); **35**, 1284 (1961).

85. N. C. Blais and D. L. Bunker, *J. Chem. Phys.*, **41**, 2377 (1964); H. H. Mok and J. C. Polanyi, ibid., **53**, 4588 (1970); K. Morukama and M. Karplus, ibid., **55**, 63 (1971); G. W. Koeppl and M. Karplus, ibid., **55**, 4667 (1971).

86. R. Wolfgang, *Prog. React. Kinet.*, **3**, 97 (1965).

87. R. M. Noyes, *J. Chem. Phys.*, **48**, 323 (1968).

88. L. M. Raff, D. L. Thompson, L. B. Sims, and R. N. Porter, *J. Chem. Phys.*, **56**, 5998 (1972).

89. C. Eckart, *Phys. Rev.*, **35**, 1303 (1935).

90. R. P. Bell, *Trans. Faraday Soc.*, **55**, 1 (1959); H. S. Johnston and D. Rapp, *J. Am. Chem. Soc.*, **83**, 1 (1961).

91. R. C. Tolman, *Phys. Rev.*, **23**, 699 (1924).

92. R. L. Burwell, Jr. and W. S. Briggs, *J. Amer. Chem. Soc.*, **74**, 5101 (1952); F. G. Gault, J. J. Rooney, and C. Kemball, *J. Catal.*, **1**, 255 (1962).

CHAPTER 3

# MOLECULAR ORBITALS AND SHAPES OF SIMPLE MOLECULES

The purpose of this chapter is to examine the valence-shell MOs of simple molecules. This will be done for various possible geometric arrangements of the nuclei, conforming in general to known structures. Only molecules capable of symmetry will be considered and only possible symmetric structures will be examined. The reason for this limitation, of course, is to enable the use of symmetry-based arguments. Any conclusions drawn for symmetric molecules can be carried over to unsymmetric molecules with due caution.

It is desirable to become familiar with the MOs of certain molecules, in a pictorial sense as well as in a mathematical one, so that one can visualize which orbitals are the key ones in reactions with reagents of a given type, for example, nucleophiles (electron donors) or electrophiles (electron acceptors). Also one needs to visualize the key MOs in the unimolecular isomerization or decomposition of a molecule by itself. Whenever possible, the rigorous selection rules derived from PT in Chapter 1 will be used. The key role, or alleged key role, played by the HOMO and the LUMO, or by the lowest excited states of a molecule, will be examined. By looking at the MO sequences generated by several possible structures, it will perhaps become clear why one structure is preferred.

If a general structural theory can be developed using MO considerations, it becomes possible to predict the structures of activated complexes for the elementary reactions of simple molecules. From the MO sequence of the unstable activated complex, it should be possible to predict its mode of decomposition. This is equivalent to predicting the path whereby it is formed from the reactants, since one needs only to traverse the reaction coordinate in the reverse direction. We have seen from PT that the potential energy of a molecule is *always* at an extremum when the nuclei are arranged in such a way as to create an element of symmetry. If the energy is a minimum, the nuclear configuration is that of a stable, or at least metastable, molecule. If the energy is a maximum, the structure is that of a TS or is very close to it. Since our knowledge of the structures of activated complexes is indirect, it is better to consider the structures of stable molecules, where our information is excellent. A key question is why is it that symmetry exists in molecules? It is safe to say that, even before the structure of $H_2O$ was determined, scientists expected both O—H bond distances to be the same. The question still is: "Why was this expected"? Of course, over the long run both H atoms of water must behave the same because they are indistinguishable. However this impartiality could equally well be achieved by a structure with one short O—H bond and one long O—H bond. The long and short bonds could then exchange, quite rapidly, by passage over a small potential hill.

The situation is illustrated in Fig. 1, where the potential energy of the water molecule is considered as a function of two of its three normal vibrational modes. The actual variation of energy with bond angle (the bending mode) is shown. The only theoretical requirement is that $E$ be either a maximum or minimum at $\theta = 180°$, since at this value of $Q_0$ the symmetry elements

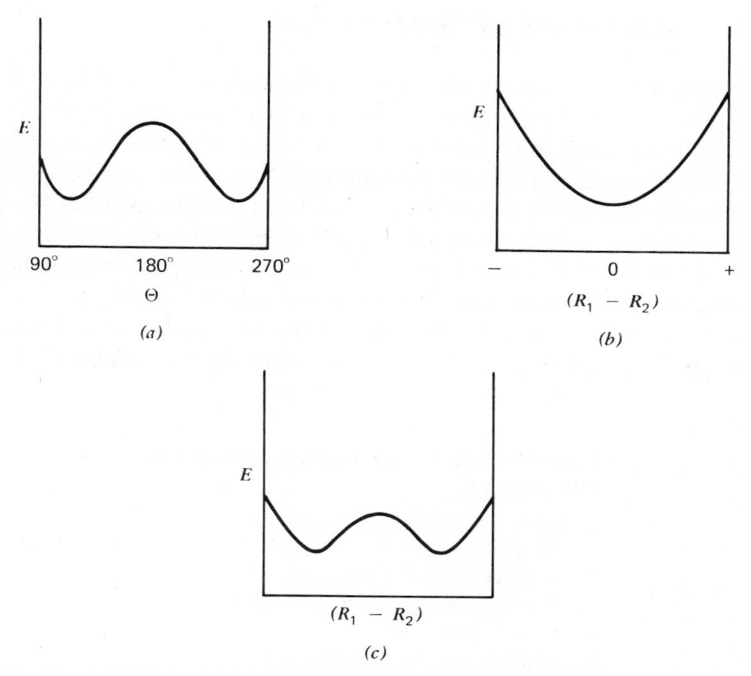


**Figure 1.** (*a*) Variation of potential energy of $H_2O$ with bending angle, $\theta$; (*b*) actual variation of potential energy of $H_2O$ with the asymmetric stretch coordinate $(R_1 - R_2)$; (*c*) possible variation of potential energy with $(R_1 - R_2)$ for $H_2O$.

of either $C_{\infty v}$ or $D_{\infty h}$ are created. The energy is found to be a maximum. The minima are found at values of $\theta$ (and $360°$-$\theta$) which are not related to the creation of new symmetry elements, and which symmetry arguments cannot predict.

Both the actual (equal bond lengths) and the possible (unequal bond lengths) situations are shown for the asymmetric stretch. In the actual case, the energy is a minimum at $Q_0 = 0$, corresponding to either $C_{2v}$ or $D_{\infty h}$. In the possible case, the energy at $Q_0 = 0$ is a maximum and two equal minima exist at some value of $\pm Q_0$. The height of the energy barrier between them would determine the rate of interconversion of the two structures.

The variation of $E$ with the symmetric stretch is not shown. This nuclear motion does not change the point group, except in the limit of complete dissociation. Arguments based on symmetry cannot tell us what the equilibrium bond distances will be. The hope will be, nevertheless, that such arguments can tell us whether a particular symmetric structure corresponds to a potential energy maximum or minimum.

## STRUCTURES OF MOLECULES $XY_n$

First consider the structures of molecules $XY_n$, characterized by a central atom X, surrounded by $n$ atoms Y. The number of electrons is variable and indeed plays an equal role, along with the number of nuclei, in determining the stable geometries. It turns out that there is a very large number of theories that correctly predict the experimental structures, at least as far as the point groups are concerned. Prediction of equilibrium angles and bond distances is much more difficult. Table 1 gives a list of the better-known methods of predicting simple structures. It is by no means a complete list.

The simple electrostatic theory treats atoms as simple charged spheres.[1] If X is a positive sphere and Y is negative, electrostatic repulsion of the Y

**Table 1 Methods for Predicting the Shapes of Molecules $XY_n$**

1. Simple electrostatic
2. Valence-bond theory
3. Valence-shell electron-pair repulsion
4. Crystal field theory
5. Electrostatic force theory
6. Molecular orbital theory (Walsh)
7. Perturbation theory (Jahn–Teller effects)

spheres produces the shapes found. Nonbonding electrons are not considered specifically.

The valence-bond approach uses hybridized AOs on the central atom.[2] These hybrid orbitals have directional properties that determine shapes. Provision is made for nonbonding electrons to be placed in unhybridized AOs of the several atoms.

The valence-shell electron-pair repulsion (VSEPR) theory considers the repulsive interaction of all electron pairs, bonding or not, in the valence shell of X.[3] Lone pairs are assumed to be more repulsive than bonding ones. The method works well for molecules where the overall distribution of electrons around the central atom is nearly spherically symmetrical. Table 2 gives the shapes predicted by the VSEPR theory for various numbers of electron pairs.

Crystal-field theory is useful in predicting molecular structures when there is a nonspherically symmetric charge distribution in the molecule. The atoms Y are treated as charged spheres that create a potential acting on the electrons. These are assumed to be in atomic orbitals on X. The theory was originally developed for compounds of the transition metals,[1,4] in which incomplete $d$ shells produce nonspherical symmetry. It has recently been extended to molecules of the nontransition elements in which incomplete $p$ shells of the central atom are important.[5]

The electrostatic force method considers the forces acting on the several nuclei of a molecule.[6] The forces are due to other nuclei and to electron clouds assumed to be in various AOs. It is necessary to use MOs, in fact, to partition the electronic charges. The interesting variation is that forces, rather than energetics, are used as a basis for prediction.

## The Walsh Method

The classical MO theory of molecular shapes is due to Walsh.[7] His method is based on construction of an approximate MO sequence, and on consideration of the way in which orbital energies would change as the molecular geometry is varied. The prime bases for orbital-energy changes are those of increasing or decreasing orbital overlap and of increasing or decreasing $s$ and $p$ mixing of the orbitals of the central atom. Depending on the number of MOs occupied and their energy changes on structure variation, the stablest shape is arrived at.

To demonstrate Walsh's procedure, and for later use, let us generate the MOs of a linear $H_2X$ molecule of $D_{\infty h}$ symmetry. X is any atom which supplies an $s$ and three $p$ AOs. Only the 1$s$ orbital of each H atom is considered. If the nuclear axis is the $y$ axis, the MOs that are formed depend on the species of

**Table 2 Stereochemistry by VSEPR Theory**

| Total Number of Electron Pairs in Valence Shell | Predicted Arrangement of Electron Pairs in Space | Number of Internuclear ($\sigma$-Bonding) Pairs | Number of Lone Pairs (Nonbonding) | Example | Shape |
|---|---|---|---|---|---|
| 2 | Linear | 2 | 0 | $HgCl_2$ | linear |
| 3 | Triangular Plane | 3 | 0 | $BCl_3$ | triangular plane |
| | | 2 | 1 | $SnCl_2$ (g) | V shaped |
| 4 | Regular tetrahedron | 4 | 0 | $CH_4$ | regular tetrahedron |
| | | 3 | 1 | $NH_3$, $AsCl_3$ | pyramidal |
| | | 2 | 2 | $H_2O$ | V shaped |
| 5 | Trigonal bipyramid | 5 | 0 | $PCl_5$ | trigonal bipyramid |
| | | 4 | 1 | $TeCl_4$ | irregular tetrahedron |
| | | 3 | 2 | $ClF_3$ | T shaped |
| | | 2 | 3 | $[ICl_2]^-$ | linear |
| 6 | Regular octahedron | 6 | 0 | $SF_6$ | regular octahedron |
| | | 5 | 1 | $IF_5$, $XeOF_4$ | square pyramid |
| | | 4 | 2 | $[ICl_4]^-$, $XeF_4$ | square planar |
| 7 | Pentagonal bipyramid | 7 | 0 | $IF_7$ | pentagonal bipyramid |
| 8 | Square antiprism or dodecahedron | 8 | 0 | $[TaF_8{}^-]^{3-}$ | square antiprism |

the four AOs of X, and of the symmetry-adapted linear combinations of $s_1$ and $s_2$, as shown in Table 3.

Mixing together AOs of the same symmetry, the MOs formed are, in order of increasing energy,

$$(\sigma_g)(\sigma_u)(\pi_u)(\sigma_g^*)(\sigma_u^*)$$

The $\sigma_g$ orbital and $\sigma_u$ orbital are lowest in energy because they are bonding MOs. Similarly $\sigma_g^*$ and $\sigma_u^*$ are highest in energy because they are antibonding. The $\pi_u$ orbitals are intermediate because they are nonbonding. The $\sigma_g$ orbital is lower than $\sigma_u$ because an atomic *ns* orbital is lower in energy than the

**Table 3 Symmetry Species of Atomic Orbitals of $H_2X$ in Linear ($D_{\infty h}$) and Bent ($C_{2v}$) Forms**

| X | H | $\Gamma_{D_{\infty h}}$ | $\Gamma_{C_{2v}}$ |
|---|---|---|---|
| $s$ | $s_1 + s_2$ | $\sigma_g$ | $a_1$ |
| $p_y$ | $s_1 - s_2$ | $\sigma_u$ | $b_2$ |
| $p_z$ | — | $\pi_u$ | $a_1$ |
| $p_x$ | — | $\pi_u$ | $b_1$ |

corresponding $np$ orbital. It is not so obvious whether $\sigma_g^*$ or $\sigma_u^*$ is lower in energy. Actually $\sigma_g^*$ is lower for two reasons. It contains some $s$ character of atom X, making it more stable. Also $\sigma_u$ is lowered in energy and $\sigma_u^*$ is raised in energy very much because the $(s_1 - s_2)$ atom combination is usually similar in energy to $p_y$. This leads to strong interaction. A very stable $s$ orbital on X will interact only slightly with the $(s_1 + s_2)$ function.

The way in which the orbital energies change on bending the molecule is shown in Fig. 2. The reasons for the indicated changes are easily seen in Table 3, which gives the symmetry species of the AOs in the $C_{2v}$ point group. The $\sigma_g$ and one of the $\pi_u$ orbitals now become of the same symmetry, $a_1$, and mix together. The $\sigma_g$ and $\sigma_g^*$ are raised in energy because $p_y$ character is mixed in. One $\pi_u$ orbital is sharply lowered both because $s$ character is added and because the orbital changes from nonbonding to bonding. The other $\pi_u$ orbital remains nonbonding. Its energy is assumed to change very little on bending.

The reason for the raising in energy of $\sigma_u$ is best seen in Fig. 3, which shows the MOs for both linear and bent $H_2X$ in schematic form. Bending reduces the bonding overlap of the $b_2$ orbital (formerly $\sigma_u$) and hence raises its energy. At the same time, the fact that it is antibonding between the terminal H atoms begins to manifest itself. For the latter reason the $\sigma_u^*$ orbital increases in energy somewhat on bending.

With Fig. 2 as a guide, the stable structures of $H_2X$ are predicted as a function of the number of valence-shell electrons. With four electrons, or fewer, only orbitals which increase in energy on bending will be occupied. Accordingly such molecules should be linear. This would be the case for $BeH_2$ and $BH_2^+$. With a fifth or sixth electron added, as in $BH_2$ or $CH_2$, a strong factor favoring bending comes in, since the $a_1$ orbital (formerly $\pi_u$) is occupied. Also with seven or eight valence electrons, a bent structure is preferred, since filling the $b_1$ orbital imposes no preference for linearity. Molecules such as $NH_2$ and $H_2O$ will also be bent.

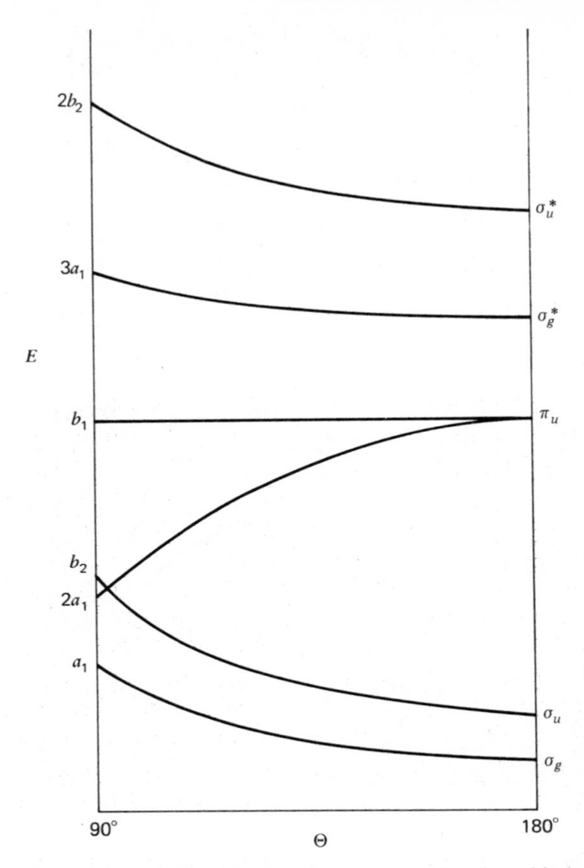


**Figure 2.** Walsh diagram for $H_2X$ molecules showing variation in orbital energies upon bending; $\theta = 180°$ is linear.

Ten electron molecules such as $NeH_2$ are unknown. However by analogy to $XeF_2$ and for other reasons to be discussed later, it is quite certain that a linear configuration would be more stable than a bent one. Figure 2 shows some justification for this, since the $\sigma_g^*$ orbital increases in energy on bending. In general, the Walsh diagram must be considered as very successful. Diagrams similar to Fig. 3 have been constructed for molecules with formulas HXY, $XY_2$, $XY_3$, $H_2X_2$, and $H_2XY$ with equal success in structure prediction.[7]

It is by no means obvious that the Walsh diagrams should work. Clearly some important factors have been omitted from consideration. This includes repulsion of the nuclei, and repulsion of the end groups in very polar mole-

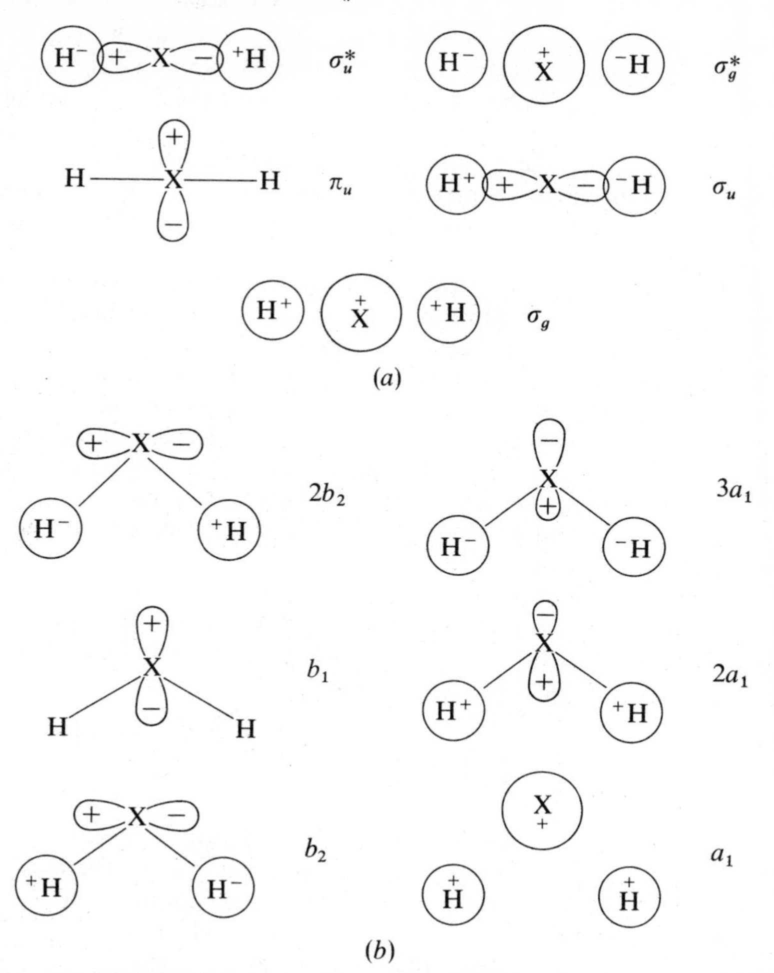


**Figure 3.** Schematic representation of the valence-shell MOs of: (*a*) linear $H_2X$, and (*b*) bent $H_2X$ molecule. Only one of the pair of $\pi_u$ orbitals is shown.

cules. Also detailed calculations of the orbital energies as a function of bending angle show that Fig. 2 is not valid for all $H_2X$ molecules.[8] For example, $Li_2O$ is isoelectronic with water and is predicted to be bent by Walsh arguments. Hartree–Fock SCF calculations show that it is instead linear. This results from the high ionicity, and resultant repulsion of the two Li atoms.[8]

Criticism of incompleteness can be leveled at all of the methods shown in Table 1. Nevertheless, they are all more or less useful in rationalizing the structures of the molecules to which they apply. It is no surprise to find that more elaborate, but still very approximate, MO theories are very good at predicting molecular geometries. These include the Wolfsberg–Helmholtz, or extended Hückel, method[9] and the floating orbital method.[10] A major problem is that all of these methods, simple or elaborate, become much more difficult as the molecular complexity increases. Just as we come to molecules whose structures are still unknown, our methods for a priori prediction become impractical.

Structural theory based on the FOJT and SOJT effects suffers equally in this regard. One may well wonder if we need another method for telling us about structures we already know. The justification is that FOJT and SOJT exercises are an excellent way of becoming familiar with MO theory, particularly with the all important symmetry properties. We learn from simple examples, such as the rearrangement of a molecule from an unstable geometry, how filled and empty MOs interact during the course of changing nuclear configurations. We see which orbital energy schemes are conducive to stable structures, and we learn which elements of symmetry are likely to exist in a given molecule.

The basis of SOJT theory has been given earlier (p. 13). For convenience, the essential equation is reproduced. We assume a particular nuclear configuration for a molecule, $Q_0$, and then distort by a small amount $Q$ along one of the normal modes. The energy becomes

$$E = E_0 + \frac{Q^2}{2}\left\langle \Psi_0 \left| \frac{\partial^2 U}{\partial Q^2} \right| \Psi_0 \right\rangle + \frac{\sum_k \left[ Q \left\langle \Psi_0 \left| \frac{\partial U}{\partial Q} \right| \Psi_k \right\rangle \right]^2}{(E_0 - E_k)} \tag{1}$$

We assume a nondegenerate ground state and we probe only nonsymmetric nuclear displacements (those that change the point group). Therefore there is no term in (1) that is linear in $Q$. We are at a maximum or minimum in potential energy and the task at hand is to determine which. The first quadratic term in (1) is positive, being the classical restoring energy. The second term in negative, being the relaxability along the coordinate $Q$.

Clearly if the potential energy is a maximum, the first term must be larger than the second. If the energy is a minimum, the second term must be larger than the first. This is because the curvature is positive for a maximum but negative for a minimum. If the two terms are comparable in magnitude, the assumed structure is not rigid. There is a very small force constant for the $Q$ vibrational mode, leading to a wide amplitude of displacement. With small activation, the molecule will go over into another structure differing little in energy from the first.

It is not practical to evaluate the infinite sum over all excited states given by (1). Instead we consider only the first few excited states.[12] The symmetry rule is, as before,

$$\Gamma_{\psi_0} \times \Gamma_{\psi_k} \subset \Gamma_Q \quad (2)$$

Instead of relying on spectroscopically identified states, MO theory is used. Even a rough energy sequence will identify the excited states likely to be major contributors to the infinite sum. The argument is that unless a low-lying state of the correct symmetry exists for the molecule, the sum is unlikely to have a substantial value. Only excited states generated by the valence-shell MOs are considered.

The application of this procedure to linear molecules $H_2X$ gives results that are almost astonishing.[13,14] The MO energy levels previously deduced are shown in Fig. 4 for $BeH_2$ and $H_2O$. The lowest excited state for $BeH_2$ should be due to the excitation $(\sigma_u) \rightarrow (\pi_u)$. As explained in Chapter 1, the mixing in of an excited-state wavefunction into the ground state wave function produces a change in electron density, $\rho_{0k}$, the transition density. The spatial symmetry of $\rho_{0k}$ is that of the direct product of the molecular orbitals which generate the excited state, $\varphi_i$ and $\varphi_f$.

$$\Gamma_{\psi_0} \times \Gamma_{\psi_k} = \Gamma_{\varphi_1} \times \Gamma_{\varphi_f} = \Gamma_{\rho_{0k}} \quad (3)$$

In this case $\rho_{0k}$ has a symmetry of $\sigma_u \times \pi_u = \Pi_g$. Figure 4 shows the transition density of $\Pi_g$ symmetry in a pictorial fashion. A plus sign means an increased electron density and a negative sign, a decreased electron density.

If the nuclei now follow these changes in electron density, the result will be a simple rotation of the $BeH_2$ molecule. Indeed the two rotations of a $D_{\infty h}$ molecule are of $\Pi_g$ species. The vibrations are $\Sigma_g$, $\Sigma_u$, and $\Pi_u$. An excited state of the right symmetry to cause a bending of the molecule would require a $(\sigma_g) \rightarrow (\pi_u)$ excitation. But this would be quite a high-energy process because $\sigma_g$ is such a stable orbital. We conclude that $BeH_2$ is stable in the linear form.

The situation in linear $H_2O$ is quite different. The lowest excited state comes from $(\pi_u) \rightarrow (\sigma_g^*)$. This gives a $\rho_{0k} = \Pi_u$, which matches up with the bending mode of the molecule as shown in Fig. 4. Linear water would be unstable and a bending motion converts it into the $C_{2v}$ point group. Note that after an initial small bend to break the symmetry, the excitation becomes $(2a_1) \rightarrow (3a_1)$, since only excited states of the same symmetry as the ground state can mix.

Notice that although the language used is quite different, the perturbation method is equivalent to the Walsh method in results. This means not only that the structural predictions are the same, but also the changes in orbital energies that are responsible. The mixing of $2a_1$ and $3a_1$ (as well as $a_1$) is described by the convention of saying that excited states are mixed with the ground-state wavefunction. The mixing lowers $2a_1$ and raises $3a_1$. Since the two levels are moving away from each other (Fig. 2), the mixing will stop eventually, and so will the bending.

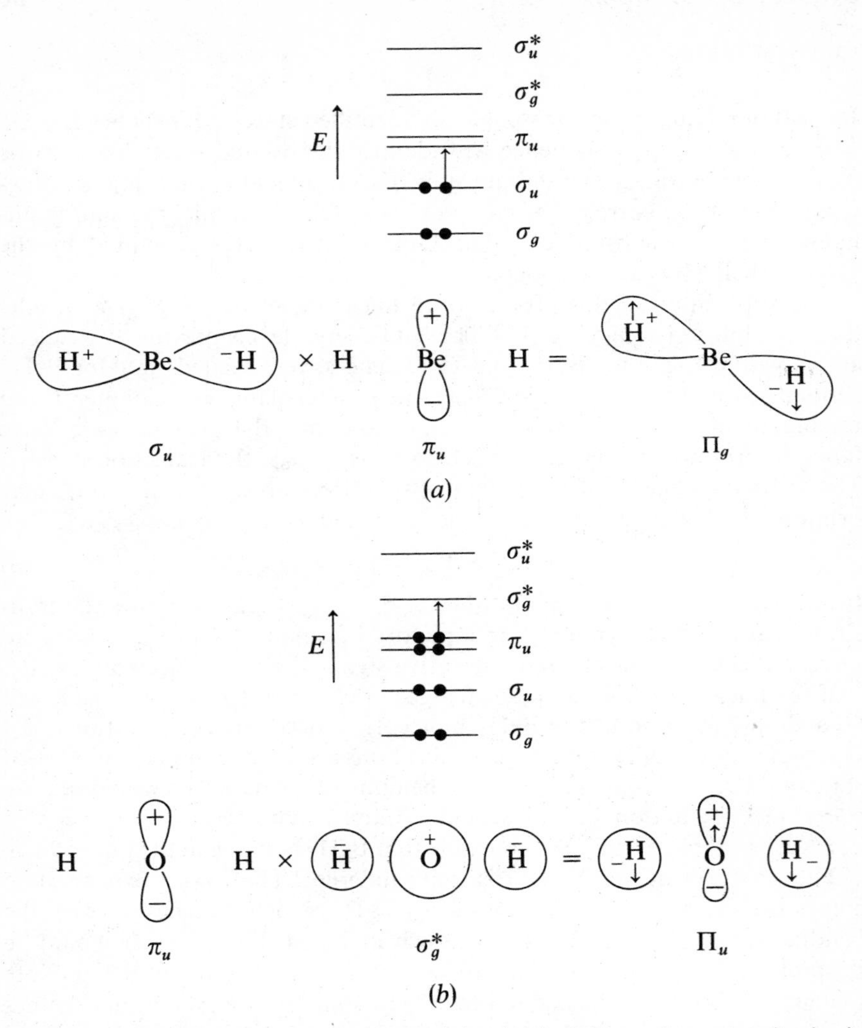


**Figure 4.** (*a*) Mixing of excited state due to $(\sigma_u) \rightarrow (\pi_u)$ in $BeH_2$ produce a transity density, $\rho_{ok}$, which corresponds to rotation of molecules; (*b*) mixing of excited state due to $(\pi_u) \rightarrow (\sigma_g{}^*)$ in linear water produces a transity density that causes the molecule to bend.

We can use PT more easily than Walsh's method, to answer another question. Should a $B_2$ distortion also occur for bent water? That is, is there any reason to expect that an asymmetric stretch should occur, to make the two bond lengths unequal at equilibrium as shown in Fig. 1? Reference to Fig. 2 shows that this is unlikely to be a spontaneous process. The only relevant excited state would be that from the $(b_2) \rightarrow (3a_1)$ excitation. This is more unfavorable than the $(2a_1) \rightarrow (3a_1)$, even after the molecule has bent.

We can predict that bending will occur for molecules $H_2X$ containing 5–8 valence electrons, as found experimentally. With a little imagination, further detail may be guessed at. For example, $BH_2$, with only one electron in the $\pi_u$ orbital, should be harder to excite to the $\sigma_g^*$ level than $NH_2$ or $H_2O$, which have doubly occupied $\pi_u$ orbitals. The bending angle in $BH_2$ is 131°, and those of $NH_2$ and $H_2O$ are 103.4° and 105.2°, in agreement with expectation.[15] Similarly $CH_2$ in the linear triplet state, $^3\Sigma_g$, would be harder to bend than singlet $CH_2$, $^1\Delta_g$, because the latter has double occupancy of one of the $\pi_u$ orbitals. The bond angles are 136° for the triplet state and 102.4° for the singlet state.

Consider molecules with 9–10 valence electrons according to Fig. 4. These would be $H_2F$ and $NeH_2$, for example. The $\sigma_g^*$ is singly and doubly occupied. Symmetric $H_2F$ would be the transition state for the exchange reaction

$$HF + H \longrightarrow H + FH \tag{4}$$

The bending excitation $(\pi_u) \rightarrow (\sigma_g^*)$ is much less favored now because the $\sigma_g^*$ orbital already has an electron in it. However the $(\sigma_g^*) \rightarrow (\sigma_u^*)$ excited state is easily accessible. This leads to a transition density of $\Sigma_u$ symmetry. This relates to the asymmetric stretch, which is exactly the nuclear motion needed to convert linear $H_2F$ into the more stable (HF + H). We conclude that the transition state for (4) is probably linear, but can easily be 10–30° off linearity with little change in energy. This accords well with ab initio calculations[16] of the potential energy surface for the closely related reaction

$$F + H_2 \longrightarrow HF + H \tag{5}$$

The unknown molecule $NeH_2$ is predicted to be stable to bending, but very unstable to the $\Sigma_u$ reaction coordinate. This, of course, leads to its decomposition. If we consider the simpler molecule $HeH_2$, the MO sequence

$$(\sigma_g)^2(\sigma_u)^2(\sigma_g^*)^0$$

is generated, since only the *s* orbital of He is used. This molecule is also unstable to a $\Sigma_u$ mode.

## Structures of $XY_2$ Molecules

The more complex molecules to be considered next are linear, symmetric $HX_2$ molecules, as well as $HeX_2$. Now the $p$ AOs of the end atoms, as well as the $s$, must be included. Table 4 shows the MOs that are formed and their energies for the specific case of $HF_2^-$.[17] The virtual orbital, $3\sigma_g$, is unoccupied and its energy is calculated much too high (see p. 27). Nevertheless there is a large energy gap between the filled $\pi_g$ orbital and the empty $3\sigma_g$, and $HF_2^-$ is expected to be linear and stable, as indeed it is. However $HeF_2$, which would have the same MO configuration, is unstable both experimentally and according to ab initio calculations.[17]

To understand why this is so, let us examine the MOs of Table 4 more closely. The $s$ and $p_z$ orbitals of the terminal X atoms generate the same $\sigma_g$ and $\sigma_u$ species by taking linear combinations. The lowest energy $\sigma_g$ and $\sigma_u$ orbitals in fact are largely the 2$s$ orbitals of fluorine in composition. This is reflected by the low energy values in Table 4. They may be considered as nearly nonbonding. Only the $2\sigma_g$ orbital is a good bonding orbital, the $p_y$ orbitals of F mixing well with the $s$ orbital of H. There are no bonding orbitals of $\pi_u$, $\pi_g$ or $\sigma_u$ type since the central H atom does not mix in these symmetries. The $\pi$ orbitals, in particular, are simply lone-pair orbitals on each F atom.

**Table 4 Symmetry Species of Atomic and Molecular Orbitals of $HX_2$ (or $HeX_2$)[a]**

| X | H | $\Gamma_{D_{\infty h}}$ |
|---|---|---|
| $s_1 + s_2$ | $s$ | $\sigma_g$ |
| $s_1 - s_2$ | — | $\sigma_u$ |
| $p_y - p_y$ | — | $\sigma_g$ |
| $p_y + p_y$ | — | $\sigma_u$ |
| $p_x + p_x$, $p_z + p_z$ | — | $\pi_u$ |
| $p_x - p_x$, $p_z - p_z$ | — | $\pi_g$ |

Energies of MOs Formed in Atomic Units (1 AU = 27.21 eV) for $HF_2^-$

| | | | |
|---|---|---|---|
| $\sigma_g$ | −1.256 | $3\sigma_u$ | −0.274 |
| $\sigma_u$ | −1.199 | $\pi_g$ | −0.295 |
| $2\sigma_g$ | −0.464 | $3\sigma_g$ | 0.957 |
| $\pi_u$ | −0.325 | | |

[a] From reference 17.

In going from H to He as a central atom, the ionization potential of the central atom is changed from 13.60 eV to 24.58 eV. The ionization potential of fluorine is 17.42 eV, for an electron in a $2p$ orbital. Thus the $2\sigma_g$ orbital becomes largely the $s$ orbital of He. It no longer is a good bonding orbital, and the molecule is unstable. Atomic orbitals must have similar energies, as well as proper symmetries, to give good overlap and lead to bonding. If He is replaced by the more easily ionized Xe (IP = 12.13 eV) or Kr (IP = 13.40 eV), then overlap occurs and stable, linear, $XeF_2$ and $KrF_2$ are formed.

These molecules have more complicated MO schemes, since all three atoms contribute $s$ and $p$ orbitals. There are bonding orbitals, now of both $\sigma_g$ and $\sigma_u$ type. In addition $\pi_u$ orbitals are formed that are bonding. These represent the first examples of double bonding of the familiar $\pi$ type. The $\pi_g$ orbitals are nonbonding, and are localized on the terminal Y atoms of $XY_2$. Figure 5 shows schematically the shape of these orbitals for $CO_2$, along with some orbital energies.[18,19] The configuration would be, for 16 electrons,

$$(\sigma_g)^2(\sigma_u)^2(2\sigma_g)^2(2\sigma_u)^2(\pi_u)^4(\pi_g)^4(2\pi_u)^0(3\sigma_g)^0(3\sigma_u)^0$$

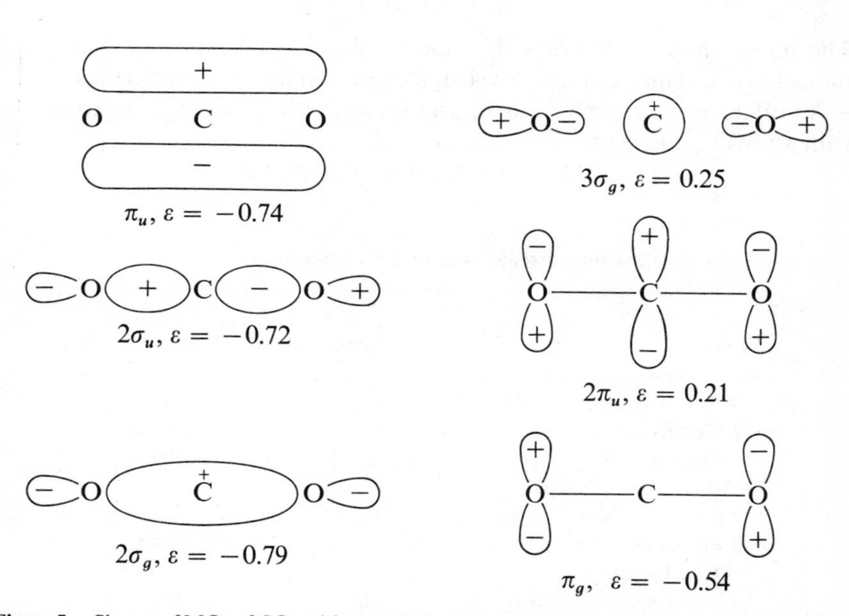


**Figure 5.** Shapes of MOs of $CO_2$ with energies in atomic units; note that these are not intended to be accurate contour diagrams—the object in this, and similar figures, is to show the symmetry properties, the bonding characteristics, and the approximate atomic coefficients ($\sigma_g$ and $\sigma_u$ orbitals, not shown, are chiefly the sum and difference of the $2s$ orbitals on each oxygen atom) (energies from reference 19).

The energy sequence is the same as that estimated by Walsh in 1953, and used by him to predict the structures, bent or linear, for a number of $XY_2$ molecules. The order is not invariant, as shown by ab initio calculations on other linear $XY_2$ molecules, but the variations are small and do not affect the arguments to be made. They relate to the exact position of the $2\sigma_u$ orbital, which is close in energy to $\pi_u$ in any case. Using the SOJT method, the lowest excited state would be the $(\pi_g) \rightarrow (2\pi_u)$ transition. This gives a $\rho_{0k}$ of $\pi_g \times \pi_u = \Sigma_u$ or $\Delta_u$. The easiest molecular process would be a dissociation into CO + O, or possibly a distortion to a structure with unequal bond lengths. The energy gap between $\pi_g$ and $\pi_u$ is large, and the molecule is stable in the linear symmetric form.

If one or two electrons are added to give a 17 or 18 valence-electron system, these must go in the $2\pi_u$ orbital. This allows easy mixing of the $(2\pi_u) \rightarrow (3\sigma_g)$ excited state. This gives $\Pi_u$ symmetry for $\rho_{0k}$ and favors bending of the molecule. Thus $NO_2$ and $NO_2^-$ will be bent, the latter more than the former. This will continue for 19- and 20-electron molecules as well, as shown in Table 5. However for 22 electrons, the configuration becomes

$$\ldots(\pi_g)^4(2\pi_u)^4(3\sigma_g)^2(3\sigma_u)^0$$

The linear shape becomes stable and only a $\Sigma_u$ normal mode is favored, indicating marginal stability toward dissociation for these molecules.

An interesting example is the $C_3$ molecule, with 12 valence electrons and with an MO sequence[20]

$$\ldots(2\sigma_g)^2(\pi_u)^4(2\sigma_u)^2(\pi_g)^0$$

**Table 5 Structural Predictions for $XY_2$ Molecules**

| System | $\rho_{0k}$ Symmetry | Structure (Experimental) |
|---|---|---|
| 8 Electrons, $Li_2O$ | $\Pi_g$ | linear |
| 12 Electrons, $C_3$ | $\Pi_u$ | linear |
| 16 Electrons, $CO_2$ $NO_2^+$, $N_3^-$, $BeF_2$ | $\Sigma_u$, $\Delta_u$ | linear |
| 17 Electrons, $NO_2$ | $\Pi_u$ | bent |
| 18 Electrons, $NO_2^-$ $O_3$, $CF_2$, $SO_2$ | $\Pi_u$ | bent |
| 19 Electrons, $NF_2$, $ClO_2$ | $\Pi_u$ | bent |
| 20 Electrons, $F_2O$, $Cl_2O$, $TeCl_2$, $SCl_2$, $ICl_2^+$ | $\Pi_u$ | bent |
| 22 Electrons, $XeF_2$, $I_3^-$, $ICl_2^-$ | $\Sigma_u$ | linear |

This suggests $\rho_{0k}$ to be of $\Pi_u$ symmetry, predicting a bent structure. This is not so, the molecule being linear. The bending frequency for $C_3$, however, is very low,[21] only 63 $cm^{-1}$, which means a small force constant. This may be compared with the corresponding bending frequency for $CO_2$, which is 667 $cm^{-1}$. The SOJT effect shows up here as a case where the first and second quadratic terms of (1) are nearly equal. The observed first excited state of $C_3$ is indeed $\Pi_u$ and only 3 eV above the ground state.[21]

All of the alkaline earth dihalides should be linear, since like $CO_2$, they have 16 valence electrons. It has been known for some time that those of the light alkaline earths, Be and Mg, are indeed linear, but that the dihalides of the heavier atoms tend to be bent.[22] All of the barium halides are bent. This has been explained by invoking a mixing of the empty *d* orbitals of the metal atom, to modify the Walsh diagrams.[23]

The SOJT theory may also be used. We need to add the manifold of *d* orbitals to the MO sequence for $XY_2$. The order expected from crystal-field theory is

$$(2\sigma_g)^2(2\sigma_u)^2(\pi_u)^4(\pi_g)^4 \;\vdots\; (\delta_g)^0(2\pi_g)^0(3\sigma_g)^0 \;\vdots\; (4\sigma_g)^0(2\pi_u)^0$$

with the *d* orbitals enclosed by the dashed lines. The $\delta_g$ orbitals would be the $d_{xy}$ and $d_{x^2-y^2}$ AOs, if we now call the internuclear axis the *z* axis, as is customary. The $2\pi_g$ would be $d_{xz}$ and $d_{yz}$, and the $3\sigma_g$ would be the $d_{z^2}$ orbital, presumably highest in energy. Transition-metal dihalides would have a number of low-lying excited states due to *d*–*d* transitions. However all of these necessarily lead to a transition density of *gerade* symmetry. Hence they favor neither a bending of the linear molecule, nor an asymmetric stretch.

The alkaline earth halides will have the $\pi_u$ and $\pi_g$ orbitals filled and the *d* orbitals nominally empty. Since $\Pi_u \times \Delta_g = \Pi_u + \Phi_u$, it is quite possible that the $(\pi_u) \rightarrow (\delta_g)$ excitation occurs easily and causes the molecule to bend. Only Ca, Sr, and Ba have empty *d* orbitals in their valence shells, and they would then have the lowest excitation energies. Note that $1\pi_u$ and $\delta_g$ mixing is equivalent to saying that *pd* hybrid orbitals are formed. These overlap at angles different from 180°.

An ab initio calculation for $ZnF_2$ gives the MO sequence[24]

$$\vdots\; \underline{(2\pi_g)^4(\delta_g)^4(3\sigma_g)^2} \;\vdots\; (2\sigma_g)^2(1\pi_u)^4(2\sigma_u)^2(1\pi_g)^4(4\sigma_g)^0$$

The fact that the *d* orbitals lie so low in energy need not be disturbing. For zinc, these are inner-shell orbitals and are more stable than the valence shell by a substantial amount. It is more disturbing that the simple crystal field ordering within the *d* shell is not found. Actually all of the *d* orbitals in

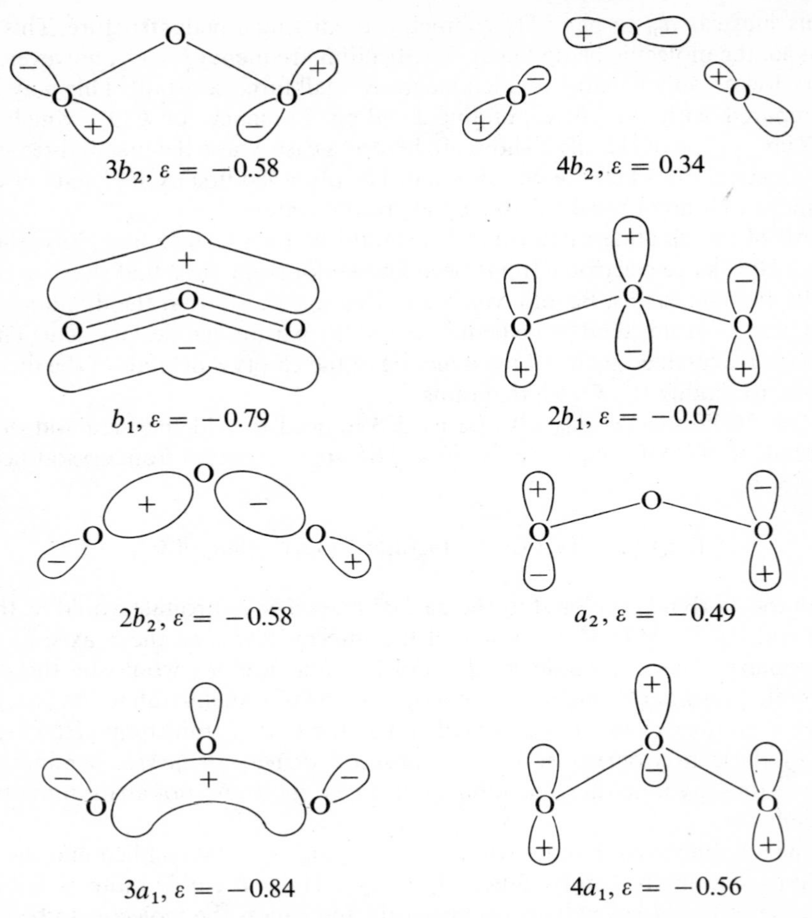


**Figure 6.** The MOs of ozone, with energies in atomic units (1 AU = 27.21 eV); $a_1$, $b_2$, and $2a_1$ orbitals (not shown) are linear combinations of the 2*s* orbitals on each O atom (energies from reference 26).

$ZnF_2$ have about the same energy in this calculation, as indicated by the underlining above. A similar calculation on $CuF_2$ also does not give the expected order of *d* levels.[25] Both $ZnF_2$ and $CuF_2$ are predicted to be linear, as found experimentally, and also as predicted by SOJT theory.

For the post-transition elements, electrons are added first to the $4\sigma_g$ orbital, essentially an *s* orbital on the central atom. In $GeCl_2$, $SnCl_2$, and $PbCl_2$ the two electrons in this orbital are called the "inert" pair. However they are not stereochemically inert. They cause bending by allowing an easy excitation from $(4\sigma_g) \rightarrow (2\pi_u)$ giving a $\rho_{0k}$ of $\Pi_u$ species. In these cases an *sp* hybrid orbital is formed which also favors bent bonding (see Fig. 3).

After bending the symmetry becomes $C_{2v}$. A typical MO sequence is that for $O_3$.[26]

$$(a_1)^2(b_2)^2(2a_1)^2(3a_1)^2(2b_2)^2(b_1)^2(3b_2)^2(4a_1)^2(a_2)^2(2b_1)^0(4b_2)^0(5a_1)^0$$

The approximate shapes of some of these orbitals are shown in Fig. 6, along with the orbital energies. Only the higher occupied and lower empty MOs are shown, since they are the chemically interesting ones. There is a low-lying excited state from the $(a_2) \rightarrow (2b_1)$ transition. This raises the question of a further distortion of $O_3$ to one of $C_s$ symmetry, with unequal bond lengths. Such a structure is reached by the $B_2$ vibrational mode, and $(A_2 \times B_1) = B_2$.

Indeed the inability to write a single Lewis diagram for $O_3$, and the related $SO_2$ molecule, seems to suggest two equilibrium structures which are rapidly equilibrating, for instance,

Each structure would have one long (single) bond and one short (double) bond. The average bond distances would be equal if rapid averaging occurred. The situation could be given by Fig. 1*c*, with a small barrier. However such a situation would be detectable because of a doubling of some of the lines of the infrared (vibrational) or the microwave (rotational) spectrum of the ozone molecule. No such splitting is observed, and we must conclude that the barrier either does not exist (Fig. 1*b*) or is extremely small.[27] The bond distances in $O_3$ and $SO_2$ are equal to within 0.003 Å, and are so at all times except for the distortions due to the $B_2$ asymmetric stretch.

## STRUCTURES OF $XY_3$ AND $XY_4$ MOLECULES

We take up next the structures of $XH_3$ molecules, starting with the planar symmetric geometry of $D_{3h}$ symmetry. Table 6 shows the species of the valence shell AOs. The $e'$ set of H AOs are symmetry adapted linear combinations. They are not a unique set, since 1, 2, and 3 are indistinguishable. They are selected in this case to give maximum overlap with $p_y$ and $p_x$ respectively, in the molecular plane. The energy ordering of the resulting MOs is obvious. The sequence is

$$(a_1')(e')(a_2'')(2a_1')(2e')$$

**Table 6 Symmetry Species for Atomic Orbitals of Planar and Pyramidal $XH_3$**

| X | H | $\Gamma_{D_{3h}}$ | $\Gamma_{C_{3v}}$ |
|---|---|---|---|
| $s$ | $s_1 + s_2 + s_3$ | $a_1'$ | $a_1$ |
| $p_z$ | — | $a_2''$ | $a_1$ |
| $p_y$ | $2s_1 - s_2 - s_3$ | $e'$ | $e$ |
| $p_x$ | $s_2 - s_3$ | $e'$ | $e$ |

The $a_1'$ orbital is lowest because it contains the $s$ orbital of X, more stable than $p_x$ and $p_y$, which also form bonding MOs. The $a_2''$ orbital is in the middle because it is nonbonding, and the antibonding $2a_1'$ and $2e'$ orbitals lie highest.

The energy levels are shown in Fig. 7 for $BH_3$ and for $NH_3$, containing 6- and 8-valence electrons respectively. For the former, the lowest excitation is $(e') \rightarrow (a_2'')$. This gives rise to a transition density of $(E' \times A_2'') = E''$ symmetry. There is no vibrational mode of $E''$ symmetry in planar $XY_3$ molecules. In fact, the electron-density changes are those that cause the molecule to rotate about the $x$ and $y$ axes. $BH_3$ and other six-electron molecules are stable in the planar form. These include $BeH_3^-$ and $CH_3^-$, the methyl carbenium ion.

For $NH_3$, there is an easy excitation from $(a_2'') \rightarrow (2a')$. This is favorable to a reaction coordinate of $A_2''$ symmetry. The $A_2''$ mode of planar $XY_3$ is the out-of-plane deformation that converts the planar $D_{3h}$ form into the pyramidal $C_{3v}$ form. We predict that $NH_3$ will take up the pyramidal structure. So will other 8-electron molecules such as $PH_3$, $H_3O^+$, and $CH_3^-$. The $A_2''$ vibration is shown in Fig. 8, along with the other normal modes of

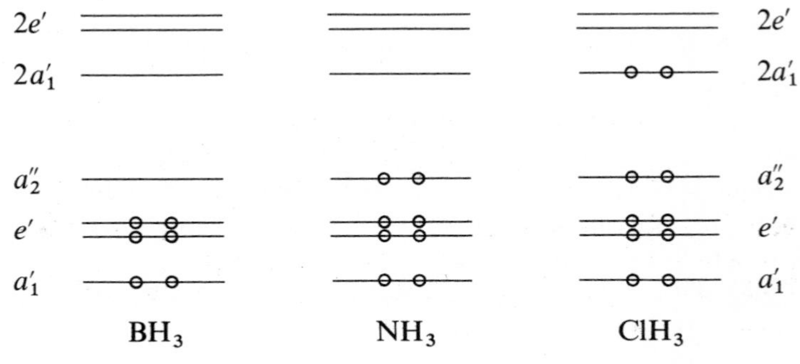


**Figure 7.** Molecular orbital occupations for $BH_3$, $NH_3$, and $ClH_3$ in planar form.

both planar and pyramidal $XY_3$. Observe how the $A_2''$ vibration becomes a totally symmetric mode ($\nu_2$) in the lower symmetry.

Figure 7 also shows the hypothetical 10-electron molecule, $ClH_3$. There is an easily attained $\rho_{0k}$ of $(A_1' \times E') = E'$ symmetry. This is seen to be (Fig. 8) consistent with a reaction coordinate that converts the planar symmetric form into either a Y-shaped or T-shaped form of $C_{2v}$ symmetry. The T shape is that found experimentally for $ClF_3$ and $BrF_3$, as indicated in Table 2. It is useful to recall that the same structures are almost always found for corresponding hydride and halide molecules. We can expect that $ClH_3$ will be more stable as a Y- or T-shaped molecule than in the equiangular form. Alternatively we can interpret an $E'$ reaction coordinate as being the one that produces the facile reaction

$$ClH_3 \longrightarrow ClH + H_2 \tag{6}$$

The case of a seven-electron molecule, the methyl radical, is interesting. Only one electron is in the critical $a_2''$ orbital in the planar form. This raises the $(a_2'') \rightarrow (2a_1')$ transition energy compared to $NH_3$, and reduces the out-of-plane bending tendency. Since $BH_2$, also with one electron in the HOMO, is found to be bent, we might still expect $CH_3$ to be nonplanar. Experimentally it is found to be planar. The SOJT effect still shows up in a strong reduction of the force constant for the $A_2''$ mode in $CH_3$. The ratio of the $E'$ (stretching) to $A_2''$ (bending) force constants is about 9 for $BH_3$ and about 30 for $CH_3$. These results are from matrix-isolation studies of these unstable species.[28] The $NH_3^+$ molecule is also much more planar than the $NH_3$ molecule, as deduced from the vibrational fine structure of the photoelectron spectrum.[29]

The seven-electron radicals, $SiH_3$, $GeH_3$, and $SnH_3$, are probably all pyramidal.[30] This has been explained by Pauling on the basis of varying electronegativity of X in $XH_3$, causing changes in orbital overlaps.[31] It can also be easily explained by the SOJT effect. As the central atom X becomes less electronegative, the energy of $a_2''$ is raised relative to that of $2a_1'$. This makes the required mixing of these two orbitals easier and favors the pyramidal distortion. The $a_2''$ orbital is, of course, a pure $p$ orbital of X. The same result can be achieved if the electronegativity of the Y groups in $XY_3$ is increased, since the $2a_1'$ contains a substantial amount of the AOs of Y, and will be lowered in energy. In fact $CH_2F$, $CHF_2$, and $CF_3$ are all pyramidal.[32]

The MO sequence for $NH_3$ in its stable $C_{3v}$ form becomes

$$(a_1)^2(e)^4(2a_1)^2(3a_1)^0(2e)^0$$

The shapes of these MOs are shown in Fig. 9, along with the orbital energies. It is tempting to try to use this sequence to prove that pyramidal $BH_3$ would be unstable. However this is not permissible because the energy

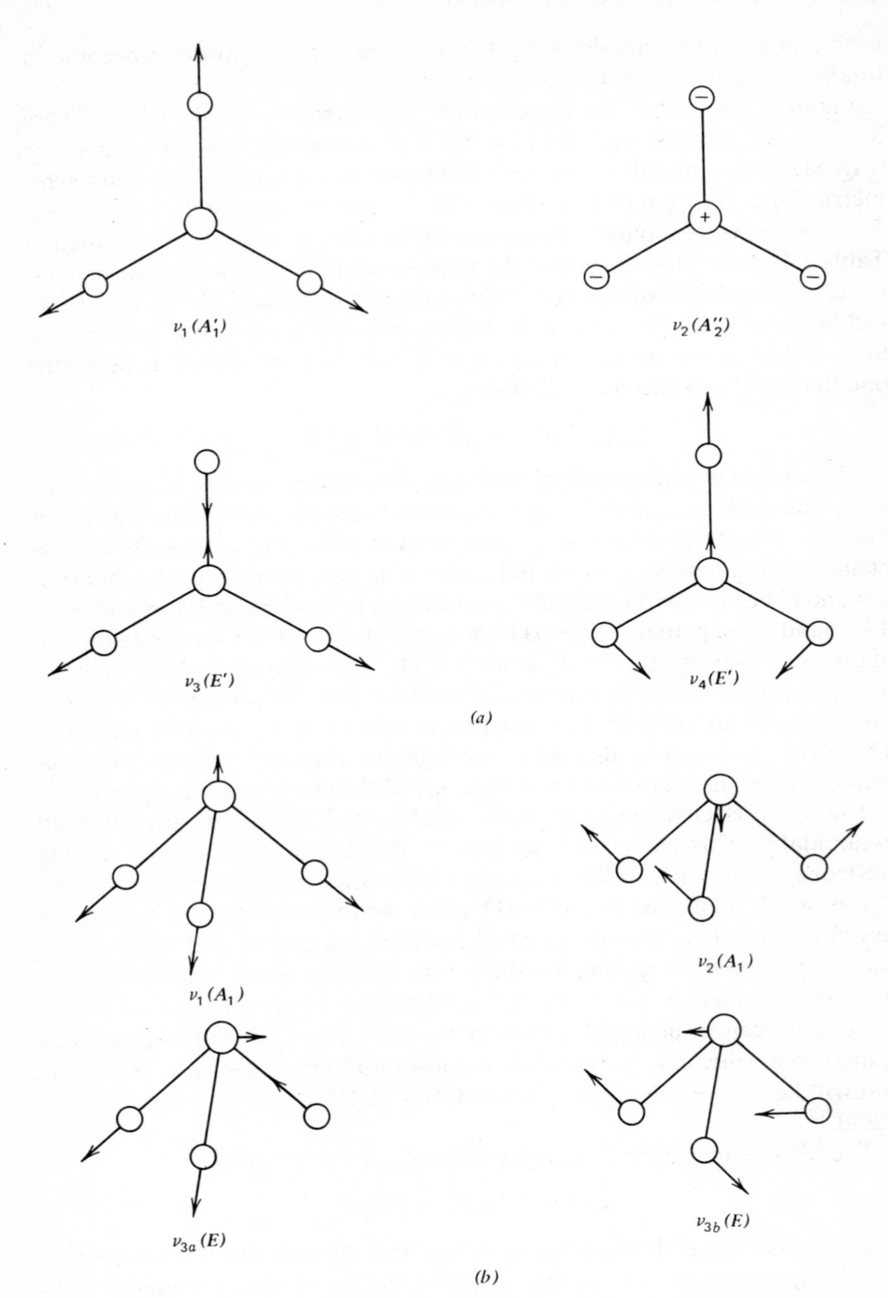


**Figure 8.** Normal modes of vibration for: (*a*) planar and (*b*) pyramidal, $XY_3$ molecules.

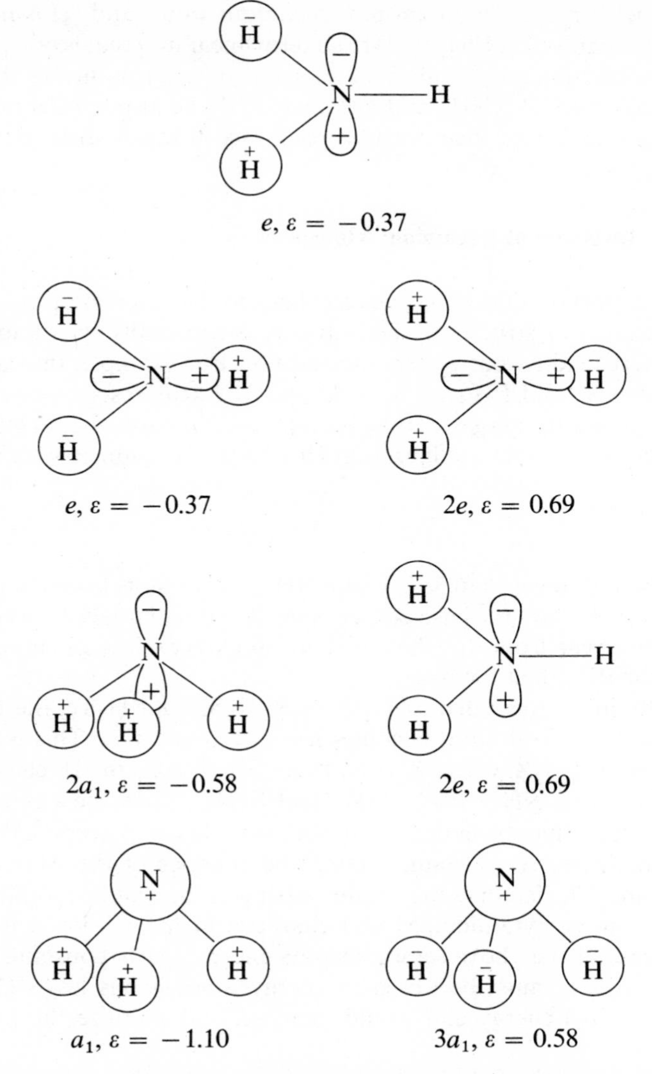


**Figure 9.** Schematic representation of MOs of $NH_3$, with their energies in AU [energies from W. E. Palke and W. N. Lipscomb, *J. Am. Chem. Soc.*, **88**, 2384 (1966)].

would be neither at a minimum nor maximum value, and (1) is not valid. The correct equation (in Chapter 1) has a term linear in $Q$ that is nonvanishing because the reaction coordinate is of $A_1$ symmetry. It is $\nu_2$ in Fig. 8. We can consider molecules like $NH_3$ and $PH_3$, which do lie at potential minima in $C_{3v}$, and try to assess their relative resistance towards distortion to the planar $D_{3h}$ form.

## Inversion of Pyramidal Molecules

This is an important question, since it relates to the inversion phenomenon, of great interest in structural chemistry. Experimentally it is known that amines invert readily, but phosphines do not. For example, the calculated barriers for $NH_3$ and $PH_3$ are 5.2 kcal and 37.2 kcal, respectively.[33] These are the differences in energies of the planar forms compared to the equilibrium configurations. The MO configuration for $PH_3$ in $C_{3v}$ symmetry is calculated to be[34]

$$(a_1)^2(e)^4(2a_1)^2(2e)^0(3a_1)^0$$

The $2e$ orbital is lower than $3a_1$, unlike $NH_3$. This results from the inclusion of $3d$ orbitals of P into the basis set, since the $2e$ MO is largely $3d$ in character. It is still not clear from this that the $(2a_1) \rightarrow (3a_1)$ excitation energy would be larger for $PH_3$ than for $NH_3$.

Normally in a series such as $H_2O$, $H_2S$, $H_2Se$, and $H_2Te$, the IP drops steadily as the central atom becomes less electronegative. This is not true for the series $(CH_3)_3X$, where X = N, P, As, Sb.[39b] Also the IP changes very little in the series $NH_3$, $PH_3$, $AsH_3$ and $SbH_3$. Since the electron most easily removed comes from the $2a_1$ orbital, some factor is keeping the energy of this orbital nearly constant, though the energies of the AOs of X are steadily rising. The factor is the greater mixing of the $s$ and $p$ atomic orbitals of X on becoming pyramidal. More $s$ character in $2a_1$ will lower its energy, or in the case of the above series, keep its energy nearly constant. The $3a_1$ orbital conversely must be raised in energy, since it has less $s$ character. The $(2a_1) \rightarrow (3a_1)$ energy gap would increase, and stabilize the pyramidal form.*

Another clue is provided by the amount of bending from the planar form for the several $H_3X$ molecules. The bond angle for $NH_3$ is about 108°, and that for $PH_3$ is only 93°. Clearly the phosphine molecule $2a_1$ orbital becomes relatively more stable on bending, than does the corresponding orbital for $NH_3$. This is why greater bending occurs. The reverse process would then be resisted more strongly.[35]

* See C. C. Levin, *J. Am. Chem. Soc.*, **97**, 5649 (1975).

## $XY_3$ Molecules

Nonhydride molecules of formula $XY_3$ contain enough electrons so that accurate MO functions are difficult to obtain. The large number of MOs is indicated by the valence shell AOs shown in Table 7 for the planar structure. For the simplest case, $BF_3$, an ab initio calculation gives the MO sequence in order of increasing energy with the three

$$(a_1')^2(e')^4(2a_1')^2(2e')^4(a_2'')^2\underline{(3e')^4(e'')^4(a_2')^2}(2a_2'')^0(3a_1')^0(4e')^0$$

underlined orbitals very close in energy. This ordering has been substantiated by photoelectron spectroscopy, both for $BF_3$ and the isoelectronic $SO_3$.[37] The lowest empty MO is certainly the $2a_2''$ orbital, largely a $p_z$ orbital on the boron atom. Note that there is $\pi$ bonding in the occupied $a_2''$ orbital. Figure 10 shows the shapes of the HOMOs and LUMOs of $BF_3$.

The lowest excitation would be the $(a_2') \rightarrow (2a_2'')$ giving rise to $\rho_{0k}$ of $A_1''$ symmetry. Figure 8 shows there is no vibration of this species. The next transition $(e'') \rightarrow (2a_2'')$ could promote an $E'$ distortion of the molecule, but the energy gap is now quite large. In any case, 24 electron molecules such as $BF_3$, $SO_3$, $NO_3^-$, $CO_3^{2-}$, $BO_3^{3-}$, $GaCl_3$, and $AlCl_3$ all are stable in the planar, symmetric form. If two electrons are added to form 26-electron molecules such as $NF_3$, $PCl_3$, $AsBr_3$, $SbI_3$, $XeO_3$, $SO_3^{2-}$, and $BrO_3^-$, the symmetry of $\rho_{0k}$ becomes $A_2'' \times A_1' = A_2''$. This causes the molecule to become pyramidal, and of $C_{3v}$ symmetry.

**Table 7 Symmetry Species for Atomic Orbitals of $XY_3$ in Planar and Pyramidal Forms**

| X | Y | $\Gamma_{D_{3h}}$ | $\Gamma_{C_{3v}}$ |
|---|---|---|---|
| $s$ | $s_1 + s_2 + s_3$ | $a_1'$ | $a_1$ |
| | $p_{y_1} + p_{y_2} + p_{y_3}$ | $a_1'$ | $a_1$ |
| — | $p_{x_1} + p_{x_2} + p_{x_3}$ | $a_2'$ | $a_2$ |
| $p_z{}^a$ | $p_{z_1}{}^a + p_{z_2} + p_{z_3}$ | $a_2''$ | $a_1$ |
| $p_y{}^b$, $p_x$ | $p_{y_1}, p_{y_2}, p_{y_3}$ | $e''$ | $e$ |
| | $p_{x_1}, p_{x_2}, p_{x_3}$ | | |
| — | $\left.\begin{matrix} 2p_{z_1} - p_{z_2} - p_{z_3} \\ p_{z_2} - p_{z_3} \end{matrix}\right\}$ | $e''$ | $e$ |

[a] The $p_z$ orbitals are perpendicular to the molecular plane.
[b] The $p_y$ orbitals of F point in to the B atom.
[c] The $e'$ orbitals of F are linear combinations of $p_x$ and $p_y$.

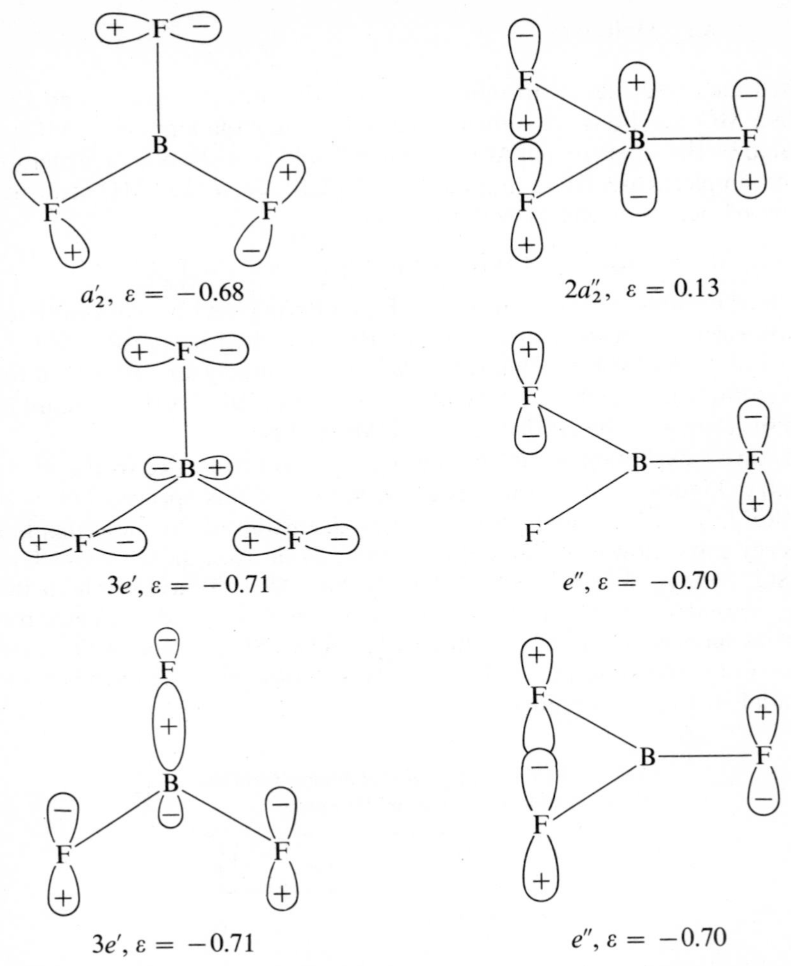


**Figure 10.** Shapes of frontier orbitals of $BF_3$ (energies in atomic units from reference 26).

Molecules with 28 valence electrons, $BrF_3$ or $ClF_3$, would have $\rho_{0k} = (A'_1 \times E') = E'$. This gives the in-plane distortion to the T shape found for these molecules. Of course an $E'$ mode can also give rise to a Y-shaped molecule, by distorting along a $-E'$ coordinate rather than $+E'$. A disadvantage of SOJT theory is that it cannot distinguish between these two possibilities, since the energy according to (1) is quadratic in $Q$.

Molecules with 22–23 electrons, such as $CO_3$, $NO_3$, or $CO_3^-$ are, in fact, calculated to be stable as Y-shaped.[38] In these cases, the SOJT instability arises from the $(3e') \rightarrow (a'_2)$ transition, which requires very little energy. It is also possible that the underlined orbitals of the $BF_3$ MO sequence shift about. In this case the 22–23-electron molecules would be unstable because of a FOJT. There would be an incomplete $3e'$ or $e''$ orbital, giving rise to a degenerate state. The reaction coordinate would still be $E'$. This coordinate might also be considered as the one corresponding to dissociation of these unstable molecules[12]

$$CO_3 \longrightarrow CO + O_2 \tag{7}$$

$$NO_3 \longrightarrow NO + O_2 \tag{8}$$

The Y shape would be a precursor to the final products in which two oxygen atoms have drawn together. The symmetry elements of $C_{2v}$ would be preserved during the complete reaction as given in (7) or (8). However an analysis in terms of bond symmetries shows that (7) and (8) are forbidden (p. 264).

On becoming pyramidal, the point group for $XY_3$ molecules becomes $C_{3v}$. The most important change is that the $2a''_2$ orbital becomes much more stable in the pyramidal form. The $(3a'_1)$ becomes much less stable. Both are of the same symmetry in $C_{3v}$, namely totally symmetric, and there is strong mixing between them. The $a'_2$ orbital becomes $a_2$, and remains nonbonding since no orbital of the central atom has this symmetry (Table 7). For molecules such as $NF_3$ and $SO_3^{2-}$, for which Lewis diagrams can be drawn with a simple octet about each terminal atom, the interesting orbitals are the same as for $NH_3$. That is, they have $\sigma$ bonding and $\sigma^*$ antibonding MOs, and a lone pair of electrons in an $a_1$ orbital concentrated on the central atom.

### $XH_4$ Molecules

We come next to molecules with the formula $XY_4$, considering first the hydrides such as $CH_4$. The symmetries of the AOs in methane are given in Table 8. Accurate calculations give the MO energies shown below.[39] The same sequence holds for other 8-valence electron

$$\begin{array}{llll} (a_1) & -0.944\ AU & (2a_1) & +0.305 \\ (t_2) & -0.542 & (2t_2) & +0.313 \end{array}$$

molecules, such as $NH_4^+$ and $BH_4^-$. The lowest energy transitions are either $(t_2) \rightarrow (2a_1)$ or $(t_2) \rightarrow (2t_2)$. Since $T_2 \times T_2 = A_1 + E + T_1 + T_2$ several possible distortions are suggested. Figure 11 shows the vibrational modes of tetrahedral and planar $XY_4$ molecules. In particular, the $E$ mode of a tetrahedral molecule converts it into the planar form.

**Table 8 Symmetry Species for Atomic Orbitals of $XH_4$ Molecules in Tetrahedral and Square Planar Forms**[a]

| X | H | $\Gamma_{T_d}$ | $\Gamma_{D_{4h}}$ |
|---|---|---|---|
| $s$ | $s_1 + s_2 + s_3 + s_4$ | $a_1$ | $a_{1g}$ |
| $p_z$ | $s_1 + s_4 - s_2 - s_3$ | $t_2$ | $a_{2u}, b_{1g}$ |
| $p_x$ | $s_1 + s_2 - s_3 - s_4$ | $t_2$ | $e_u$ |
| $p_y$ | $s_1 + s_3 - s_2 - s_4$ | $t_2$ | $e_u$ |

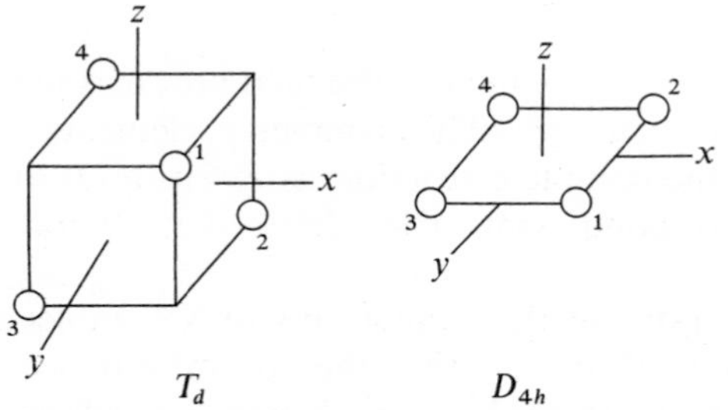

$T_d$ $D_{4h}$

[a] Numbering and coordinate system used.

However there is no need to fear that $CH_4$ will be other than its familiar shape. The saving feature is the large energy gap between the filled $t_2$ orbital and the empty, antibonding ones. The calculated difference of 0.847 AU cannot be taken seriously, because virtual orbitals are always calculated too high in energy by SCF methods. Experimentally the gap is about 10 eV. Methane is one of the most transparent molecules known, to UV light. Any structure in which all the stable MOs are occupied and all the unstable MOs are empty, with a large gap between, must be a stable one.

With eight valence electrons so ideally suited to the $T_d$ structure, it is no surprise to find all other numbers of valence electrons unstable in this form. For example, $CH_4^+$, a seven-electron molecule would be in a $T_2$ ground state. This creates a FOJT instability. The indicated distortion is either $T_2$, leading to a $C_{3v}$ or $C_{2v}$ structure, or $E$ to a $D_{2d}$ structure. Again we cannot tell from symmetry arguments alone which change will occur. However we can deduce the direction of distortion by referring to the changes in the orbital energies for each point group. These are shown in Fig. 12 for $CH_4^+$ in $T_d$, $D_{2d}$, $D_{4h}$ and $C_{3v}$ point groups.

All of these other geometries remove the ground-state degeneracy of the $T_d$ form, as required by the Jahn–Teller principle. The $C_{3v}$ distortion primarily produces a high energy orbital for the odd electron, without significant

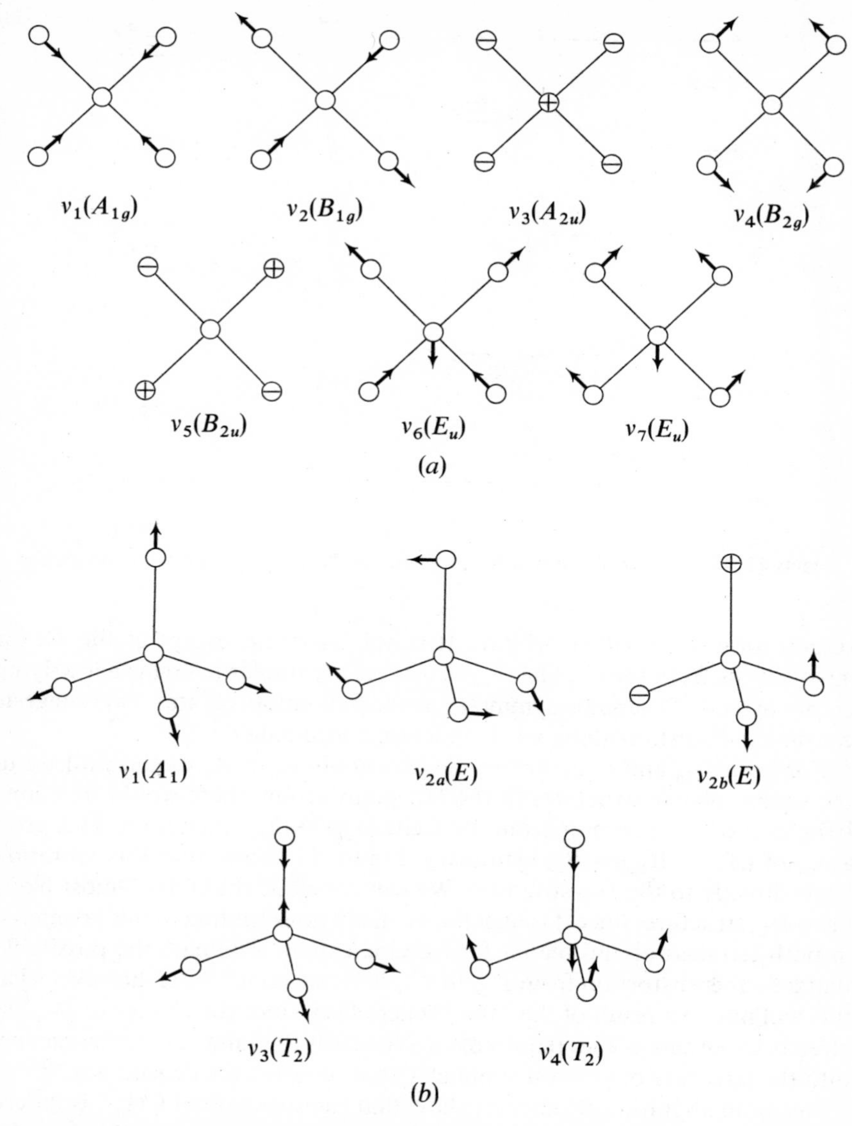


**Figure 11.** Normal modes of vibration for $XY_4$ molecules in: (*a*) square planar and (*b*) tetrahedral, form. Only one component of the triply degenerate $T_2$ modes is shown.

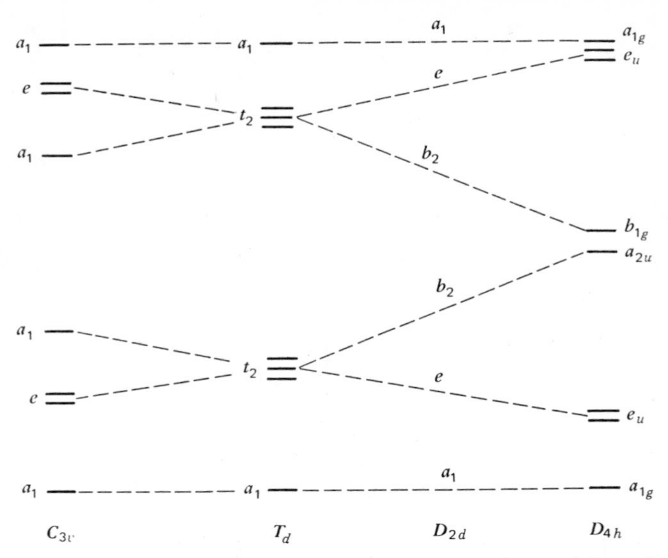


**Figure 12.** Correlation of MOs of $XH_4$ molecule in $T_d$, $C_{3v}$, $D_{2d}$, and $D_{4h}$ point groups.

stabilization of the other orbitals. It is not favorable, except as the TS for dissociation. Both the $C_{2v}$ and $D_{2d}$ structures are similar in giving a low-lying empty orbital of the same symmetry as the half-empty orbital. This suggests continued distortion along an $A_1$ reaction coordinate.

For both $D_{2d}$ and $C_{2v}$, further movement along an $A_1$ mode can lead to the square planar structure. In the $D_{4h}$ point group, there would be a low-lying excited state coming from the facile $(a_{2u}) \rightarrow (b_{1g})$ excitation. This gives a $\rho_{0k}$ of $(A_{2u} \times B_{1g}) = B_{2u}$ symmetry. Figure 11 shows that this vibration leads directly to the $D_{2d}$ structure. We can conclude that $CH_4^+$ most likely has a $D_{2d}$ structure, since it is unstable to distortions leading to this geometry, in both tetrahedral and planar forms. One cannot eliminate the possibility that a further distortion from $D_{2d}$ to $C_{2v}$ will not occur. Note, however, that this will not be a result of the Jahn–Teller effect, since the change to $D_{2d}$ has already taken care of that requirement. Normally the Jahn–Teller changes are into the structure of greatest symmetry that removes the degeneracy.[40]

Accurate ab initio calculations show that the structure of $CH_4^+$ is indeed $D_{2d}$.[41a] The dihedral angle is 53°, about halfway between the 90° angle for $T_d$ and 0° angle for $D_{4h}$ structures. The experimental evidence is from the photoelectron spectrum of $CH_4$.[41b] The fine structure clearly shows that the distortion in $CH_4^+$ is along an $E$ mode, which means a $D_{2d}$ structure. The virtue of the $D_{2d}$ structure in this case is that the half-empty orbital is raised in energy (the $b_2$ orbital), and the filled orbitals ($e$) are lowered in energy. One $p$ orbital of carbon has poorer overlap, but the other two $p$ orbitals have better overlap, compared to the tetrahedral form.

A 6-electron molecule such as $BH_4^+$ would be unstable in $T_d$ also. It would be stable in $D_{4h}$ symmetry, as seen in Fig. 12. The lowest excited state would be of $(E_u \times A_{2u}) = E_g$ symmetry. There is no vibrational mode of $E_g$ type. Methane itself and other eight electron molecules would be unstable in $D_{4h}$ since the $B_{2u}$ mode would be even more easily excited than for $CH_4^+$. Square planar $CH_4$ is calculated to be 250 kcal higher in energy than the tetrahedral form.[42] No wonder that racemization of optically active organic molecules does not occur by this route! Some of MOs of planar and tetrahedral $CH_4$ are shown in Fig. 13. The others not shown can easily be constructed from the information in Table 8.

### $XY_4$ Molecules

$XY_4$ molecules have a large number of electrons and numerous MOs. An ab initio calculation has been carried out for $CF_4$, in the known tetrahedral structure.[43] The energy sequence of the MOs of the valence shell is

$$(a_1)^2(t_2)^6(2a_1)^2(2t_2)^6(1e)^4(3t_2)^6(t_1)^6(3a_1)^0(4t_2)^0$$

It is of interest that an almost identical ordering was found in an early semiempirical calculation for $CCl_4$.[44] This was the first application of the Wolfsberg–Helmholz method to organic molecules. Compared to $CH_4$, we find a doubling of the filled $a_1$ and $t_2$ orbitals because both the $s$ and $p_y$ orbitals of F or Cl generate these symmetries. The $e$, $3t_2$, and $t_1$ orbitals are $\pi$-type orbitals made by the $p_z$ and $p_x$ AOs of the halogen. For many purposes, the critical orbitals are the bonding $2a_1$ and $2t_2$ orbitals of $\sigma$ character and their antibonding counterparts, $3a_1$ and $4t_2$. The remaining filled orbitals are essentially equivalent to lone pairs of electrons on F or Cl.

$CF_4$ is stable by symmetry, but also because of the large energy gap between the $\pi$ orbitals and the antibonding $3a_1$ orbital. Experimentally this is about 7 eV, from the UV spectrum. Other 32-electron molecules such as $SO_4^{2-}$, $ClO_4^-$, $SnF_4$, or $SiCl_4$ would also be stable in $T_d$. Adding two electrons as in $SF_4$, would create a low-lying excited state from the $(3a_1) \rightarrow (4t_2)$ transfer. The corresponding $T_2$ reaction coordinate carries the tetrahedral $SF_4$ into the $C_{2v}$ structure, which is found experimentally.[45] $XeF_4$, with 36 valence electrons would have an incomplete $(4t_2)^2$ shell. This would create a FOJT, or SOJT effect leading to a distortion toward a planar structure.

$XeF_4$, with 108 electrons altogether, is not within the reach of accurate SCF MO calculations at present. Nevertheless an ab initio calculation with a limited basis set has been made for the known square planar structure. The

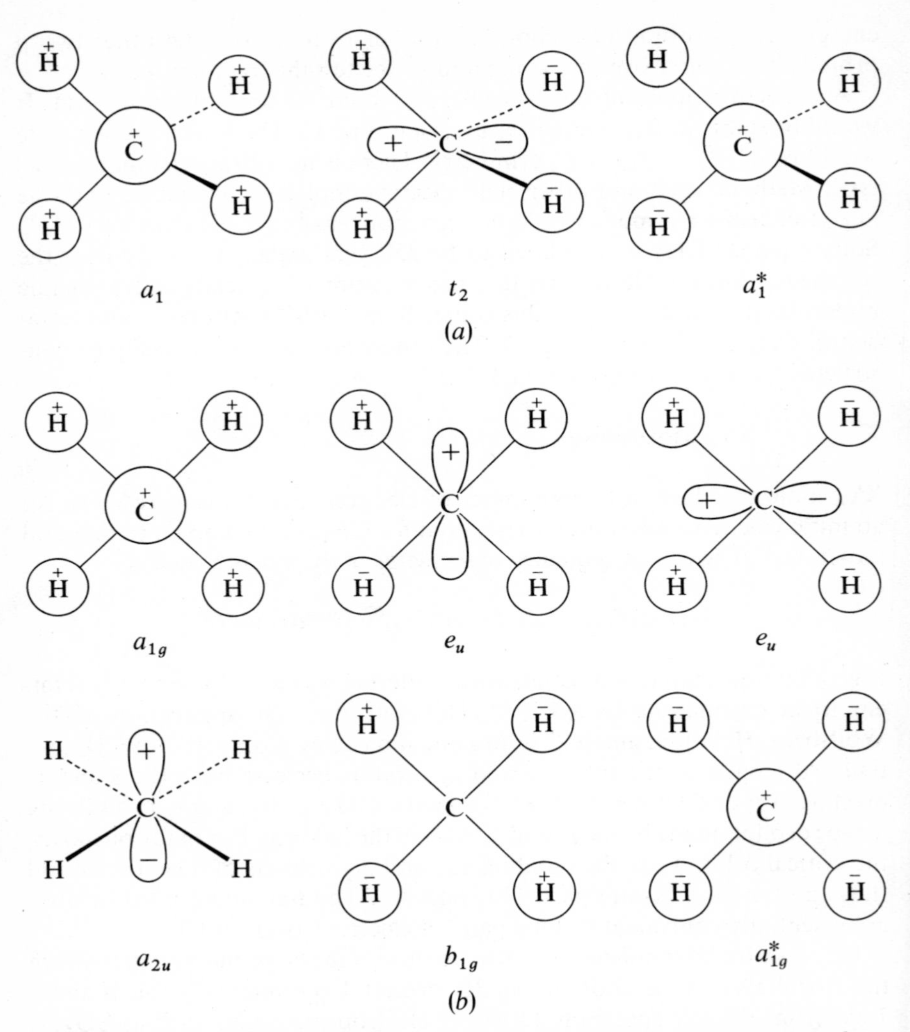


**Figure 13.** Some molecular orbitals of $CH_4$ in: (*a*) tetrahedral form and (*b*) square planar form, the $t_2$ orbital using the $p_x$ orbital of carbon is shown; The $y$ and $z$ axes are both twisted 45° out of the plane of the page to agree with Table 8.

MOs are, in order of increasing energy,

$$\ldots(a_{1g})^2(e_u)^4(a_{2u})^2(b_{2g})^2(e_g)^4(b_{2u})^2(2e_u)^4(a_{2g})^2(b_{1g})^2(2a_{2u})^2(2a_{1g})^2(3e_u)^0$$

The transition density of the $(2a_{1g}) \rightarrow (3e_u)$ excitation is of $E_u$ symmetry. This corresponds to the reaction mode (Fig. 11), leading to dissociation

$$XeF_4 \longrightarrow XeF_2 + F_2 \tag{9}$$

However $XeF_4$ is quite stable with respect to this reaction, consistent with a reasonably large gap between $2a_{1g}$ and $3e_u$. In any case $XeF_4$ is predicted to be stable toward bending out of the plane.

$CCl_4$ or $SO_4^{2-}$ would have the configuration, in $D_{4h}$,

$$\ldots(2e_u)^4(a_{2g})^2(b_{1g})^2(a_{2u})^0(2a_{1g})^0$$

with a small energy gap between the HOMO and LUMO. This creates an easy movement along a reaction coordinate of $B_{1g} \times A_{2u} = B_{2u}$ symmetry. This takes the molecule into the $D_{2d}$ structure, and by continuing, into the stable tetrahedral form. The $SF_4$ molecule would have 34 valence electrons, so that the excited state would be from $(a_{2u}) \rightarrow (2a_{1g})$. This $A_{2u}$ mode (Fig. 11) converts into a $C_{4v}$ structure, whereas $C_{2v}$ is found. The excited state of $B_{2u}$ symmetry, arising from the configuration

$$\ldots(a_{2g})^2(b_{1g})^1(a_{2u})^1(2a_{1g})^2$$

would also be needed, since $A_{2u} + B_{2u}$ converts planar $SF_4$ into its correct structure. It may be noted that the $b_{1g}$, $a_{2u}$, $2a_{1g}$, and $3e_u$ orbitals of $SF_4$ are essentially the same as the corresponding highest orbitals of $CH_4^+$ shown in Fig. 12.

The nonexistent molecule $SH_4$ would have the configuration

$$(a_{1g})^2(e_u)^4\underline{(a_{2u})^2(b_{1g})^2(2a_{1g})^0(2e_u)^0}$$

One indicated instability is towards an $E_u$ mode, which would convert the molecule into $H_2S + H_2$. The combination of $A_{2u}$ and $B_{1g}$ nuclear motions suggested by the MO sequence leads to an incorrect $C_{2v}$ structure. The average structure of $SH_4$, or $SF_4$, would be the $C_{4v}$ form predicted by the $A_{2u}$ mode with all H atoms equivalent. This is the TS for the rapid interconversion of axial and equatorial atoms. Note that it is not necessary to pass through the planar form to achieve this pseudorotation.

### Transition-Metal $ML_4$ Complexes

Transition metal complexes often occur in square planar form. Compared to the hypothetical planar $CF_4$ molecule, the biggest change is that an orbital of $b_{1g}$ species becomes available on the central atom. This is the

$d_{x^2-y^2}$ orbital. The effect is to lower the energy of $b_{1g}$ greatly, and also to create an antibonding $b_{1g}^*$ orbital. The first effect automatically makes planar complexes more plausible for transition-metal complexes, since the formerly facile $(b_{1g}) \rightarrow (a_{2u})$ transition becomes difficult. Alternatively, we now have $dsp^2$ hybrid orbitals.

Considering only the $\sigma$ orbitals and the $d$ orbitals of the complex, the MO sequence is probably

$$(a_{1g})^2(b_{1g})^2(e_u)^4 \;\vdots\; (b_{2g})(e_g)(a_{1g}^*)(b_{1g}^*) \;\vdots\; (a_{2u})^0(2a_{1g})^0$$

The positioning of the $d$ orbitals in order has been the subject of much controversy. We now recognize that attempts to use the visible-UV absorption spectrum as a guide are rather hopeless. Instead the results of several ab initio calculations are available. These include planar $NiF_4^{2-}$,[47] $CuCl_4^{2-}$,[48] $PtCl_4^{2-}$,[49] $CuF_4^{2-}$,[50] and $Ni(CN)_4^{2-}$.[51] Surprisingly, these all agree on the ordering

$$d_{xy} < d_{xz}, d_{yz} < d_{z^2} \ll d_{x^2-y^2}$$

given in the sequence above. The calculations all agree also in that the $d$ levels are scrambled among the $\pi$ and nonbonding levels of the ligands.

Distortion toward a $D_{2d}$, or $T_d$, structure requires a $B_{2u}$ reaction coordinate. The most effective transition to produce this structural change will be the $(e_u) \rightarrow (e_g)$ transition. This is a way of saying that a tetrahedral transition metal complex uses $sd^3$ hybrid orbitals, rather than $sp^3$. The $e_u$ orbitals are based on the $p_x$ and $p_y$ orbitals of the central atom.

It can be seen that the low-lying states arising from $d$–$d$ transitions are not effective since they produce only *gerade* transition densities. As we fill up the $d$ shell, the planar form is stabilized by blocking transitions from $e_u$ to $e_g$, or others, such as $b_{2u}$ to $a_{1g}^*$. At $d^8$ there is very efficient blocking since the $b_{1g}^*$ orbital is much higher in energy than $a_{1g}^*$. Hence $NiX_4^{2-}$ complexes will be square planar, if they are low-spin. However high-spin $d^5$ to $d^{10}$ complexes will be unstable in $D_{4h}$, since there will be a $(b_{1g}^*) \rightarrow (a_{2u})$ excitation available to them.

No other predictions can be made with much confidence. However if we consider the other extreme possibility of a tetrahedral structure, further insight can be gained. The order of orbitals expected from ligand field theory is

$$\ldots \underline{(t_2)(t_1)} \;\vdots\; (e)(t_2) \;\vdots\; \underline{(a_1)(t_2)}$$

with the underlined orbitals close in energy. There are ab initio calculations to support this sequence for the $d^0$ complexes, $MnO_4^-$ and $CrO_4^{2-}$,[51a] for $d^5$ $FeCl_4^-$ and for the $d^6$ $FeCl_4^{2-}$,[51b] and the $d^{10}$ complex, $Ni(CO)_4$.[51b] It must be admitted that the latter is not a simple case because the CO ligand is a strong $\pi$ acceptor.

The $MnO_4^-$ and $CrO_4^{2-}$ ions are also characterized by strong $\pi$-bonding effects in which the ligands (oxide ions) act as electron donors. Compared to $CF_4$, where the bonding is chiefly in the $a_1$ and $t_2$ orbitals of $\sigma$ character, the bonding in $MnO_4^-$ is in the $e$ and $t_2$ orbitals.[51a] The $e$ orbitals are strictly $\pi$-bonding, and the $t_2$ orbitals are mixed $\sigma$ and $\pi$. The chief AOs used are the $3d$ orbitals of Mn and the $2p$ orbitals of O. The crystal-field orbitals of $e$ and $t_2$ symmetry are the antibonding partners of the bonding orbitals.

Since $(T_2 \times E) = (T_1 \times E) = (T_1 + T_2)$, it appears that there are no low-lying excited states that can favor the $E$ reaction coordinate needed to convert $T_d$ into $D_{2d}$. This is true for all systems from $d^0$ to $d^{10}$. However FOJT distortions are expected for many of these complexes. Only $d^0$, $d^2$, high-spin $d^5$, and $d^7$, and $d^{10}$ complexes have $A_1$ or $A_2$ ground states and are stable. All others are $E$, $T_1$, or $T_2$ states (p. 78). These all allow a distortion in the direction leading toward a planar structure. Since a $D_{2d}$ structure is enough to remove the degeneracy in most cases, it is possible that the distortion stops there.

Figure 14 shows the splittings expected for the $d$ orbitals in a field of $D_{2d}$ symmetry. The number of levels and their symmetries are rigorously fixed. However the direction and magnitude of the splittings depend on the detailed interactions and cannot be predicted. Those shown are expected for a flattened tetrahedron using crystal-field type arguments. The $d$ levels are also shown for $D_{4h}$ symmetry, using the same naive approach.

If we now combine the results from a consideration of both $D_{4h}$ and $T_d$ structures, we form the structural predictions given in Table 9 for metal

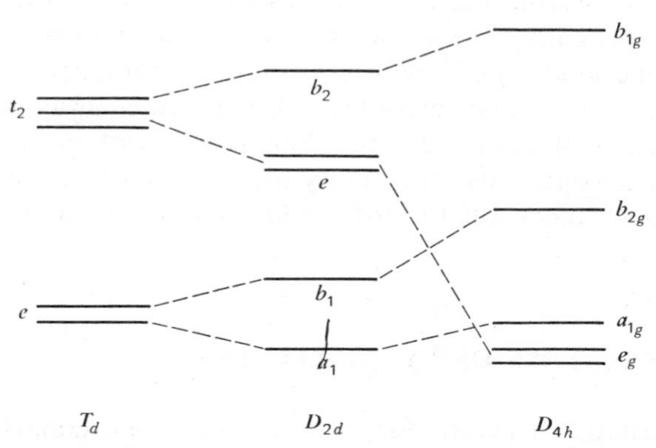


**Figure 14.** Splittings of the set of $d$ orbitals in the central atom in $ML_4$ when structure is tetrahedral, flattened, or planar; Crystal-field model used for splittings.

**Table 9 Stable Structures for $ML_4$ Molecules or Complexes**

| | High Spin | Low Spin | | High Spin | Low Spin |
|---|---|---|---|---|---|
| $d^0$ | $T_d$ | $T_d$ | $d^6$ | $D_{2d}$ | $D_{4h}$ |
| $d^1$ | $D_{2d}$ | $D_{2d}$ | $d^7$ | $T_d$ | $D_{4h}$ |
| $d^2$ | $T_d$ | $D_{2d}$ | $d^8$ | $D_{2d}$ | $D_{4h}$ |
| $d^3$ | $D_{2d}$ | [a] | $d^9$ | $D_{2d}$ | $D_{2d}$ |
| $d^4$ | $D_{2d}$ | $D_{2d}$ | $d^{10}$ | $T_d$ | $T_d$ |
| $d^5$ | $T_d$ | [a] | | | |

[a] Less than $D_{2d}$ symmetry.

complexes, $ML_4$, with various numbers of $d$ electrons.[52] The $d^0$ and $d^{10}$ systems are considered stable as tetrahedral complexes because the four L ligands each supply two electrons to be used in $\sigma$ bonding. An $8\sigma$-electron molecule would be stable in $T_d$, just as methane is. Low-spin $d^6$ and $d^7$ complexes are considered as stable in $D_{4h}$ by the assumption that the filled ligand orbital of $b_{2u}$ symmetry is far enough removed in energy from the empty $a_{1g}^*$ orbital of the $d$ shell.

The predictions are in excellent agreement with known structures, with one or two exceptions. $VCl_4$, a $d^1$ system, is found to be regular tetrahedral, even though the ground state is a $^2E$, and the Jahn–Teller effect predicts a distortion to $D_{2d}$.[53] $NiCl_4^{2-}$, a high-spin $d^8$ complex is sometimes found to be $D_{2d}$, as predicted,[54] and sometimes as $T_d$. In the latter case it appears that spin–orbit coupling removes the degeneracy of the $T_1$ ground state.[55]

It is often thought that four-coordinated cupric complexes are square planar, instead of having the $D_{2d}$ structures predicted in Table 9. This is not true, no simple $CuX_4^{2-}$ is known which is not of $D_{2d}$ structure, either in solution or in the solid state.[56] What is often found for $Cu^{2+}$, of course, is a structure having four ligands close by and two further away, but these are not $XY_4$ molecules. It can be seen that interactions with groups above and below the plane would raise the empty $a_{2u}$ level markedly in the planar form. This would block off the $(b_{1g}^*) \rightarrow (a_{2u})$ excitation and stabilize the planar structure.

## STRUCTURES OF $XY_5$ MOLECULES

The major alternative structures for $XY_5$ molecules are trigonal bipyramidal, $D_{3h}$, and square pyramidal, $C_{4v}$. Both structures are found experimentally with similar frequency, although $D_{3h}$ is somewhat more common.[57] There is much evidence which shows that many $XY_5$ molecules interconvert rapidly between $D_{3h}$ and $C_{4v}$ structures.[58] Also there is evidence that some species exist as a mixture of both forms in similar proportions.[59] These are the phenomena that must be accounted for by structural theory.

The only known $XH_5$ molecule is protonated methane, $CH_5^+$. Ab initio MO calculations show the energy level ordering in $D_{3h}$ is

$$(a_1')^2(a_2'')^2(e')^4(2a_1')^0(2e')^0$$

with $a_2''$ and $e'$ very close in energy.[60] A similar calculation on the hypothetical $PH_5$ molecule gives the ordering[61]

$$(a_1')^2\underline{(e')^4(a_2'')^2}(2a_1')^2(2e')^0(2a_2'')^0$$

The forms of these MOs are shown in Fig. 15, which also clarifies the AOs used as a basis set. Inclusion of the $d$ orbitals of phosphorous has little effect.

For both $CH_5^+$ and $PH_5$ the transition density of the lowest excited state is of $E'$ symmetry. As Fig. 16 shows, this is the normal mode that takes the $D_{3h}$ structure initially in to a $C_{2v}$ form. If continued, it will convert into the $C_{4v}$ structure. This is the well-known Berry pseudorotation mechanism for interconversion of $XY_5$ structures.[62] For $PH_5$, at least, the gap between $2a_1'$ and $2e'$ is fairly large. This indicates that trigonal bipyramidal $PH_5$ exists at an energy minimum, but that isomerization to a square pyramid form would not be difficult.

The $D_{3h}$ structure for $PH_5$ is calculated to be 3.9 kcal more stable than $C_{4v}$.[61] The MOs for the square pyramidal form are in order of increasing energy

$$(a_1)^2(e)^4(2a_1)^2(b_1)^2(3a_1)^0(2e)^0(4a_1)^0$$

These are also shown in Fig. 15. The transition density is of $B_1$ symmetry, which is the normal mode needed to take the $C_{4v}$ structure back into $D_{3h}$ (Fig. 16). Figure 17 shows the correlation diagram for the overall interconversion.[63] The intermediate configuration has $C_{2v}$ symmetry, whose elements are preserved. Since there is so little disturbance of the MOs on carrying out the $D_{3h} \rightleftharpoons C_{4v}$ isomerization, it is not surprising that the change is facile.

The structure of $CH_5^+$ is not known experimentally, but the calculations agree on a structure of $C_s$ symmetry (see p. 301). This is a three–two arrangement of hydrogen atoms in which the energy is essentially invariant to rotation of the two parts with respect to each other. This is of considerable interest because such an intermediate and a corresponding "turnstile" process has been suggested for the mechanism of the $D_{3h} \rightleftharpoons C_{4v}$ interconversion.[64]

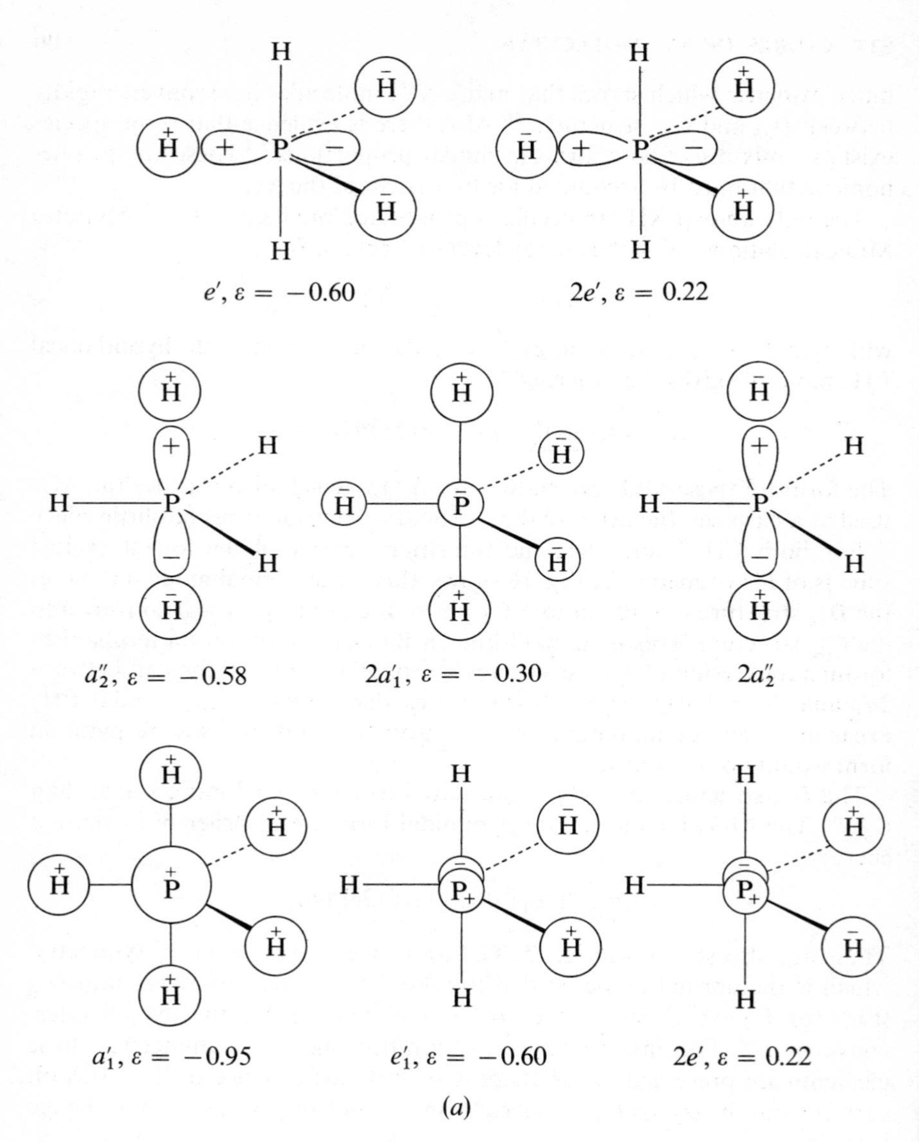


**Figure 15.** (*a*) Molecular orbitals of $PH_5$ in trigonal bipyramid form, $D_{3h}$.

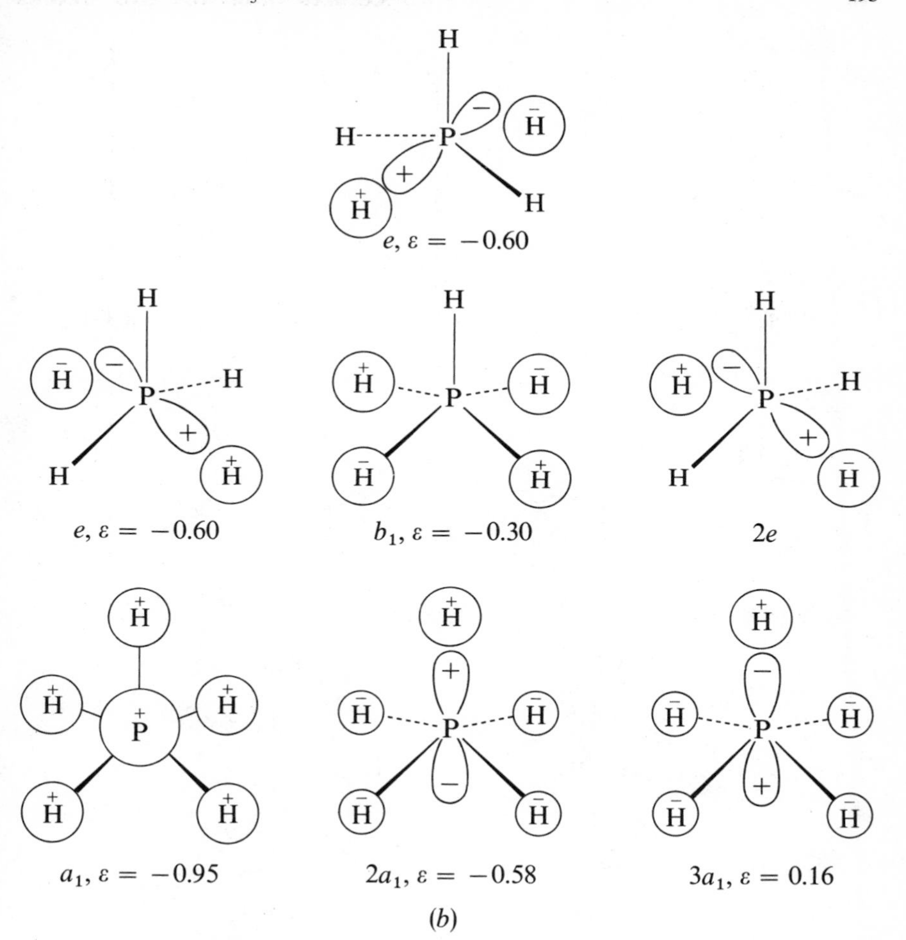


**Figure 15.** (*b*) MOs of $PH_5$ in square pyramid form, $C_{4v}$ (energies from reference 61).

The MO sequence for $CH_5^+$ in $C_{4v}$ is

$$(a_1)^2(e)^4(2a_1)^2(b_1)^0(3a_1)^0$$

which gives the correct transition density of $B_1$ symmetry to convert into $C_{2v}$. Instead of continuing on to $D_{3h}$, a further distortion to $C_s$ must be favorable. This, of course, would occur whether the molecule starts as $D_{3h}$ or as $C_{4v}$. The way in which this would occur is shown in Fig. 17. In $C_{2v}$, the occupied $e$ orbitals split into $b_2$ and $b_1$ (or the $e'$ orbitals split into $a_1$ and $b_2$). This allows a transition density of either $B_1$ or $B_2$ symmetry, by excitation into the

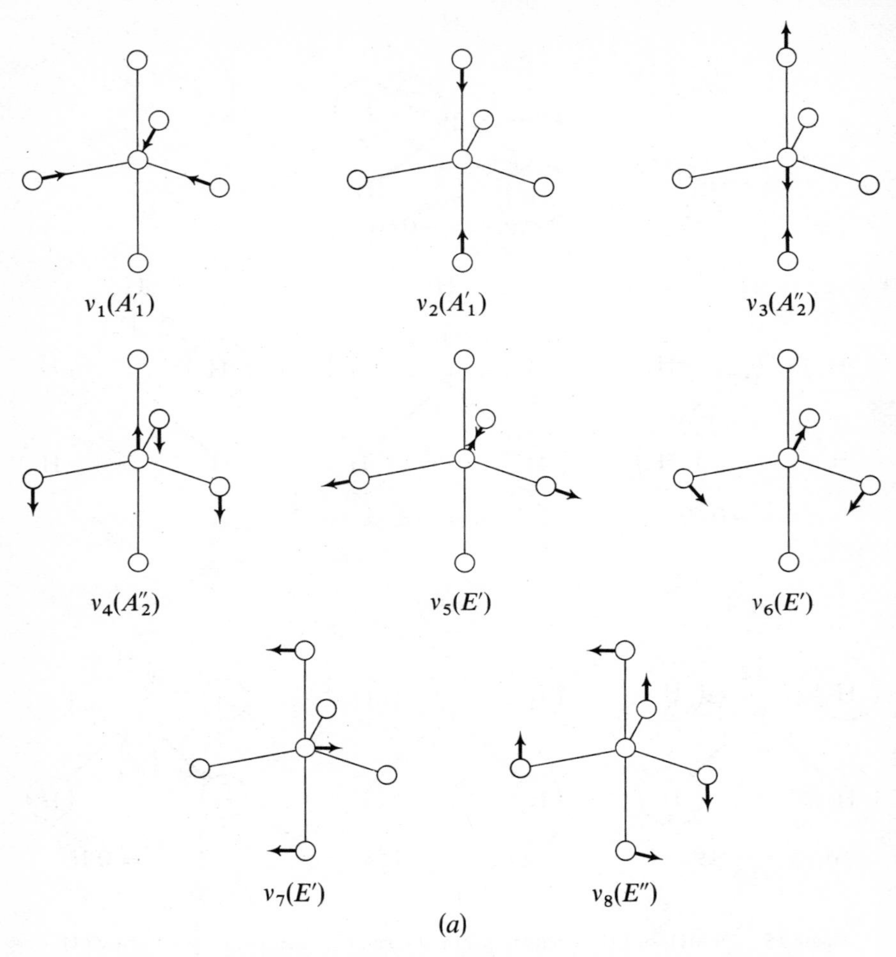


**Figure 16.** Normal modes for $XY_5$ molecule in: (*a*) trigonal bipyramid form.

lowest empty orbital, which is of $a_1$ symmetry. The $B_1$ or $B_2$ motions convert $C_{2v}$ into one or the other of the two $C_s$ forms shown in Fig. 16.

These excitations are very easy for $CH_5^+$ because the excited state still uses an orbital which is a net bonding one. In $PH_5$ a similar splitting of the $2e'$ or $2e$ orbital occurs in $C_{2v}$, but now excitation must be from a filled bonding MO to an empty antibonding MO. Hence the $C_s$ structure can be stable for $CH_5^+$, but would be, at best, a TS for $PH_5$. Various MO calculations show the Berry pseudorotation to be favored over the turnstile rotation for $PH_5$ and $PF_5$.[63,64]

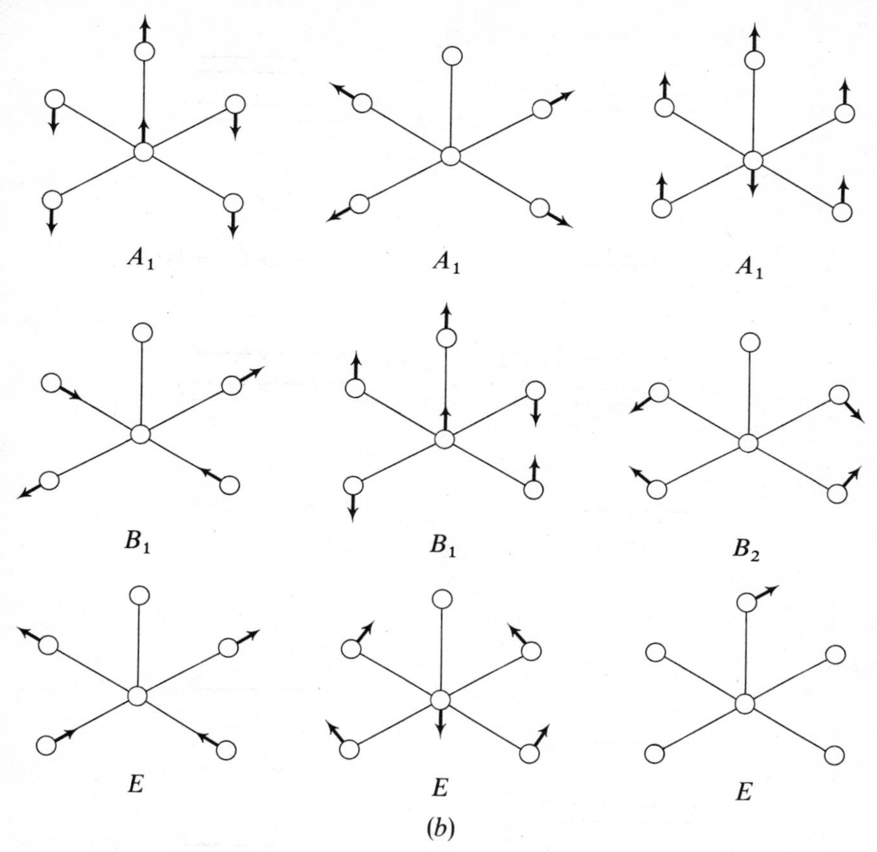


(*b*)

**Figure 16.** Normal modes for $XY_5$ molecule in: (*b*) square pyramid form.

An ab initio calculation has been made for $PF_5$ in both $C_{4v}$ and $D_{3h}$ geometries.[65a] Approximate calculations have been made for $AsF_5$ and $BrF_5$.[65b] There are a large number of MOs because of the $p$ orbitals on F. If the lone pairs on F are ignored, the $\sigma$ orbitals for $PF_5$ or $AsF_5$ will give the same results as for $PH_5$. Inclusion of the lone pairs is equivalent to asking whether changes in $\pi$ bonding can be significant in determining the relative stabilities of trigonal bipyramid and square pyramid forms.

The structures of $BrF_5$, and also those of $IF_5$ and $SbF_5^{2-}$, are found to be square pyramid. This can be justified by looking at the $\sigma$ orbitals only. For $IF_5$, for example, the MO sequence in $C_{4v}$ would be

$$(a_1)^2(e)^4(2a_1)^2(b_1)^2(3a_1)^2(2e)^0(4a_1)^0$$

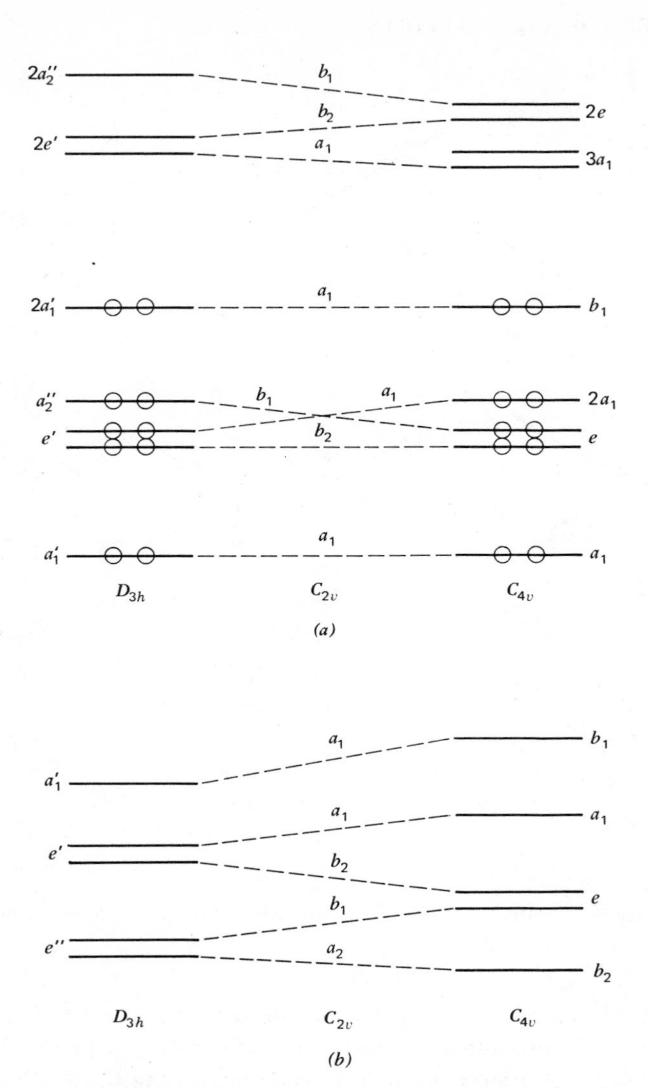


**Figure 17.** (*a*) Correlation diagram for MOs of $PH_5$ during conversion from $D_{3h}$ to $C_{4v}$ structures. The $C_{2v}$ point group is intermediate in this rearrangement; (*b*) correlation diagram for *d* orbitals of $ML_5$ during the $D_{3h} \rightleftarrows C_{4v}$ interconversion.

which is stable to the $B_1$ distortion leading to $D_{3h}$. In the latter structure the configuration would be

$$(a_1')^2(e')^4(a_2'')^2(2a_1')^2(2e')^2(2a_2'')^0$$

Because of the incomplete $e'$ shell there would be a FOJT or SOJT effect leading to an $E'$ reaction mode.

$PF_5$, $AsF_5$, and $PCl_5$ would exist at energy minima at least, in $D_{3h}$, but ready interconversion would be possible between square pyramidal and trigonal bipyramidal forms. Since $PCl_5$ does not absorb light until 2500 A, or 5 eV, the excited states must lie fairly high and isomerization of the more stable $D_{3h}$ structure would not be spontaneous. The experimental barrier for $PCl_5$ is estimated to be only 3.7 kcal.[58] This low barrier makes the interconversion rapid, even on the n.m.r. time scale.

Barriers may be estimated from the force constants for the lowest frequency fundamental. This is the equatorial-in-plane bending mode of $E'$ species. In solution and in the gas phase, the barriers are all very low. In the solid state, for instance, for $InCl_5^{2-}$, they can be as large as 9.2 kcal.[58] $InCl_5^{2-}$ and $TlCl_5^{2-}$ are reported as $C_{4v}$ structures, whereas $D_{3h}$ is found for $VF_5$, $SbCl_5$, $NbCl_5$, $PCl_5$, $SnCl_5^-$, $CuCl_5^{3-}$, and $CdCl_5^{3-}$.

### $ML_5$ Complexes

Five-coordinated complexes are very common for the transition metals.[57] While many semiempirical MO calculations have been made, there are no ab initio calculations. The simple crystal-field approach gives the $d$ orbital sequence in $D_{3h}$

Filled ligand orbitals ¦ $(e'')(e')(a_1')$ ¦ empty orbitals

Similarly in $C_{4v}$ the ordering would be

Filled ligand orbitals ¦ $(e)(b_2)(a_1)(b_1)$ ¦ empty orbitals

Since $\pi$-bonding interactions are likely to be important in determining structure,[66] it becomes rather complicated to make predictions for various numbers of $d$ electrons. It is more helpful to look at the correlation diagram for the $d$ orbitals only for the $D_{3h} \leftrightarrow C_{4v}$ conversion. These are shown in Fig. 17$b$, as calculated by the extended Hückel method.[66a] There are no orbital crossings, and the Berry pseudorotation can occur for any number of $d$ electrons, if the spin stays constant.

Ignoring the effect of incomplete shells, Fig. 17$b$ shows that low-spin $d^3$–$d^4$ will prefer a $D_{3h}$ structure and $d^5$–$d^6$ a $C_{4v}$ structure. For a $d^8$ low-spin complex the MO configurations would be

$$(e'')^4(e')^4(a_1')^0 \qquad D_{3h}$$

$$(b_2)^2(e)^4(a_1)^2(b_1)^0 \qquad C_{4v}$$

which are both ideal for SOJT instability via the $(e') \rightarrow (a_1')$ and $(a_1) \rightarrow (b_1)$ excitations. In fact it is just for low-spin $d^8$ complexes such as $Fe(CO)_5$ and $Os(PF_3)_5$ that a dynamic equilibrium between the two structures has been demonstrated.[67] In many other $d^n$ complexes the evidence is that five-coordinate complexes have been found both as $D_{3h}$ and $C_{4v}$ forms. In fact, the structures found are often between $D_{3h}$ and $C_{4v}$.[57]

It may be mentioned that finding structures of any kind which are the idealized geometric forms is rather a rare event, at least in the solid state. In the solid it is, of course, possible to measure structures accurately by x-ray diffraction methods. Small deviations from idealized symmetries may be due to crystal-packing effects. A systematic procedure has been given for referencing the structures of real molecules to the idealized shapes.[67b]

Low-spin $d^7$ complexes should be more stable than $d^8$ in the square pyramidal form. This follows from the more difficult excitation from a half-filled $d$ orbital than from a filled one. In fact $Co(CN)_5{}^{3-}$ and $Co(CNCH_3)_5{}^{2+}$ have $C_{4v}$ structures.[68a] In the $D_{3h}$ structure, low-spin $d^7$ is unstable because of the ground state would be $E'$, from the $(e')^3$ configuration.

There is strong evidence that low spin $d^6$ complexes are square pyramids, as suggested by Fig. 17*b*. For example, $Cr(CO)_5$, and also $Mo(CO)_5$ and $W(CO)_5$, have $C_{4v}$ symmetry.[68b] These unstable species are prepared photochemically from the hexacarbonyls, and preserved in a low-temperature argon matrix. In the $D_{3h}$ form, a low-spin $d^6$ system would be $(e'')^4(e')^2$ and have a FOJT instability.

Low-spin $d^5$ complexes are also predicted to be FOJT unstable in both structures by Fig. 17*b*. They would be stable in $C_{4v}$, if the ordering were $(e)^4(b_2)$ instead of $(b_2)^2(e)^3$. Experimentally such complexes are found as square pyramids, though few simple $ML_5$ examples are known, where all ligands are the same. $V(CO)_5$ is reported to be $C_{4v}$.[68c] Fortunately some data from another source can be added for both low-spin $d^5$ and $d^6$.

In the substitution reactions of octahedral cobalt(III), rhodium(III), and iridium(III) complexes, it is often postulated that a five-coordinate intermediate is formed.[69] The intermediates $Co(NH_3)_5{}^{3+}$, $Co(NH_3)_4(H_2O)^{3+}$, and $Co(en)_2NH_3{}^{+}$ do seem to maintain a $C_{4v}$ structure.[70] However other intermediates such as $Co(NH_3)_4NH_2{}^{2+}$ and $Co(en)_2Cl^{2+}$ rearrange, presumably by way of a trigonal–bipyramid form. These latter species are outside the scope of our symmetry arguments, because some of the ligands are chemically too different, $NH_2{}^{-}$ compared to $NH_3$. Technically, of course, symmetry is destroyed when one en, $NH_2CH_2CH_2NH_2$, replaces two $NH_3$ molecules. Chemically there is little change. In addition to the $d^6$ cobalt(III) examples, Rh(III), Ir(III), both $d^6$, and Ru(III), which is $d^5$, also appear to be $C_{4v}$ when forming five-coordinated intermediates.[71]

Substitution reactions of Cr(III) complexes are also very stereoretentive. From Fig. 17*b*, a high-spin $d^3$ complex would be predicted to be a square pyramid, when five-coordinated. However the evidence is that species such as $Cr(NH_3)_5{}^{3+}$ are *not* formed in the reactions of octahedral chromium(III) complexes. High-spin $d^4$ complexes, and $d^9$, should behave like low-spin $d^8$ complexes, according to Fig. 17*b*. High spin $d^5$, as well as $d^{10}$, should behave like $PF_5$. That is, a $D_{3h}$ structure should be preferred, but ready interconversion to $C_{4v}$ should be possible.

## $XY_6$ MOLECULES

In six-coordinated systems the problem is to explain the almost universal occurrence of octahedral structures. Figure 18 shows the normal vibrations of an $XY_6$ molecule, indicating the kinds of distortion that might occur. While there are no $XH_6$ molecules, there are complexes such as $Ca(NH_3)_6{}^{2+}$, containing only $\sigma$-donor ligands and metal ions with no $d$ electrons. In Table 10 are listed the valence-shell AOs that would be used for bonding in

**Table 10 Symmetries of Atomic Orbitals of $XY_6$ Molecules in $O_h$ Point Group[a]**

| X | Y | $\Gamma_{O_h}$ |
|---|---|---|
| $s$ | $\sigma_1 + \sigma_2 + \sigma_3 + \sigma_4 + \sigma_5 + \sigma_6$ | $a_{1g}$ |
| $p_z$ | $\sigma_1 - \sigma_4$ | |
| $p_y$ | $\sigma_3 - \sigma_6$ | $t_{1u}$ |
| $p_x$ | $\sigma_2 - \sigma_5$ | |
| $d_{x^2-y^2}$ | $\sigma_1 - \sigma_2 + \sigma_4 - \sigma_5$ | $e_g$ |
| $d_{z^2}$ | $2\sigma_3 + 2\sigma_6 - \sigma_1 - \sigma_2 - \sigma_4 - \sigma_5$ | |
| $d_{xy}$ | | |
| $d_{xz}$ | Ligand $\pi$ combinations | $t_{2g}$ |
| $d_{yz}$ | | |
| — | Ligand $\pi$ combinations | $t_{1g}$ |
| — | Ligand $\pi$ combinations | $t_{2u}$ |
| $p_x, p_y, p_z$ | Ligand $\pi$ combinations | $t_{1u}$ |

[a] Ligand $\sigma$ orbitals are combinations of $s$ and $p_z$, The $\pi$ orbitals are $p_x$ and $p_y$.

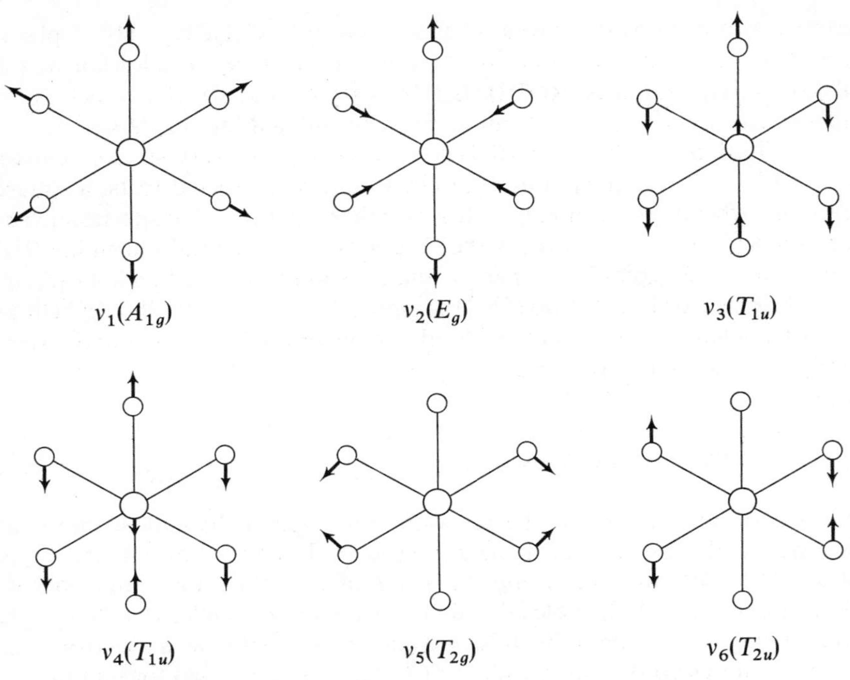


**Figure 18.** Normal modes of vibration of octahedral $XY_6$ molecules.

such a case. The $\sigma$ bonds would be of $a_{1g}$, $e_g$ and $t_{1u}$ species. Figure 19 shows the MOs of an $XY_6$ molecule.

The photoelectron spectrum of $W(CH_3)_6$ shows three low lying IPs at 10.0, 10.35, and 10.8 eV.[72] These have been assigned as corresponding to the $e_g$, $t_{1u}$, and $a_{1g}$ orbitals in order. While the assignment is dubious, being based on matching the PES intensity with multiplicity (2:3:1), the spectrum shows the expected three peaks. Ionization from the C—H bonding MOs of the methyl groups occur at higher energies. The energies obviously are all comparable for the $\sigma$ MOs binding to the central atom. While W(VI) has no $d$ electrons, it does have low-energy empty $d$ orbitals, which are needed for $e_g$ bonding. The empty $t_{2g}$ orbitals can be used for $\pi$ bonding with suitable ligands.

Consideration of hexahalides, such as $SF_6$, brings in a large number of $\pi$-type MOs, some of which are bonding and some nonbonding. An ab initio calculation for $SF_6$ gives the valence shell, excluding the $2s$ electrons of fluorine which are much lower in energy,[73]

$$(t_{1u})^6(a_{1g})^2(t_{2g})^6(e_g)^4(2t_{1u})^6(t_{2u})^6(t_{1g})^6(2a_{1g})^0(3t_{1u})^0$$

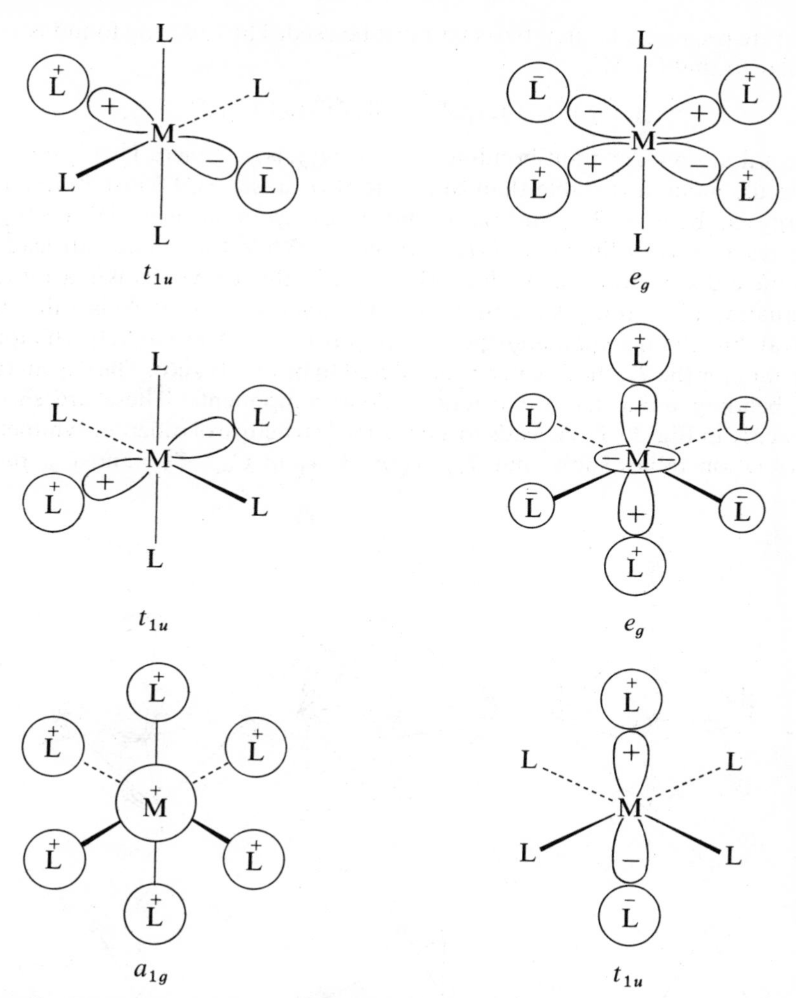


**Figure 19.** Molecular orbitals of an $ML_6$ octahedral molecule in which only $\sigma$ bonding to the ligands occurs; the antibonding orbitals $a_{1g}^*$, $t_{1u}^*$, and $e_g^*$ are formed by changing one set of signs on the corresponding bonding orbitals.

The $t_{2g}$ orbitals are $\pi$ bonding, the $t_{1u}$ orbitals contribute to both $\sigma$ and $\pi$ bonding. The $t_{2u}$ and $t_{1g}$ orbitals are nonbonding. $SF_6$ is not only colorless down to 2200 A, but also extremely inert. There is a large gap between the HOMO and LUMO.

A remarkable ab initio calculation has been made for $XeF_6$, which contains 108 electrons.[46] Of course, for such a molecule the results are not very accurate because a limited basis set must be used. The ordering found is very similar to that for $SF_6$

$$(t_{2g})^6(t_{2u})^6(2t_{1u})^6(t_{1g})^6(e_g)^4(2a_{1g})^2(3t_{1u})^0$$

except that now the $\sigma$-antibonding orbital, $2a_{1g}$, is occupied. This makes the molecule much less stable than $SF_6$, but also causes SOJT instability. The energy gap between $2a_{1g}$ and $3t_{1u}$ is only 3.7 eV, from the near-UV spectrum. The reaction coordinate is of $T_{1u}$ symmetry. While this mode can lead to complete dissociation in the limit (Fig. 18), in this case it causes a rapidly fluctuating distortion of the molecule. The average symmetry is still octahedral, but the instantaneous point group is $C_{3v}$.[45] Alternatively, the force constant for the $T_{1u}$ mode can be considered to be nearly zero. The degenerate $T_{1u}$ bending mode has three independent components. These are shown explicitly in Fig. 20. Each leads to a deformed structure of different symmetry. Correlation tables show that $T_{1u}$ becomes $A_1$ in $C_{4v}$, $C_{3v}$, and $C_{2v}$ point

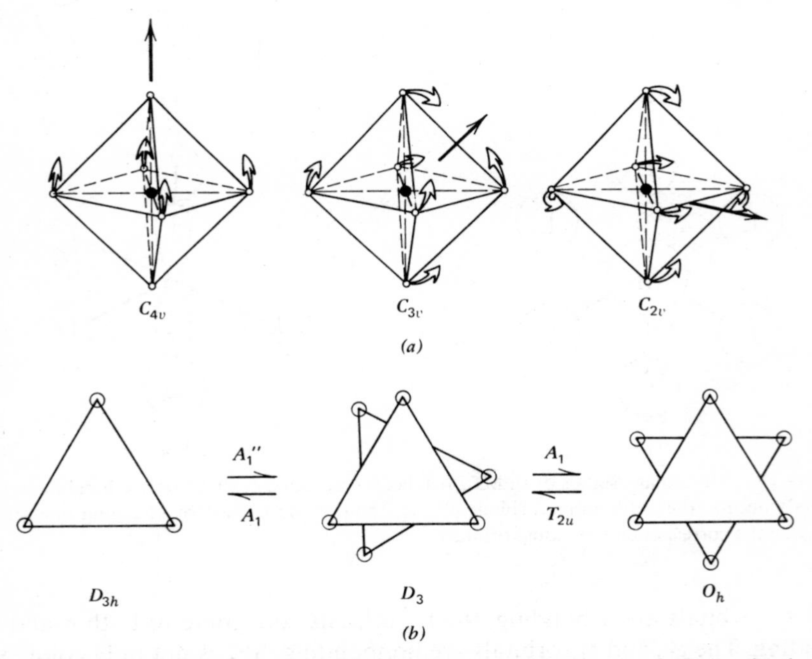


**Figure 20.** (*a*) The three components of the $T_{1u}$ bending mode of an octahedral molecule; the point groups generated are shown, and the dark arrow is the principal axis in the new point group [after S. Y. Wang and L. L. Lohr, Jr., *J. Chem. Phys.*, **60**, 3908 (1974)]; (*b*) the twisting modes $A_1''$ and $T_{2u}$ that interconvert trigonal prismatic and octahedral structures by way of a $D_3$ intermediate.

groups, which then correspond to the structures formed. One cannot say a priori which deformation is favored, but other effects combine to stabilize the $C_{3v}$ form.

Complex ions such as $SiF_6^{2-}$ and $PF_6^-$ are isoelectronic with $SF_6$. They would be stable in regular octahedral forms. However $SbCl_6^{3-}$ and $TeCl_6^{3-}$ are isoelectronic with $XeF_6$, and should show fluctuating structures or permanent distortions from $O_h$ symmetry. Instead in the solid state all of the above ions are found to be regular octahedral. This must be attributed to crystal packing forces which freeze the latter two ions into the average structure. Evidence for the SOJT effect comes from the infrared and Raman spectra.[74] The $T_{1u}$ modes are very broad and of unusually low force constants, depending markedly on the counter ion in the solid state.

### Transition Metals

We turn next to transition metal complexes. The MO scheme for $\sigma$ bonded systems such as $Co(NH_3)_6^{3+}$ is expected to be

$$(a_{1g})^2(e_g)^4(t_{1u})^6 \; \vdots \; (t_{2g})^6(2e_g)^0 \; \vdots \; (2a_{1g})^0(2t_{1u})^0$$

with the $d$ orbitals enclosed by dashed lines. The lowest-lying transitions are within the $d$ manifold and give rise to $(T_{1g} + T_{2g})$ symmetries for the transition densities, $\rho_{0k}$. There is no $T_{1g}$ mode (Fig. 19) and the $T_{2g}$ mode does not lead to an alternative structure, but to a distortion. There is some evidence that the Raman $T_{2g}$ mode has a low force constant for some transition-metal complexes.[75]

If $\pi$ bonding ligands, such as the halide ions are present, orbitals of $t_{1g}$, $t_{1u}$, and $t_{2u}$ symmetry are expected just before the $d$ manifold. An ab initio calculation of $NiF_6^{4-}$ verifies this expectation, giving a sequence[76]

$$\ldots\underline{(t_{2u})^6(t_{1u})^6(t_{1g})^6} \; \vdots \; (t_{2g})^6(e_g)^2 \; \vdots$$

The calculation extends over the ground state, $^3A_{2g}$, and the two lowest excited states, $^3T_{1g}$ and $^3T_{2g}$. This amounts to a calculation of the parameter $10D_q$ of crystal-field theory. The value calculated was 6089 $cm^{-1}$, compared to 7250 $cm^{-1}$, experimental. Even though the complex is highly ionic, the charge on nickel being calculated as +1.82, the small amount of covalency accounts for most of the value of 10 $D_q$.

Excitation from the $\pi$ orbitals of $t_{1g}$, $t_{1u}$, and $t_{2u}$ symmetry into the $d$ shell, gives rise to charge-transfer bands. For $NiF_6^{4-}$ and many other complexes, these bands occur at high energies.[77] If the oxidation state of the central atom becomes very high, it is expected that the separation between the

$t_{1u}$, $t_{2u}$, and $t_{2g}$ orbitals will become small. Since $T_{1u}$ (or $T_{2u}$) $\times$ $T_{2g}$ gives rise to $T_{1u}$ and $T_{2u}$ symmetry, structural changes should also result. Orgel has given some interesting examples of this effect.[78] The examples are from the solid state and must be viewed with some reservations, since solids are subject to other forces that can distort structure. Metal ions of high oxidation state such as $Hf^{4+}$, $Zr^{4+}$, $Nb^{5+}$, $Mo^{6+}$, and $V^{5+}$ often form oxide lattices with distorted octahedral structures in which the metal ion is off center. It can be seen from Fig. 19 that the $T_{1u}$ vibration does draw the central atom away from the center of the octahedron.

The oxides, sulfides, and selenides of metal ions with an inert pair of electrons such as $Tl^{+}$, $Pb^{2+}$, and $Bi^{2+}$, also form distorted octahedral structures.[78] Since the inert pair is in an orbital of $a_{1g}$ symmetry, the distortion is explained by an excitation from this orbital into a higher $t_{1u}$ orbital. This is identical with the explanation given earlier for $XeF_6$.

A third example discussed by Orgel[79] concerns sixfold coordination of $d^{10}$ ions such as $Hg^{2+}$, $Au^{+}$, $Ag^{+}$, and $Cu^{+}$, which give badly distorted octahedra in the solid state, whereas other $d^{10}$ ions such as $Zn^{2+}$, $Cd^{2+}$, and $Tl^{3+}$ do not. Orgel has shown that a small *s–d* energy gap in the metal ion leads to the distortion, whereas a large gap does not. Such an *s–d* mixing will be of $(A_{1g} \times E_g) = E_g$ symmetry and the corresponding vibration (Fig. 17) will be excited. This causes a distortion in which two *trans*-groups are closer to the metal than the other four, or, just as probable, two groups are further than the other four. The former distortion is characteristic of $Hg^{2+}$, $Au^{+}$, $Ag^{+}$, and $Cu^{+}$. In solution, the effect occurs to such an extreme that usually only two ligands are held tightly by these metal ions.

The best known examples of tetragonal distortion are the high-spin $d^4$ and $d^9$, and low-spin $d^7$ complexes. These have $E_g$ ground states in $O_h$ and are subject to a FOJT distortion. Even here rather modest forces, such as crystal packing, are enough to suppress the distortion in some cases. Also the observed effects may well be due to the *s–d* mixing described above, or at least are augmented by such mixing.

### Trigonal Prismatic Structures

The possibility of exciting the $T_{2u}$ vibration is particularly interesting since, as shown in Fig. 18, this vibration twists an octahedron into a trigonal prism structure. Excitation from $t_{1u}$ or $t_{2u}$ into the $t_{2g}$ orbital should favor the $T_{2u}$, as well as the $T_{1u}$ vibrations; the latter distortion is much more common than the former. The only molecular examples of the trigonal–prismatic structures are the well-known dithiolate complexes, $MS_6C_6R_6$.[80] Considering the dithiolate ligand, $R_2C_2S_2{}^{2-}$, as formally divalent, the metal ions which give trigonal–prismatic structures are Mo(VI), W(VI), Re(VI), Cr(VI), and V(VI). These are all $d^0$ or $d^1$ metal ions. Metal ions of lower oxidation state, or more $d$ electrons, are octahedral.

In these examples conditions have been optimized to make the $(t_{1u})$, $(t_{2u}) \rightarrow (t_{2g})$ transitions of low energy and hence to favor the $T_{2u}$ vibration. The conditions are as follows: (1) $\pi$-bonding ligands with filled $\pi$ orbitals, (2) at least one empty $t_{2g}$ orbital on the central metal ion, (3) a high oxidation state for the central metal ion (to lower the $t_{2g}$ energy), (4) a donor atom of the ligand that is easily oxidized (to raise the $t_{1u}$ and $t_{2u}$ levels), and (5) strong ligand–ligand interactions due to overlap. Condition (5) originates from the work of Schmidtke,[81] who has considered the effect of ligand–ligand interactions in detail. A significant conclusion is that the $t_{1u}$ orbital is raised with respect to the $t_{2g}$ orbital. There is evidence from bond distances of strong sulfur–sulfur interactions in the trigonal–prismatic structures.[80]

To complete the argument it is necessary to show when the trigonal–prismatic structure is stable to rearrangement into an octahedral structure. Only approximate MO or crystal field calculations are available.[80,82] These give the energy-level order

$$(e')^4(a_2')^2 \;\vdots\; (2e')(a_1')(e'') \;\vdots\; (a_1'')^0(2e'')^0$$

The required twisting mode to convert a $D_{3h}$ structure into $D_3$ and eventually into $O_h$ is of $A_1''$ symmetry (see Fig. 20). Complexes with $d^0$ or $d^{10}$ configurations are predicted to be stable in trigonal–prismatic form by the above sequence. However $d^2$ to $d^9$ systems are predicted unstable because $(E' \times E'') = (A_1'' + A_2'' + E'')$. A $d^1$ complex is marginally stable. Experimentally Re(VI) is trigonal prismatic and V(IV) is in between trigonal and octahedral.[83]

While trigonal structures are rare, there is much evidence that octahedral molecules can pass through such a structure, sometimes with little activation energy.[84] This is a mechanism for the interconversion of isomers of certain *tris* chelates, where the symmetry is not truly $O_h$, but rather $D_3$ at most. Just as expected from the MO analysis of octahedral complexes, the activation energy for the trigonal twist mechanism is much larger for a filled $(t_{2g})^6$ shell (assuming essential $O_h$ symmetry) than for a complex with a hole in the $t_{2g}$ shell.[85]

## $XY_7$ AND $XY_8$ MOLECULES

A coordination number of seven is relatively rare, but some examples are known, mainly fluorides and fluoride complexes. The molecule $IF_7$ has been much discussed since there is evidence that it is stereochemically nonrigid, somewhat like $XeF_6$.[86] The most recent conclusion is that it has a rigid $D_{5h}$ pentagonal–bipyramidal structure on the infrared and Raman time scale. On the much longer n.m.r. time scale, $IF_7$ and $ReF_7$ have nonrigid structures, since all seven fluorine atoms have become equivalent by an intramolecular process. Presumably this occurs by a pseudorotation process similar to that postulated for $PCl_5$, $AsCl_5$, and so on. An $E'_1$ vibration could convert the $D_{5h}$ structure into a capped octahedron of $C_{3v}$ symmetry, losing the identity of axial and equatorial positions.[87]

Only one MO calculation exists for $IF_7$, and it ignores $\pi$ bonding.[88] The MO sequence in $D_{5h}$ symmetry is

$$(a'_1)^2(e'_1)^4(a''_2)^2(2a'_1)^2(e'_2)^4\underline{(2a''_2)^0(2e'_1)^0}$$

Since $\underline{2a''_2}$ and $\underline{2e'_1}$ are close together in energy, there is a $\rho_{0k}$ of $(E'_2 \times E'_1) = (E'_1 + E'_2)$ symmetry. However the calculated energy gap is very large (10 eV). While the calculations cannot be taken too seriously, there is probably a fairly substantial gap between $e'_2$ and $2e'_1$. For example, the molecule is colorless. This suggests a moderately stable structure, at least. It is of interest that $ReF_7$, which is pale yellow, shows evidence of being nonrigid even on the infrared time scale.[88] At low temperatures, for example, $ReF_7$ is polar, whereas $IF_7$ is not.

A coordination number of eight is somewhat more common than seven, but still quite rare. The most symmetrical structure possible would be a simple cube of $O_h$ point group. Strangely enough, this structure is never observed. Instead either an Archimedean antiprism, $D_{4d}$, or a dodecahedral structure, $D_{2d}$, is found.[90] These structures are displayed in Fig. 21 which also shows the $E_u$ modes needed to produce the less symmetric shapes from the cubic one.

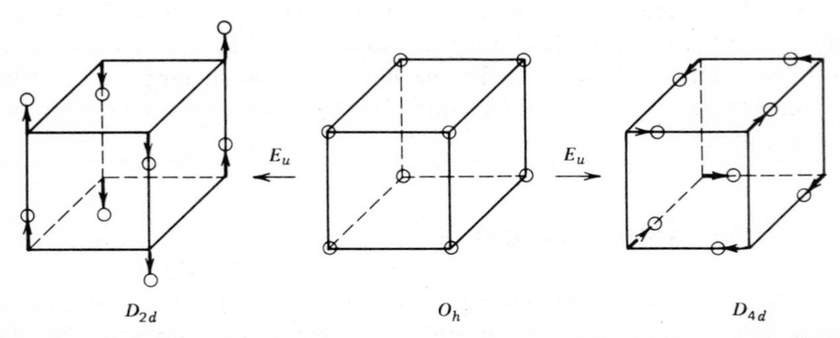


**Figure 21.** The $E_u$ modes that convert a regular cubic $XY_8$ molecule, $O_h$ into a dodecahedron, $D_{2d}$, or Archimedean antiprism, $D_{4d}$.

Note that the cubic molecules has two interpenetrating groups of four ligands arranged tetrahedrally. The molecular orbitals can easily be found by taking sums and differences of the LCAO used in the $T_d$ case. Consider a molecule $MH_8$, or any case where only $\sigma$ bonds are formed. Figure 22 shows the proper combinations of ligand $\sigma$ orbitals to use in the $O_h$ point group. They are of $a_{1g}$, $a_{2u}$, $t_{1u}$, and $t_{2g}$ symmetry. The central atom contributes the same basis sets as shown in Table 10, that is, $a_{1g}$, $t_{1u}$, $e_g$, and $t_{2g}$.

Without any detailed calculations it follows that the order of orbital energies generated is roughly

$$(a_{1g})^2\underline{(t_{1u})^6(t_{2g})^6}\,\underline{(a_{2u})^2 \,\vdots\, (e_g)}(2t_{2g}) \,\vdots\, (2a_{1g})(2t_{1u})$$

Both $a_{2u}$ and $e_g$ are nonbonding. As Fig. 22 shows, the $a_{2u}$ orbital is actually antibonding if ligand–ligand overlap is considered. If there are 16 valence electrons (no $d$ electrons), there should be a low energy excitation $(a_{2u}) \rightarrow (e_g)$, which is possible. This favors the $E_u$ vibrational mode. This mode is

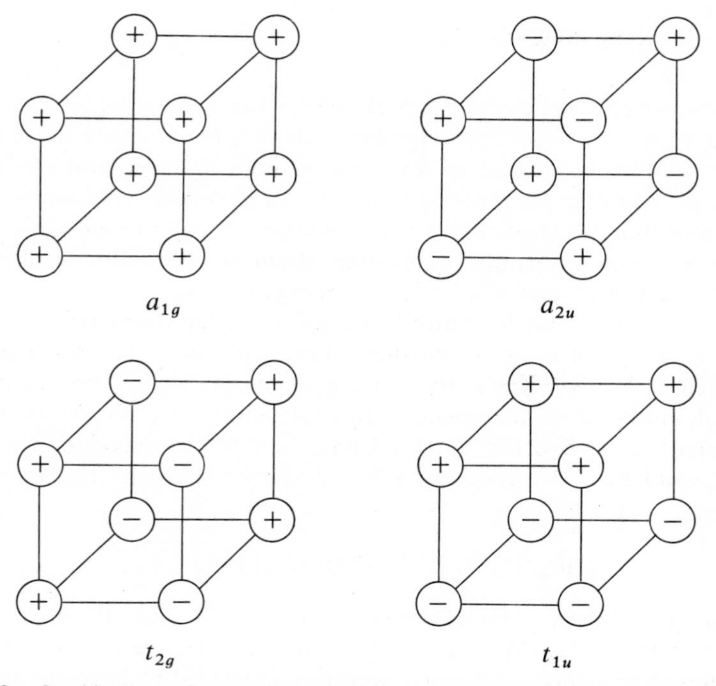


**Figure 22.** Combinations of $\sigma$ orbitals on the ligands in an $MH_8$ molecule with proper symmetries in $O_h$ point group.

obtained by taking the out-of-phase sum of the $E$ modes of a tetrahedral molecule (Fig. 11). This doubly degenerate mode has both a twisting component and a scissoring component. The former produces the $D_{4d}$ structure and the latter, the $D_{2d}$ form.[91]

In this way the instability of the $O_h$ point group is explained for $d^0$ metals and $\sigma$ donor ligands. A spin-paired $d^4$ system might well be stable in the $O_h$ form, from this analysis. Of course, $\pi$-bonding ligands increase the number of possible excited states of the correct symmetry to cause rearrangement. Experimentally $d^0$ complexes and molecules are almost always antiprisms and $d^2$ complexes are apparently dodecahedra.[93,94] Some explanation might be attempted for this by generating MO schemes for $D_{4d}$ and $D_{2d}$. However it is probably more pertinent to realize that all eight-coordinated molecules studied in solution have been found to be structurally unstable.[94] Potential energy calculations on models show very little energy difference between $D_{2d}$ and $D_{4d}$ structures, and no barrier between them.[95] It is not likely that symmetry arguments alone can tell anything about relative preferences in such cases.

## $X_2Y_2$ MOLECULES

We consider next more complex molecules of the $X_n Y_m$ type. Such molecules have a variety of structures experimentally. There are many more normal modes possible compared to $XY_n$, and a more stringent test of theory is possible.[96] Among the possible structures of molecules containing at least four atoms bonded sequentially, are those that are conformers of each other. These are interconvertible by rotation about single bonds. Such changes usually occur with very low activation energy (1–7 kcal).

For molecules with four atoms we will consider linear ($D_{\infty h}$), *cis* ($C_{2v}$), *trans* ($C_{2h}$), and skew ($C_2$) structures. The sequence of bonded atoms will be taken as Y—X—X—Y, though other possibilities obviously exist. The normal modes that interconvert these structures are shown in Fig. 23. Consider first the hydrides, $X_2H_2$. A reliable SCF MO calculation for acetylene is available.[97] Considering only the valence electrons, the order of the MOs is

$$(\sigma_g)^2(\sigma_u)^2(2\sigma_g)^2(\pi_u)^4(\pi_g)^0(2\sigma_u)^0(3\sigma_g)^0(3\sigma_u)^0$$

The shapes of these orbitals are shown in Fig. 24. The transition density, $\rho_{0k}$, is of $\Pi_u \times \Pi_g = \Delta_u + \Sigma_u^+ + \Sigma_u^-$ symmetry. Hence $C_2H_2$ is stable in the linear form since a $\Pi_g$ or $\Pi_u$ vibration is needed to bend the molecule. Also there is a large energy gap between the $\pi_u$ and $\pi_g$ orbitals.

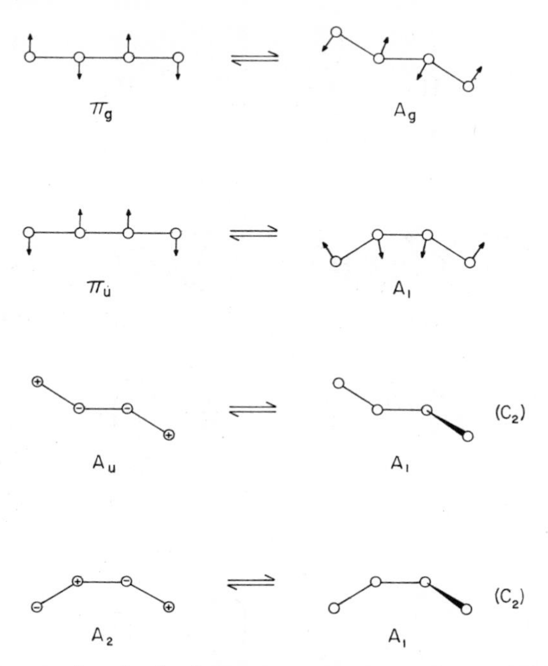


**Figure 23.** The normal modes that interconvert common structures of $X_2Y_2$ molecules.

Assuming the same MO sequence is valid for related molecules, diimide, $N_2H_2$, has the configuration

$$\ldots(\pi_u)^4(\pi_g)^2(2\sigma_u)^0$$

We see that diimide would be a triplet in the lowest-energy linear form. Also there is only a small energy gap between the $\pi_g$ and $2\sigma_u$ orbitals. The symmetry of $\rho_{0k}$ is $\Pi_u$, and $N_2H_2$ is predicted to exist in the *cis* bent configuration. After bending it would no longer be a triplet in its lowest-energy form since the degeneracy of the $\pi_g$ orbitals is lifted. Actually diimide exists in both *cis* and *trans* forms, with the *trans* form the more stable. Self-consistent field MO calculations have been made for $N_2H_2$ in both the *cis* and *trans* forms.[98] For the *cis* ($C_{2v}$) structure the sequence is

$$\ldots(3a_1)^2(b_1)^2(2b_2)^2(4a_1)^0(3b_4)^0,$$

the symmetry of $\rho_{0k}$ is $B_2$ or $A_1$. Since an $A_2$ vibration is needed to convert into the *trans* form, the molecule is predicted to be stable. Also in the *trans* ($C_{2h}$) form the MO sequence is

$$\ldots(2b_u)^2(a_u)^2(3a_g)^2(b_g)^2(4a_g)^0(3b_u)^0,$$

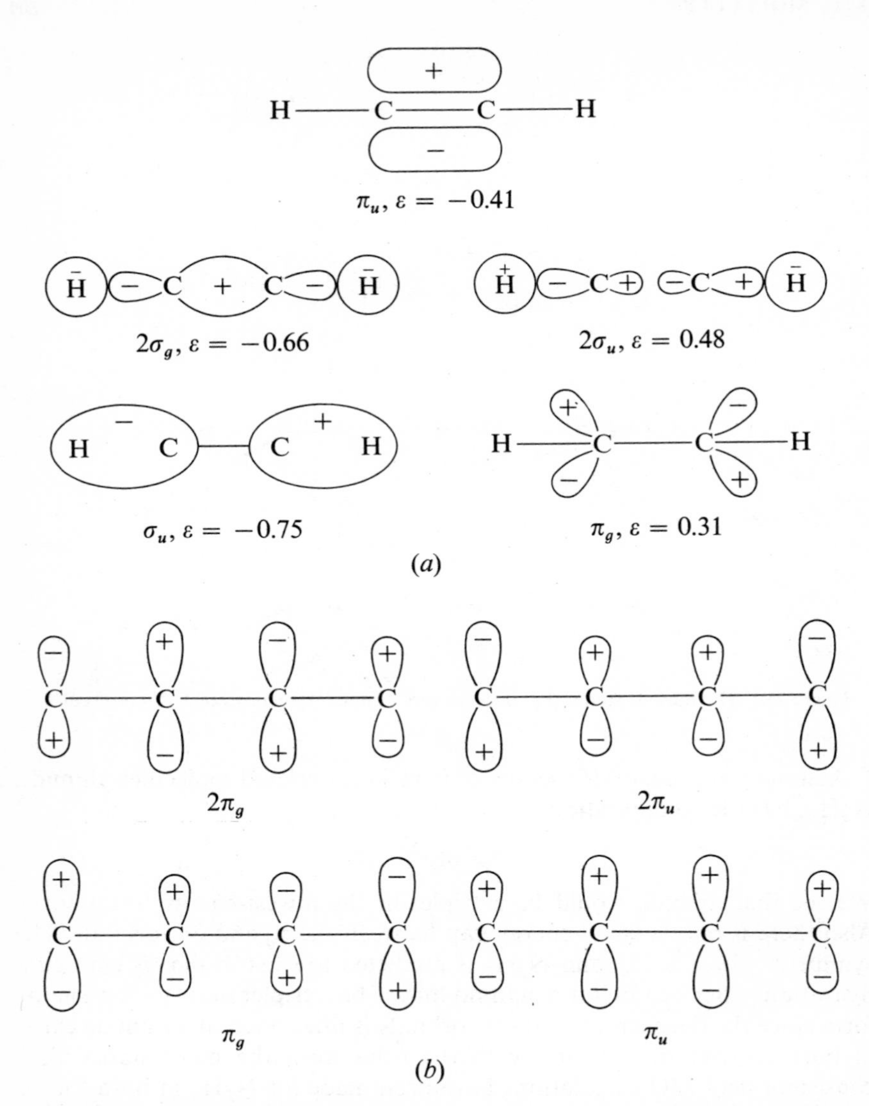


**Figure 24.** (*a*) Molecular orbitals of $C_2H_2$, with energies in atomic units; the $\sigma_g$ orbital, not shown, is strongly bonding for all four atoms [energies from W. S. Palke and W. N. Lipscomb, *J. Am. Chem. Soc.*, **88**, 2384 (1966)]; (*b*) the $\pi$-type orbitals of linear $C_4$; these are the HOMO and LUMO orbitals and resemble the $\pi$ orbitals of butadiene.

so that $\rho_{0k}$ is $A_g$ or $B_g$. Since an $A_u$ vibration is needed to rotate the molecule into the *cis* form, again stability is predicted. These predictions of stable *cis* and *trans* isomers are expected since alkyl diazenes, R—N=N—R, exist as *cis* and *trans* isomers, with substantial activation energies for interconversion. For these molecules, the *trans* isomers are more stable than the *cis*, probably for steric reasons.

Turning to $H_2O_2$, the MO configuration in the linear form would be

$$\ldots(\pi_u)^4(\pi_g)^4(2\sigma_u)^0,$$

the transition density is of $\Pi_u$ symmetry, and the molecule would bent into the *cis* structure. Actually hydrogen peroxide has a skew $C_2$ structure with a dihedral angle of about 100°.

An SCF MO calculation for $H_2O_2$ in the *cis* and *trans* forms is available.[99] In $C_{2v}$ symmetry the MO configuration is

$$\ldots(3a_1)^2(2b_2)^2(1a_2)^2(3b_2)^0(4a_1)^0,$$

and $\rho_{0k}$ is of $B_1$ symmetry. Furthermore there is a large energy gap between the $(1a_2)$ and $(4a_1)$ orbitals so that the $A_2$ vibration is not easily excited. The rotation needed to get the correct structure is not favored.

In $C_{2h}$ symmetry the MO sequence for $H_2O_2$ is

$$\ldots(1a_u)^2(4a_g)^2(1b_g)^2(4b_u)^0.$$

In this case $\rho_{0k}$ is of the correct $A_u$ symmetry to aid in the rotation of the molecule. However the energy gap is large between the $b_g$ and the $b_u$ levels and $\rho_{0k}$ would be small. Experimentally there is almost free rotation about the oxygen—oxygen bond, the barrier being about 3 kcal. Accordingly, the SOJT effect has failed when applied to rotation about a single bond.

### Nonhydrides

An accurate SCF calculation has been made for the molecule $C_4$, in the linear form in which it occurs experimentally.[100] The ordering of MO energies is

$$(\sigma_g)^2(\sigma_u)^2(2\sigma_g)^2(2\sigma_u)^2(3\sigma_g)^2(\pi_u)^4(\pi_g)^2(2\pi_u)^0(2\pi_g)^0(3\sigma_u)^0$$

Since $\Pi \times \Pi = \Sigma^+ + \Sigma^- + \Delta$ with the $g$ or $u$ subscript added, the molecule is stable to bending. Other molecules with 16 valence electrons would also be linear *provided* the same MO sequence can be assumed. Since molecules like the sodium chloride dimer, $Na_2Cl_2$, are not linear but rhomboid, it is clear that a very polar molecule will not have the same orbital pattern as a nonpolar one.[96] Some of the MOs for linear $X_2Y_2$ are shown in Fig. 24.

An 18-electron molecule, like $C_2N_2$, would have the configuration

$$\ldots(\pi_u)^4(\pi_g)^4(2\pi_u)^0$$

This would be stable to bending also, and indeed cyanogen is linear. There are no stable $X_2Y_2$ molecules known with 20 electrons. The only transity density to match with a vibrational mode would be $\Sigma_g^+$. This would correspond to the dissociations

$$N_4 \longrightarrow 2N_2 \quad (10)$$

$$C_2O_2 \longrightarrow 2CO \quad (11)$$

A molecule with 22-valence electrons would have the configuration

$$\ldots(\pi_u)^4(\pi_g)^4(2\pi_u)^4(2\pi_g)^0$$

This would be a stable closed-shell configuration and the molecule would be linear. An example would be $C_2F_2$. However $N_2O_2$ also has 22 electrons and is not linear but bent. Its most striking feature is its instability towards dissociation

$$N_2O_2 \longrightarrow 2NO \quad (12)$$

which is also not predicted. The 22-electron molecule $S_4{}^{2+}$, or $Se_4{}^{2+}$, is not linear but exists in a square planar form. This can be considered an extreme $C_{2v}$ case. Clearly not all isoelectronic $X_2Y_2$ molecules have the same structure, and hence Walsh's rules would not apply.

A 24-electron molecule would be $N_2F_2$. This would have the orbital pattern

$$\ldots(2\pi_u)^4(2\pi_g)^2(3\sigma_u)^0$$

with $2\pi_g$ and $3\sigma_u$ close in energy (both antibonding). This leads to a $\pi_u$ bending mode, and a bent $C_{2v}$ structure. A 26-electron molecule would have the $2\pi_g$ shell filled, but the excitation to $3\sigma_u$ would still be easy, and the molecule would bend to $C_{2v}$. Examples would be $O_2F_2$, $S_2Cl_2$, and $S_4{}^{2-}$.

An SCF calculation for *trans*-$N_2F_2$ gives the MO order[101]

$$\ldots(b_u)^2(a_u)^2(a_g)^2(b_g)^0(b_u)^0$$

The symmetry of the transition density is $(A_g \times B_g = B_g)$, and the molecule is predicted to be stable with respect to rotation about the double bond. A less accurate calculation for *cis*-$N_2F_2$ gives the ordering[101b]

$$\ldots(b_2)^2(a_2)^2(a_1)^2(b_1)^0(a_1)^0$$

Hence $\rho_{0k}$ is of $B_1$ symmetry and the molecule is again stable to twisting. Experimentally both *cis* and *trans* $N_2F_2$ are known, and there is a high activation energy (32 kcal) needed to isomerize the two forms.[102] The *cis* form is more stable.

Using the same MO sequence for $O_2F_2$, gives the configurations in $C_{2v}$ and $C_{2h}$ geometries

$$\ldots(a_1)^2(b_1)^2(a_1)^0 \quad \text{and} \quad (a_g)^2(b_g)^2(b_u)^0$$

This correctly predicts the skew $C_2$ structure of such molecules from the *trans* side, but does not predict it from the *cis* side.

This is another example of the inability of SOJT to explain rotation about single bonds. The failure is not unexpected. Going back to (1), it simply means that both the positive and negative quadratic terms are small. Rotation of groups about single bonds, keeping the electron density fixed at that given by $\psi_0$, costs little energy. This results from the high degree of axial symmetry of the electron density.

If X and Y are the same, a possible structure is the tetrahedral one found for white phosphorus, $P_4$. Even for $X_2Y_2$ a pseudotetrahedral structure of $C_{2v}$ point group is plausible. An ab initio calculation on $P_4$ gives the MO ordering[103]

$$(a_1)^2(t_2)^6(a_1)^2\underline{(2t_2)^6(e)^4(t_1)^0(3t_2)^0}$$

There is a large gap between $e$ and $t_1$, and this shape is particularly favorable for 20 valence electrons. All other numbers of electrons would give configurations unstable to FOJT or SOJT effects. If $N_4$ existed, it would be tetrahedral.

## $X_2Y_4$ MOLECULES

A molecule with six atoms contains a considerably larger number of vibrational modes. The nonsymmetric vibrations of ethylene, a planar molecule of $D_{2h}$ symmetry were shown earlier (p. 23). Figure 25 shows some of the possible structures of $X_2Y_4$ molecules. We will consider $X_2H_4$ molecules by first assuming that they have the structure of $C_2H_4$. The choice of axes used places the $z$ axis perpendicular to the plane, and the $x$ axis passes through the two X atoms.

A very reliable SCF calculation of the MO level scheme in $C_2H_4$ is available.[104] The relevant MOs are as follows:

$$\ldots(2a_g)^2(b_{1g})^2(b_{1u})^2(b_{2g})^0(3a_g)^0(2b_{2u})^0(2b_{3u})^0$$

with the last three being within 1 eV of each other and some 6 eV above the $b_{2g}$ level. For $C_2H_4$ the symmetry of $\rho_{0k}$ is $(B_{1u} \times B_{2g}) = B_{3u}$. It can be seen on page 23 that the $B_{3u}$ vibration distorts the molecule but does not lead to any other structure which could be considered as a stable alternative to the symmetrical one. Furthermore, the energy gap between $b_{2g}$ and $b_{1u}$ is experimentally equal to 7.5 eV, so the SOJT effect is small in any case.

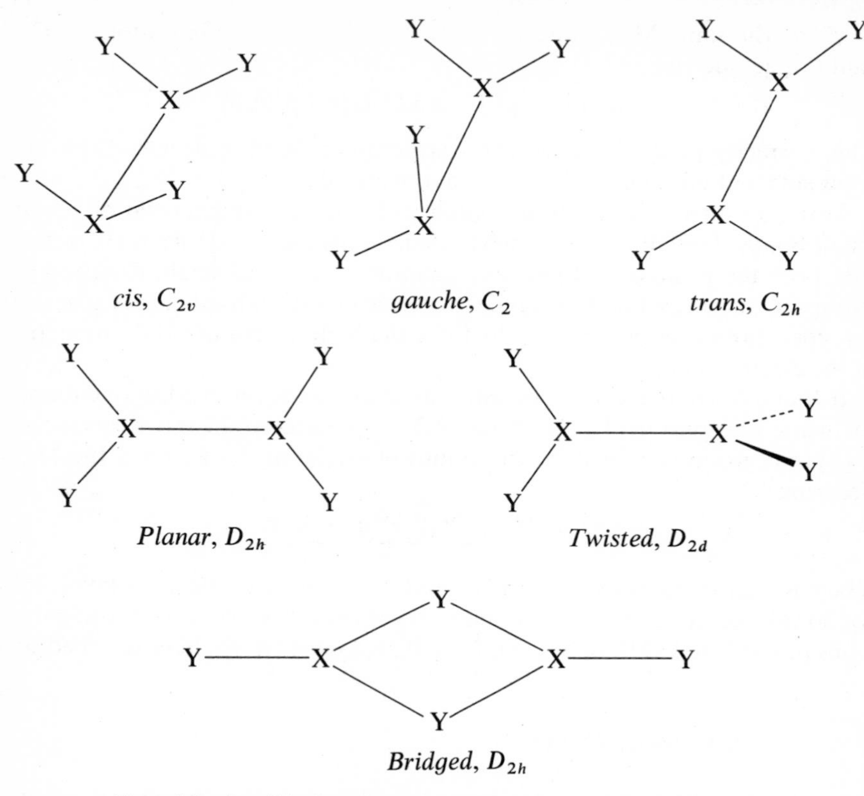


**Figure 25.** Possible structures taken up by $X_2H_4$ and $X_2Y_4$ molecules.

There are numerous symmetry elements in $D_{2h}$, and no degenerate species. Accordingly many MOs of different symmetries are possible for $X_2H_4$. A simple trick makes it possible to visualize them easily. Imagine the MOs to be formed by combining the familiar MOs of a homonuclear diatomic molecule with the group MOs of four H atoms. These are formed from SALC (p. 5), which correspond to $a_g$, $b_{3u}$, $b_{2u}$, and $b_{1g}$ in $D_{2h}$.

From correlation tables we find that the $D_{\infty h}$ species reduce into $D_{2h}$ as follows

$$\Sigma_g^+ \longrightarrow A_g \qquad \Pi_u \longrightarrow B_{1u} + B_{2u}$$

$$\Sigma_u \longrightarrow B_{3u} \qquad \Pi_g \longrightarrow B_{2g} + B_{1g}$$

Now simply combine the MOs of the same species. The bonding combinations are shown in Fig. 26; there will also be antibonding combinations,

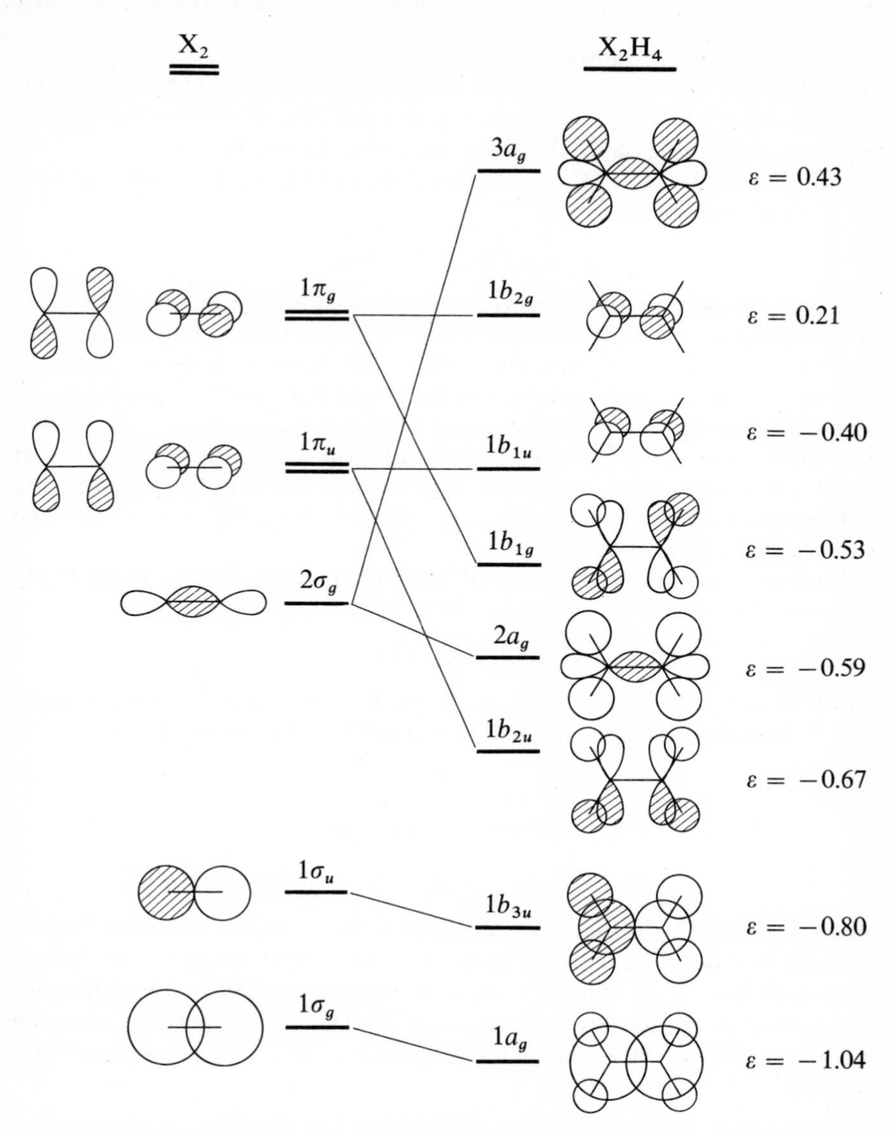


**Figure 26.** Molecular orbitals of $X_2H_4$ formed by combining the MOs of $X_2$ with those of $H_4$; the shaded areas refer to a negative sign for the wavefunction and the unshaded areas, to a positive sign (energies from reference 104).

of which only the two lowest in energy are shown. The $b_{1u}$ and $b_{2g}$ orbitals of $X_2$ have no counterparts in the hydrogen-atom system. They correspond to the $\pi$ and $\pi^*$ orbitals of ethylene. It is also possible to generate the MOs of $X_2H_4$ by combining the MOs of two bent $XH_2$ molecules.

If the MO ordering for $C_2H_4$ is also valid for $B_2H_4$, we have two electrons fewer to accommodate,

$$\ldots(2a_g)^2(b_{1g})^2(b_{1u})^0(b_{2g})^0.$$

The only transition in the required energy range is $(b_{1g}) \rightarrow (b_{1u})$. This gives the symmetry product $A_u$, which is the symmetry of the twisting mode giving rise to a $D_{2d}$ structure. Actually $B_2H_4$ is not known; however, $B_2Cl_4$ is known to have the staggered $D_{2d}$ structure, and $B_2F_4$ exists in the staggered form in the gas phase, and in the planar form in the solid state.[105] In both of the latter cases there is a nearly free rotation about the boron—boron bond, the barrier being only a few kilocalories. For simple molecules, the halides and hydrides usually have the same structure. This seems to be the case for $B_2Y_4$ molecules.

We may also go on to the dimer of beryllium hydride, $Be_2H_4$. In the $C_2H_4$ structure the MO-level scheme would be

$$\ldots(2a_g)^2(b_{1g})^0(b_{1u})^0.$$

The symmetry of $\rho_{0k}$ is $B_{1g}$. Figure 25 shows that this is the normal mode that rotates the two $BeH_2$ fragments into the bridged structure.

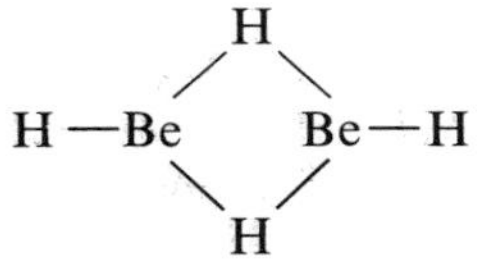

```
          H
        /   \
H—Be         Be—H
        \   /
          H
```

Again the structure of $Be_2H_4$ is not known; however, the beryllium halides and the beryllium dialkyls probably have the above bridged structure as dimers in the vapor state.[106] There is available an SCF calculation for ethylene in a twisted $D_2$ configuration.[107] The results may be extended slightly to give an MO scheme for staggered, or $D_{2d}$, symmetry. For $C_2H_4$ we have

$$\ldots(e)^4(2a_1)^2(2e)^2(3a_1)^0(2b_2)^0(3e)^0,$$

with a large energy gap between $2a_1$ and $2e$ and with $1e$ and $2a_1$ very close. The ground state would be the triplet $^3A_2$.

The lowest excited state is the $^3E$, which leads to $\rho_{0k}$ having $E$ symmetry. The $E$ vibrations include the $B_{2g}$ and $B_{1u}$ wagging vibrations of planar ethylene, which can now act independently for each $CH_2$ group. The structure for triplet ethylene is not known, but there are reasons to believe it has the staggered structure with a pyramidal arrangement of bonds about each carbon atom.[108]

If we arbitrarily keep the electron spins paired to produce singlet but twisted ethylene, we would obtain the $^1B_1$ state. Close-lying states also in the $(2e)^2$ configuration would be $^1A_1$ and $^1B_2$. The $B_1$ vibrational mode is the twisting mode that converts the staggered structure into the correct planar one.

For $B_2H_4$ the MO configuration would be, in $D_{2d}$,

$$\ldots(1e)^4(2a_1)^2(2e)^0$$

Because of the large energy gap, this would be stable in the staggered form. The MO scheme for $N_2H_4$ would be $\ldots(2a_1)^2(2e)^4(3a_1)^0$, which would give $\rho_{0k}$ of $E$ symmetry. As already mentioned, this includes the wagging modes that carry staggered $N_2H_4$ toward its stable structure. However a rotation around the single N—N bond of about 90° is still needed to get the most stable geometry. Similarly, for $Be_2H_4$ the MO configuration would be $\ldots(1e)^4(3a_1)^0$. The $E$ vibration would be strongly favored. This includes the $B_{1g}$ and $B_{2u}$ rocking modes of ethylene now acting independently for each $XH_2$ group. These take $Be_2H_4$ toward its stable bridged structure. However a rotation would again be needed to complete the transformation. In spite of the many possibilities, a single MO scheme (for $C_2H_4$) seems to do a remarkable job of predicting the structural features of $X_2H_4$ molecules. The exception is that rotations about single bonds are poorly predicted.

### Nonhydride Molecules

The situation with $X_2Y_4$ molecules is much less promising. Two ab initio calculations are available, for planar $N_2O_4$ and $C_2F_4$.[109] The MO sequence for $N_2O_4$ is, in part,

$$\ldots(b_{2g})^2(b_{3u})^2(b_{2u})^2(b_{1g})^2(b_{3g})^2(a_u)^2(a_g)^2(b_{1u})^0(b_{3u})^0(b_{2g})^0(a_g)^0$$

There are 34 valence electrons, two fewer than for $C_2F_4$. For this molecule the MO order is

$$\ldots(b_{2g})^2(b_{1u})^2(a_u)^2(b_{2u})^2(b_{3g})^2(a_g)^2(b_{3u})^2(b_{2g})^0(b_{2u})^0(b_{1u})^0$$

The transition density is $B_{1u}$ in both cases but with a large energy gap, especially for the ethylene derivative. Actually $N_2O_4$, like $C_2F_4$, has a planar structure. However there is almost free rotation about the N—N bond.[110] It would appear, that unlike $C_2F_4$, this is a single bond, and again SOJT cannot predict such conformational changes.

More distressing is the observation that the two MO sequences are quite different, even though the molecules are fairly similar. When all the atoms in a molecule are changed, the same MO ordering cannot be assumed, even if the formulas and structures are the same. For example, the $N_2O_4$ sequence, if used for $C_2F_4$ by adding two electrons, would predict an unstable structure. A $B_{2g}$ distortion would be produced, favoring a nonplanar structure of $C_{2h}$ symmetry.

Constructing Lewis diagrams for the two molecules is instructive

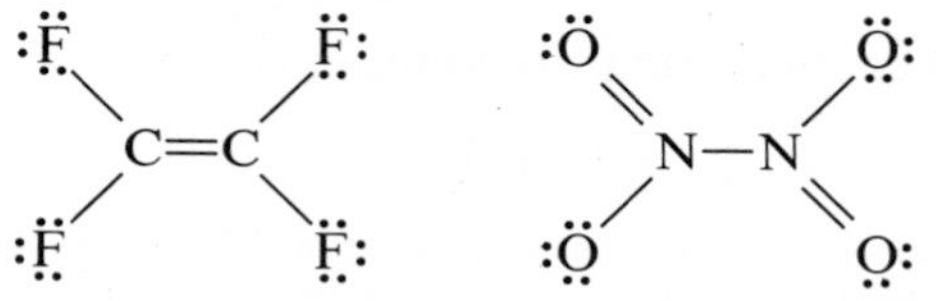

```
 :F̤̈          F̤̈:        :Ö          Ö:
    \        /             \\       /
     C=C                    N—N            (etc.)
    /        \             /       \\
 :F̤̈          F̤̈:        :Ö̤          O̤:
```

Clearly the $\pi$ bonding in $C_2F_4$ and $N_2O_4$ is quite different, a single Lewis diagram being inadequate for the latter. The resemblance of $C_2F_4$ to $C_2H_4$ is strong in that a single Lewis diagram suffices. The MO sequence of $C_2F_4$ must contain the same structural predictions that were so successful for $C_2H_4$. However the predictions are obscured by the large number of essentially extraneous MOs produced by the extra-valence orbitals of fluorine. It would appear that, unless there is a drastic change in $\pi$ bonding produced by some structural or chemical change, the $\pi$ electrons of many systems may be ignored and attention focused on the $\sigma$ bonding.

## $X_2Y_6$ MOLECULES

There are three principal structures for $X_2Y_6$ molecules: (1) the staggered ethane structure of $D_{3d}$ symmetry, (2) the bridged structure of diborane of $D_{2h}$ symmetry, and (3) the all-planar, bridged structure, also of $D_{2h}$ symmetry. Figure 27 shows the normal modes that interconvert these structures. In addition, ethane can exist in the eclipsed form of $D_{3h}$ symmetry. An $A_1''$ or an $A_{1u}$ vibration are needed to interconvert the eclipsed and staggered forms.

A number of accurate SCF calculations of $C_2H_6$ are available. In the stable $D_{3d}$ form the MO sequence is[111]

$$(a_{1g})^2(a_{2u})^2(e_u)^4(2a_{1g})^2(e_g)^4(2a_{2u})^0(2e_u)^0$$

considering only the valence electrons. The symmetry of the transition density is $E_u$, but there is a very large energy gap between the $(e_g)$ and $(a_{2u})$ orbitals. Accordingly the molecule is structurally stable. The forms of the molecular orbitals of ethane are shown in Fig. 28. In this and in Fig. 27, the $z$ axis in the C—C bond axis, to coincide with the threefold axis of ethane. A convenient way to visualize the MOs of ethane is to form them by taking

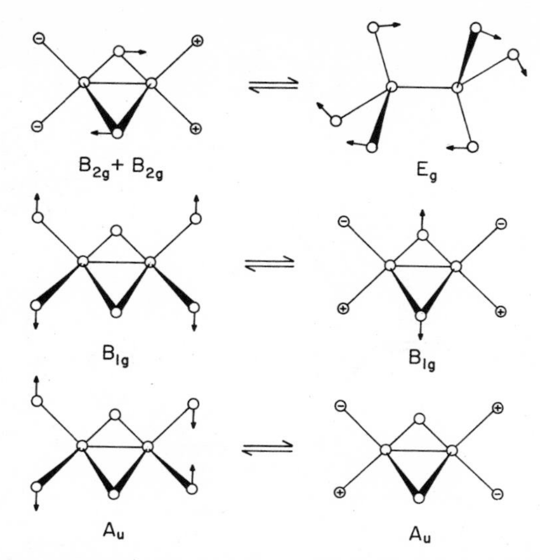


**Figure 27.** The normal modes that interconvert various structures of $X_2H_6$ molecules.

the sum and differences of the simpler MOs of two planar methyl radicals (see Fig. 28).

The MO sequence is also known for $B_2H_6$ in the ethane structure.[112] The order is

$$(a_{1g})^2(a_{2u})^2(e_u)^4(e_g)^4(2a_{1g})^0(2a_{2u})^0.$$

The symmetry of $\rho_{0k}$ is seen to be $E_g$. Also the energy difference between $e_g$ and $2a_{1g}$ must be small. Clearly $B_2H_6$ is unstable in $D_{3d}$ symmetry and will convert to its stable bridged structure. Other 14-electron systems such as $Si_2H_6$ and $N_2H_6^{2+}$ will have the ethane structure, whereas $Be_2H_6^{2-}$, a 12-electron system, will have the diborane structure.[113]

The MO sequence of $C_2H_6$ in the diborane structure has been accurately calculated.[112] The order is

$$\ldots(b_{2u})^2(a_g)^2(b_{3g})^2(b_{2g})^2(2a_g)^0(b_{1u})^0.$$

A $B_{2g}$ vibration is needed to convert the diborane structure into the $D_{3d}$ ethane structure. Since the highest occupied orbital, $b_{2g}$, is an antibonding one, the excitation energy to $2a_g$ must be small, and the $B_{2g}$ vibration is easily excited.

For diborane itself the configuration, in the correct structure, is[114]

$$\ldots(b_{2u})^2(2a_g)^2(b_{3g})^2(b_{2g})^0(2b_{1u})^0(2b_{3u})^0$$

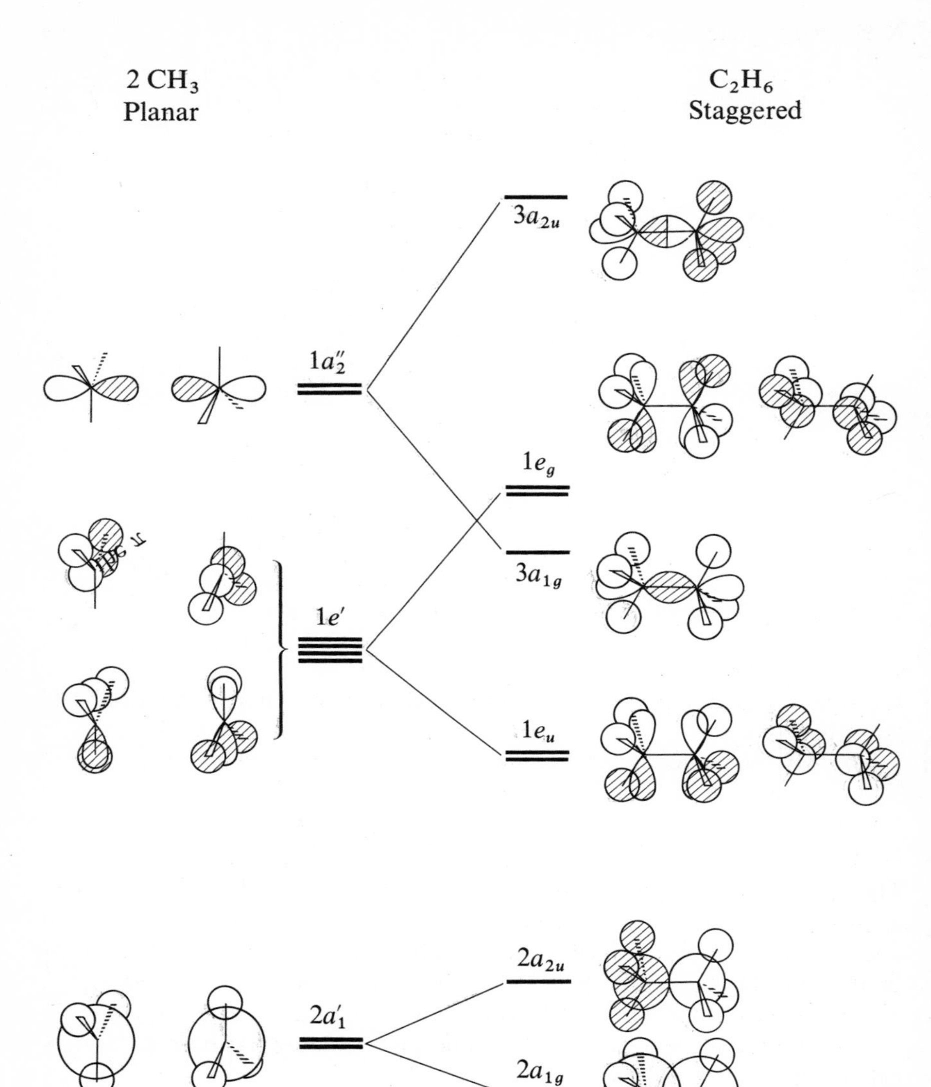


**Figure 28.** Molecular orbitals and relative energies for ethane in staggered $D_{3d}$ conformation; the orbitals are shown as if formed by combining the orbitals of two planar $CH_3$ radicals; the shaded and unshaded areas refer to opposite signs for the wavefunction. The $1a_{1g}$ and $1a_{2u}$ orbitals (not shown) are the 1*s* orbitals of carbon. [Reprinted with permission from B. M. Gimarc, *J. Am. Chem. Soc.*, **95**, 1417 (1973). Copyright by the American Chemical Society.]

so that only bonding orbitals are filled. The symmetry of $\rho_{0k}$ is $B_{1g}$, which converts the staggered bridged structure into the all-planar bridged structure. However the first absorption band of $B_2H_6$ has its maximum at 1800 A, so that the energy gap between $b_{3g}$ and $b_{2g}$ is 6.8 eV. This makes the structure stable in the staggered form. The diborane MOs in Fig. 29 are shown as if they were generated by starting with the ethane structure, $D_{3d}$, and converting into the $D_{2h}$ form by an $E_g$ vibration. In this case the elements of $C_{2h}$ are preserved.

Ethane in the eclipsed form of $D_{3h}$ symmetry has the accurate MO sequence[115]

$$(a_1'')^2(a_2'')^2(e')^4(a_1')^2(e'')^4(2a_2'')^0(2e')^0$$

with a very large energy gap between the filled and empty orbitals. The symmetry of the $(1e'') \rightarrow (2e')$ transition would be $A_1''$ needed to rotate the eclipsed form into the stable staggered form. Because of the high energy, this would be an ineffective transition. Similarly, the barrier to rotation from the staggered to the eclipsed form is only 2–3 kcal, yet the transition in normal ethane of $A_{1u}$ symmetry must correspond to a very large energy gap. Rotations about single bonds are not predicted correctly by SOJT theory.

It is of interest to consider the hypothetical molecule $F_2H_6$. Assume first that it has the diborane structure, and that the MO sequence for $B_2H_6$ is still valid. This gives the configuration

$$\ldots(2a_g)^2(b_{1u})^2(2b_{2u})^2(2b_{3u})^0(2b_{1u})^0$$

The symmetry of the transition density is $(B_{2u} \times B_{3u}) = B_{1g}$, and the molecule would become planar by the conrotatory twist mode. The energy gap would be small according to the calculations.

The known molecule $I_2Cl_6$ does exist in the all-planar form. A semi-empirical MO sequence for all-planar $F_2H_6$ is available.[116] The ordering is

$$\ldots(2b_{3u})^2(2b_{1u})^3(3a_g)^2(3b_{1u})^0(2b_{2g})^0(3b_{3u})^0(4a_g)^0$$

so that no easy excitations favoring either the $B_{1g}$ or $A_{1u}$ (disrotatory) twist modes are possible. The molecule should be stable in the planar structure. Many metal halides and complexes are known of formula $M_2L_6$. Some are planar, like $Pt_2Br_6{}^{2-}$, and some have the diborane structure, like $Al_2Cl_6$. The last structure can be rationalized since in terms of $\sigma$ bonds, $Al_2Cl_6$ is isoelectronic with $B_2H_6$. It would be a mistake to think that simple Walsh rules apply for $X_2Y_6$ molecules, however. For example, $Cl_2O_6$ is isoelectronic with $C_2Cl_6$. The structure is much more likely to be of the diborane variety *A*, rather than the ethane one, B.

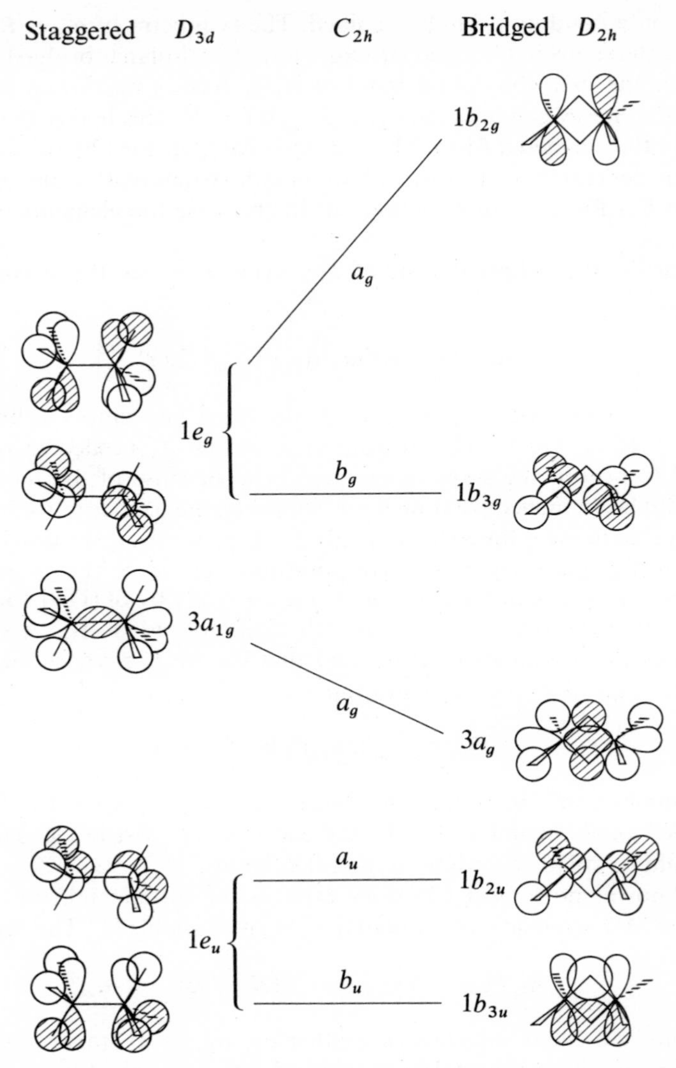


**Figure 29.** The MOs of $B_2H_6$ (on the right) shown as if formed by a rotation from the ethane structure, $D_{3d}$, by the way of an $E_g$ mode, generating intermediates with a $C_{2h}$ structure. [Reprinted with permission from B. M. Gimarc, *J. Am. Chem. Soc.*, **95**, 1417 (1973). Copyright by the American Chemical Society.]

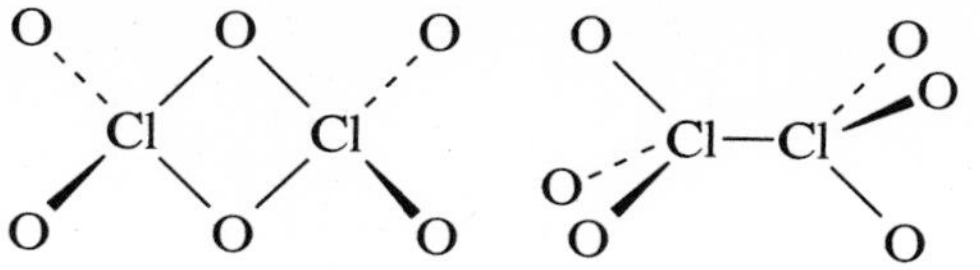

A B

The difference may be attributed to the much more positive charge on Cl in $ClO_3$, than on C in $CCl_3$. Hence two chlorine atoms repel each other and prefer to bond to oxygen.

## HXY, $H_2XY$, AND $HXY_2$ MOLECULES

An HXY molecule can only belong to the $C_{\infty v}$ point group, if linear, or to $C_s$, if not. An ab initio MO scheme for linear HCN is[117]

$$(\sigma)^2(2\sigma)^2(3\sigma)^2(\pi)^4(2\pi)^0(4\sigma)^0$$

which predicts $\rho_{0k}$ as $\Sigma$ or $\Delta$, and a stable linear structure. The same MO sequence is valid for HOF, with 14 valence electrons instead of 10.[118] This creates a $\rho_{0k}$ of $\Pi$ symmetry and the molecule bends. Molecules or radicals with 11, 12 or 13 electrons, such as HCO, HNO, $HO_2$, will also be bent.

$H_2XY$ molecules, like formaldehyde, are either planar, $C_{2v}$, or nonplanar, $C_s$. An accurate MO sequence for $CH_2O$ in the planar form is[119]

$$(a_1)^2(2a_1)^2(b_2)^2(3a_1)^2(b_1)^2(2b_2)^2(2b_1)^0(4a_1)^0$$

The transition density is $(B_1 \times B_2) = A_2$, with a large energy gap. In any case the $A_2$ vibration does not exist for $CH_2O$, only a rotation having this symmetry. Formaldehyde has 12 valence electrons, as has $BH_2F$, which is also planar. The radical $CH_2N$, with 11 electrons also should be planar. However $CH_2F$, with 13 electrons, or $NH_2F$ with 14 electrons, have one or two electrons in the $2b_1$ orbital. This allows for any easy $(2b_1) \rightarrow (4a_1)$ excitation. The resultant reaction coordinate of $B_1$ symmetry is the out-of-plane deformation leading to a $C_s$ structure.

The planar formate ion has the MO energy-level ordering

$$\ldots(3a_1)^2(2b_2)^2(b_1)^2(3b_2)^2(4a_1)^2(a_2)^2(2b_1)^0(5a_1)^0$$

The transition density is of $(A_2 \times B_1) = B_2$ symmetry, with a large energy difference between $a_2$ and $2b_1$. The $B_2$ mode is an asymmetric in-plane distortion. The radical $CHF_2$ has one more electron, and the molecule $NHF_2$ has two more electrons than $HCO_2^-$. This allows for the $(2b_1) \rightarrow (5a_1)$ excitation. The symmetry of the reaction coordinate is $B_1$, which is the out-of-plane bending mode. These molecules become pyramidal, of $C_s$ symmetry.

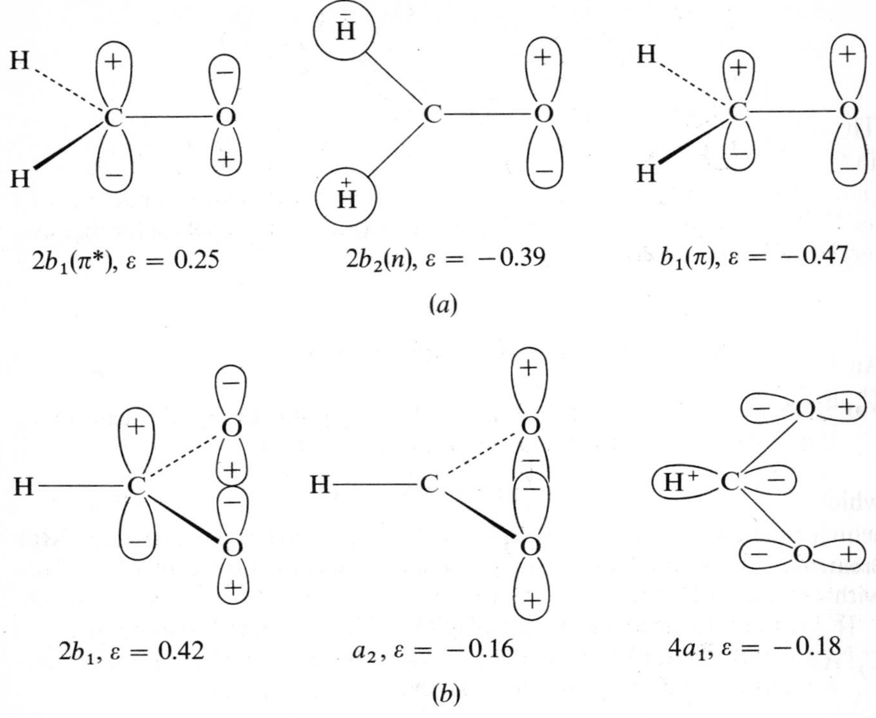


**Figure 30.** (*a*) The two HOMOs and the LUMO of formaldehyde; the common $\pi$, $\pi^*$ and $n$ labels are shown in addition to the $C_{2v}$ symmetry labels; (*b*) the two HOMOs and the LUMOs of the formate ion (energies from reference 26).

The higher MOs of formaldehyde and of the formate ion are displayed in Fig. 30.

## CONJUGATED SYSTEMS

The highly conjugated $\pi$ systems of organic chemistry are rich in examples of both FOJT and SOJT effects.[121] The reason for this can be seen in the following way. Consider a long chain of carbon atoms, arranged linearly. The $\pi$ orbitals are alternately of $\pi_u$ and $\pi_g$ symmetry in order of increasing energy (see Fig. 24). Furthermore the HOMO and LUMO orbitals approach each other in energy as the chain gets longer. Thus butadiene has a $(\pi) \rightarrow (\pi^*)$ transition at lower energy than ethylene; hexatriene is at lower energy still, and so on, until a limit is reached for very long conjugated systems.

Now the finite limit for the frequency of the longest wavelength absorption, as the chain length goes to infinity, must arise because some other factor is pushing apart the HOMO and the LUMO. Otherwise simple $\pi$ theory of the Hückel type would predict a zero-frequency limit. The other factor at work is the SOJT effect. As the gap between the HOMO and LUMO decreases, eventually a distortion will occur which pushes these levels apart again. The symmetry of the distortion can only be $(\Pi_g \times \Pi_u) = (\Sigma_u^+ + \Sigma_u^- + \Delta_u)$, of the $\Sigma_u^+$ type, which lengthens and shortens alternate bonds. Accordingly bond-length alternation is found in long chain linear polyenes, and in large cyclic polyenes.[121]

The smaller cyclic systems are characterized by the occurrence of degenerate orbitals with a definite pattern (see p. 119). If these orbitals are completely filled, the molecular shape is stable and indeed the molecule has the aromatic stability associated with the Hückel $(4n + 2)$ rule. If the degenerate orbitals are incompletely filled, then FOJT or SOJT effects occur. To illustrate these phenomena, Fig. 31 shows the MOs of the $\pi$ system of planar, symmetric $C_4H_4$, $C_6H_6$, and $C_8H_8$.

Benzene, point group $D_{6h}$, has a filled $e_{1g}$ subshell and is stable. However $C_6H_6{}^+$ would have an ${}^2E_{1g}$ ground state and would be subject to the FOJT distortion. $C_6H_6{}^-$ would have an ${}^2E_{2u}$ ground state and would also distort

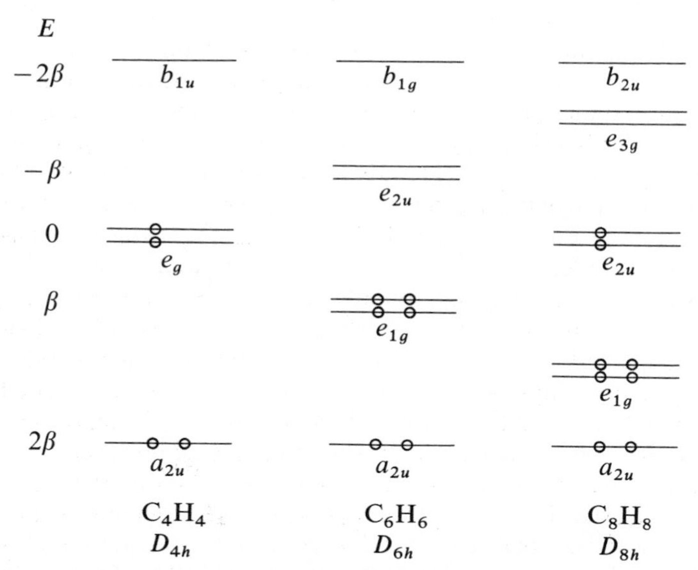


**Figure 31.** Energies and symmetries of the $\pi$ MOs of cyclobutadiene, benzene, and cyclooctatetraene; $\beta$ is the exchange integral.

from $D_{6h}$. The excited states of benzene would also be unstable structurally (see p. 481). The symmetric direct products, $(E_{1g} \times E_{1g})^+ = (E_{2u} \times E_{2u})^+ = (A_{1g} + E_{2g})$, are the same for both $C_6H_6^+$ and $C_6H_6^-$. The direction of distortion would be that of an $E_{2g}$ vibration. From the correlation tables. $E_{2g}$ becomes $(A_g + B_{1g})$ in the $D_{2h}$ point group. Hence structures of $D_{2h}$ symmetry would result.

The direct product of two identical degenerate functions is the sum of the symmetric product and the antisymmetric product. The symmetric product would be the species of the electronic states of two electrons of opposite spin. Thus the total wavefunction for two such electrons would be antisymmetric, as required, since the spin function $[\alpha(1)\beta(2) - \beta(1)\alpha(2)]$, is antisymmetric. However for two unpaired electrons, the antisymmetric direct product is needed, since the spin function is symmetric. In tables, the antisymmetric product is usually enclosed in brackets.

This information is useful in discussing the situation in cyclobutadiene. The configuration $(a_{2u})^2(e_g)^2$ gives rise to four states, $A_{1g}$, $B_{1g}$, $B_{2g}$, and $A_{2g}$. Of these $A_{2g}$ is the antisymmetric direct product. Hence it must correspond to a triplet state $^3A_{2g}$. The other states must be singlets. The triplet state would be geometrically stable as $D_{4h}$, since no low-lying state of the same multiplicity exists. If we force the electrons to pair up, then there must necessarily be three states, $^1A_{1g}$, $^1B_{1g}$, and $^1B_{2g}$, which are close in energy. Accordingly the molecule is unstable as $D_{4h}$, no matter which singlet state lies lowest. This is a pseudo JT effect.

The possible distortions are of $B_{1g}$, $B_{2g}$, and $A_{2g}$ symmetry. Referring back to Fig. 11, we see there is no skeletal mode of $A_{2g}$ species. The $B_{1g}$ mode converts into a diamond-shaped $D_{2h}$ structure, and the $B_{2g}$ mode converts into a rectangular $D_{2h}$ shape. Symmetry alone will not tell us which distortion is favored, but elementary considerations of bonding can. The object of the distortion is to remove the degeneracy of the $e_g$ orbitals, that is, to lower one component and to raise the other. The $B_{2g}$ vibration produces a first-order splitting since it affects nearest-neighbor distances and overlaps. The $B_{1g}$ mode causes a smaller splitting, initially at least, since it affects only next-nearest neighbor distances. It is predicted that singlet cyclobutadiene will have a rectangular $D_{2h}$ structure, with two long and two short bonds.

Theoretical calculations verify this prediction, with calculated carbon—carbon bond lengths of about 1.34 Å and 1.51 Å, characteristic of isolated double and single bonds.[122] The calculations, however, are unable to agree on whether cyclobutadiene is much more stable than acetylene, or much less stable. They cannot be regarded as very reliable at this stage, especially since the experimental evidence strongly suggests that singlet cyclobutadiene has a square planar $D_{4h}$ structure.[123] The disappearance of predicted Jahn–Teller distortions is by no means a rare event.

The decomposition of cyclobutadiene to acetylene, in any case, is forbidden by orbital symmetry in a $D_{2h}$ reaction mode.

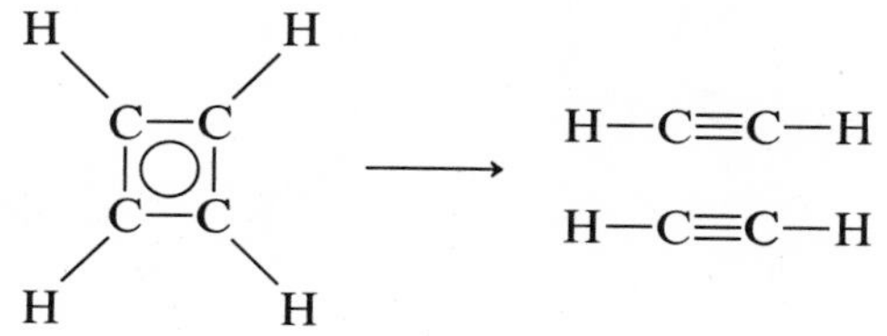

$$\text{C}_4\text{H}_4 \longrightarrow \begin{matrix} \text{H—C}\equiv\text{C—H} \\ \text{H—C}\equiv\text{C—H} \end{matrix} \tag{13}$$

$$a_g + b_{2u} + b_{1u} + b_{2g} \qquad a_g + b_{3u} + b_{1u} + b_{2g}$$

Only the $\pi$ orbitals of acetylene are labeled. The reaction that the molecule actually undergoes is an allowed Diels–Alder condensation with itself to form tricyclooctadiene. Only a plane of symmetry is conserved.

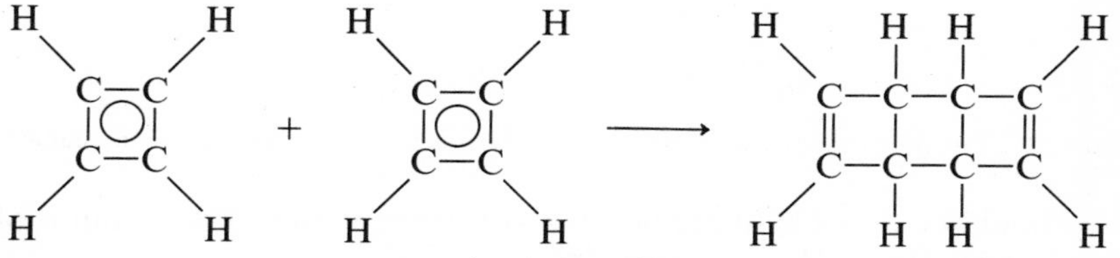

$$\text{C}_4\text{H}_4 + \text{C}_4\text{H}_4 \longrightarrow \text{C}_8\text{H}_8 \tag{14}$$

$$2a' \qquad a' + a'' \qquad 3a' + a''$$

Since $e_g$ in $D_{4h}$ splits into $a' + a''$ in $C_s$, we can localize the highest-energy pair of electrons in an orbital of either symmetry during reaction.

Planar symmetric cyclooctatetraene in the $D_{8h}$ point group would have the configuration $(a_{2u})^2(e_{1g})^4(e_{2u})^2$. Again there would be three states of similar energy, ${}^1B_{1g}$, ${}^1B_{2g}$, and ${}^1A_{1g}$. Correlation tables do not usually include the $D_{8h}$ point group. The symmetry of the point group generated by either a $B_{1g}$ or $B_{2g}$ mode can easily be found by looking at the character of the species for each symmetry element in the character table. If the character is $+1$, the element remains, if it is $-1$, the element disappears. In this way it is found that both $B_{1g}$ and $B_{2g}$ generate the $D_{4h}$ point group.

Ab initio calculations have shown that the most stable *planar* form of $C_8H_8$ is the $D_{4h}$ structure with alternating long and short bonds.[124] In other words, the molecule has four isolated double bonds. The actual structure of cyclooctatetraene is a puckered, nonplanar form of $D_{2d}$ symmetry, shown in Fig. 32, along with other possible shapes. This actual shape is some 18 kcal more stable than the planar $D_{4h}$ form. It is derived from the planar form by an additional distortion of $B_{1u}$ type.

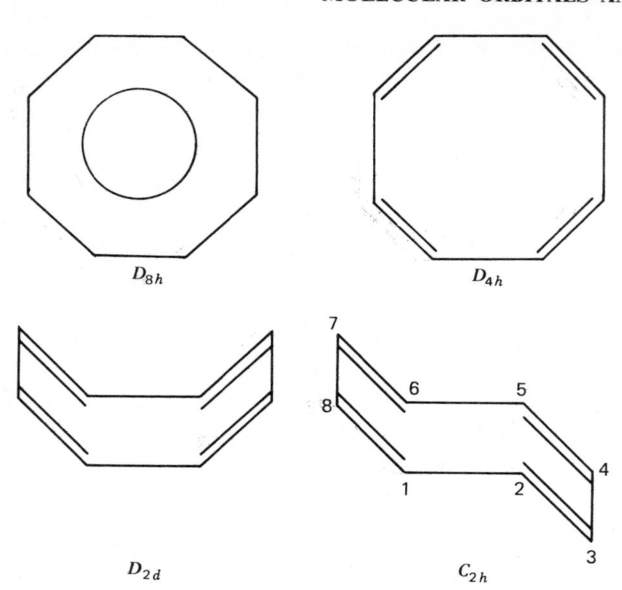


**Figure 32.** Possible structures for $C_8H_8$, cyclooctatetraene; the $D_{2d}$ structure is the stable one.

It should be possible to predict this distortion as well. The ab initio MO sequence for $D_{4h}$ $C_8H_8$ is, in part,[124b]

$$(e_u)^4(b_{1u})^2(b_{2u})^0(e_u)^0$$

so the indicated distortion is of $(B_{1u} \times B_{2u}) = A_{2g}$ type. This leads to a $C_{2h}$ or chair form, rather than the stable $D_{2d}$ boatform (Fig. 32). This failure of PT has a partial explanation. The $C_{2h}$ form is strongly destabilized by the twisting of the central double bonds. In fact the $\pi$ orbitals of carbon atoms 1, 2, 5 and 6 are nearly at right angles to each other in the chair form, so the overlap is close to zero. Starting with the planar $D_{4h}$ form, where the overlap is maximal, it is possible to twist the double bond several degrees without markedly changing the energy (the overlap goes as $\cos\theta$, where $\theta$ is the dihedral angle and the variation of the overlap with changing $\theta$ goes as $\sin\theta$). Accordingly it may be possible *initially* to distort by an $A_{2g}$ mode with less energy requirement than for a $B_{1u}$ mode.

This illustrates a fundamental difficulty with PT, namely that only small displacements are considered. Energy changes that come into play only for large displacements can be overlooked. An example of this kind of error would be found by applying the theory to a molecule like $S_4^{2+}$, assuming it to be linear. Actually it has square planar structure, $D_{4h}$,[125] which is not predicted correctly (p. 212). Clearly the effect of bonding the two terminal sulfur atoms cannot be felt for small displacements from the linear structure.

### Nonalternant Hydrocarbons

A remarkable example of the successful PT application is in the work of Nakajima on the bond-distance changes in nonalternant hydrocarbons, their ions, and their excited states.[126] Figure 33 shows the structures of some of the molecules considered, and Table 11 shows the predictions made as to reduction in symmetry caused by the SOJT effect. The predictions are based on SCF calculations of the $\pi$ molecular orbitals assuming the most symmetric planar structure, and using the Pariser–Parr–Pople approximation. This gives the energy of excitation to the first excited state and its symmetry.

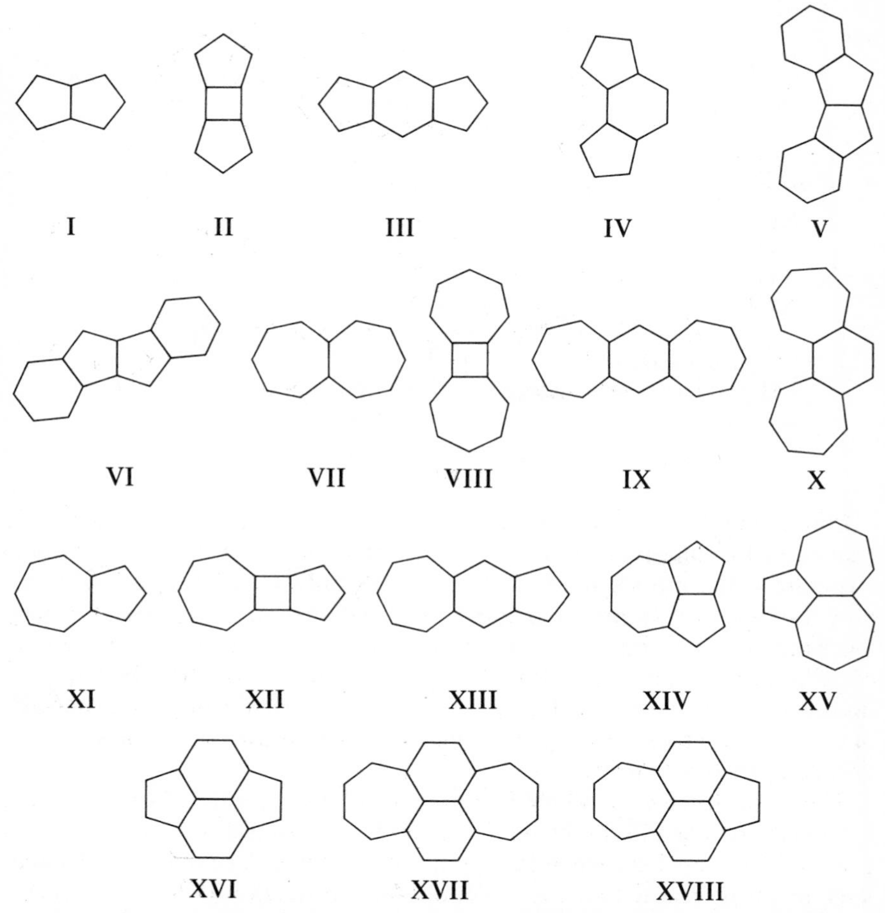


**Figure 33.** Structures of polycyclic aromatic hydrocarbons (referred to in Table 12).

**Table 11 Symmetries of $\psi_1$ and First Excitation Energies $\Delta E_1$ of Nonalternant Hydrocarbons Calculated Assuming the Full Molecular Symmetries and Possible Molecular-symmetry Reductions**[a]

| Molecule | Symmetry of $\psi_1$ | $\Delta E_1$ (eV) | Symmetry Reduction |
|---|---|---|---|
| I | $B_{3g}$ | 0.35 | $D_{2h} \rightarrow C_{2h}$ |
| II | $B_{2u}$ | 1.22 | $D_{2h} \rightarrow C_{2v}$ |
| III | $B_{3g}$ | 1.00 | $D_{2h} \rightarrow C_{2h}$ |
| IV | $B_2$ | 1.47 | $(C_{2v} \rightarrow C_s)$[b] |
| V | $B_2$ | 0.41 | $C_{2v} \rightarrow C_s$ |
| VI | $A_g$ | 2.54 | |
| VII | $B_{3g}$ | 0.26 | $D_{2h} \rightarrow C_{2h}$ |
| VIII | $B_{2u}$ | 0.81 | $D_{2h} \rightarrow C_{2v}$ |
| IX | $B_{3g}$ | 0.83 | $D_{2h} \rightarrow C_{2h}$ |
| X | $B_2$ | 1.46 | $(C_{2v} \rightarrow C_s)$ |
| XI | $B_2$ | 2.05 | $(C_{2v} \rightarrow C_s)$ |
| XII | $B_2$ | 1.07 | $C_{2v} \rightarrow C_s$ |
| XIII | $B_2$ | 0.81 | $C_{2v} \rightarrow C_s$ |
| XIV | $B_2$ | 1.73 | $(C_{2v} \rightarrow C_s)$ |
| XV | $B_2$ | 1.52 | $(C_{2v} \rightarrow C_s)$ |
| XVI | $B_{3g}$ | 2.24 | $(D_{2h} \rightarrow C_{2h})$ |
| XVII | $A_1$ | 2.98 | |
| XVIII | $B_{3g}$ | 2.15 | $(D_{2h} \rightarrow C_{2h})$ |

[a] From reference 126.
[b] The symmetry reductions in parentheses do not occur because $\Delta E_1$ is too great.

Empirically it is found that if $(E_1 - E_0)$ is less than 1.3 eV, the predicted distortion does take place. If $(E_1 - E_0)$ is greater than 1.3 eV, the experimental structure is the most symmetrical one. Since only $\pi$ MOs are considered, only in-plane distortions are predicted. It is possible to go one step further. Since the MOs are simple functions, it is possible to estimate the second-order corrections to the energy and predict the actual bond distances. In order to do this it is necessary to include a force constant for changing the distances of the $\sigma$ bonds as well. In any case, this refinement is outside of the scope of symmetry theory.

The critical value of 1.3 eV is a very small energy gap. Many previous examples discussed in this chapter showed SOJT instability for much larger energy differences between ground and excited states. A rough limit of some 4 eV might be selected for the structural interconversions of $XY_n$ and $X_2Y_n$ molecules. However it should be noticed that these changes refer chiefly to changes in bond angles, not changes in bond distances or in the nature of the bonds themselves. Such rearrangements in which bonds are neither made nor broken, but their direction in space is changed, are called "polytopal."[127] Conformational changes are usually not included in this category, but are similar. Since force constants for bond angle changes are much less than for bond distance changes, it is obvious that quite different critical values of the excited state energy gap may apply. The structural changes shown in Table 12 are all bond distance changes primarily. Hence the critical energy gap is small.

## Planarity of Conjugated Systems

Conjugated $\pi$ systems make nice examples because simple Hückel theory can be used. This makes it much easier to follow the changes that occur. Let us consider two possible structures for trimethylenemethane, $C(CH_2)_3$. One is an all planar form of $D_{3h}$ point group, and the other has one methylene group twisted out of the plane to a perpendicular position. The point group now is $C_{2v}$. Figure 34 shows the two forms, and also the MO energy diagrams calculated by simple Hückel theory. Only the $\pi$ orbitals are considered,

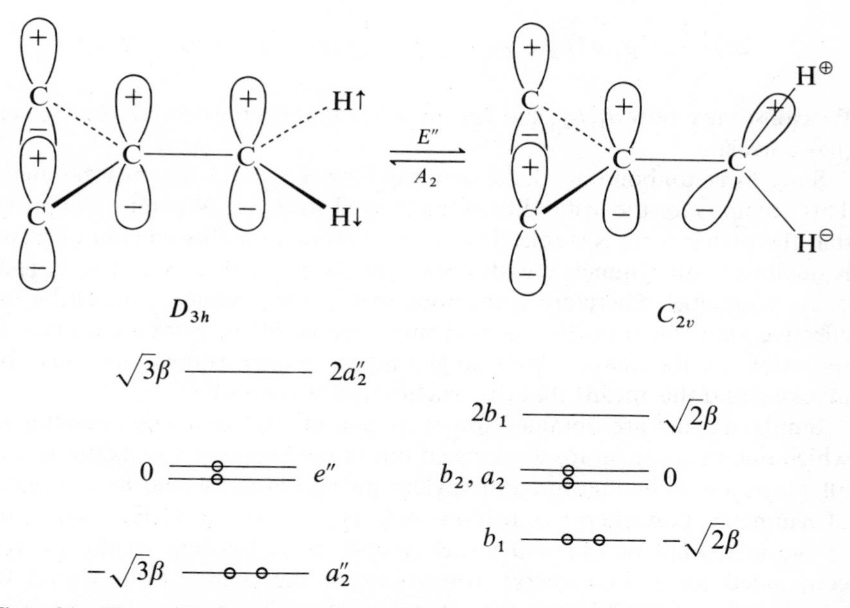


**Figure 34.** The $\pi$ orbitals and energies of $C(CH_2)_3$ in planar, $D_{3h}$, and twisted, $C_{2v}$, forms; the motions of the H atoms needed to interconvert the two forms are shown.

but accurate all-electron calculations completely verify the conclusions drawn.[128]

The all-planar form has a ground-state configuration $(a_2'')^2(e'')^2$, leading to a $^3A_2''$ state. This has a lower energy than the twisted form by an amount $2(\sqrt{3} - \sqrt{2})\,\beta$ in the Hückel approximation. The $C_{2v}$ structure is a $^3B_1$ state arising from $(b_1)^2(b_2)^1(a_2)^1$. The reaction coordinate needed to convert twisted trimethylemethane into the planar form is of $A_2$ species. This is shown by the arrows of Fig. 34.

After the $CH_2$ group begins to twist, the point group becomes $C_2$. The $b_1$, $b_2$, and $2b_1$ orbitals all become of $B$ species and mix together until the twisting is complete. The $a_2$ orbital must remain unchanged, since it cannot mix with any other $\pi$ orbital. From Hückel theory, the orbitals change as follows:

$$a_2'' \quad \frac{1}{\sqrt{6}}(3p_1 + p_2 + p_3 + p_4) \rightarrow \frac{1}{2}(2p_1 + p_3 + p_4) \quad b_1$$

$$e'' \begin{cases} \dfrac{1}{\sqrt{6}}(2p_2 - p_3 - p_4) \rightarrow p_2 & b_2 \\ \dfrac{1}{\sqrt{2}}(p_3 - p_4) \rightarrow \dfrac{1}{\sqrt{2}}(p_3 - p_4) & a_2 \end{cases}$$

$$2a_2'' \quad \frac{1}{\sqrt{6}}(3p_1 - p_2 - p_3 - p_4) \rightarrow \frac{1}{2}(2p_1 - p_3 - p_4) \quad 2b_1$$

We can see just how $b_1$, $b_2$, and $2b_1$ must mix to form the orbitals with which they correlate.

Since $b_2$ is nonbonding, the energy gap between it and $2b_1$ must be small. This accounts for the instability of the twisted structure. We must still prove that the planar form is stable. The twisting mode of methylene out of plane is obviously antisymmetric with respect to the plane (Fig. 34). It is, in fact, of $E''$ symmetry. Therefore transitions within the $\pi$ system will all be ineffective, since the transition density must necessarily be symmetric towards reflection in the plane. Accordingly higher-energy transitions must be invoked, and this means that the reaction is not favored.

Similar results are obtained for other potentially conjugated systems in which one or more atoms are twisted out of conjugation. For example, the allyl group with one terminal methylene group twisted would have a plane of symmetry. Low-energy transitions in $C_3H_5^+$, $C_3H_5$, or $C_3H_5^-$ would be of the $(a'') \rightarrow (a')$ or $(a') \rightarrow (a'')$ variety and cause twisting to the planar, conjugated form. Low-energy transitions in the planar form would be $(a'')$ to $(a'')$ and would not allow twisting. The allyl anion would be most liable to become nonplanar.

## GEOMETRIES OF COORDINATED LIGANDS

Another application of SOJT is in predicting changes in geometry that occur in small molecules upon complexation by a metal ion.[129] Included in this would be considerations of the mode of attachment of such a molecule when several possibilities exist. An immediate effect on structure results from the adding of electrons to certain orbitals of the molecule, and the removal of electrons from other orbitals. This subject will be discussed in Chapters 4 and 6. The usual assumption will be that the HOMO of the molecule will donate electrons to the metal and the LUMO will accept electrons. The most symmetric mode of bonding consistent with this is then considered first.

Second-order Jahn–Teller theory may then be used to estimate if this is a stable bonding scheme. The strength of bonding per se is not a dominant feature of such an analysis. The bonding itself requires interaction between filled and empty orbitals of the same symmetry (net positive overlap). The intensity of the bonding can only produce changes in the geometry of the coordinated ligand within the same point group as for a weakly bonded molecule. For example, ethylene coordinated to a metal atom is subject to symmetric distortions such as lengthening the C—C bond or equally moving all four hydrogen atoms away from the metal atom.[130]

The most symmetrical bonding scheme can be unstable if a suitable pairing of filled and empty MOs occurs in the complex. To illustrate, we take the case of five-coordinated complexes containing the nitrosyl ligand, $ML_4NO$. The nitrosyl bonding can either be linear, 180° bond angle, or bent, 120° bond angle, approximately.[131] The same situation exists for $ML_5NO$. We assume initially a linear coordination with $C_{4v}$ symmetry, as shown in Fig. 35. An approximate MO scheme is also shown. The normal mode which converts the linear NO group to a bent one is of $E$ species. It can either occur to maintain a dihedral plane, as shown, or a vertical plane. In any event, the point group becomes $C_s$.

For convenience consider the nitrosyl ligand to be $NO^+$, isoelectronic with CO, (if one starts with NO neutral, allow one electron to transfer to the metal). If the resulting metal atom is $d^6$, the configuration in the complex will be

$$(e)^4(b_2)^2(a_1)^0(2e)^0$$

The transition density is $B_2$, and the energy gap is large. The linear form is stable. If the metal atom is $d^8$, the configuration becomes

$$(e)^4(b_2)^2(a_1)^2(2e)^0$$

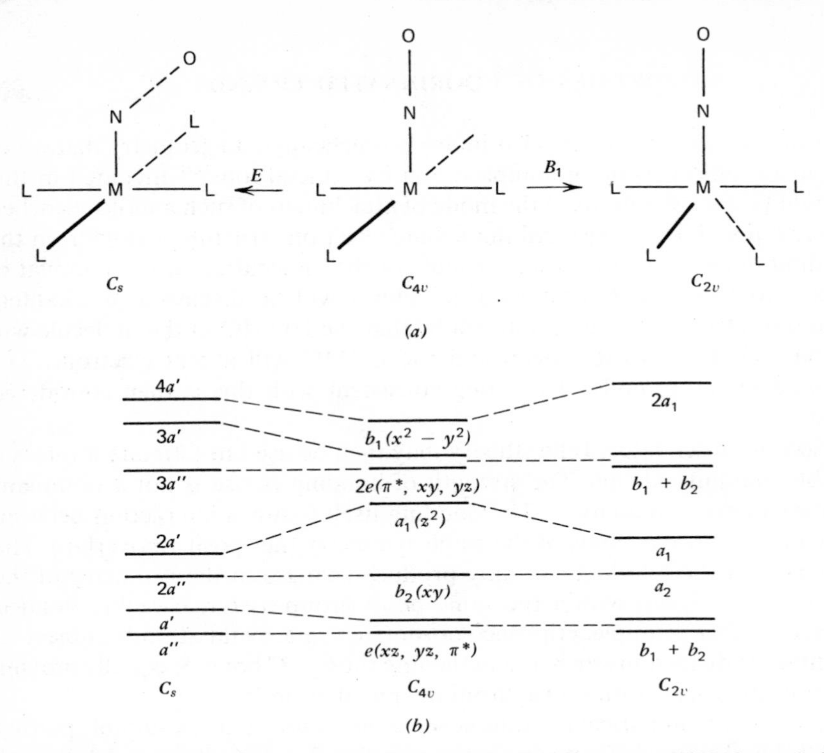


**Figure 35.** (*a*) The two distortions that can stabilize a $C_{4v}$ $ML_4NO$ complex; (*b*) MO correlation showing how a $d^8$ complex is stabilized by distortion. In each case the $a_1$, $(z^2)$ orbital is lowered in energy.

Now the energy gap is small, the transition density is $E$, and the coordinated NO will bend. Other numbers of $d$ electrons can also be considered and other coordination numbers and geometries. Note that the bending results from a mixing of the $a_1$ orbital (primarily $d_{z^2}$ on the metal) with one of the $2e$ orbitals which is $\pi$ antibonding between metal and ligand. Thus a strong $\sigma$-bond is formed at the expense of the $\pi$ bond that existed in the linear model. The strong $\sigma$ bonding can be considered to result from the partial donation of a pair of electrons from the metal to the ligand, which formally becomes $NO^-$ as a result.[132]

In the preceding analysis a square pyramidal structure was assumed. Sometimes $d^8$ complexes are found with linear nitrosyl ligands, but then the structure is a trigonal bipyramid of $C_{2v}$ symmetry. This could be explained by assuming that the $2e$ orbitals now lie above the $b_1$ orbital ($b_1$ is the anti-bonding $d_{x^2-y^2}$ orbital concentrated on the metal atom). The electron configuration becomes

$$(e)^4(b_2)^2(a_1)^2(b_1)^0(2e)^0$$

The transition density is $B_1$, which is the reaction coordinate giving rise to a trigonal bipyramid. The correlations of the MOs in $C_{4v}$, $C_{2v}$, and $C_s$ point groups is shown in Fig. 35.

### Structures of $M(C_5H_5)_2$

Another example comes from the structures of the cyclopentadienides of divalent metal ions, $M(C_5H_5)_2$, or $MCp_2$. The parent molecule ferrocene, $FeCp_2$ has the well known sandwich structure of either $D_{5h}$ (eclipsed) or $D_{5d}$ (staggered) point group. Essentially there is free rotation of the Cp rings, so that either point group can be used, or even the $D_{\infty h}$ group. Such a structure is said to be *pentahapto*,[133] bonded to five sites in the ring. There is strong $\pi$ bonding. An alternative structure is the $\sigma$-bonded one, in which only one carbon of each ring is bound to metal, the *monohapto* case. A third structure, *trihapto*, is possible, but not known for $MCp_2$ molecules.

Figure 36 shows schematically the *pentahapto* structure and the two metal-ring vibrations which can convert it in to the *monohapto* form. These are $E_{1u}$ and $E_{1g}$ species. An ab initio calculation on ferrocene gives the MO sequence (in $D_{5d}$)[134]

$$(a_{1g})^2(e_{2g})^4(a_{2u})^2(e_{2u})^4(2e_{2g})^4(e_{1g})^4(e_{1u})^4(2e_{1g})^0(2e_{2u})^0(2e_{2g})^0$$

The empty orbitals are not calculated ab initio, but result from a combination of semiempirical calculations and experimental evidence on other metallocenes.[135]

The closed-shell configuration of ferrocene is stable, with a large energy gap. Nickelocene, $NiCp_2$, has two more electrons and is known to be in the $^3A_{2g}$ state arising from

$$\ldots(e_{1u})^4(2e_{1g})^2(2e_{2u})^0(3e_{2g})^0$$

The average structure is the same as for $FeCp_2$. However the structure is much less rigid; all vibrations having to do with the rings moving with respect to the metal are of much lower frequency.[136] In particular the $E_{1u}$ and $E_{1g}$ modes are very soft. This can be attributed to the easy $(2e_{1g}) \rightarrow (2e_{2u})$ excitation, since $(E_{1g} \times E_{2u}) = (E_{1u} + E_{2u})$, and to the $(2e_{1g}) \rightarrow (3e_{2g})$.

If the same MO sequence can be assumed, then $HgCp_2$ should be

$$\ldots(e_{1u})^4(e_{1g})^4(e_{2u})^0(e_{2g})^0$$

This should be much more easily deformed than $NiCp_2$, by way of an $E_{1u}$ mode. This can account for the fact that mercurocene is a $\sigma$-bonded, monohapto, compound.[137] It is not known whether the rings are in the $C_{2v}$ or $C_{2h}$

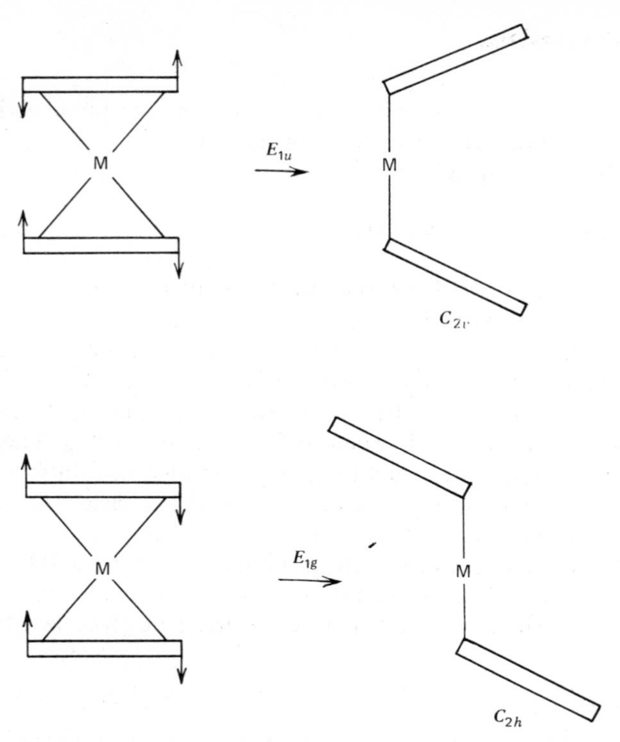


**Figure 36.** The two vibrational modes that convert a $MCp_2$ molecule from a $D_{5d}$ sandwich structure to $C_{2v}$ or $C_{2h}$ $\sigma$-bonded structures.

form. $PbCp_2$ and $SnCp_2$ have two more electrons than $HgCp_2$. The configuration would be

$$(e_{1u})^4(e_{1g})^4(e_{2u})^2(e_{2g})^0$$

Since $(E_{2u} \times E_{2g}) = (A_{1u} + A_{2u} + E_{1u})$, the ring-tipping mode is enhanced again, but less strongly than for $HgCp_2$ (excitation from a half-filled shell is more difficult than from a filled shell). The structures are found to be $C_{2v}$,[136] as expected from Fig. 36.

It has also been postulated that titanocene has a $C_{2v}$ structure because of the SOJT effect.[138] However a different MO sequence would be needed to predict this. Of course it is quite possible that titanocene would have a different ordering of orbital energies. While no *trihapto* examples are known for $MCp_2$, the related compounds in which a benzene molecule replaces $C_5H_5^-$ show *tetrahapto* bonding in some cases. An explanation based on the SOJT effect has been given.[129a] One of the earliest applications of SOJT theory was to bis(hexamethylbenzene)cobalt, which was shown to be unstable to $A_{2u}$ or $E_{1u}$ distortions (in $D_{6h}$).[139] The molecular shape has not yet been determined for this interesting case.

## SOME COMMENTS ON MOLECULAR ORBITALS

The point has been made that MO theory has many advantages. The orbitals have correct symmetry properties, they are calculated in a straightforward manner (at least for closed shells) by current ab initio methods, and orbital energies are closely related to experimental properties, such as IPs and absorption spectra. Indeed the simple assumptions are often made that the IPs can be equated to the orbital energies, and that UV and visible absorption bands correspond directly to differences in orbital energies. While the errors in these assumptions, have already been mentioned (p. 28), it is worth while to examine more closely the relationship between observable quantities, computed orbital energies, and the true wave functions of the molecular system. Errors due to the Born–Oppenheimer approximation and to relativistic effects are not considered.

Most ab initio calculations to date are of the SCF variety. For small molecules, at least, the basis set of AOs taken is large enough to approach the Hartree–Fock limit. However only a single configuration is considered in the HF method, and so there is a large error in the interelectronic repulsion, called the "correlation error." The remedy is inclusion of configuration interaction.[140] In considering a chemical reaction, it might be hoped that the correlation error would cancel between reactants and products. The evidence is that this does not often happen. In a dissociation reaction, for example, the correlation error is always larger in the molecule than in the atoms.

The problem of estimating energy changes in chemical reactions is complicated by the fact that all of the electrons find themselves at a different energy level after the change. Table 12 shows the orbital energies for $ZnF_2$, calculated by an SCF single-configuration method.[24] With a single exception, all of the MOs can be identified with a single AO on either Zn or F, which makes comparison easy. All of the Zn atom-based MOs are lowered in energy and all of the F atom MOs are raised in energy by about 0.1 AU, compared to the free atom. This results because the final wavefunction corresponds to a charge of $+1.39$ on zinc, and a negative charge half of this on each fluorine. The $7\sigma_g$ orbital is a bonding MO, containing 17.2% Zn 4*s*, 7.9% Zn 3*d*, and the rest F 2*p*, plus 1.1% F 2*s*. The energy of the 4*s* orbital on Zn atom (not shown) is experimentally equal to 9.39 eV, or $-0.347$ AU.

**Table 12**[a]

| Molecular-orbital Energy | MO | Corresponding AO | Atomic orbital energy |
|---|---|---|---|
| −353.364 | $1\sigma_g$ | Zn 1s | −353.264 |
| −44.434 | $2\sigma_g$ | Zn 2s | −44.334 |
| −39.003 | $1\pi_u$ | Zn 2p | −38.899 |
| −39.003 | | Zn 2p | −38.899 |
| −38.997 | $1\sigma_u$ | Zn 2p | −38.899 |
| −26.216 | $2\sigma_u$ | F 1s | −26.383 |
| −26.216 | $3\sigma_g$ | F 1s | −26.383 |
| −5.713 | $4\sigma_g$ | Zn 3s | −5.607 |
| −3.919 | $2\pi_u$ | Zn 3p | −3.808 |
| −3.919 | | Zn 3p | −3.808 |
| −3.915 | $3\sigma_u$ | Zn 3p | −3.808 |
| −1.493 | $5\sigma_g$ | F 2s | −1.567 |
| −1.483 | $4\sigma_u$ | F 2s | −1.567 |
| −0.870 | $1\pi_g$ | Zn 3d | −0.755 |
| −0.870 | | Zn 3d | −0.755 |
| −0.869 | $1\delta_g$ | Zn 3d | −0.755 |
| −0.869 | | Zn 3d | −0.755 |
| −0.860 | $6\sigma_g$ | Zn 3d | −0.755 |
| −0.581 | $7\sigma_g$ | bonding orbital | |
| −0.582 | $3\pi_u$ | F 2p | −0.726 |
| −0.582 | | F 2p | −0.726 |
| −0.582 | $5\sigma_u$ | F 2p | −0.726 |
| −0.565 | $2\pi_g$ | F 2p | −0.726 |
| −0.565 | | F 2p | −0.726 |

[a] Orbital energies of linear $ZnF_2$; bond distance = 1.76 A; energies in AU; data from reference 24.

If we look at the reaction

$$ZnF_{2(g)} \longrightarrow Zn_{(g)} + 2F_{(g)} \tag{15}$$

the biggest change is that the two 4s electrons of zinc are now in the $7\sigma_g$ orbital, which is much more stable. The experimental dissociation energy for (12) is 191 kcal = 8.3 eV = 0.305 AU.[141] The dissociation energy from the theoretical calculation is 5.3 eV = 122 kcal. An error of 69 kcal is rather depressing, considering that this is a rather accurate calculation.

Molecule formation from atoms would be expected to lower some orbital energies and raise others, as a general phenomenon. If we take ionization potentials as a measure of orbital energy (vide infra), we find results such as the following:

| | | | | | |
|---|---|---|---|---|---|
| H | 13.6 eV | 1*s* | Cl | 13.0 eV | 3*p* |
| $H_2$ | 15.4 | $\sigma_g$ | $Cl_2$ | 11.3 | $\pi_g$ |
| N | 14.0 | 2*p* | | 13.9 | $\pi_u$ |
| $N_2$ | 15.6 | $2\sigma_g$ | | 15.7 | $2\sigma_g$ |
| | 16.9 | $\pi_u$ | | | |

As expected, electrons in bonding MOs have their binding energy lowered upon molecule formation. Electrons in antibonding MOs, such as the $\pi_g$ orbital of $Cl_2$, have their energies raised.

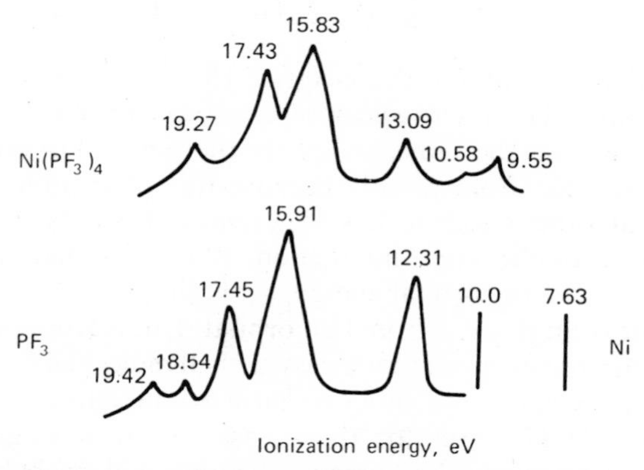


**Figure 37.** Ionization potentials of Ni, $PF_3$, and $Ni(PF_3)_4$; the vertical lines are for the 3*d* and 4*s* orbitals of nickel atoms, with 3*d* the more stable (after reference 142).

A more complicated example is shown in Fig. 37. The IPs for Ni atoms, $PF_3$ molecules, and the complex $Ni(PF_3)_4$ are shown.[142] The two IPs for Ni shown are for removal of an electron from the 4*s* orbital (7.63 eV) and from a 3*d* orbital (10 eV). The two lowest IPs for the complex are from orbitals localized heavily on the metal atom 3*d* orbitals. They presumably are from $t_2$ and *e* orbitals, similar to $Ni(CO)_4$.[143] The figures of 9.6 and 10.6 eV show little bonding character. The lone pair of electrons on $PF_3$ have an IP of 12.3 eV in the free ligand. In the complex, this is reduced to 13.0 eV, showing that a $\sigma$ bond has been formed. The lower ligand MOs seem little affected by complexation.

### Koopmans's Theorem

As pointed out earlier (p. 41) the total energy of a molecule is given by the expression

$$E_0 = \sum_n \varepsilon_n + \sum_{A,B} I_{AB} - \sum_{i,j} I_{ij} \tag{16}$$

where $\varepsilon_n$ refers to the energies of the occupied MOs, $I_{AB}$ refers to nuclear–nuclear, and $I_{ij}$ to electron–electron, potential energies. The last term is subtracted because in an SCF calculation each electron–electron repulsion has been counted twice. Suppose we now remove one electron; the energy becomes

$$E_1 = \sum \varepsilon'_n + \sum I_{AB} - \sum I'_{ij} \tag{17}$$

We assume that, except for the electron that is removed, the molecule remains the same. The internuclear separations are not changed and the wavefunction remains the same, except for the removal of one spin orbital. Then all the $\varepsilon'_n$ differ from $\varepsilon_n$ only because the interaction of the missing electron with all other electrons has been removed. Similarly $I'_{ij}$ differs from $I_{ij}$ only because of the same interaction. It follows that the $E_1 - E_0 = -\varepsilon_m = \mathrm{IP}_m$. That is, the orbital energy, $\varepsilon_m$, is simply the negative of the IP needed to remove an electron from that orbital. This is Koopmans's theorem.

The above argument is valid for a closed-shell molecule. For an open-shell molecule there are some difficulties because orbital energies are not always clearly defined.[144] However for those cases where a single determinant wavefunction is adequate, Koopman's theorem will probably be as valid as for closed shell molecules. This will still leave considerable room for failure.

It is obvious that, if the wavefunction reorganizes (relaxes) to a better electron density distribution, as the electron ionizes, then the measured IP will be *less* than the orbital energy. On the other hand, the greater correlation energy in the molecule compared to the ion will give an error that tends to compensate. Changes in the internuclear distances on ionization are related to the difference between the vertical and adiabatic IPs. In fact nuclei do not have time to rearrange during ionization so that vertical IPs are measured.

For many molecules Koopmans's theorem is approximately obeyed. Actually it is found that the experimental IP is some 8 % lower than the orbital energies computed by SCF methods. For example, for $CF_4$ the following comparison has been found[145]

| Orbital | $IP_{exp}$ | $0.92\ \varepsilon_{calc}$ |
|---|---|---|
| $1t_1$ | $\geq$15.35 eV | 17.85 eV |
| $3t_2$ | 17.1 | 18.08 |
| $1e$ | 18.3 | 19.63 |
| $2t_2$ | 21.7 | 22.90 |
| $2a_1$ | 25.1 | 25.90 |

These orbitals were discussed earlier (p. 185). It is remarkable that the same calculation gives orbital energies for the C and F 1*s* orbitals (in $CF_4$), which are in good agreement with IPs measured by the ESCA technique.[146] A factor of 0.95 $\varepsilon_{calc}$ fits the 1*s* binding energies.

Koopmans's theorem works best for relatively hard, or nonpolarizable, molecules, where electronic rearrangement upon ionization is small. For soft, polarizable, molecules, failure is to be expected. It has been known for some time that aromatic hydrocarbons do not agree with Koopmans's theorem as well as do aliphatic hydrocarbons.[147a] Complexes of the transition metals are also particularly susceptible to failures because of the polarizabilities of the *d* electrons.[147b] The failures here refer to the predictions that: (1) the lowest IP will be for removal of an electron from the HOMO and (2) the various measured, or calculated by ab initio methods, IPs can be set equal to the various orbital energies in order.

For example, the SCF calculations on $Ni(CN)_4^{2-}$ referred to earlier[51b] were extended to the $Ni(CN)_4^-$ formed by removal of an electron from each of five different orbitals. The two lowest IPs were found to be from the $a_{1g}$ and $e_g$ MOs made up mainly of metal 3*d* AOs. However these were not the highest occupied MOs, which were ligand $\pi$ in character. For the latter, it was approximately true that the orbital energy was equal to the IP. For the metal *d* orbitals, this was not at all true. The same situation has been found for ferrocene.[134] The two lowest IPs were calculated to be for the removal of the *d*-like $e_{2g}$ and $a_{1g}$ electrons, but these orbitals lie much lower in energy than some of the ligand $\pi$ orbitals.

Experimentally the opposite situation is also found. The ionization potentials of the acetylacetonates, or hexafluoroacetylacetonates, of various metal ions are found to be nearly the same, independent of the metal ion and the number of *d* electrons contained.[148] Since the enol form of acetylacetone itself gives the same IP, it is clear that ionization is from the $\pi$ orbitals of the ligand and not from the *d* shell. Yet naive ligand-field theory always assumes that the *d* manifold lies in between the filled ligand orbitals and the empty antibonding orbitals of the system. This assumption accounts for the fact that *d*–*d* transitions occur more easily than charge transfer processes. Only in the cases of $Ti(hfa)_3$ and $V(hfa)_3$, where hfa = $CF_3COCHCOCF_3^-$, does the photoelectron spectrum show ionization of $d$ electrons before ligand $\pi$ electrons.[149]

A good indication of the erratic behavior of $d$ electrons is found in a recent SCF calculation on planar, trigonal $FeF_3$.[150] The ground state is found to be the high-spin $^6A_1'$ state. The MO sequence is

$$(4e')^4 \;\vdots\; (e'')^2(6a_1')^1(5e')^2 \;\vdots\; (7a_1')^2(6e')^4(7e')^4(2e'')^2(3a_2'')^2(a_2')^2$$

with the $d$ orbitals, as usual, enclosed. The half-filled $d$ shell is found on the average to be 0.35 AU = 10 eV *lower* in energy than the six completely filled shells above them. Yet removing the electrons from the higher MOs and putting them into the $d$ shell gives higher total energies than the configuration shown. This clearly shows the inadequacies of simple one-electron MO procedure.

Koopmans's theorem is not always poor for transition-metal complexes. For reasons which are not very clear, it works well for $Ni(CO)_4$.[51b] Often many IPs can be measured for a single molecule. Semiempirical MO calculations can then be made that "identify" each of these with a particular orbital energy. Such identifications must be regarded with great suspicion, since accurate SCF calculations often show quite a different interpretation.[151] Assignments of the various IPs is often made on the basis of intensities: an $e$ orbital is assumed to give twice the intensity of a nondegenerate orbital, for example. Such procedures are not justified. Photoionization cross sections depend on the atomic compositions of MOs, as well as on their degeneracies.

Semiempirical schemes are by no means without merit. They still serve to develop a framework for understanding many chemical and physical phenomena. Even a rough idea of the energy of a particular MO can be very useful. This applies to the vast amount of semitheoretical analysis of absorption spectra as well. The major goals of such analyses has been to develop an overall understanding and to predict the spectra of closely related molecules. These goals have been met, in general. An absolute requirement is that the basic imperfections of such methods be well understood.[152]

The only reliable theoretical prediction of absorption spectra is a good-quality ab initio calculation on the ground state of the molecule and separate calculations of the several excited states. Changes in orbital energies upon excitation must be evaluated. These result chiefly from changes in interelectronic repulsions, which cannot be ignored, but changes in geometry can also occur that affect the positions of the 0,0 bands. Correlations of UV and visible spectra with ionization potentials is very hazardous.

Orbital energies, both experimentally and theoretically, are more difficult for ionic solids than for gaseous molecules, and SCF calculations for an isolated $CrO_4^{2-}$ ion will produce positive eigenvalues for many occupied orbitals.[51a] Thus such electrons should not be bound. This agrees with the observation that no mononegative ion can bind another electron exoenergetically. In the solid state, $K_2CrO_4$, for example, the electrons would be bound because of the positive potential produced by the cations. To get an absolute binding energy (orbital energy) requires a Madelung calculation of an infinite array of cations and anions.

Experimentally photoelectron spectroscopy cannot be used for insulating solids. However the ESCA method using an x-ray source is workable. The resolution is low, but it is possible to measure the kinetic energy of electrons emitted from the valence shell.[153] However the binding energies are not absolute values, but are referenced to the Fermi level of the solid. Presumably measuring work functions of the solid would permit absolute values to be calculated. There would still be a problem of the Madelung potential to consider.

At any rate it is possible to get relative IPs, which can then be compared to various MO-level schemes. For example, $ClO_4^-$ gives a series of bands at 6.3, 9.0, 13.4, 16.5, 27.0, and 34.4 eV.[153a] It is rather disturbing to find that the same values are found independent of the cation that is used. Some idea of the Madelung and work-function corrections may be obtained by noting that the $3p$ ionization of $K^+$ is found at 17 eV in $KMnO_4$. The gas-phase value is 31.8 eV.

## REFERENCES


1. F. Basolo and R. G. Pearson, *Mechanisms of Inorganic Reactions*, 2nd ed., Wiley, New York, 1967, Chapter 2.
2. L. Pauling, *The Nature of the Chemical Bond*, 3rd ed., Cornell U. P., Ithaca, N. Y., 1960.
3. R. J. Gillespie, *Molecular Geometry*, Van Nostrand, Princeton, N. J., 1972; H. B. Thompson, *J. Am. Chem. Soc.*, **93**, 4609 (1971).
4. L. E. Orgel, *An Introduction to Transition Metal Chemistry*, Wiley, New York, 1960.
5. G. W. Schnuelle and R. G. Parr, *J. Am. Chem. Soc.*, **94**, 8974 (1972).
6. H. Nakatsuji, *J. Am. Chem. Soc.*, **95**, 345, 354 (1973); B. M. Deb, ibid., **96**, 2030 (1974); B. M. Deb, P. N. Sen, and S. K. Bose, ibid., 2044.
7. A. D. Walsh, *J. Chem. Soc.*, 2260–2317 (1953).
8. S. D. Peyerimhoff, R. J. Buenker, and L. C. Allen, *J. Chem. Phys.*, **45**, 734 (1966); R. J. Buenker and S. D. Peyerimhoff, ibid., 3682.
9. R. Hoffmann, *J. Chem. Phys.*, **39**, 1397 (1963); **40**, 2047, 2474 (1964).
10. B. M. Gimarc, *J. Am. Chem. Soc.*, **92**, 266 (1970).
11. A. A. Frost, *J. Chem. Phys.*, **40**, 3530 (1964).
12. R. F. W. Bader, *Can. J. Chem.*, **40**, 1164 (1962).
13. L. S. Bartell, *J. Chem. Ed.*, **45**, 754 (1968).
14. R. G. Pearson, *J. Am. Chem. Soc.*, **91**, 4947 (1969).

15. Literature references to structural data for a large number of simple molecules are contained in references 3, 5, 6, and 14.
16. P. K. Pearson, S. V. O'Neil, H. E. Schaefer, III, and C. F. Bender, *J. Chem. Phys.*, **56**, 4626 (1972).
17. P. N. Noble and R. N. Kortzeborn, *J. Chem. Phys.*, **52**, 5375 (1970).
18. W. L. Jorgensen and L. Salem, *The Organic Chemists' Book of Orbitals*, Academic, New York, 1973, p. 130.
19. S. D. Peyerimhoff, R. J. Bunker, and J. L. Whitten, *J. Chem. Phys.*, **46**, 1707 (1967).
20. R. E. Clementi, *J. Chem. Phys.*, **36**, 45 (1962).
21. G. Herzberg, *Dis. Faraday Soc.*, **35**, 113 (1963).
22. A. Büchler, J. L. Stauffer, and W. Klemperer, *J. Am. Chem. Soc.*, **86**, 4554 (1964).
23. E. F. Hayes, *J. Phys. Chem.*, **70**, 3740 (1966).
24. D. R. Yarkony and H. F. Schaefer, III, *Chem. Phys. Lett.*, **15**, 1514 (1972).
25. H. Basch, C. Hollister, and J. W. Moskowitz, *Chem. Phys. Lett.*, **4**, 79 (1969).
26. L. C. Snyder and H. Basch, *Molecular Wave Functions and Properties*, Wiley, New York, 1972, p. 194.
27. R. H. Hughes, *J. Chem. Phys.*, **24**, 131 (1956).
28. A. Snelson, *J. Phys. Chem.*, **74**, 537 (1970); A. Kaldor and R. F. Porter, *J. Am. Chem. Soc.*, **93**, 2140 (1971).
29. J. W. Rabalais, et al., *J. Chem. Phys.*, **58**, 3370 (1973).
30. R. L. Morehouse, J. J. Christiansen, and W. Gordy, J., *Chem. Phys.*, **45**, 1751 (1966).
31. L. Pauling, *J. Chem. Phys.*, **51**, 2767 (1969).
32. K. Morukama, L. Pedersen, and M. Karplus, *J. Chem. Phys.*, **48**, 4801 (1968).
33. J. M. Lehn and B. Munsch, *Chem. Commun.*, 1327 (1969).
34. (a) I. H. Hillier and V. R. Saunders, *J. Chem. Soc.*, **A**, 2475 (1970); (b) S. Elbel, H. Bergmann and W. Ensslin, *J. Chem. Soc., Faraday Trans.*, **II**, 555 (1974).
35. A. Rauk, J. D. Andose, W. G. Frick, R. Tong, and K. Mislow, *J. Am. Chem. Soc.*, **93**, 6507 (1971).
36. See reference 26, p. 320. The $a_2''$ orbitals have been mislabeled as $a_1''$.
37. (a) P. J. Bassett and D. R. Lloyd, *Chem. Commun.*, **36** (1970); (b) R. L. DeKock and D. R. Lloyd, *J. Chem. Soc. Dalton*, 526 (1973).
38. J. F. Olson and L. Burnelle, *J. Am. Chem. Soc.*, **92**, 3659 (1970).
39. Reference 26, p. 10.
40. A. D. Liehr, *J. Phys. Chem.*, **69**, 389 (1963).
41. (a) J. Arents and L. C. Allen, *J. Chem. Phys.*, **53**, 73 (1970); (b) J. W. Rabalais et al., *Phys. Scripta*, **3**, 113 (1971).
42. H. J. Monckhurst, *Chem. Commun.*, 1111 (1968).
43. See reference 26, p. 362.
44. J. D. Robinson, Ph.D. Thesis, Washington University, St. Louis, Mo. 1956.
45. L. S. Bartell and R. M. Gavin, Jr., *J. Chem. Phys.*, **48**, 2466 (1968).
46. H. Basch, J. W. Moskowitz, C. Hollister, and D. Hankin, *J. Chem. Phys.*, **55**, 1922 (1971).
47. H. Basch, C. Hollister, and J. W. Moskowitz, *Sigma Molecular Orbital Theory*, O. Sinanoglu and K. Wiberg (eds.), Yale U. P., New Haven, 1970, p. 449.
48. J. Demuynck and A. Viellard, *Chem. Phys. Lett.*, **6**, 204 (1970).

49. W. T. Van der Lugt, *Chem. Phys. Lett.*, **10**, 117 (1971).
50. J. A. Tossell and W. N. Lipscomb, *J. Am. Chem. Soc.*, **94**, 1505 (1972).
51. (a) I. H. Hillier and V. R. Saunders, *Chem. Phys. Lett.*, **9**, 219 (1971); (b) D. E. Ellis and F. W. Averill, *J. Chem. Phys.*, **60**, 2856 (1974); (c) J. Demuynck and A. Viellard, *Theor. Chim. Acta.*, **28**, 241 (1973).
52. R. G. Pearson, *J. Am. Chem. Soc.*, **91**, 1252 (1969).
53. P. A. Cox, S. Evans, A. Hamnett, and A. F. Orchard, *Chem. Phys. Lett.*, **7**, 414 (1970).
54. G. D. Stucky, J. B. Folkers, and T. J. Kistemacher, *Acta Cryst.*, **23**, 1064 (1967).
55. L. Sacconi, *Transition Metal Chem.*, **4**, 199 (1968).
56. D. Forster, *Chem. Commun.*, **113** (1967); O. Mønsted and J. Bjerrum, *Acta Chem. Scand.*, **21**, 1116 (1967); B. J. Hathaway and A. A. G. Tomlinson, *Coord. Chem. Rev.*, **5**, 1 (1970).
57. (a) B. A. Frenz and J. A. Ibers, *MTP International Review of Science, Physical Chemistry, Series One*, Vol. 11, Butterworths, London, 1972, p. 33; (b) J. S. Wood, *Prog. Inorg. Chem.*, S. J. Lippard, (ed.), Interscience, New York, Vol. 16, 1972, p. 354.
58. R. R. Holmes, *Acc. Chem. Res.*, **5**, 296 (1972).
59. K. N. Raymond, P. W. R. Corfield, and J. A. Ibers, *Inorg. Chem.*, **7**, 1362 (1968); C. P. Brock, et al., ibid., **12**, 1304 (1973).
60. V. Dyczmons, V. Staemmler, and W. Kutzelnigg, *Chem. Phys. Lett.*, **5**, 361 (1970).
61. A. Rauk, L. C. Allen, and K. Mislow, *J. Am. Chem. Soc.*, **94**, 3035 (1972).
62. R. S. Berry, *J. Chem. Phys.*, **32**, 933 (1960).
63. R. Hoffmann, J. M. Howell, and E. L. Muetterties, *J. Am. Chem. Soc.*, **94**, 3047 (1972).
64. I. Ugi, D. Marquarding, H. Klusacek, P. Gillespie, and F. Ramirez, *Acc. Chem. Res.*, **4**, 288 (1971).
65. (a) R. S. Berry, M. Tamres, C. J. Ballhausen, and H. Johansen, *Acta Chem. Scand.*, **22**, 231 (1968); (b) A. Strich and A. Viellard, *J. Am. Chem. Soc.*, **95**, 5574 (1973).
66. (a) A. R. Rossi and R. Hoffmann, *Inorg. Chem.*, **14**, 365 (1975); (b) J. K. Burdett, ibid., 375.
67. (a) P. Meakin, J. P. Jesson, F. N. Tebbe, and E. L. Muetterties, *J. Am. Chem. Soc.*, **93**, 1797 (1971); ibid., **94**, 5271 (1972); (b) E. L. Muetterties and J. Guggenberger, *J. Am. Chem. Soc.*, **96**, 1748 (1974).
68. (a) A. O. Caride, H. Panepucci, and S. J. Zanette, *J. Chem. Phys.*, **55**, 3651 (1971); (b) R. N. Perutz and J. J. Turner, *Inorg. Chem.*, **14**, 262 (1975); (c) E. P. Kundig and G. A. Ozin, *J. Am. Chem. Soc.*, **96**, 1820 (1974).
69. F. Basolo and R. G. Pearson, *Mechanisms of Inorganic Reactions*, 2nd ed., Wiley, New York, 1967, Chapters 3 and 4.
70. D. A. Buckingham, I. I. Olsen, and A. M. Sargeson, *Austr. J. Chem.*, **20**, 597 (1967); *J. Am. Chem. Soc.*, **90**, 6654 (1968).
71. J. A. Broomhead and L. A. P. Kane-Maguire, *Inorg. Chem.*, **7**, 2519 (1968).
72. S. Cradock and W. Savage, *Inorg. Nucl. Chem. Lett.*, **8**, 753 (1972).
73. F. A. Gianturco, C. Guidotti, and U. Lamanna, *J. Chem. Phys.*, **57**, 840 (1972).
74. D. M. Adams and D. M. Morris, *J. Chem. Soc.*, **A**, 2067 (1967); ibid., 878 (1971).
75. K. Nakamoto, *Infrared Spectra of Inorganic Compounds*, Wiley, New York, 1963, p. 119.
76. J. W. Moskowitz, C. Hollister, C. J. Hornbach, and H. Basch, *J. Chem. Phys.*, **53**, 2570 (1970).

77. C. K. Jørgensen, *Orbitals in Atoms and Molecules*, Academic, New York, 1962, Chapter 8.
78. L. E. Orgel, *Trans. Faraday Soc.*, **26**, 138 (1958); *J. Chem. Soc.*, 3815 (1958).
79. L. E. Orgel, *J. Chem. Soc.*, 4186 (1958).
80. H. B. Gray, R. Eisenberg and E. I. Stiefel, *Advances in Chemistry Series*, No. 62, American Chemical Society, Washington, 1966; R. Eisenberg and H. B. Gray, *Inorg. Chem.*, **6**, 1844 (1967).
81. H. H. Schmidtke, *J. Chem. Phys.*, **45**, 3920 (1966); *Theor. Chim. Acta*, **9**, 199 (1968).
82. G. N. Schrauzer and V. P. Mayweg, *J. Am. Chem. Soc.*, **88**, 3235 (1966).
83. E. Stiefel, Z. Dori, and H. B. Gray, *J. Am. Chem. Soc.*, **89**, 3353 (1967).
84. Reference 1, Chapter 4; S. S. Eaton and R. H. Holm, *J. Am. Chem. Soc.*, **93**, 4913 (1971).
85. M. C. Palazzoto, D. J. Duffy, B. L. Edgar, L. Que, Jr., and L. H. Pignolet, *J. Am. Chem. Soc.*, **95**, 4537 (1973).
86. W. J. Adams, H. B. Thompson, and L. S. Bartell, *J. Chem. Phys.*, **53**, 4041 (1970).
87. R. D. Burbank and N. Bartlett, *Chem. Commun.*, 645 (1968).
88. H. H. Claasen, E. L. Gasner, and H. Selig, *J. Chem. Phys.*, **49**, 1803 (1968).
89. E. W. Kaiser, J. S. Muenter, W. Klemperer, and W. E. Falconer, *J. Chem. Phys.*, **53**, 53 (1970).
90. S. J. Lippard, *Prog. Inorg. Chem.*, **8**, 109 (1967); J. L. Hoard and J. V. Silverton, *Inorg. Chem.*, **2**, 235 (1963).
91. S. F. A. Kettle and R. V. Parish, *Spectrochim. Acta*, **21**, 1087 (1965).
92. R. Krishnamurthy and W. B. Schaap, *J. Chem. Ed.*, **46**, 799 (1969). This paper shows how to easily calculate crystal field energies for numerous simple structures.
93. T. V. Long and G. A. Vernon, *J. Am. Chem. Soc.*, **93**, 1919 (1971).
94. E. L. Muetterties, *J. Am. Chem. Soc.*, **93**, 5261 (1971).
95. D. G. Blight and D. L. Kepert, *Theor. Chim. Acta*, **11**, 51 (1968).
96. R. G. Pearson, *J. Chem. Phys.*, **52**, 2167 (1970).
97. A. D. McLean, *J. Chem. Phys.*, **32**, 1595 (1960).
98. See reference 26, p. 62; M. B. Robin, R. R. Hart, and N. A. Kuebler, *J. Am. Chem. Soc.*, **89**, 1564 (1967).
99. W. H. Fink and L. C. Allen, *J. Chem. Phys.*, **46**, 2261 (1967).
100. E. Clementi, *J. Am. Chem. Soc.*, **83**, 4501 (1961).
101. (a) See reference 26, p. 332; (b) J. J. Kaufman, L. A. Burnelle, and J. R. Hamann, *Advances in Chemistry Series*, No. 54, 8 (1966).
102. J. Binenboym, A. Burcat, A. Lifshitz, and J. Shamir, *J. Am. Chem. Soc.*, **88**, 5039 (1966).
103. C. R. Brundle, N. A. Kuebler, M. B. Robin, and H. Basch, *Inorg. Chem.*, **11**, 20 (1972).
104. T. H. Dunning and V. McKoy, *J. Chem. Phys.*, **47**, 1735 (1967).
105. K. Hedberg and R. R. Ryan, *J. Chem. Phys.*, **41**, 2214 (1964); J. N. Gayles and J. Self, ibid., **40**, 3530 (1964).
106. A. Büchler and W. Klemperer, *J. Chem. Phys.*, **29**, 121 (1958); D. B. Chambers, F. Glockling, and J. R. C. Light, *Quart. Rev.* (*Lond.*), **22**, 317 (1968).
107. J. W. Moskowitz and M. C. Harrison, *J. Chem. Phys.*, **42**, 1726 (1965); A. J. Buenker, ibid., **48**, 1368 (1968).
108. A. D. Walsh, *J. Chem. Soc.*, 2325 (1953).
109. See reference 26, pp. 368 and 378.

110. R. G. Snyder and I. C. Hisatsune, *J. Mol. Spectrom.*, **1**, 139 (1957).
111. See reference 26, p. 74.
112. R. J. Buenker, S. D. Peyerimhoff, L. C. Allen, and J. L. Whitten, *J. Chem. Phys.*, **45**, 2835 (1966).
113. $Be_2H_6^{2-}$ is unknown, but its structure should be the same as that of $(C_2H_5)_4Be_2H_2^{2-}$, which has a hydride bridged diborane structure. See G. W. Adamson and H. M. M. Shearer, *Chem. Commun.*, 240 (1965).
114. See reference 26, p. 48.
115. W. H. Fink and L. C. Allen, *J. Chem. Phys.*, **46**, 2261 (1967).
116. B. M. Gimarc, private communication.
117. D. C. Pan and L. C. Allen, *J. Chem. Phys.*, **46**, 1797 (1967).
118. R. J. Buenker and S. D. Peyerimhoff, *J. Chem. Phys.*, **45**, 3682 (1966).
119. S. Aung, R. M. Pitzer, and S. I. Chan, *J. Chem. Phys.*, **45**, 3457 (1966).
120. See reference 26, p. 230.
121. For general reviews, see L. Salem, the *Molecular Orbital Theory of Conjugated Systems*, Benjamin, New York, 1966; T. Nakajima, *Fortschr. Chem. Forsch.*, **32**, 1 (1972).
122. R. J. Buenker and S. D. Peyerimhoff, *J. Chem. Phys.*, **48**, 354 (1968); M. J. S. Dewar, M. C. Kohn and N. Trinajstic, *J. Am. Chem. Soc.*, **93**, 3437 (1971).
123. O. L. Chapman, D. De La Cruz, R. Roth, and J. Pacansky, *J. Am. Chem. soc.*, **95**, 1338 (1973); A. Krantz, C. Y. Lin, and M. D. Newton, ibid., 2744; G. Maier, *Angew. Chem. Int. Ed.*, **13**, 425 (1974).
124. (a) G. Wipff, U. Wahlgren, E. Kochanski, and J. M. Lehn, *Chem. Phys. Lett*, **11**, 350 (1971); (b) J. M. Lehn, private communication.
125. R. J. Gillespie and J. Passmore, *Acc. Chem. Rev.*, **4**, 413 (1971).
126. T. Nakajima, *Pure Appl. Chem.*, **28**, 219 (1971); T. Nakajima, A. Toyota, and S. Fujii, *Bull. Chem. Soc. Jap.*, **45**, 1022 (1972).
127. E. L. Muetterties, *J. Am. Chem. Soc.*, **91**, 1636 (1969).
128. D. R. Yarkony and H. F. Schaefer, III, *J. Am. Chem. Soc.*, **96**, 3754 (1974).
129. (a) D. M. P. Mingos, *Nature Phys. Sci.*, **229**, 193 (1971); (b) ibid., **230**, 154 (1971).
130. Lj. Manojlovic-Muir, K. W. Muir, and J. A. Ibers, *Disc. Faraday Soc.*, **47**, 84 (1969).
131. C. G. Pierpoint and R. Eisenberg, *J. Am. Chem. Soc.*, **93**, 4905 (1971); D. M. P. Mingos, *Inorg. Chem.*, **12**, 1209 (1972); J. H. Enemark and R. D. Feltham, *J. Am. Chem. Soc.*, **96**, 5004 (1974); *Theor. Chem. Acta*, **34**, 165 (1974); R. Hoffmann, M. Chem, A. Rossi, and D. Mingos, *Inorg. Chem.*, **13**, 2666 (1974).
132. D. J. Hodgson, N. C. Payne, J. A. McGinnety, R. G. Pearson, and J. A. Ibers, *J. Am. Chem. Soc.*, **90**, 4486 (1968).
133. F. A. Cotton, *J. Am. Chem. Soc.*, **90**, 6230 (1968).
134. M. M. Coutiere, J. Demuynck, and A. Viellard, *Theor. Chim. Acta*, **27**, 281 (1973).
135. R. Prins, *J. Chem. Phys.*, **50**, 4804 (1969).
136. L. Hedberg and K. Hedberg, *J. Chem. Phys.*, **53**, 1228 (1970).
137. E. Maslowsky and K. Nakamoto, *Inorg. Chem.*, **8**, 1108 (1969); F. A. Cotton and T. J. Marks, *J. Am. Chem. Soc.*, **91**, 7281 (1969).
138. H. H. Brintzinger and L. S. Bartell, *J. Am. Chem. Soc.*, **92**, 1105 (1970).
139. B. J. Nicholson and H. C. Longuet-Higgins, *Mol. Phys.*, **9**, 461 (1965).

140. H. F. Schaefer, III, *The Electronic Structure of Atoms and Molecules*, Addison-Wesley, Reading, Mass., 1972. This gives a good account of methods for obtaining rigorous quantum mechanical results for molecules.

141. R. C. Feber, *Los Alamos Report*. LA-3164, 1965.

142. J. C. Green, D. I. King, and J. H. D. Eland, *Chem. Commun.*, 1121 (1970); J. H. Hillier, V. R. Saunders, M. J. Ware, P. J. Bassett, D. R. Lloyd, and N. Lynaugh, ibid., 1316.

143. D. R. Lloyd and E. W. Schlag, *Inorg. Chem.*, **8**, 2544 (1969).

144. C. C. J. Roothaan, *Rev. Mod. Phys.*, **32**, 179 (1960); J. L. Dodds and R. McWeeny, *Chem. Phys. Lett.*, **13**, 9 (1972).

145. C. R. Brundle, M. B. Robin, and H. Basch, *J. Chem. Phys.*, **53**, 2196 (1970).

146. K. Siegbahn et al., ESCA Applied to Free Molecules, North Holland, Amsterdam, 1969.

147. (a) J. R. Hoyland and L. Goodman, *J. Chem. Phys.*, **33**, 946 (1960); (b) C. K. Jørgensen, *Chimia*, **27**, 203 (1973).

148. S. M. Schildcrout, R. G. Pearson, and F. E. Stafford, *J. Am. Chem. Soc.*, **90**, 4006 (1968); G. M. Bancroft, C. Reichert, and J. B. Westmore, *Inorg. Chem.*, **7**, 870 (1968).

149. S. Evans, A. Hamnett, and A. F. Orchard, *Chem. Commun.*, 1282 (1970); *J. Coord. Chem.*, **2**, 57 (1972).

150. R. W. Hand, W. J. Hunt, and H. F. Schaefer, III, *J. Am. Chem. Soc.*, **95**, 4517 (1973).

151. M. B. Hall, M. F. Guest, and I. H. Hillier, *Chem. Phys. Lett.*, **15**, 592 (1972).

152. D. A. Brown, W. J. Chambers, and N. J. Fitzpatrick, *Inorg. Chim. Acta Rev.*, **6**, 7 (1972). This paper reviews various methods used in ligand-field theory.

153. (a) R. Prins and T. Novokov, *Chem. Phys. Lett.*, **9**, 593 (1971); (b) ibid., **16**, 86 (1972); (c) A. Calabrese and R. G. Hayes, *J. Am. Chem. Soc.*, **95**, 2819 (1973).

# CHAPTER 4

# REACTION MECHANISMS FOR SIMPLE MOLECULES

In this chapter orbital symmetry is used as a guide to predict the most favorable reaction path for a number of reactions of simple molecules. Consideration is given to $XY_n$ and $X_2Y_n$ molecules and other very symmetric molecules. Hopefully the results can be extrapolated to the same reactions for less symmetric molecules. The kinds of reaction discussed are few in number but of fundamental importance. They include

(a) isomerizations,
(b) dissociations,
(c) reactions with electrophiles, nucleophiles and free radicals,
(d) reactions with diatomic molecules.

The reagents under (c) will usually be considered to be monatomic species which form bonds by way of $\sigma$ orbitals. That is, they are $\sigma$ donors and acceptors.

The reactions of free radicals, such as Cl·, are similar to those of nucleophiles in some respects and to those of electrophiles in others, which makes prediction more difficult. Also dissociation reactions may yield free radical products, or products which are themselves nucleophiles or electrophiles. For example,

$$CH_3Cl \longrightarrow CH_3\cdot + Cl\cdot \tag{1}$$

$$CH_3Cl \longrightarrow CH_3^+ + Cl^- \tag{2}$$

Symmetry rules as such usually do not distinguish between these several alternatives. However total energies and external reaction conditions do.

Solvation effects obviously play a major role in all reactions in which ions are involved. For the most part these effects will not be specifically discussed, except when critical to the argument. Theoretical calculations by ab initio methods are invariably made for the dilute gas phase, and this must be borne in mind when making comparisons with experiment.

## THE PRINCIPLE OF LEAST MOTION

For any reaction the problem to be considered is the finding of the reaction coordinate which leads to the minimum value of $E_0$, the potential energy barrier to reaction. We start with a knowledge of the structures of the reactant molecules and with the symmetries and approximate energies of the valence-shell MOs. The reactants are brought together to form an initial "product." We assume that reasonable theories exist for predicting the stable shapes of these "product" molecules.

In many cases the initial "products" will be very unstable, corresponding to activated complexes or reactive intermediates en route to final, stable products. Knowing the geometries of the reactants and the "products" can we determine which path is followed in getting from one to the other, and how orbital symmetries dictate this path?

Clearly not every possible path can be tested, nor can we tell much about very unsymmetrical reaction paths by using symmetry arguments. Fortunately, there is a very powerful postulate that can be used as a guide. This is the principle of least motion (PLM). As enunciated by Rice and Teller, this principle states that the lowest activation energy for a reaction is the one that requires the least motion of the nuclei and the least disruption of the original electronic distribution. A similar statement had been made earlier by Franck and Rabinowitsch.[2]

The PLM seems intuitively very reasonable. Unfortunately it is not very clearly stated as to how one measures the motion of the nuclei or the disruption of the electron distribution. Furthermore no proof of the principle has ever been given. It is a pure postulate whose validity must be tested by application. As it turns out, it has been assumed by most chemists concerned with the intimate details of reaction mechanisms. This goes back to the work of organic chemists such as Ingold, Robinson, and Hughes and physical chemists such as Eyring, London, and Polanyi in the early 1930s.

The elucidation of reaction mechanisms consists of two rather distinct parts: (1) the identification of the several elementary processes (unimolecular, bimolecular, etc.) that cooperate to bring about the observed reaction and (2) a development of the detailed stereochemical picture of each of the above elementary processes. At the moment, we are concerned more with the latter aspect. That is, we wish to know the exact path for each elementary process, and the geometries of the activated complexes and intermediates formed.

The usual procedure in the past has been to make an initial assumption about the detailed path of the elementary step, based in some way on the PLM. If this assumption is contradicted by experimental facts, then it must be abandoned.[3] In the great majority of cases the mechanism indicated by the PLM has seemed to be consistent with the facts. This does not prove that the assumed mechanism is true, of course.

A number of attempts have been made to see if certain commonly accepted mechanisms are in accord with the PLM.[4] The difficulty is in weighting the nuclear displacements, which are easy to estimate, with the disturbance of the electron cloud, which is more difficult. The usual procedure is to use weighting factors based on force constants for bond stretching and bond bending. Alternative mechanisms are then compared by some minimization procedure. Such calculations showed that usually, but not always, the PLM is in fact obeyed.

A much more satisfactory test of an assumed reaction path for an elementary reaction would be a series of ab initio energy calculations. Such calculations would require a very large basis set of atomic functions and the inclusion of many configurations, besides the predominant one, to have any value. Also a very large range of possible nuclear coordinates would have to be considered to find the reaction coordinate of lowest energy barrier. In spite of the difficulties, such calculation are just starting to appear.[5] Examples include ($H + H_2$), ($H^- + CH_4$), ($F + H_2$), ($H_2 + CH_2$), ($2CH_2 \rightarrow C_2H_4$), and others. The immediate future will yield many more such calculations, but there will still be a need for less expensive and time-consuming methods.

Before going further, it is necessary to have a practical definition of the least-motion path. The requirement of minimum nuclear motion will usually be easy to follow, or at least not to violate seriously. To give minimum disruption of the electron distribution, we will adopt the following: "The least-motion (LM) path for an elementary reaction is the one which creates and preserves the greatest number of symmetry elements to be found in the final products." If two or more molecules react or are formed, the total point group that they generate must be taken as a basis for classification. Thus the PLM states that the reactant molecules approach each other in the most symmetric way leading to the product.

This statement seems to be consistent with the meaning of a least nuclear motion and a minimum disturbance of the electronic configuration. Since all reaction coordinates are totally symmetric, except at maxima and minima of potential energy (see p. 15), symmetry elements introduced at the start will usually remain up to the activated complex configuration. At this point new symmetry elements may be created, and on passing through the TS, symmetry elements may be destroyed. Demands on the reaction coordinate at this stage can be minimized by creating in advance as much of the finally required symmetry as possible.

No proof is available that the potential energy is a minimum rather than a maximum for the asymmetric normal mode that destroys a symmetry element. However since it is a minimum in the final product, it is likely to be so at the TS. Note also that a symmetric structure minimizes unfavorable overlaps, by making them exactly zero, and tends to maximize favorable overlaps.

Sometimes there will be more than one path that are equal in terms of the number of symmetry elements preserved. An example would be conrotatory and disrotatory ring opening in cyclobutene, where either a mirror plane or a two-fold axis persists. In such cases both paths must be treated as of equal a priori probability.

A simple example of the above procedure is the assumption of a colinear approach of the hydrogen atom in the reaction

$$H + H_2 \longrightarrow H\text{---}H\text{---}H \longrightarrow H_2 + H \quad (3)$$

The point group generated is $C_{\infty h}$ rather than the lower group $C_s$. Note that it is the known linear structure of the TS, $H_3$, which selects the reaction coordinate. The procedure, of course, must only be used for a single elementary process, not for a reaction which is the sum of several elementary steps.

The assumption is that the LM path is the energetically favored one, unless it is forbidden by orbital symmetry, or other factors. If it is, then other paths must be tested, in order of decreasing symmetry. At this point, however, there is no very sure guide as to which alternate path is favored.[6]

The LM path must be tested by as many criteria as possible. A substantial activation energy is indicated by any of the following

(a) correlation of orbitals leading to excited states of products,
(b) failure to match symmetries in the bonds that are made and the bonds that are broken,
(c) absence of low lying excited states of correct symmetry in reactants,
(d) zero, or poor, overlap of HOMO and LUMO, or alternate orbitals, after start of interaction,
(e) creation of antiaromatic TS,
(f) excessive strain in bond distances or bond angles en route to product,
(g) excessive repulsion due to overlap of filled orbitals, or close approach of nuclei en route to product.

The use of the PLM in the manner described automatically leads to a TS of rather high symmetry, but there are cases where it can be shown that a highly symmetric structure cannot be the TS. One case was discussed earlier (p. 11), where it was shown that a structure that can decompose into equivalent products in more than one way cannot be an activated complex.

Another set of examples can be found by considering cycloaddition or sigmatropic shifts.[7] Take the Cope rearrangement, a [3,3] sigmatropic shift.

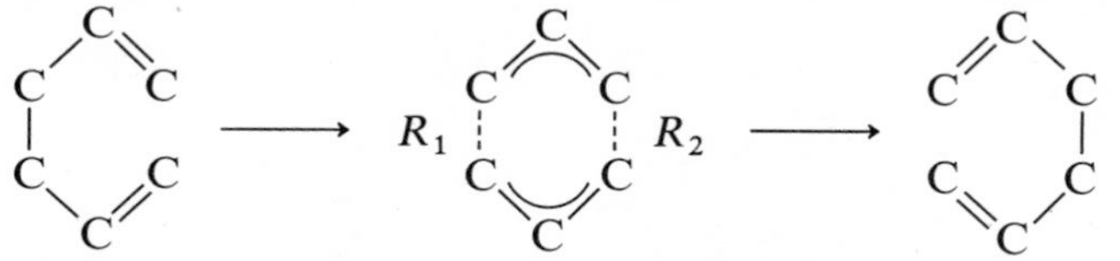

$$\text{C}_6 \text{ (1,5-hexadiene skeleton)} \longrightarrow R_1 \; \text{C}_6 \text{ (symmetric TS)} \; R_2 \longrightarrow \text{C}_6 \text{ (1,5-hexadiene skeleton)} \tag{4}$$

The TS is assumed to be symmetric, and changes in $R_1$ and $R_2$ dominate the reaction coordinate. There will be a symmetric stretch for which $(R_1 + R_2)$ is the symmetry coordinate, and an asymmetric stretch with $(R_1 - R_2)$ as the coordinate. The force constants are $(f + f')$ for the symmetric and $(f - f')$ for the asymmetric mode. The constant $f$ is the C—C bond-stretching constant, and $f'$ is the interaction constant. Now the force constant for the reaction coordinate must be negative, and there can be only one negative force constant in the total force-constant matrix of the activated complex.

Now $(f + f')$ must be positive, since dissociation into two allyl radicals does not occur (symmetric stretch), and $(f - f')$ must then be negative. This is certainly possible for the [3,3] shift, since the two bonds are still strongly coupled. However it becomes progressively less possible as we go to [5,5] shifts, [7,7] shifts, and so on. Eventually the symmetric structure must become an unstable intermediate, existing at a potential minimum.

In spite of these difficulties, it will still be advantageous to discuss the formation of highly symmetric intermediates. The hope in these cases will be that the activated complexes will differ from the symmetric structures by only a small amount. For example, they may differ only by nuclear displacements of the magnitude of vibrational amplitudes. It must be admitted that there is always the possibility that the displacements are much larger. The symmetric structures will then not be good models for the activated complexes.

## Dimerization of $H_2X$

To illustrate the procedure, let us consider the general case of dimerization of two $H_2X$ molecules to form planar $X_2H_4$. The LM path is a coplanar approach of $D_{2h}$ point group. Figure 1 shows the orbital-correlation diagram for this process.[6,8] Triplet ground-state carbene $CH_2$ has the configuration $(a_1)^2(b_2)^2(2a_1)^1(b_1)^1$, which correlates smoothly with the ground state of ethylene in this LM path.[11]

$$2CH_2(^3B_1) \longrightarrow C_2H_4(^1A_g) \tag{5}$$

However singlet carbene has the configuration $(a_1)^2(b_2)^2(2a_1)^2$. This can be

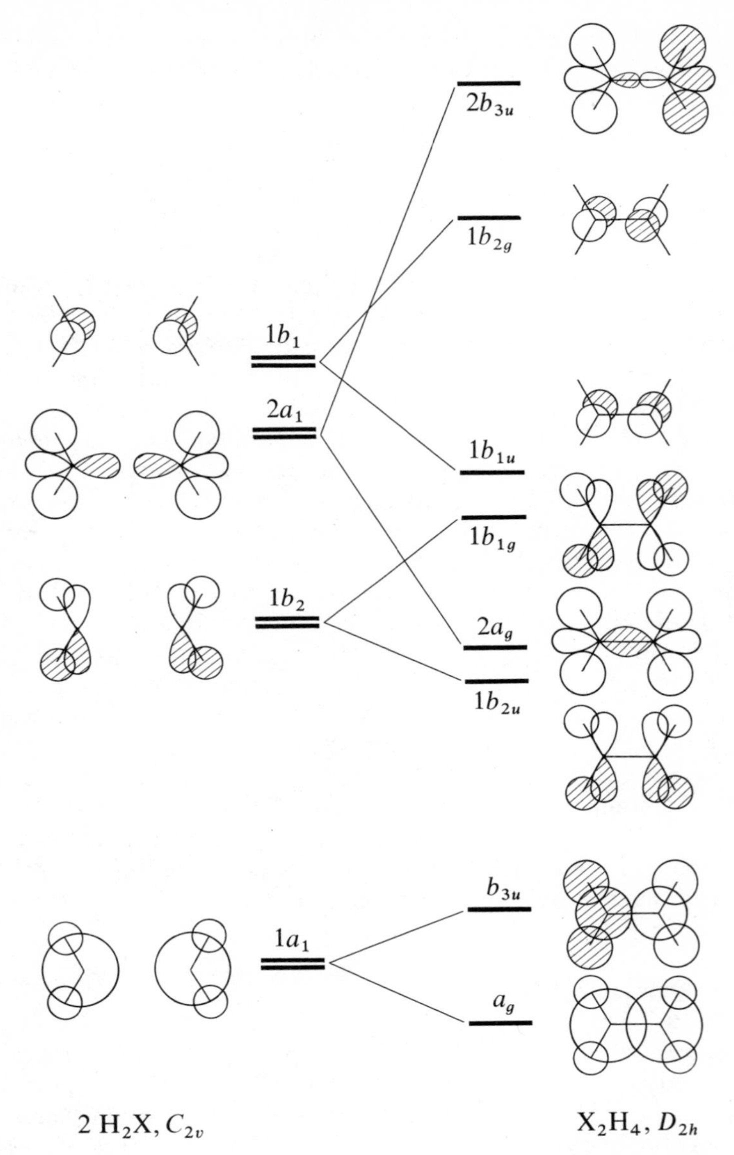


**Figure 1.** Correlation diagram for $2H_2X \rightarrow X_2H_4$ in a coplanar approach, $D_{2h}$ symmetry (after reference 8).

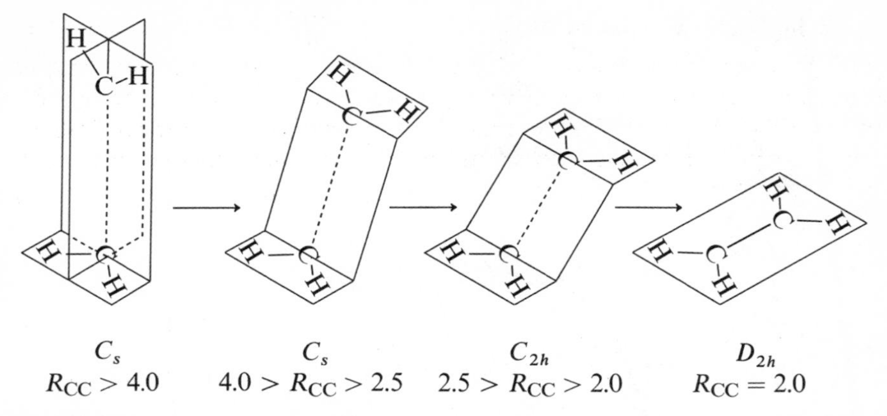


**Figure 2.** The non-LM approach of two $CH_2$ singlet molecules to form ground-state ethylene (after reference 6).

seen to correlate with doubly excited ethylene, with two electrons in $b_{3u}$, a $\sigma$ orbital, rather than in $b_{1u}$, the $\pi$ orbital. Hence the colinear approach of two singlet $CH_2$ molecules gives rise to a purely repulsive state.[9]

However a non-LM approach of two singlet $CH_2$ molecules does allow for the formation of ground-state ethylene without an energy barrier.[6] According to extended Hückel calculations, which leave something to be desired, the lowest-energy path starts with a perpendicular approach of $C_s$ symmetry (Fig. 2). As the carbon—carbon distance decreases, one $CH_2$ molecule bends relative to the other, and at about 2.5 Å, the point group becomes $C_{2h}$. This changes smoothly into $D_{2h}$ at C—C bond distances less than 2.0 Å. There is no energy barrier at any point along the way to product.

The reason for the forbidden coplanar approach and the virtue of the $C_s$ approach are easily seen by looking at simple Lewis diagrams for the two reaction coordinates.

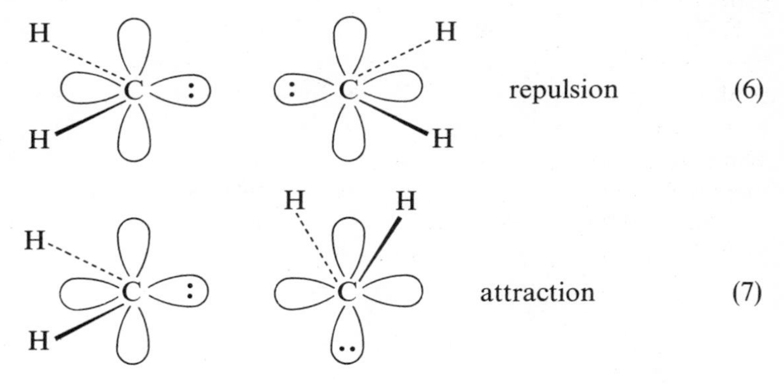

In the first case two filled $a_1$ orbitals are approaching each other and each is orthogonal to the empty $b_1$ orbitals. There is a net repulsion. In the less symmetric approach, there is a maximum overlap of one filled and one empty orbital, and some overlap of the remaining pair. Agreeing with the calculation, a $C_{2h}$ approach from the start should be of very similar energy, since both pairs overlap equally.

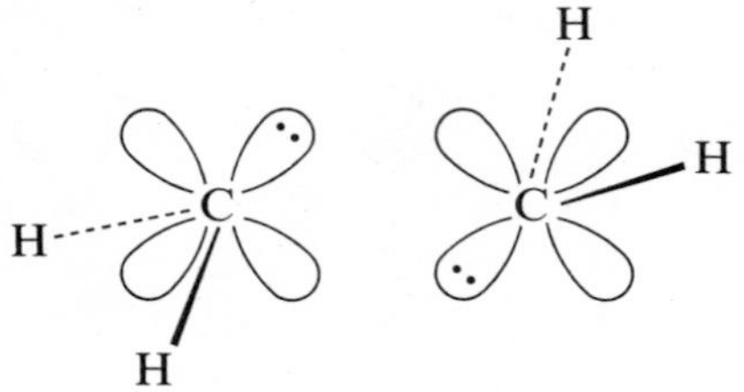

attraction (8)

In this example the LM path is forbidden by orbital symmetry. Reduction to a $C_{2h}$ point group removes the symmetry restriction, since there is now good overlap between filled and empty orbitals, and also the correlation diagram does not show crossing. Actually there is little evidence that singlet carbene dimerizes at all, whereas triplet carbene undoubtedly does.[10]

Consider the hypothetical reverse process

$$C_2H_4(^1A_g) \longrightarrow 2CH_2(^1A_1) \tag{9}$$

using Fig. 2 as a guide. The reaction coordinate would initially be an $A_g$ stretch of the C—C bond, combined with a $B_{2g}$ mode to rock the $CH_2$ groups into the $C_{2h}$ position. The $B_{2g}$ mode becomes $A_g$ in the lower point group. At $R = 2.5$ Å or so, a $B_u$ mode would change the point group to $C_s$, where $B_u$ becomes $A'$. Note that whenever an asymmetric reaction coordinate is needed, it exists only at a point and only because the potential energy has become a maximum with respect to that particular set of nuclear motions. This does not mean that a maximum will be seen in a plot of $E$ against the generalized reaction coordinate $Q$.

$NH_2$ radicals dimerize to form hydrazine. The electron configuration of $NH_2$ is $(a_1)^2(b_2)^2(2a_1)^2(b_1)^1$. From Fig. 1, it again can be seen that the LM coplanar approach is forbidden, This is irrelevant in any case since planar $N_2H_4$ would be of high energy. The actual structure is a *gauche* form of $C_2$ symmetry. The least motion approach in this point group is allowed since there would be good overlap between the two half-filled $b_1$ orbitals, as shown in (10). Two $BH_2$ radicals, $(a_1)^2(b_2)^2(2a_1)^1$, could react by the staggered $D_{2d}$ LM approach to give ground state $B_2H_4$.

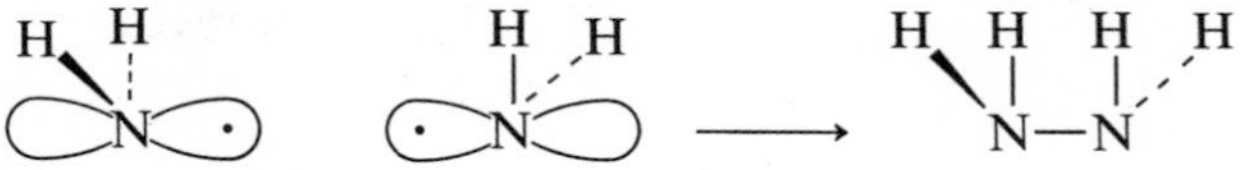

(10)

$$H_2B\cdot \quad \cdot BH_2 \longrightarrow H_2B{-}BH_2 \qquad (11)$$

## POLYTOPAL REARRANGEMENTS

In Chapter 3 PT was used to probe the relative stabilities of various polytopal forms of simple molecules. The question of orbital symmetry barriers to polytopal rearrangements was not explored in any detail. One polytope may be much less stable than another without any intervening barrier for their interconversion. Stereoisomerization of the more stable form by passage through the less stable form would not occur. An example would be the tetrahedral and planar forms of $CH_4$ and its derivatives. The planar form of methane is over 200 kcal mole$^{-1}$ less stable than the tetrahedral form.

This difference in stability is created by the symmetry properties of the valence-shell AOs used. Figure 12 of this chapter gives the correlation diagram for $T_d$ methane going to $D_{4h}$. The instability of planar $CH_4$ is due almost entirely to a bonding MO ($t_2$) becoming a nonbonding MO ($a_{2u}$). The conversion $T_d \rightarrow D_{2d} \rightarrow D_{4h}$ requires an $E$ reaction coordinate, which can be created only by excitation from a bonding $t_2$ orbital into an antibonding $a_1^*$ or $t_2^*$ orbital. The large energy gap for this transition is the warning that this is a very energetic path.

Indeed the racemization of optically active methane derivatives has never been observed to occur via an intramolecular path. Instead dissociation into radicals or ions occurs on thermal activation. This is accompanied by rapid racemization of the radicals or ions, followed by recombination to restore the original compound in the racemic condition.

Another extreme would be two polytopal isomers of similar energy, but where a symmetry barrier prevents the ready isomerization of one into the other. An example of this would be the *cis* and *trans* forms of azo compounds, such as $N_2F_2$ or $N_2R_2$. There is always a vibrational mode, or sometimes two, which constitutes the LM path for polytopal rearrangement. These vibrational modes are shown in Chapter III for $XY_n$ and $X_2Y_n$ molecules. They represent the reaction coordinate that carries out the desired change with the destruction of the fewest elements of symmetry. We will assume that the PLM selects these as favored paths unless forbidden by symmetry.

Turning first to $XY_n$ molecules, it appears that there is no theoretical example of a strongly forbidden polytopal rearrangement. As discussed elsewhere (p. 129), there are cases where a forbiddenness of second order appears. Excited states corresponding to excitation of nonbonding electrons, such as $d$ electrons, may be the correlation indicated products of certain isomerizations. The barriers created in some of these reactions are found to be small, for example, square planar and tetrahedral nickel(II) complexes. The corresponding equilibria between planar and tetrahedral platinum(II) complexes have never been observed. This may be the result of simple thermodynamic instability of tetrahedral $PtL_4$. It may also result from the undoubtedly larger symmetry barrier due to the $d$ electrons.

The more complex $X_2Y_n$ molecules show several cases of symmetry-restricted rearrangements. The simplest example would be *cis–trans* isomerization in diimide, $N_2H_2$.

$$\text{H}-\ddot{\text{N}}=\underset{..}{\text{N}}-\text{H} \;(\textit{trans}) \longrightarrow \ddot{\text{N}}=\ddot{\text{N}} \;(\text{H},\ \text{H};\ \textit{cis}) \qquad (12)$$

There are two LM paths: (1) rotation about the double bond and (2) movement of one or both H atoms in the plane. The first preserves the $C_2$ axis, and the other the mirror plane. Applying the bond-symmetry rule to (12) shows that the $A_u$ rotation is forbidden. The two N—H bonds are, and remain, A and B in $C_2$, and the two lone pairs of electrons on nitrogen are also A and B. These offer no problem. However the $\pi$ bond is A in the *trans* isomer and B in the *cis* isomer. Note that the twofold axis rotates from being perpendicular to the plane to lying in the plane.

The $B_u$ mode, which gives the $C_s$ point group, is allowed. The bonds and lone pairs are all A′ and the double bond is A″. This procedure is somewhat dangerous since a plane that does not intersect the bonds is often not diagnostic (p. 89). However orbital correlation show that the above analysis is correct.[11] Experimentally azo compounds, $R_2N_2$, isomerize with considerable activation energy. Semiempirical calculations indicate the planar mechanism with one hydrogen atom moving.[11]

$$\text{H}-\ddot{\text{N}}=\underset{..}{\text{N}}-\text{H} \;(\textit{trans}) \longrightarrow \text{H}-\ddot{\text{N}}=\underset{..}{\text{N}}-\text{H} \;(\text{linear}) \longrightarrow \ddot{\text{N}}=\ddot{\text{N}} \;(\text{H},\ \text{H};\ \textit{cis}) \qquad (13)$$

*Cis–trans* isomerization of olefins by rotation about the double bond is also symmetry forbidden.

$$\text{RHC}=\text{CHR} \;(\textit{trans}:\ \text{R},\ \text{H}\ /\ \text{H},\ \text{R}) \longrightarrow \text{RHC}=\text{CHR} \;(\textit{cis}:\ \text{R},\ \text{R}\ /\ \text{H},\ \text{H}) \qquad (14)$$

If R is different from H, then the point group is the same as for azo compounds and the same analysis applies. For ethylene itself, Fig. 3 shows the orbital correlation for the $b_{1u}$ and $b_{2g}$ orbitals. These are the $\pi$ or HOMO, and $\pi^*$, or LUMO, of ethylene. The elements of the $D_2$ point group are preserved. Since the $z$ and $y$ $C_2$ axes interchange for 180° rotation, the subscripts 1 and 2 are interchanged. Hence the filled, bonding $b_1$ orbital becomes an empty, antibonding orbital in the product. The intended crossing is prevented because $b_1$ and $b_2$ become degenerate at the halfway point.

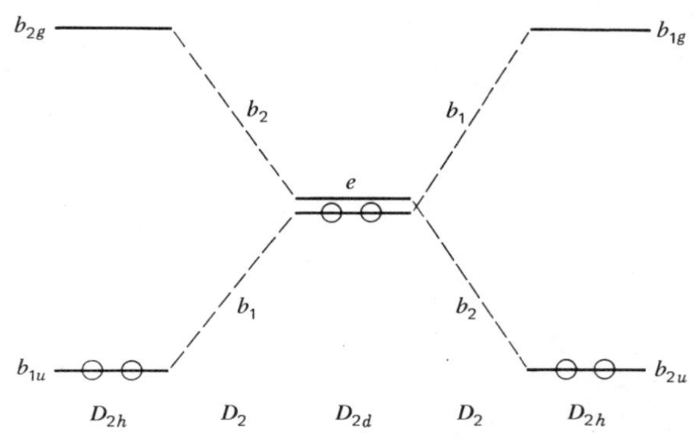


**Figure 3.** Correlation diagram for the $\pi$ and $\pi^*$ orbitals of ethylene during twisting about the double bond (*cis–trans* isomerization); the reaction is forbidden.

In the olefin case, the in-plane mechanism is energetically impossible. Only the forbidden twisting process is available. For ethylene the activation energy is 65 kcal, from the *cis–trans* isomerization of 1,2-dideuterioethylene.[12] It was at one time thought that isomerization of olefins could occur by thermal excitation of one electron from the $\pi$ to the $\pi^*$ states.[13] This would circumvent the symmetry barrier. Such a process is important in photochemical isomerization of olefins (see p. 489), but it is unlikely for the thermal process, at least for simple olefins.[14]

As a check on procedures, it is desirable to make sure that twisting about the single bond of ethane is allowed. In $D_{3d}$, the staggered form of ethane, the twisting mode is $A_{1u}$. This converts the point group into $D_3$. The six C—H bonds are $A_1$, $A_2$, and 2E in this point group. The C—C bond is $A_1$. Now rotation by 120° simply changes each of the three $C_2$ axes into another. The $C_3$ axis is unaffected, but the character of each species is the same for each of the $C_2$ axes. Therefore their interchange causes no change in symmetry labeling. The bonds remain of the same species throughout and the process is allowed.

The twisting mode of diborane of $A_u$ species is forbidden by orbital symmetry. In principle this could be a mechanism for isomerizing the (unknown) *cis–trans* isomers of $R_2B_2H_4$.

(15)

The conrotatory twisting mode of $B_{1g}$ species is allowed (see p. 219). However this cannot cause isomerization of *cis–trans* isomers.

As the molecular formula becomes more complex, the possibilities for forbidden polytopal rearrangements increase. Several examples from organic chemistry have been discussed.[15] A case in point is the $C_5H_5^+$ cation. The

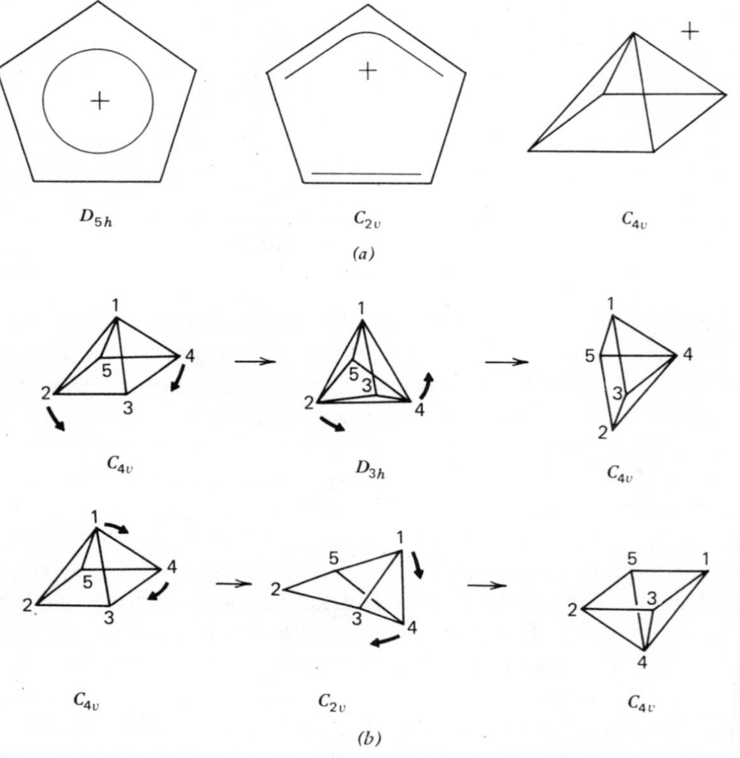


**Figure 4.** (*a*) Structures for $C_5H_5^+$ ions; (*b*) the high-energy path via a $D_{3h}$ intermediate, and the low-energy path via a $C_{2v}$ intermediate, for polytopal rearrangement of pyramidal $C_5H_5^+$ (after reference 17).

ground state structure for this ion as a singlet is probably planar with $C_{2v}$ symmetry.[16] The planar structure of $D_{5h}$ point group is unstable because of a Jahn–Teller effect. Another singlet form of $C_{4v}$ symmetry can exist in which one CH group is apical and the other four are in a plane.

The square pyramidal form is capable of being readily changed into its polytopal isomers.[15] That is, any of the five C—H groups can become the unique apical group. This does *not* occur by the LM path $C_{4v} \rightarrow C_{2v} \rightarrow D_{3h} \rightarrow C_{2v} \rightarrow C_{4v}$ shown in Fig. 4. Instead rearrangement goes by a less symmetric path $C_{4v} \rightarrow C_s \rightarrow C_{2v} \rightarrow C_s \rightarrow C_{4v}$, also in Fig. 4. The MO sequence for pyramidal $C_5H_5^+$ explains this choice of reaction coordinate. The configuration is[15]

$$(a_1)^2(e)^4(b_1)^0(2e)^0(2a_1)^0$$

for the MOs involved in bonding the apical CH group. The lowest excited state gives a transition density of $(E \times B_1) = E$ species. The $E$ mode takes $C_{4v}$ into the $C_s$ but *not* into the $C_{2v}$ point group. The next higher excited state, $(e) \rightarrow (2e)$, would be needed for the $B_2$ mode needed to produce a $C_{2v}$ intermediate. The LM path that is by-passed is essentially the Berry pseudorotation path for pentavalent phosphorus (p. 191). In $C_5H_5^+$ the corresponding path is forbidden because the $D_{3h}$ intermediate (Fig. 4) is not stable, but is a very unstable species.

## DISSOCIATION REACTIONS

Another kind of unimolecular elementary process is dissociation of a molecule into smaller fragments.[17] For complex molecules, the number of possible fragments becomes very large, and prediction is difficult. We start by considering the symmetry features of dissociation of $XY_n$ molecules. Even here at least two possibilities exist, for instance,

$$H_2O \longrightarrow H + OH \qquad (16)$$

$$H_2O \longrightarrow H_2 + O \qquad (17)$$

$$NH_3 \longrightarrow H + NH_2 \qquad (18)$$

$$NH_3 \longrightarrow H_2 + NH \qquad (19)$$

More extensive dissociation into three or more atoms can be ruled out on energetic grounds.

Table 1 is a compilation of predicted reaction coordinates for dissociations of these two types for $XY_n$ molecules. The assumption is that the PLM is obeyed, unless demonstrated to be unlikely. This assumption then fixes the

**Table 1 Reaction Coordinates for Least-Motion Dissociation of $XY_n$ Molecules**

| Reaction | Initial Point Group | Point Group for Reaction | Symmetry of Reaction Coordinate |
|---|---|---|---|
| $XY_2 \rightarrow XY + Y$ | $D_{\infty h}$ | $C_{\infty v}$ | $\Sigma_u$ |
| $XY_2 \rightarrow X + Y_2$ | $D_{\infty h}$ | $C_{2v}$ | $\Pi_u$ |
| $XY_2 \rightarrow XY + Y$ | $C_{2v}$ | $C_s$ | $B_2$ |
| $XY_2 \rightarrow X + Y_2$ | $C_{2v}$ | $C_{2v}$ | $A_1$ |
| $XY_3 \rightarrow XY_2 + Y$ | $D_{3h}$ | $C_{2v}$ | $E'$ |
| $XY_3 \rightarrow XY + Y_2$ | $D_{3h}$ | $C_{2v}$ | $E'$ |
| $XY_3 \rightarrow XY_2 + Y$ | $C_{3v}$ | $C_s$ | $E'$ |
| $XY_3 \rightarrow XY + Y_2$ | $C_{3v}$ | $C_s$ | $E'$ |
| $XY_4 \rightarrow XY_3 + Y$ | $T_d$ | $C_{3v}$ | $T_2$ |
| $XY_4 \rightarrow XY_2 + Y_2$ | $T_d$ | $C_{2v}$ | $E$ or $T_2$ |
| $XY_4 \rightarrow XY_3 + Y$ | $D_{4h}$ | $C_{2v}$ | $E_u$ |
| $XY_4 \rightarrow XY_2 + Y_2$ | $D_{4h}$ | $C_{2v}$ | $E_u$ |
| $XY_4 \rightarrow XY_3 + Y$ | $C_{2v}$ | $C_s$ | $B_1$ or $B_2$ |
| $XY_4 \rightarrow XY_2 + Y_2$ | $C_{2v}$ | $C_{2v}$ | $A_1$ |
| $XY_5 \rightarrow XY_4 + Y$ | $D_{3h}$ | $C_{2v}$ | $E'$ |
| $XY_5 \rightarrow XY_4 + Y$ | $D_{3h}$ | $C_{2v}$ | $A''_2$ |
| $XY_5 \rightarrow XY_3 + Y_2$ | $D_{3h}$ | $C_s$ | $E'$ |
| $XY_5 \rightarrow XY_4 + Y$ | $C_{4v}$ | $C_{2v}$ | $B_1$ |
| $XY_5 \rightarrow XY_4 + Y$ | $C_{4v}$ | $C_s$ | $E$ ? |
| $XY_5 \rightarrow XY_3 + Y_2$ | $C_{4v}$ | $C_{2v}$ | $B_1$ |
| $XY_5 \rightarrow XY_3 + Y_2$ | $C_{4v}$ | $C_s$ | $E$ ? |
| $XY_6 \rightarrow XY_5 + Y$ | $O_h$ | $C_{4v}$ | $T_{1u}$ |
| $XY_6 \rightarrow XY_4 + Y_2$ | $O_h$ | $C_{2v}$ | $T_{1u}$ |
| $XY_7 \rightarrow XY_6 + Y$ | $D_{5h}$ | $C_{2v}$ | $E'_1$ or $E'_2$ |
| $XY_7 \rightarrow XY_5 + Y_2$ | $D_{5h}$ | $C_{2v}$ | $E'_1$ or $E'_2$ |

elements of symmetry that are maintained as one or two Y atoms are removed. For example, linear $XY_2$ would change from $D_{\infty h}$ to $C_{\infty v}$ as one Y atom is removed, and to $C_{2v}$ as two Y atoms are removed.

$$\text{Y—X—Y} \longrightarrow \text{Y}\cdots\text{X—Y}\ C_{\infty v} \tag{20}$$

$$\text{Y—X—Y} \longrightarrow \text{Y}\cdots\text{Y (with X above, dashed bonds)}\ C_{2v} \tag{21}$$

From the correlation tables, the reaction coordinate in the first case would be $\Sigma_u$, and in the second case, $\Pi_u$.

For bent $XY_2$ molecules, the point group would become $C_s$ for a reaction such as (16) and would remain $C_{2v}$ for (17). The reaction coordinate would be

$B_2$ in the first case and $A_1$ in the second. The dissociation of bent and linear $XY_2$ molecules has already been discussed (pp. 65 and 167), and the remaining treatment will be brief. The most serious restriction on either of the reactions

$$XY_2 \longrightarrow XY + Y$$

or

$$XY_2 \longrightarrow X + Y_2 \tag{22}$$

has to do with the spin state of the atom X or Y that is formed. There is often a change in spin required if ground-state atoms are to be formed. The alternative is to form excited states of the atoms.

The latter alternative usually makes the overall energy requirement too large. An example would be (17) if an O atom is formed in the excited $^1D$ state, 22 kcal above the $^3P$ ground state. The alternative of forming ground state O atom, imposes a spin restriction that makes the process unlikely. Nevertheless this is the path that is usually selected, if the energy requirement is otherwise reasonable. Some examples would be

$$N_2O \longrightarrow N_2 + O(^3P) \tag{23}$$

$$CO_2 \longrightarrow CO + O(^3P) \tag{24}$$

Molecules $XY_3$ in the planar $D_{3h}$ form would require an $E'$ reaction mode for either kind of dissociation

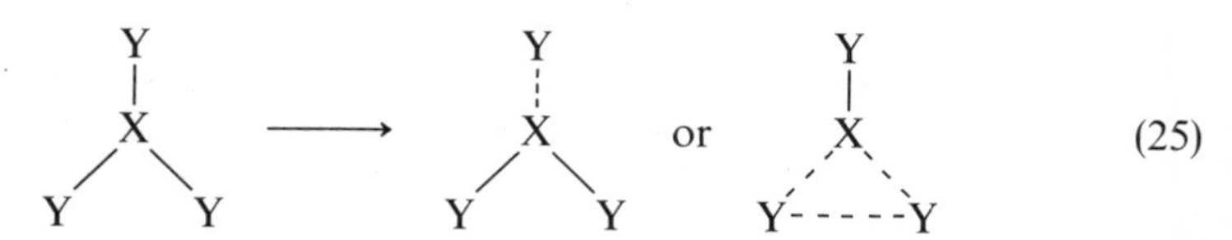


since the point group becomes $C_{2v}$ in both cases. Looking at the simplest example, $BH_3$ has the configuration

$$(a_1')^2(e')^4(a_2'')^0(2a_1')^0(2e')^0$$

From ab initio calculations, there is a large energy gap between $e'$ and $2a_1'$, which is the lowest useful excitation. The molecule $BH_3$ should be very stable to dissociation. $BF_3$ has several nonbonding MOs concentrated on the fluorine atoms (p. 179). This offers the possibility of lower-lying excited states because of the high electronegativity of F; however, all of these nonbonding MOs require at least as much excitation energy as does $BH_3$.

The meaning of exciting electrons from a nonbonding MO of $BF_3$ to promote dissociation is straightforward. It corresponds to $\pi$ bonding in the products. To simplify matters, consider that $F^-$ dissociates instead of F.

Then we have

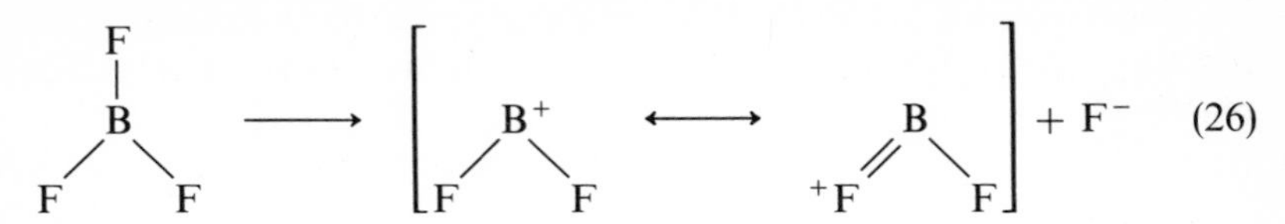


A suitable excitation would be from a bonding $e''$ orbital to the antibonding $a_2''$ orbital. One might think that $BCl_3$ could benefit more from such $\pi$ bonding, since the energy gap between $e''$ and $a_2''$ is less.[13] However it must be remembered that, because of poor overlap, $\pi$ bonding is rarely as effective in the heavier elements as it is for first-row elements.

$BF_3$ has 24 valence electrons. The possible dissociation of planar 22- and 23-electron molecules like $CO_3$ and $NO_3$ has already been discussed (p. 181). The reaction coordinate is $E'$ for dissociation into $CO + O_2$, or $NO + O_2$, respectively. However such processes are forbidden. This is seen most easily by considering the reaction

$$\text{H—BH}_2 \ (\text{bent}) \longrightarrow \text{H—B:} + \text{H—H} \qquad (27)$$

$$2a_1 + b_2 \qquad\qquad 2a_1 + a_1$$

The four valence electrons of the BH molecule are in $\sigma$ orbitals, which transform as $a_1$ in $C_{2v}$. Hence the symmetry of the bonds that are broken does not match with the symmetry of the bonds that are made. The same forbidden character exists for $NO_3$ and $CO_3$, in spite of their instability. However planar 28-electron molecules, such as $BrF_3$, are allowed by symmetry to dissociate into $BrF + F_2$ (see p. 281).

Pyramidal molecules such as $NH_3$ require an $E$ species reaction coordinate for both (18) and (19). Nitrene, NH, would be formed in its singlet state, which lies somewhat higher in energy than the ground-state triplet. The overall energy requirements of (18) and (19) are rather similar. In the point group $C_s$ for the LM removal of $H_2$ or H from $NH_3$, both reactions are allowed. However it can be shown that reaction (19) is still partially forbidden.[19] The reason is that in the planar configuration, (19) is forbidden in the same way that reaction (27) is, except that a lone pair in a $b_1$ orbital appears in $NH_3$ and a lone pair in either a $b_1$ or $b_2$ orbital in NH. Bending out of plane converts both $a_1$ and $b_1$ to $a'$, so that a formal correlation can be made without violating the symmetry rules. However the correlation is only formal since the lone pair of planar $NH_3$ would actually correlate with a $b_1$ orbital and not a $b_2$.

Accordingly (18), which is allowed, is favored over (19). The MO sequence for $NH_3$ is $(a_1)^2(e)^4(2a_1)^2(3a_1)^0(2e)^0$ with a large gap between the HOMO and LUMO. Certainly an $E$ reaction path would require a large energy, some 104 kcal mole$^{-1}$ for the allowed reaction (18).

The situation is quite different for the transient molecule $H_3O$. With one more electron, there is an easy excitation from the $3a_1$ to the $2e$ orbital. The $E$ mode leads to the dissociation

$$H_3O \longrightarrow H_2O + H \tag{28}$$

According to an ab initio calculation, reaction (28) has an activation energy of 7.5 kcal.[20] The structure of $H_3O$ is pyramidal $C_{3v}$.

This instability of $H_3O$ cannot simply be ascribed to its being a free radical, and hence reactive. $CH_3$ is equally a free radical and reactive, but it is stable to the reaction

$$CH_3 \longrightarrow CH_2 + H \tag{29}$$

The explanation lies in the MO sequence for planar methyl radical

$$(a_1')^2(e')^4(a_2'')^1(2a')^0(2e')^0$$

The low-lying states now give $(E' \times A_2'') = E''$ or $(A_2'' \times A') = A_2''$ transition densities. Neither of these is the $E'$ mode needed for (29). However planar, trigonal $ClH_3$, which is $\ldots(a_2'')^2(2a')^2(2e')^0$, would be unstable to the $E'$ mode, dissociating into $H_2$ + HCl in this case. This is an allowed reaction.

For tetrahedral molecules such as $CH_4$, the two reactions to consider are

$$CH_4 \longrightarrow CH_3 + H \tag{30}$$

$$CH_4 \longrightarrow H_2 + CH_2,\ ^1A_1 \tag{31}$$

The reaction coordinate is $T_2$ for (30) and either $E$ or $T_2$ for (31), depending on whether the $CH_2$ angle is less than or greater than the tetrahedral value. For singlet methylene the bond angle is 102° and an $E$ mode is indicated. The MO sequence for $CH_4$ is

$$(a_1)^2(t_2)^6(2a_1)^0(2t_2)^0$$

with $2a_1$ and $2t_2$ close together but far from the bonding MOs. The lowest excitation is favorable to $T_2$ but the large energy gap ensures a strong bond.

Since the singlet state of carbene lies only a few kcal above the ground triplet state,[21] reaction (31) is energetically about as favorable as (30). However there is an additional symmetry barrier to be surmounted, as well as the already large thermodynamic one.

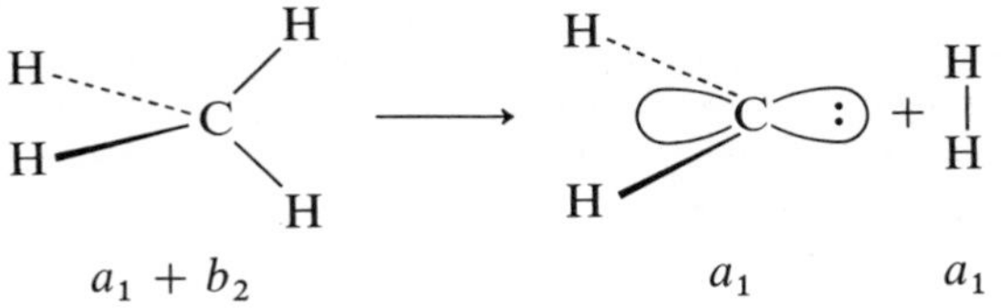

$$\mathrm{CH_4} \longrightarrow \mathrm{H_2C}{:} + \mathrm{H{-}H} \qquad (32)$$

$$a_1 + b_2 \qquad\qquad a_1 \qquad a_1$$

In the $C_{2v}$ point group, the two C—H bonds broken are $a_1 + b_2$. The H—H bond is $a_1$, and the lone pair of carbene is also $a_1$. A less symmetric dissociation of $C_s$ species is called for, and an increase in activation energy would result.[22] Note that there is no real indication in the MO sequence for $CH_4$ that (31) is forbidden. An $E$ excited state does not appear much more inaccessible than a $T_2$ state.

Removal of one electron from $CH_4$ to form the ion, results in a spontaneous dissociation.[23]

$$CH_4^+ \longrightarrow CH_3^+ + H \qquad (33)$$

The most stable structure of $CH_4^+$ is $D_{2d}$, but a $C_{3v}$ structure is almost as stable. In this shape the orbitals are shown elsewhere (p. 184). A low-energy excitation from $2a_1$ to $3a_1$ is available, which allows reaction (33) to occur. In a similar fashion a molecule such as $CH_3Br$ of $C_{3v}$ point group, would have a sequence for the $\sigma$ MOs of

$$(a_1)^2(e)^4(2a_1)^2(3a_1)^0$$

Since the $2a_1$ orbital is relatively high in energy, corresponding to weaker bonding of Br to C, an $A_1$ reaction coordinate is more easily excited

$$CH_3Br \longrightarrow CH_3\cdot + Br\cdot \qquad (34)$$

For tetrahedral complexes of the transition metals, the MO sequence in $T_d$ would be

$$\ldots(t_1)^6(t_2)^6 \,\vdots\, (e)(t_2^*) \,\vdots\, (a_1)^*$$

with the $d$ orbitals enclosed by dashed lines. Since $(T_1 \times E) = (T_2 \times E) = (T_1 + T_2)$, there are numerous ways to favor a dissociation into radicals or ions, for instance,

$$MX_4 \longrightarrow MX_3^+ + X^- \qquad (35)$$

There could be transfer of electrons from $t_1$ ($\pi$) or $t_2$ ($\sigma$ or $\pi$) orbitals into an empty $e$ orbital. Or there could be $(e) \rightarrow (t_2^*)$ transitions in the $d$ shell. Certainly such tetrahedral complexes are quite labile and appear to react by dissociation mechanisms in many cases.

The best documented examples are for $d^{10}$ complexes such as $Ni(CO)_4$, or $ML_4$, where M is Ni, Pd, or Pt and L is a phosphine or a phosphite.[24] Dissociation in these examples should occur as a result of $t_2^*$ to $a_1^*$ excitations. The $t_2^*$ orbitals are mainly $p$ and $d$ AOs on the metal, and the $a_1^*$ is mainly an $s$ orbital on the metal. The activation energies for dissociation are found to be, for $L = P(OC_2H_5)_3$, in the order Ni > Pd < Pt. This does not correlate with

| | $E_a$ | $ns - np$ | $(n-1)d - ns$ |
|---|---|---|---|
| Ni | 26 kcal | 5.3 eV | −1.8 eV |
| Pd | 22 | 4.32 | 0.81 |
| Pt | 26 | 4.76 | −0.76 |

the experimental energy gap between the valence shell $s$ and $p$ orbitals of the free metal atoms, nor between the $d$ and $s$ orbitals, shown above.

## Square Planar Molecules

Reaction of square planar $XY_4$ molecules can also occur to give both $XY_3$ + Y and $XY_2 + Y_2$. In both cases the LM path is of $C_{2v}$ symmetry and an $E_u$ mode is needed. There may be confusion because correlation tables show $E_u$ in $D_{2h}$ becoming $(B_1 + B_2)$ in $C_{2v}$. The rules require that $E_u$ become $A_1$ to be a true reaction coordinate. Actually, it does, the confusion arising from a change in the $z$ axis from perpendicular in $D_{4h}$ to in-plane in $C_{2v}$.

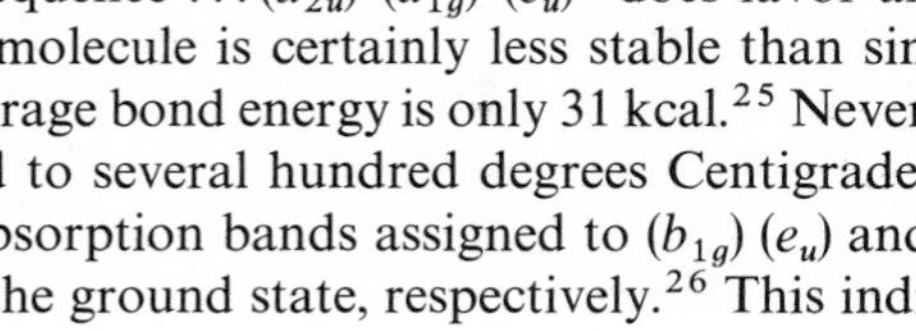

$$XY_4\ (D_{4h}) \xrightarrow{E_u} XY_2\ (C_{2v}) + Y_2$$

For $XeF_4$ the sequence $\ldots(a_{2u})^2(a_{1g})^2(e_u)^0$ does favor an $E_u$ reaction coordinate, and the molecule is certainly less stable than similar, tetrahedral molecules. The average bond energy is only 31 kcal.[25] Nevertheless the molecule can be heated to several hundred degrees Centigrade without decomposing. The UV absorption bands assigned to $(b_{1g})$ $(e_u)$ and $(a_{1g})$ $(e_u)$ are 6.8 and 9.4 eV above the ground state, respectively.[26] This indicates reasonable stability.

An examination of the concerted loss of $F_2$ from $XeF_4$ shows that only *trans* groups can be lost.

$$\text{XeF}_4\ (a_1 + b_2) \longrightarrow \text{F–Xe(:)–F}\ (a_1) + \text{F–F}\ (a_1) \tag{36}$$

$$\text{XeF}_4\ (a_1 + b_2) \longrightarrow \text{F–F}\ (a_1) + \text{F–Xe(:)–F}\ (b_2) \tag{37}$$

The loss of *cis* F atoms is forbidden by symmetry, since a lone pair appears in a *p* orbital of $a_1$ species.

The great majority of square planar complexes are of low-spin $d^8$ transition metal ions. The MO sequence, considering only $\sigma$ bonding ligands, would be

$$(a_{1g})^2(e_u)^4(b_{1g})^2 \,\vdots\, (b_{2g})^2(e_g)^4(a_{1g})^2(b_{1g}^*)^0 \,\vdots\, (a_{2u})^0(a_{1g}^*)^0$$

with the *d* orbitals enclosed by dashed lines, as usual. The only effective transition promoting dissociation by an $E_u$ mode, would be from the $\sigma$-bonding $e_u$ orbitals into the $\sigma^*$-antibonding $b_{1g}$ orbital of the *d* manifold. Transitions such as $(e_g) \rightarrow (a_{2u})$ are ineffective because $\rho_{0k}$ is zero in the plane of the complex.

It is an experimental fact, long puzzling, that nucleophilic substitution reactions of planar $d^8$ complexes occur only by $S_N2$ mechanisms.[27] An $S_N1$ mechanism has not been substantiated, and only occasionally suggested. This differs markedly from the case of similar tetrahedral complexes where $S_N1$ and $S_N2$ processes are about equally common. One might be tempted to try to explain this by considering the MO sequence above.

However there is no reason why the necessary charge transfer for planar complexes should be any less accessible than the corresponding states in tetrahedral molecules. Instead the failure of planar complexes to react by an $S_N1$ mechanism probably has a different explanation, lying in the special stabilization of a $d^8$ system in a $D_{4h}$ environment. Figure 5 shows the correlation of those orbitals that are primarily *d* orbitals on the metal in $D_{4h}$, $D_{3h}$, and $T_d$. $D_{3h}$ would be the point group if a single ligand dissociated from a planar complex, and the remaining ligands rearranged to a symmetric, planar structure.

It can be seen that there is a very large loss of ligand field-stabilization energy for a planar complex, but not for a tetrahedral one. Ligand-field

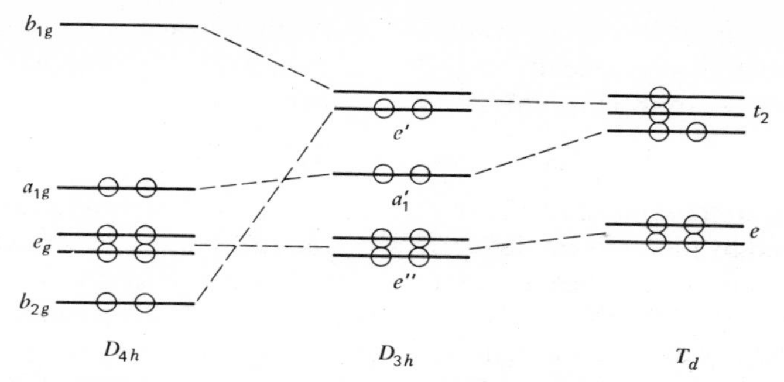


**Figure 5.** Correlation of $d$ orbitals of a $d^8$ metal complex, $ML_4$, as it loses one ligand, L, to become planar, symmetric $ML_3$; the exact ordering of the filled orbitals in $D_{4h}$ is uncertain.

stabilization energy (LFSE) results from an unequal filling of the $d$ manifold with electrons, such that the least stable orbitals remain empty.[28] Planar complexes only exist because of such stabilization. Tetrahedral complexes have little or no stabilization of this kind, and in fact may gain LFSE upon dissociation.

The two modes of dissociation of $SF_4$, with its $C_{2v}$ structure, are

$$SF_4 \longrightarrow SF_3 + F \tag{38}$$

$$SF_4 \longrightarrow SF_2 + F_2 \tag{39}$$

Reaction (38) is complicated because the leaving atom can be either apical requiring a $B_1$ mode, or equatorial, requiring a $B_2$ mode. The LM principle requires the apical group to leave since the $SF_3$ group is left very near the pyramidal structure required. If $F^-$ leaves, $SF_3{}^+$ would have a $C_{3v}$ structure finally. The apical S—F bond distance is longer than the equatorial S—F distance, and the apical force constant for S—F stretch is less than half of that for equatorial S—F stretch.[29a]

The MO sequence calculated, however, is

$$(a_1)^2(b_1)^2(b_2)^2(2a_1)^2(3a_1)^2(2b_2)^0(2b_1)^0(4a_1)^0$$

which suggests a $B_2$ reaction coordinate.[29b] This is not convincing in itself, but another observation strongly suggests that an equatorial atom will be more labile. Both theory[29b] and experiment[29c] show that the $3a_1$ orbital is quite antibonding toward the apical ligands. The transition $(3a_1) \rightarrow (2b_2)$ would not only weaken the equatorial bonds, but also strengthen the apical S—F bonds. Dissociation should be much easier than for $CF_4$, for example. $SF_4$ is a very reactive molecule, hydrolyzing instantly and selectively fluorinating many molecules. Structurally related molecules, $R_2SX_2$, exist in polar solvents as $R_2SX^+$ and $X^-$, at least to a substantial degree.[30]

The concerted loss of $F_2$ from $SF_4$ is a fully allowed process, although thermodynamically difficult. Many $R_2SX_2$ molecules exist in an equilibrium, which is very rapid. This is also true for related

$$R_2SX_2 \rightleftharpoons R_2S + X_2 \tag{40}$$

molecules containing Se and Te in place of S. This kind of equilibrium may be established by initial formation of $R_2SX^+$ and $X^-$, but it is very likely that in nonpolar solvents a single concerted reaction is energetically easier.

In the case of $R_2SCl_2$, which have $C_{2v}$ structures,[30] it is the two apical halogen atoms that are lost. However symmetry considerations show that in $SF_4$, two apical F atoms can be lost as $F_2$, but two equatorial atoms also can be

$$SF_4 \longrightarrow SF_2 + F_2 \tag{41}$$

$$a_1 + b_1 \qquad\qquad b_1 \qquad a_1$$

Apical loss is shown in (41), where it is seen that the bond-symmetry rule is obeyed. A lone pair of electrons in a $b_1$ orbital is formed on sulfur. In equatorial loss of two F atoms, the bonds that are broken have $a_1 + b_2$ symmetry. An F—F bond of $a_1$ species is formed, and a lone pair on S in an orbital of $b_2$ type is created. A lone pair on S of $a_1$ symmetry is not shown in (41) since it remains constant.

Apical–equatorial loss of $F_2$ is a very asymmetric process. No elements of symmetry are conserved, and symmetry arguments cannot be used. However the orbital following method shows that the process is essentially forbidden. One can look at the reverse of the various reactions of $SF_4$ as well, adding $F_2$ to $SF_2$, In this way simple overlap arguments can be used for allowedness.

## Dissociation of $XY_5$ Molecules

Molecules $XY_5$ also have a variety of ways in which the reactions

$$XY_5 \longrightarrow XY_4 + Y \tag{42}$$

and

$$XY_5 \longrightarrow XY_3 + Y_2 \tag{43}$$

can occur. Table 1 shows that a trigonal bipyramidal molecule, such as $PCl_5$, can lose either an apical Cl by way of an $A_2''$ reaction coordinate, or an equatorial Cl by an $E'$ reaction mode. The MO sequence for $PH_5$ is

$$(a_1')^2(a_2'')^2(e')^4(2a_1')^2(2e')^0(2a_2'')^0$$

which predicts that an $E'$ mode is favored, that is, loss of an equatorial group. The $A_2''$ mode comes in through a higher-energy transition.

This is quite a surprising conclusion, since a great deal of evidence on the behavior of five-coordinated phosphorus compounds (phosphoranes) is best interpreted on the basis that apical groups are lost more easily than equatorial.[31] Also experimentally apical bond distances are somewhat longer than equatorial bond distances in molecules like $PCl_5$. This suggests weaker bonds. Several comments are appropriate. One is that conclusions drawn from less symmetric molecules, such as $PX_4Y$ where Y is a weakly held group, can be misleading. The most weakly bonded Y indeed does appear in the apical position as a rule,[32] and hence must appear to leave from there. The other comment is that the reactions that $PCl_5$ and so on actually do undergo are indeed of $E'$ symmetry. One is the facile Berry pseudorotation, converting into a $C_{4v}$ structure. The second is the concerted loss of $Cl_2$, as in (41). This reaction will be discussed later as the reverse reaction, called "oxidative–addition."

The most important factor favoring loss of an apical group, however, is the principle of least motion. Consider the specific reaction

$$PCl_5 \longrightarrow PCl_4^+ + Cl^- \qquad (44)$$

This reaction occurs readily in the presence of a chloride-ion acceptor. $PCl_5$ itself can play such a role since in the solid state it exists as $PCl_4^+$, $PCl_6^-$. The phosphonium ion has a tetrahedral structure, since it has eight valence electrons. Now Fig. 17 of this chapter shows that an $A_2''$ vibration of an $XY_5$ molecule converts smoothly into the tetrahedral structure. Furthermore it is the mode that preserves the greater number of symmetry elements, $C_{3v}$ compared to $C_{2v}$ for the $E'$ mode. Hence PLM predicts apical loss of a group.

This analysis is supported by another set of observations. Low-spin $d^8$ metal ions form a number of five-coordinate complexes of the $D_{3h}$ point group. These are structurally labile, rapidly interconverting back and forth into the square pyramidal isomers. The $d$ manifold would be

$$\vdots (e'')^4(e')^4(a_1')^0 \vdots$$

which favors the needed $E'$ mode. Complexes such as $Ni(PR_3)_3(CN)_2$ have two cyanide ions in the axial positions. These are tightly held and nonlabile. The phosphine ligands are weakly held and dissociate easily.[33]

$$Ni(PR_3)_3(CN)_2 \rightleftharpoons Ni(PR_3)_2(CN)_2 + PR_3 \qquad (45)$$

There is clearly dissociation from an equatorial position in these examples.

This is also consistent with the PLM. Figure 17 shows that the $E'$ mode can lead easily to a square planar structure, but not to a tetrahedral one. The reverse is true for $A_2''$. Since the $Ni(PR_3)_2(CN)_2$ formed in (43) is square planar, it follows that an $E'$ mode is required by PLM. Incidentally, it is not likely that transitions within the $d$ manifold are solely responsible for dissociation. Instead the MOs more closely related to bonding must be involved as well.

The loss of F from $IF_5$ can occur, or in a polar solvent, $F^-$ can be released

$$IF_5 \longrightarrow IF_4{}^+ + F^- \tag{46}$$

This can happen either from an apical position, or from the basal plane of the square pyramid structure. An examination of the reverse of (46) suggests loss of a basal group, contrary to the PLM (p. 321). The reaction coordinate is $E$ for basal loss and $B_1$, for apical loss.

Similarly the concerted loss of $F_2$ from $IF_5$ can occur either axially–equatorially or equatorially–equatorially. The former requires an $E$ mode, a plane of symmetry being kept, and the latter a $B_1$ mode, the elements of $C_{2v}$ being conserved. The latter path is the LM one, but again an analysis of the reverse reaction for the process

$$IF_5 \rightleftharpoons IF_3 + F_2 \tag{47}$$

favors the non-LM axial–equatorial path (p. 289).

### Octahedral Molecules

Molecules $XY_6$ are almost always of the $O_h$ point group. The dissociation process

$$XY_6 \rightleftharpoons XY_5 + Y \tag{48}$$

requires a $T_{1u}$ reaction mode if $XY_5$ is formed, initially at least, in the $C_{4v}$ point group. This is the LM path for (48). The molecule $SF_6$ has the MO sequence

$$(2t_{1u})^6(t_{2u})^6(t_{2g})^6(2a_{1g})^0(3t_{1u})^0$$

so that a $T_{1u}$ mode is not easily excited. The molecule is very stable, with a strong S—F bond. $XeF_6$ has the configuration

$$(e_g)^4(2a_{1g})^2(3t_{1u})^0$$

with a small energy gap between $(2a_{1g})$ and $(3t_{1u})$. Accordingly the molecule is unstable to dissociation, the Xe—F bond energy being only 30 kcal.[25] $XeF_6$ often acts as a fluorinating agent, but the mechanisms are complex. An illuminating study has been made[34] of the relative tendencies of the following reactions:

$$XeF_2 \longrightarrow XeF^+ + F^- \tag{49}$$

$$XeF_4 \longrightarrow XeF_3{}^+ + F^- \tag{50}$$

$$XeF_6 \longrightarrow XeF_5{}^+ + F^- \tag{51}$$

The procedure is to add $F^-$ acceptors of varying strengths. The results show that $XeF_6$ is the most prone to dissociate and $XeF_4$, the least. This correlates with the fact that $XeF_6$ has the lowest-energy excited state of the three molecules. However $XeF_2$ and $XeF_4$ have similar absorption spectra.[26]

If $\pi$-bonding ligands are not included, the MO sequence for octahedral complexes of the transition metals is usually[35]

$$\underline{(a_{1g})^2(t_{1u})^6(e_g)^4} \,\vdots\, (t_{2g})(e_g^*) \,\vdots\, (a_{1g}^*)(t_{1u}^*)^0$$

These orbitals are shown in Fig. 6, giving some idea of the separation between them. Since $(T_{1u} \times T_{2g}) = (A_{2u} + E_u + T_{1u} + T_{2u})$, and $(T_{1u} \times E_g) = (T_{1u} + T_{2u})$, there are numerous transitions of $T_{1u}$ symmetry possible.

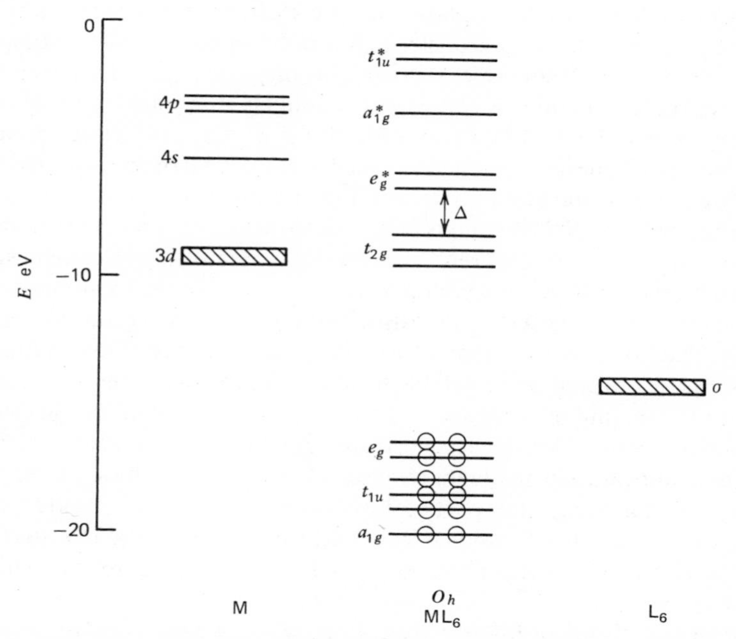


**Figure 6.** Molecular orbitals of an octahedral $ML_6$ complex where L has only $\sigma$-bonding orbitals; approximate energy values are shown in eV; the *d* electrons are accommodated in the $t_{2g} - e_g^*$ set of orbitals, and $\Delta = 10$ Dq, and the approximate energies of the valence orbitals of M and the $\sigma$ orbitals of the ligands are also shown.

Several experimental facts must be accounted for. For example, the dissociation energy of a series of similar complexes, say $M(NH_3)_6{}^{2+}$ where M is a metal ion of the first transition series, shows a steady increase as the nuclear charge on M increases. There are perturbations due to the filling in of the metal $d$ orbitals $t_{2g}$ and $e_g^*$, but these are relatively small (this is not to say that they are not of great chemical significance). Also as the positive charge on M increases, the bond energy increases. The process under discussion is heterolytic bond cleavage, for instance,

$$Ni(NH_3)_6{}^{2+} \longrightarrow Ni(NH_3)_5{}^{2+} + NH_3 \qquad (52)$$

These observations suggest that the main excitations responsible for dissociation are between the filled $\sigma$-bonding orbitals and the empty $\sigma^*$-antibonding orbitals. The gap between these would presumably increase as the electronegativity of the metal increases. Since the $\sigma$ orbitals are concentrated on the ligands and the $\sigma^*$ orbitals on the metal, it seems initially inconsistent that mixing of the latter would promote even greater charge transfer to ligand, as in (52). Actually this offers no problem, since the sign of mixing of excited-state wavefunctions will always be adjusted to fit the situation (p. 14).

One might look then for a relationship between bond-dissociation energies and the positions of charge-transfer bands in the spectra of metal complexes. These bands would correspond to $(t_{1u}) \rightarrow (a_{1g}^*)$ or $(t_{1u}) \rightarrow (e_g^*)$ excitations. However there is a good reason why it is unlikely that much correlation will be found. It is well known that an ionic model works very well for predicting the dissociation energies of reactions such as (52).[28] All of the expected effects of varying charge and size of the metal ion and the ligands are found, as long as the donor atom of the ligand is fairly electronegative. Also the variations in bond strengths due to different numbers of $d$ electrons are rather well explained by crystal-field theory, which also uses a simple electrostatic model.[36]

Consider a molecule $MY_6$ made up of one spherical cation and six spherical anions. The bonding is assumed to be 100% electrostatic. Dissociation into $MY_5{}^+$ and $Y^-$ would be calculable from Coulomb's law. There would be no need for the mixing of excited states, since the wavefunction for the electrons would not change. The valence electrons remain on the same atoms, namely Y. Even if allowance is made for some covalent bonding, the situation would still remain that the greater part of the bonding energy would be independent of excited states. Only a minor part would correlate with the UV absorption spectra. The same conclusions apply to all other metal complexes which are highly polar.

For donor atoms of low electronegativity, covalency is more important, and excited states should be more contributory in reactions. For example, it has been suggested that stabilizing the $t_{2g}$ orbitals by suitable ligands is important in preventing homolytic cleavage of metal–alkyl bonds.[37] The implication is that $(t_{2g}) \rightarrow (t_{1u}^*)$ excitation is important in promoting such dissociation.

There is another circumstance in which covalency becomes important, even for polar molecules. This is when $\pi$-bonding ligands, such as the halide ions, are in $MY_6$. These contribute MOs of $t_{1u}$ and $t_{2g}$ symmetry between the $\sigma$-bonding orbitals and the $t_{2g}$ level of the metal ion. Upon dissociation of $Y^-$, $\pi$ bonding from ligand to metal becomes important. The transitions are $(t_{1u})$ or $(t_{2u}) \rightarrow (t_{2g})$. If the $t_{2g}$ level is filled, then rearrangement of the dissociating complex occurs so that the $e_g$ level can become $\pi$ accepting.[38]

The simultaneous loss of two ligands, as in

$$XY_6 \longrightarrow XY_4 + Y_2 \qquad (53)$$

is more commonly known under the name of "reductive elimination." It is a reaction of great importance as a step in catalytic cycles involving transition-metal catalysts. It will be discussed later in terms of its reverse reaction under the heading of "oxidative addition."

Table 1 also shows the LM reaction paths for $XY_7$ molecules, although little is known about their reaction mechanisms. Both processes such as

$$IF_7 \longrightarrow IF_6^+ + F^- \qquad (54)$$

$$IF_7 \longrightarrow IF_5 + F_2 \qquad (55)$$

should occur by removal of one or two groups from the equatorial plane of a $D_{5h}$ molecule. While (54) is allowed, (55) is forbidden by the bond-symmetry rule.

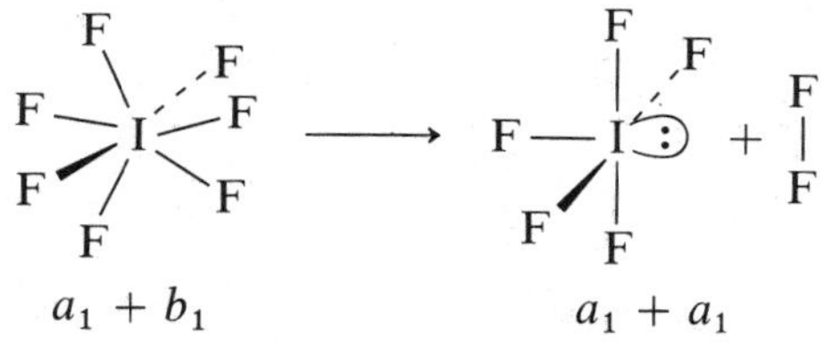

$$a_1 + b_1 \longrightarrow a_1 + a_1 \qquad (56)$$

The lone pair of $IF_5$ shows up in an $a_1$ orbital, whereas a $b_1$ orbital is needed. The kinetics of the reaction reverse to (56) have been studied. The rate law is

$$\text{rate} = k[IF_5][F_2] \qquad (57)$$

the reaction appears homogeneous, and the indications are that it is a simple bimolecular process. Overall the oxidative addition of $F_2$ to $IF_5$ is exothermic by 29 kcal. The reaction is rather slow, temperatures of 60–95°C being required. The Arrhenius activation energy $E_a = 14$ kcal and $\Delta S^\ddagger = -26$ eu. These results are perhaps not incompatible with a mechanism in which a non-LM path is followed, but the activation energy seems low.

## DISSOCIATION OF $X_2Y_n$ MOLECULES

Besides the reactions corresponding to those of $XY_n$ molecules,

$$X_2Y_n \longrightarrow X_2Y_{n-1} + Y \tag{58}$$

$$X_2Y_n \longrightarrow X_2Y_{n-2} + Y_2 \tag{59}$$

at least one additional reaction must be considered for $X_2Y_n$ molecules. This is the symmetric fragmentation process,

$$X_2Y_n \longrightarrow 2XY_{n/2} \tag{60}$$

Also reaction (59) can occur in two ways, in which each X atom loses one Y atom, or one X atom loses both Y atoms. We will consider only the former possibility, since the latter is usually of high energy.

With this restriction, (59) is rather rapidly disposed of. With rare exceptions, it is forbidden by symmetry considerations, for example,

$$C_2H_6 \longrightarrow C_2H_4 + H_2 \tag{61}$$

$$N_2H_4 \longrightarrow N_2H_2 + H_2 \tag{62}$$

$$cis\text{-}N_2H_2 \longrightarrow N_2 + H_2 \tag{63}$$

$$C_2H_2 \longrightarrow C_2 + H_2 \tag{64}$$

In all cases, an LM reaction coordinate has been assumed. One exception to the general rule is the dehydrogenation of diborane. In this case a concerted reaction is allowed in which the bridging hydrogens are eliminated as molecular $H_2$. The point group is $C_{2v}$. The reaction coordinate is $B_{1u}$, an upward motion of

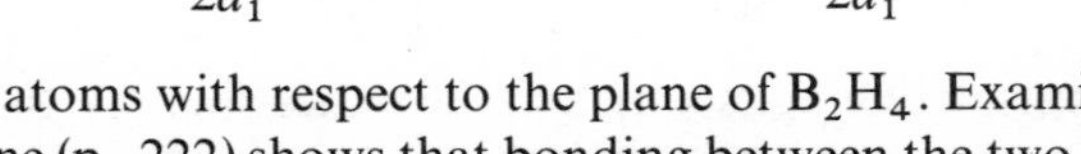

```
                                        H
                                        |
 H     H     H                    H     H     H
   \  / \   /                       \       /
    B     B          ---->           B --- B          (65)
   /  \ /   \                       /       \
 H     H     H                    H           H
      2a1                               2a1
```

the bridging H atoms with respect to the plane of $B_2H_4$. Examination of the MOs of diborane (p. 222) shows that bonding between the two borons exists only by virtue of the bridging H atoms. The two MOs, $3a_g$ and $b_{1u}$, are essentially two-electron, three-center bonds. In $C_{2v}$ with the $z$ axis remaining $C_2$, both of these are $a_1$. They become eventually the B—B $\sigma$ bond in $B_2H_4$ and the H—H bond in $H_2$.

While reaction (65) is not known to occur, its reverse is known. At least $H_2$ and $Cl_2$ add rapidly to $B_2Cl_4$, the known analog of $B_2H_4$.[40] The observed products are $BHCl_2$ and $BCl_3$, which would result from the spontaneous dissociation of $B_2H_2Cl_4$ and $B_2Cl_6$. It is tempting to also consider a broadside addition.

H—H
H B—B H
H H
⟶
H H
B—B
H H H H
(66)

utilizing the overlap of the filled $\sigma$ orbital of $H_2$ and the empty $\pi$ orbital of $B_2H_4$. This process is also allowed, but it generates the unstable ethane-like structure for $B_2H_6$. The two new B—H bonds contain only one electron each. Alternatively, by a symmetry restriction, only the $a_1$ component of the two new B—H bonds contains electrons. The $b_2$ component is empty.

The known reaction on heating $B_2H_6$ is dissociation into $2\,BH_3$.

$$B_2H_6 \longrightarrow 2\,BH_3 \qquad (67)$$

This occurs rather easily since the bond-dissociation energy is only 36 kcal. Ab initio calculations for (67) show that the reaction coordinate is initially $B_{2g}$. The reaction begins by rotation of two $BH_3$ fragments away from each other (about the $y$ axis). The point group becomes $C_{2h}$, and the two $BH_3$ groups leave each other nearly face to face.[41] Figure 7 shows the TS for (67), which is 2.6 kcal above the energy of the final products.

The relative ease of (67) may be compared with the difficulty of (68), which requires some 80 kcal of energy

$$C_2H_6 \longrightarrow 2CH_3\cdot \qquad (68)$$

The UV absorption of ethane is shifted about 2 eV toward higher energy compared to diborane. However a direct comparison is not possible since the dissociation (68) requires primarily doubly excited states (see p. 50).

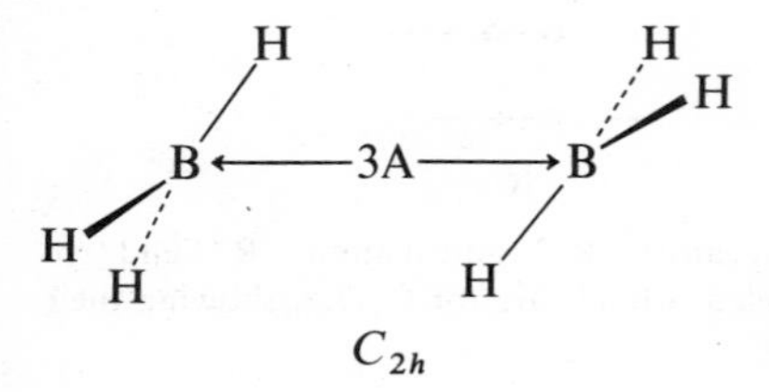


**Figure 7.** The TS for the formation of bridged $B_2H_6$ from two $BH_3$ molecules (after reference 41).

The pyrolysis of ethane gives ethylene, hydrogen, and a little methane. The mechanism is a chain reaction in which (68) is the initiating step. This is followed by[42]

$$CH_3\cdot + C_2H_6 \longrightarrow CH_4 + C_2H_5\cdot \tag{69}$$

$$C_2H_5\cdot \longrightarrow C_2H_4 + H\cdot \tag{70}$$

$$H\cdot + C_2H_6 \longrightarrow H_2 + C_2H_5\cdot \text{ (etc.)} \tag{71}$$

The novel feature is (70), the rather facile decay of ethyl radical into ethylene and a hydrogen atom. This contrasts with the stability of methyl radical toward carbene formation and a hydrogen atom, reaction (29).

Whenever a free radical is formed, low-lying excited states are automatically created. If the odd electron is in a stable bonding orbital, excitation from below is possible. If it is in an unstable orbital, excitation to orbitals above are favored. Also if the free radical is formed by removal of an atom, or group, as in (69), at least two orbitals are markedly affected. One is the stable orbital most closely related to bonding the atom which is removed. This orbital will generally be raised in energy by removal of the atom. At the same time, unstable orbitals strongly antibonding toward the atom removed, will decrease in energy. This situation is illustrated in Fig. 8, along with other circumstances which make free radicals reactive.

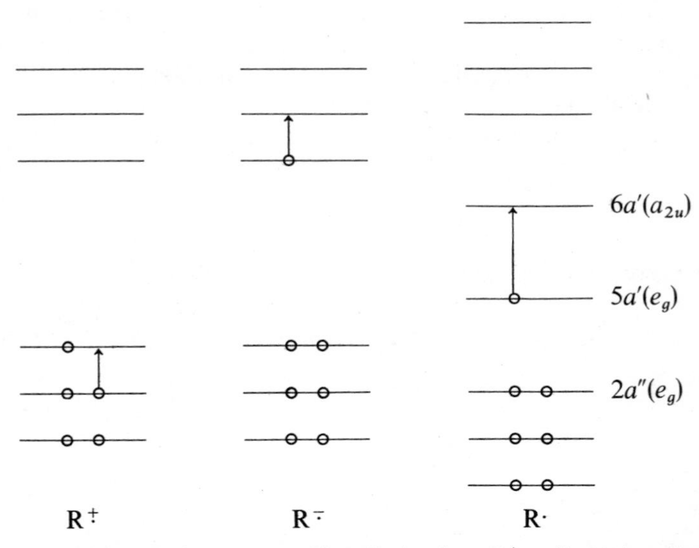


**Figure 8.** Low-lying excited states created in radical cations, $R^{+}$, radical anions, $R^{-}$, and free radicals, R·, created by removing a group; the labeled orbitals are for $C_2H_5\cdot$, showing their relationship to those of $C_2H_6$.

The situation which facilitates reaction (70) is the last one discussed. Comparing the MOs of staggered ethane (p. 220) to the MOs of the bisected ethyl radical,[43] it is found that the occupied $e_g$ orbital, which is C—H bonding, becomes the HOMO of $C_2H_5$. The LUMO of $C_2H_5$ derives from the $2a_{2u}$ orbital of $C_2H_6$, which is C—H antibonding. Removal of the hydrogen atom, brings these two orbitals quite close in energy. In the $C_s$ point group of eclipsed $C_2H_5$, they are both of $A'$ species. The reaction coordinate for removal of another H atom to form ethylene, is also $A'$.

A comparison of the bond-dissociation energies in the series

$$C_2H_6 \longrightarrow C_2H_5 + H \quad 98 \text{ kcal} \tag{72}$$

$$C_2H_4 \longrightarrow C_2H_3 + H \quad 104 \text{ kcal} \tag{73}$$

$$C_2H_2 \longrightarrow C_2H + H \sim 120 \text{ kcal} \tag{74}$$

is instructive. The reaction coordinate for (72) is $E_u$, which takes the $D_{3d}$ point group into $C_s$, and for (74) it is $\Sigma_u{}^+$, which takes $D_{\infty h}$ into $C_{\infty v}$. However reaction (73) requires the activation of at least three vibrational modes. That is, the $D_{2h}$ point group must become $C_s$, even for the LM path. Many symmetry elements must be destroyed and each vibrational mode for ethylene, all of which are nondegenerate, can destroy only a few.

This means that many excited states of ethylene are called on for contributions to the activation process. Nevertheless the energy requirement for (73) is not exceptional compared to (72) and (74). A complex reaction path can require no more energy than a simple one, at least in cases where the energy required is large thermodynamically.

A somewhat related problem is in the pyrolysis of azo compounds. The question is whether or not both alkyl groups are lost at once (75), or whether the process is stepwise, (76) and (77).

$$trans\text{-R—N=N—R} \longrightarrow R\cdot + N_2 + R\cdot \tag{75}$$

$$\text{R—N=N—R} \longrightarrow \text{R—}N_2\cdot + R\cdot \tag{76}$$

$$\text{R—}N_2\cdot \longrightarrow R\cdot + N_2 \tag{77}$$

In the first case the reaction coordinate is $A_g$ and no symmetry is lost. In the second case the reaction coordinate must be $B_u$, and only a plane of symmetry remains. The bulk of the evidence shows that the two-step mechanism is followed.[44] Thus a less symmetric path is followed in preference to a more symmetric one. This is not a violation of the PLM, however, since the principle applies independently to each of the elementary reactions (75), (76), and (77).

Similarly a vinyl radical is formed in (73), or an ethyl radical in (72) without the concerted loss of a second hydrogen atom to form acetylene or ethylene. In the former case, the reaction coordinate would be very much simpler for a concerted loss of two H atoms, but the energy needed is larger. In some cases substituted ethane derivatives pyrolyze in a way that suggests concertedness.

$$\text{RS}\!-\!\text{C}\!-\!\text{C}\!-\!\text{SR} \longrightarrow \text{C}\!=\!\text{C} + \text{RSSR} \quad (78)$$

In the above example stereospecific *trans* elimination occurs, with RS radicals definitely formed as intermediates.[45] The stereospecificity could be explained by a concerted loss of two RS radicals. More likely it simply requires that the second RS radical is lost before the intermediate can racemize. This could occur if the sulfur atom of the second RS interacts with the free radical center on carbon to hold its structure (see p. 335).

## THE OXIDATIVE-ADDITION REACTION

It is quite surprising that it is only in the last ten years that an important new elementary reaction has been recognized. This is the oxidative-addition reaction[46] which may be written as

$$\text{M} + \text{XY} \longrightarrow \text{X—M—Y} \quad (79)$$

Here M stands for an atom in a reduced oxidation state and oxidized by two units on adding the molecule XY. Other groups will be attached to M, as a rule. If X and Y are held together by a single bond, then X and Y are added to M with complete breaking of the X—Y bond. This is the case when XY is a molecule such as $H_2$, $Cl_2$, HCl, and $CH_3I$. If XY is multiply bonded, for instance, $O_2$, then the bonding is reduced in the adduct, but some residual bonding remains.[47] No stereochemistry is implied in (79) to begin with, except that X and Y must be adjacent in the latter case.

The earliest recognized examples of (79) were cases of addition of XY to transition-metal complexes of $d^8$ and $d^{10}$ configuration.[48] A typical reaction would be oxidative addition of $CH_3I$ to Vaska's compound, a planar four-coordinated iridium(I) complex.

$$trans\text{-Ir(CO)Cl[P}(C_6H_5)_3]_2 \longrightarrow \text{Ir(CO)Cl[P}(C_6H_5)_3]_2CH_3I \quad (80)$$

An octahedral six-coordinated iridium(III) complex is formed. In this case the methyl group and iodine atom are *trans* to each other in the product. Other cases are known where X and Y appear *cis* to each other in reactions with the same metal complex.

There are several mechanisms whereby reaction (79) can occur, if the product formed is used as a criterion. We are concerned with concerted mechanisms where XY is added in a single step. For symmetric reactants, this is simply the reverse of some of the unimolecular dissociations disscussed earlier. For thermodynamic reasons, the addition step is often easier to study. Also symmetry arguments are often easier to apply if the view is taken that reactions occur via a flow of electrons from an occupied MO in the substrate (M) to an empty MO in the oxidant XY. Note that even $H_2$ can be an oxidant if the product is a metal hydride (hydrogen becomes negative).

As a simpler example of oxidative addition, consider the equilibrium

$$PCl_3 + Cl_2 \rightleftharpoons PCl_5 \tag{81}$$

This reaction is exothermic by 21 kcal mole$^{-1}$. Equilibrium is established very quickly in the gas phase, too quickly to study by ordinary means. It is quite certain that chlorine atoms are not formed as intermediates and hence a concerted process is indicated. In polar solvents ionic mechanisms such as reaction (44) could be a factor, but not in the gas phase.

An even simpler example would be a reaction of two diatomic molecules. Earlier it had been stressed that two molecules of this kind could not react by a four-center planar mechanism. However a head-on collision of one molecule with the broadside of another was not considered because only diatomic products were sought. Removing this restriction we find that reactions of the kind

$$\begin{matrix} \text{F} \\ | \\ \text{F} \end{matrix} + \;:\ddot{\underset{..}{\text{Br}}}\text{—Br} \longrightarrow \begin{matrix} \text{F} \\ \diagdown \\ \end{matrix}\ddot{\underset{..}{\text{Br}}}\text{—Br} \begin{matrix} \\ \diagup \\ \text{F} \end{matrix} \tag{82}$$

$$a_1 \qquad a_1 + b_1 + b_2 \qquad\qquad a_1 + b_2 + a_1 + b_1$$

are allowed by symmetry. The $C_{2v}$ point group is assumed. The lone pairs of the central bromine atom are $(a_1 + b_1 + b_2)$ to start with. At the finish, there is a lone pair above and below the molecular plane, giving $a_1$ for their sum and $b_1$ for their difference. The two Br—F bonds give $(a_1 + b_2)$ for their sum and difference. The other lone pairs of the halogen atoms are not affected. Essentially the reverse reaction of (82) could give two molecules of BrF. Reaction of BrF with another molecule of $F_2$ would give $BrF_3$, a stable molecule. Products of the kind formed by the mechanism of (82) are only stable if the central halogen atom is of low electronegativity, and the terminal ones are of high electronegativity.

There is still an activation energy expected for (82), even when thermodynamically favorable. The bond distance between the two apical fluorine atoms in $BrF_3$ is about 3.6 Å, much greater than the F—F distance in $F_2$ which is 1.43 Å. Thus $BrF_3$ would be formed from BrF and $F_2$ with a Y-shaped structure rather than the stable T shape. From the nmr spectrum of $BrF_3$ and $ClF_3$, it is known that all three F atoms are equilibrating rapidly. The barrier is about 5 kcal for $ClF_3$,[50] which means the symmetric structure (120° angles for F—Cl—F) is that much in energy above the equilibrium form. The Y-shaped structure would be still higher in energy.

Experimentally it is found that $Br_2$ and $F_2$ react rapidly at room temperature to give $BrF_3$.[49] Similarly $I_2$ and $F_2$ react rapidly even at −78°C to give $IF_3$. Upon warming the latter molecule disproportionates to $IF_5$ and $I_2$. Surprisingly, even though $ClF_3$ is fairly stable, $Cl_2$ and $F_2$ react only at temperatures above 200°C, where free atoms are the reactants.[49] This is true for both $ClF_3$ and ClF as products. Also two heavy halogens, such as $I_2$ and $Br_2$, do not use a variation of (82) to form interhalogen compounds, such as IBr.

It appears that this reaction mode requires both a light and heavy halogen. The critical feature is probably the willingness with which the less electronegative halogen atom will give up electrons from its $\pi^*$ orbitals into the antibonding $\sigma^*$ orbital of the more electronegative halogen. Unless this occurs readily, to form the new bonds and to break the old F—F bond, the strain energy in the formation of the products is excessive.

The reaction of type (82) can succeed only if the reducing electrons are in a $\pi$-type orbital. If the electrons are in a $\sigma$-type orbital, the reaction is forbidden.

$$Cl_2 + :C{=}O \longrightarrow Cl_2C{=}O \qquad (83)$$

$a_1$ $a_1$ $a_1 + b_2$

$$Cl_2 + Tl{-}Cl \longrightarrow Cl_2Tl{-}Cl \qquad (84)$$

$a_1$ $a_1$ $a_1 + b_2$

$$O{=}O + :C{=}O \longrightarrow O_2C{=}O \qquad (85)$$

$a_1$ $a_1$ $a_1 + b_2$

The reducing electrons of carbon monoxide are in an orbital of $a_1$ symmetry. Also the thallous ion has its reducing electrons in an $s$ orbital, the so-called "inert" pair. An $s$ orbital is of $a_1$ species in the $C_{2v}$ point group. The reaction of dry chlorine gas with CO is remarkably slow, occurring only at high temperature, and by a free-radical mechanism.[51] Reaction (85) is forbidden both by symmetry and spin.

The reactions of $CH_2$, and other carbenes, are related to those of CO. Triplet methylene acts as a free radical, abstracting hydrogen atoms, typically.

$$CH_2(^3B_1) + RH \longrightarrow CH_3\cdot + R\cdot \quad (86)$$

But singlet carbene inserts into a number of single bonds.

$$CH_2(^1A_1) + RH \longrightarrow RCH_3 \quad (87)$$

Reactions like (87), and similar reactions of oxygen atoms and nitrenes, have long been called insertion reactions. However they may also be called oxidative-addition reactions in a mechanistic sense.

An examination of the simplest example of the insertion reaction shows that it is forbidden in the LM approach. This

$$CH_2 + H_2 \longrightarrow CH_4 \quad (88)$$

has already been shown for the reverse process in (32). Semiempirical calculations[52] show that a non-LM approach is favored, which is very asymmetric. The leading orbital interaction is between the empty MO of carbene of $B_1$ symmetry and the filled C—H or H—H $\sigma$-bonding orbital. Hence these reactions are unlike other oxidative additions in that electrons flow from XY rather than to XY.

In spite of the symmetry barriers to (87) and (88), the reactions occur readily. This is a consequence of their extreme exothermicity. In fact, reaction (88) is immediately followed by decomposition into methyl radicals and hydrogen atoms, since the product methane is formed in a highly excited vibrational state.[53] More complex products are stabilized by sharing this excitation energy among the numerous degrees of freedom. Reactions such as

$$CCl_2 + R^*H \longrightarrow R^*CCl_2H \quad (89)$$

where R* is an optically active group, occur stereospecifically, and with retention of configuration.[54] This is an important indication that the reaction is concerted and occurs by attack at the front side of the C—H bond.

### Oxidative Addition to $ML_2$ Molecules

If a triatomic molecule has its reducing electrons in an orbital that is essentially an $s$ orbital on the central atom, concerted addition of XY is forbidden. This is illustrated in Fig. 9$a$ by showing the zero overlap of the

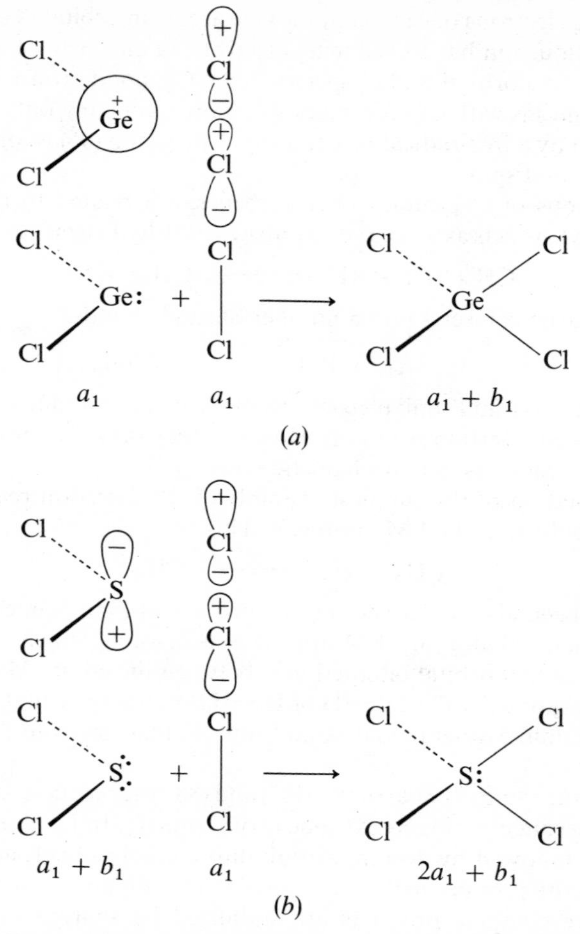


**Figure 9.** (*a*) The zero overlap between HOMO and LUMO and failure to match bond symmetries when reducing electrons are in an *s* orbital; (*b*) nonzero overlap between HOMO and LUMO and matching of bond symmetries when reducing electrons are in a *p* orbital.

HOMO and LUMO concerned. Also shown are the bond symmetries. Examples would be

$$GeCl_2 + Cl_2 \longrightarrow GeCl_4 \quad (90)$$

$$SnCl_2 + Cl_2 \longrightarrow SnCl_4 \quad (91)$$

Actually (90) is very slow, as predicted, but (91) is quite rapid in the condensed phase. It is likely that a two-step mechanism is being followed, in which ions are intermediates.

$$SnCl_2 + Cl_2 \longrightarrow SnCl_3^+ + Cl^- \quad (92)$$

$$SnCl_3^+ + Cl^- \longrightarrow SnCl_4 \quad (93)$$

In general, the more polar the bond between the metal and the halogen, the more likely this two-step sequence becomes. In a similar way, the more polar XY is, the more likely is a two-step ionic mechanism.

If the reducing electrons are in a *p* orbital on the central atom, then a concerted addition of XY is allowed. A definite stereochemistry must be followed. For a linear triatomic molecule, the addition of XY must be *trans*, as already shown in (37). For a bent molecule, the required mode of addition is shown in Fig. 9*b*. This kind of reaction has already been discussed (p. 270). The most important stereochemical prediction is that for the configuration about X and Y in the XY molecule. As mentioned earlier, XY can be an alkyl halide, or a silyl halide or hydride, in which X can be chiral.

An optically active form of XY can be the reactant. For concerted addition to molecules like $R_2S$, the prediction is that *retention of configuration* at X will be found, according to Fig. 9*b*. This is a startling conclusion since sulfides, selenides, and tellurides are well known to react with alkyl halides.

$$R_2S + R^*Br \longrightarrow [R_2SR^*Br] \longrightarrow R_2SR^{*+} + Br^- \quad (94)$$

The sulfonium ion products found with sulfides are always assumed to have *inversion of configuration* at R*. Indeed this is what has been found experimentally in many cases.[55] Such a stereochemical result is consistent with a mechanism of the same kind as (92), the familiar $S_N2$ nucleophilic substitution reaction, which occurs with inversion at R*.

$$R_2S + R^*Br \longrightarrow R_2SR^{*+} + Br^- \quad (95)$$

Of course (95) is equally an allowed reaction (see p. 303). However in certain cases, particularly in nonpolar solvents, the concerted addition should be favored. The critical test for concertedness is the configuration about the alkyl carbon atom in an alkyl halide, or other chiral molecule. Some evidence has been found that the reverse reaction to (94), occurs by both an inversion mode and a retention mode, as predicted.[56]

$$\begin{aligned} Cl^- + R_2SR^{*+} &\longrightarrow R_2S + R^*Cl \text{ (inversion)} \\ R_2SR^*Cl &\longrightarrow R_2S + R^*Cl \text{ (retention)} \end{aligned} \quad (96)$$

Compounds containing four groups attached to sulfur are called "sulfuranes." Few of these compounds have been isolated. However a great many organic reactions are found that are nicely explained by assuming sulfurane intermediates.[57] An example would be the reaction of arylsulfonium ions with organolithium compounds.

$$Ar_3S^+ + RLi \longrightarrow Li^+ + Ar_2S + Ar{-}R \qquad (97)$$

The intermediate sulfurane $Ar_3SR$ is not identified, but presumably splits off Ar—R in a concerted process. As predicted, the reactions are highly stereospecific, with retention at the carbon atom of R.[57]

Another case of concerted addition is when the reducing electrons are in an orbital that is essentially a *d* orbital on the central atom. Normally a $d^{10}$ transition-metal complex with two ligands has a linear configuration. Figure 10 shows that a *cis* addition of XY is allowed by symmetry, but that a *trans* addition is forbidden.

$$\begin{array}{llll} Au(CN)_2^- + I_2 & \longrightarrow & cis\text{-}Au(CN)_2I_2^- & \\ & & \text{allowed} & \\ & & & (98) \\ Au(CN)_2^- + I_2 & \longrightarrow & trans\text{-}Au(CN)_2I_2^- & \\ & & \text{forbidden} & \end{array}$$

Note that use of an alternative *d* orbital suggests that the *trans* addition is allowed (Fig. 10*c*). However the final product would then have a pair of electrons in a *d* orbital of high energy, so that a symmetry barrier does exist.

The experimental evidence for (98) is in disagreement with the prediction.[58] The product is definitely the *trans* isomer and appears to be formed in a simple concerted process. For example, no intermediates such as $Au(CN)_2I(H_2O)$ are detected. This intermediate would be formed in a two-step mechanism. A study of the reductive-elimination reaction of planar gold(III) complexes, however, agrees with the theoretical prediction.[58] Only *cis* elimination is found.

$$CH_3CH_2{-}\underset{\displaystyle CH_3}{\overset{\displaystyle CH_3}{\underset{|}{\overset{|}{Au}}}}{-}P(C_6H_5)_3 \xrightarrow{\Delta} CH_3CH_2CH_3 + CH_3AuP(C_6H_5)_3 \qquad (99)$$

No ethane from *trans* elimination is detected in the products.

Other *d* configurations will not allow either *cis* or *trans* oxidative addition. The correlation diagram for the *d* manifold is shown in Fig. 11. The reducing electrons for *cis* addition come from an orbital of $b_2$ symmetry in $C_{2v}$, so that at least six *d* electrons are needed (in the low spin case). However if the high energy $a_1$ orbital is not filled, the product cannot be formed in the ground state.

Another possibility is that a tetrahedral product is formed, rather than a square planar one. The correlation diagram for this case is also shown in

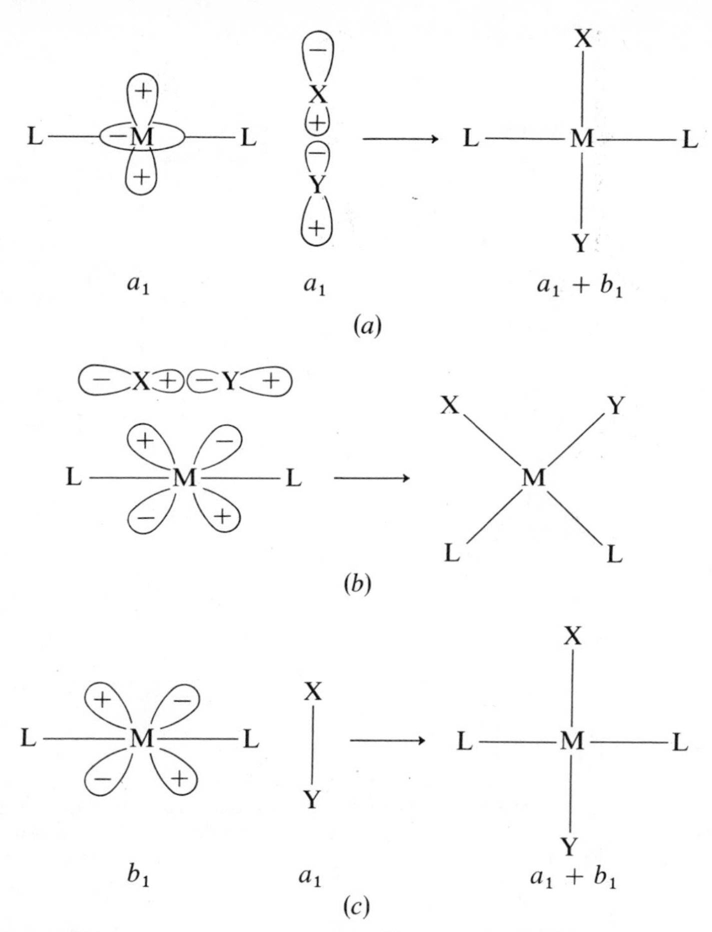


**Figure 10.** (*a*) The *trans* addition of X—Y to a $d^{10}$ linear $ML_2$ molecule is forbidden; (*b*) the *cis* addition of X—Y to $ML_2$ is allowed; (*c*) apparent allowed *trans* addition; however, electrons in the $d_{z^2}$ orbital of $a_1$ species, now become of very high energy.

Fig. 11. The reducing electrons come from an orbital of $b_1$ species in the $C_{2v}$ point group. It can be seen that no ground state $d^n$ complex could react without an orbital and/or spin problem.

### Oxidative Addition to $ML_3$

We consider next the case of oxidative addition to $ML_3$ molecules. The most important examples involve the change from trivalent to pentavalent

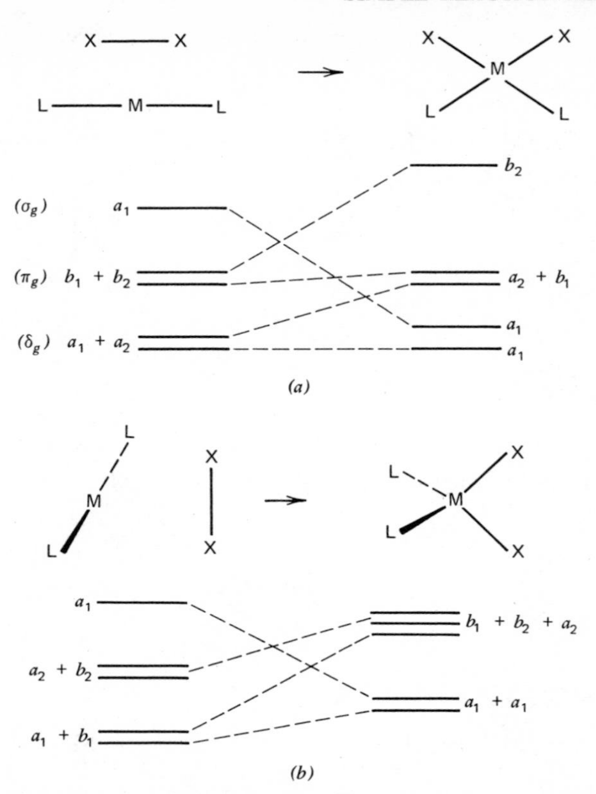


**Figure 11.** (*a*) Correlation diagram for *d* orbitals only, showing that *cis* oxidative addition to a linear $ML_2$ molecule is only allowed for $d^{10}$; (*b*) correlation diagram showing that oxidative addition of $X_2$ to $ML_2$ to form tetrahedral $ML_2X_2$ is forbidden for all ground-state $d^n$ configurations.

phosphorus as exemplified in (81). Before taking up this class of reactions, let us consider another case.

$$\begin{aligned} IF_3 + F_2 &\longrightarrow IF_5 \\ BrF_3 + F_2 &\longrightarrow BrF_5 \end{aligned} \qquad (100)$$

The first of these occurs readily at 0°C. The second requires higher temperatures, the activation energy being 16.4 kcal.[60] The kinetics are simple second-order, in line with a concerted process. Figure 12 shows that both an equatorial–equatorial addition and an axial–equatorial addition are allowed. The former is the least motion path in terms of maximum symmetry. However the latter requires less nuclear movement to get to the final product.

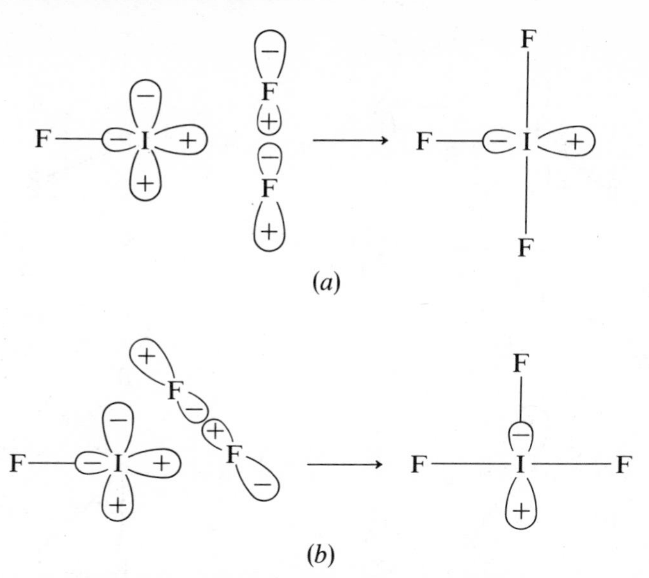


**Figure 12.** (*a*) Addition of $F_2$ to $IF_3$ with both groups adding equatorially; $C_{2v}$ point group, with 2F atoms above and below I not shown; (*b*) addition of $F_2$ to $IF_3$ with one group adding axially and one equatorially, $C_s$ point group; the two *p* orbitals of I are filled, and the $\sigma^*$ orbital of $F_2$ is empty.

Consider the symmetric addition with a $C_{2v}$ point group. An $a_1$ orbital of $IF_3$ which is filled points directly at the center of the bond of the $F_2$ molecule. There is a strong repulsive interaction. Alternatively we can see that the $F_2$ bond must be almost completely broken before much new bonding results from overlap with the filled $b_2$ orbital of $IF_3$. The axial–equatorial addition has no such disadvantage. As Fig. 12 shows, the interaction with both $a_1$ and $b_2$ is bonding. One must conclude that the less symmetric path is energetically favored.

Oxidative addition to $PCl_3$ and the large related family of organic phosphines is particularly interesting. Addition of XY can occur stereochemically in three ways. The two groups can finish up both apical, both equatorial, or one apical and one equatorial.

*a–a* *e–e* *a–e* (101)

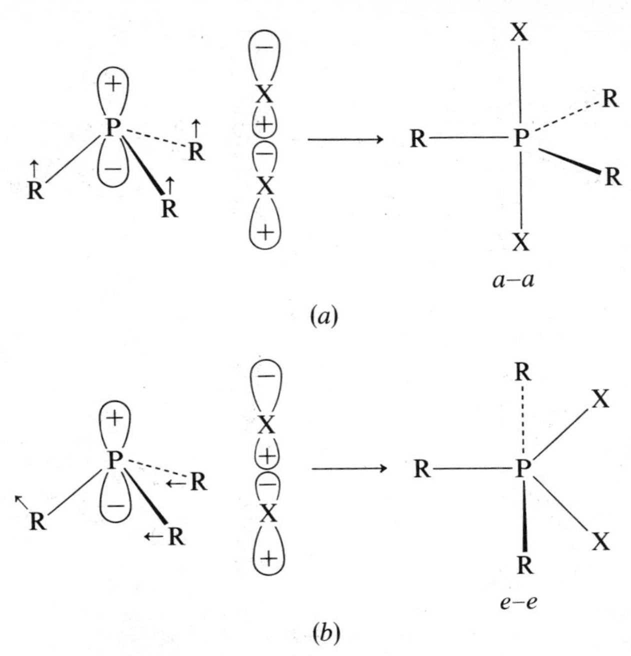


**Figure 13.** (*a*) Axial–axial addition of $X_2$ to a phosphine; (*b*) equatorial–equatorial addition of $X_2$ to a phosphine.

The latter product is actually from the LM path, in the sense that the three marker groups, R, change their initial positions the least. All three reaction paths preserve only a plane of symmetry, at most.

The orbital following method can be used, however, to show that axial–equatorial addition is forbidden.[29] The other two products can be formed in allowed modes. This is more easily seen by using the criterion of positive overlap between the HOMO and LUMO of the reactants. The details are shown in Fig. 13. The HOMO is the orbital holding the lone pair of electrons in $PR_3$ (see p. 177). The LUMO is the $\sigma^*$ orbital of XY. Since the HOMO is essentially a *p* orbital on phosphorus, and since X and Y must each use the same *p* orbital for bonding, it is necessary that they must finish up axial to each other, or both in the equatorial plane. This follows because different *p* orbitals are used to bond axial and equatorial groups.

The alternative reaction mechanism to concerted oxidative addition, is the two-step one:

$$R_3P + XY \longrightarrow R_3PX^+ + Y^- \tag{102}$$

$$R_3PX^+ + Y^- \rightleftharpoons R_3PXY \tag{103}$$

Since the second reaction is reversible, product isolation alone cannot tell us the mechanism. That is, both sets of products, the five-coordinate $R_3PXY$, called "phosphoranes," and the salts $R_3PX^+$, $Y^-$ have been isolated, but either product can result from either mechanism.

Relative rates of reaction may be used as a clue to reaction mechanism. It is well established that phosphoranes in which two of the groups attached to phosphorus form a five-membered ring are unusually stable. One point of attachment must be apical and one equatorial.[61a] Some cyclic phosphines will thus react unusually rapidly by way of concerted oxidative addition. On the other hand cyclic phosphines are known to react unusually slowly in the classical $S_N2$ reaction, (102).[61b,c] By this criterion it has shown that several phosphites react with diethylperoxide by a concerted mechanism.[62] The products are usually phosphoranes. Some alkoxyphosphonium alkoxides are also formed.

Phosphines, or phosphites, in which three different groups are attached to P, exist as optical isomers and may be resolved. This provides a possible way to study their reactions with XY molecules. It can be seen in (101) that if $X = Y$, both *a–a* and *e–e* addition produce a plane of symmetry in the phosphorane, and hence optical activity is lost. Also the pseudorotation[31,60] process would offer another route toward racemization or isomerization even if $X \neq Y$. On the other hand, the $S_N2$ mechanism of (102) would be stereospecific with retention of configuration about the phosphorus atom.[63]

If at least one of the groups attached to P is an alkoxide, OR, the final product is often $R_2PXO$, formed by the Michaelis–Arbuzov reaction.[64] An example would be

$$(CH_3O)_3P + C_2H_5I \longrightarrow (CH_3O)_3PC_2H_5^+ + I^- \qquad (104)$$

$$(CH_3O)_3PC_2H_5^+ + I^- \longrightarrow CH_3I + (CH_3O)_2P(=O)C_2H_5 \qquad (105)$$

The phosphonate ester still has a retained configuration about the phosphorus atom, the O atom occupying the position formerly held by the lone pair.

Using an optically active phosphine, it has been found that the addition of ROCl or of $R_2O_2$ gives a racemic product in nonpolar solvents like benzene or pentane, but an active product when a more polar solvent is used.[66] In these cases the product is a phosphine oxide, $R_1R_2R_3PO$, formed by the Arbuzov reaction. The solvent effect is expected to be in the observed direction. Thus a nonpolar solvent favors concerted addition, with attendant racemization. A polar solvent favors the mechanism that forms ions, (102). The observed stereochemical results could still be explained by (102), followed by (103), since the latter reaction would also be favored by a nonpolar solvent. In the concerted mechanism there should also be retention of configuration at the carbon atom of a chiral alkyl halide. This has not been observed. It is always assumed that phosphines and phosphites react with alkyl halides by an $S_N2$ mechanism with inversion at the carbon atom, (104). While this is known to be the case in many examples,[64,67] it is clearly dangerous to assume this as a general result.

### Reactions of $ML_4$ Molecules

We consider next oxidative addition to $ML_4$ molecules.

$$ML_4 + XY \longrightarrow ML_4XY \tag{106}$$

The product may be assumed to be pseudo-octahedral with the groups X and Y either *cis* or *trans* to each other. Several possible structures for $ML_4$ must be considered. There are no known examples of tetrahedral molecules undergoing oxidative addition to give six-coordinated products. Possible candidates, such as $d^{10}$ metal complexes turn out to behave differently.[68]

We can give the general symmetry requirement for oxidative addition in tetrahedral molecules, as follows:

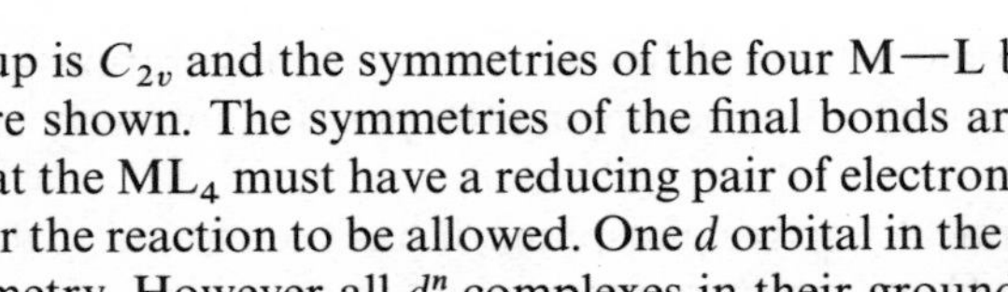

$$ML_4 + X{-}X \longrightarrow ML_4X_2 \tag{107}$$

$3a_1 + b_1 + b_2$ $\qquad$ $3a_1 + 2b_1 + b_2$

The point group is $C_{2v}$ and the symmetries of the four M—L bonds and the X—X bond are shown. The symmetries of the final bonds are also shown. This means that the $ML_4$ must have a reducing pair of electrons in an orbital of $b_1$ species for the reaction to be allowed. One $d$ orbital in the $t_2$ shell would have this symmetry. However all $d^n$ complexes in their ground states would then form a product in an excited state. For example, a $d^{10}$ molecule would form low spin $d^8$, instead of the more stable high spin.

Oxidative addition of $F_2$ to $SF_4$ is forbidden in the ground-state geometry. The lone pair is in an orbital of $a_1$ species. However $SF_4$ readily pseudorotates going through the planar form.

$$\underset{a_1}{SF_4} \quad + \quad \underset{a_1}{F\text{—}F} \longrightarrow \underset{a_1 + b_2}{SF_6} \tag{108}$$

*Trans* addition of $F_2$ to planar $SF_4$ is allowed. It is similar to the allowed addition to $XeF_4$. In both cases the reducing electrons are in a $p$ orbital of $b_1$ symmetry.

$$XeF_4 + F\text{—}F \longrightarrow XeF_6 \tag{109}$$

In spite of this prediction that the reaction of $XeF_4$ to form $XeF_6$ is allowed, as is the reaction of $XeF_2$ to form $XeF_4$ (p. 268), it appears that it requires temperatures in excess of 250°C, or photochemical activation, to form any of these substances from Xe and $F_2$.[25] Typical free-radical paths seem required. Some insight into the reasons for this can be obtained by looking at the initial reaction of xenon itself.

$$\underset{b_2}{Xe} + \underset{a_1}{F\text{—}F} \longrightarrow \underset{a_1 + b_2}{F\text{—}Xe\text{—}F} \tag{110}$$

The reaction is allowed and exothermic by 39 kcal. Yet it certainly requires very high temperatures to occur.

This provides an illustration of a statement made earlier. Even completely allowed, exothermic reactions can have a high activation energy. In this case it probably results from the tightness with which the reducing electrons are held by xenon. The new Xe—F bonds cannot form appreciably until there is a close distance of approach. At this point there are strong repulsions from overlap of filled shells. Similar factors can operate for $XeF_2$ and $XeF_4$, besides the more ordinary steric factors that *trans* addition would involve (F—F repulsions).

As already indicated, the most important examples of oxidative addition are found for the four-coordinate, square planar complexes of Ir(I), Rh(I), Pd(II), and Pt(II). These are $d^8$ metal ions, with the ordering of the $b_{2g}$, $e_g$, and $a_{1g}$ levels not being clearly established. The $b_{1g}^*$ orbital lies highest and is empty. The $b_{2g}$ orbital, while filled, is inaccessible for steric reasons. It lies in the plane of the complex. The reducing electrons must be in the $d_{z^2}$ orbital or in the $d_{xz}$, $d_{yz}$ degenerate pair.

Figure 10 can be used to discuss the symmetry requirements, by imagining two L molecules above and below the plane of the page. It can be seen that concerted *trans* addition, using the $d_{z^2}$ electrons, is forbidden, but that *cis* addition using $d_{xz}$ or $d_{yz}$ is allowed. Low-spin $d^7$ and $d^9$ could react about as well as low spin $d^8$. *Trans* addition using the $e_g$ orbitals is apparently allowed by symmetry, as shown in Fig. 10*c*, but would lead to an excited state of the product with two electrons still in the $d_{z^2}$ orbital of the six-coordinated product. Hence it must also be considered as forbidden. It is also sterically difficult to add two groups *trans* in a concerted process.

In fact both products of *cis* and *trans* addition are well known for square planar $d^8$ complexes. It must be that other, nonconcerted mechanisms are operating to account for the *trans* products. There is one very likely alternative to a concerted addition reaction. This is a two-step process in which the metal complex acts as a simple nucleophile in a classical $S_N2$ displacement reaction, as the first step.[69]

$$\begin{aligned} IrL_4{}^+ + CH_3I &\longrightarrow CH_3IrL_4{}^{2+} + I^- \\ CH_3IrL_4{}^{2+} + I^- &\longrightarrow CH_3IrL_4I^+ \end{aligned} \qquad (111)$$

Here L stands for any ligand attached to Ir(I), as an example. The sequence shown in (111) could easily lead to *trans* addition of the $CH_3{}^+$ and $I^-$ fragments, though *cis* addition is not excluded.

There are also allowed free-radical paths for oxidative-addition reactions which can be visualized. Indeed such paths have been shown to operate in a number of cases.[70] Because of the importance of these transition metal oxidative-addition reactions, a more complete discussion of this mechanism will be deferred until Chapter 5.

## ADDITION OF ELECTROPHILES AND NUCLEOPHILES TO SIMPLE MOLECULES

In this section we discuss the symmetry requirements for the addition of electrophiles, such as $H^+$ or $Cl^+$, and nucleophiles, such as $H^-$ or $Cl^-$, to molecules of formulas $XY_n$ and $X_2Y_n$. While sometimes stable adducts are formed, the more common result is that a high-energy species results. Such a species is a TS, or an intermediate, in the electrophilic or nucleophilic substitution reactions of these molecules.

We assume that the structures of the reactants are well understood, including the shapes of the valence-shell MOs. The same structural theory enables us to predict the structures of the adducts that are formed. Loss of a *different* electrophilic or nucleophilic group is required for a substitution process. While this process is essentially the microscopic reverse of the addition, the requirement of loss of a different group creates a stereochemistry for the overall process. This stereochemistry will be observable in more complex molecules.

The first examples are the reactions of triatomic molecules

$$XY_2 + Y^+ \longrightarrow XY_3^+ \tag{112}$$

$$XY_2 + Y^- \longrightarrow XY_3^- \tag{113}$$

Table 2 shows the structures of the adducts and the symmetry of the point group that is maintained during the LM approach of the reactants. Only nonradical species are considered. The reaction of $BeH_2$ with $H^+$ should lead to attack at hydride ion and not at the central atom. Also $XeF_2$ with $F^-$ should lead to reaction at fluorine and not to attachment to xenon. These last remarks follow from a consideration of the symmetries and amplitudes of the HOMO of $BeH_2$ and the LUMO of $XeF_2$.

In general an electrophile will approach the HOMO of the reactant molecule, and a nucleophile will seek out the LUMO. The expected direction of approach will be that giving maximum overlap. Many exceptions will be found to these simple rules. Even the examples of Table 2 only roughly conform. The MO configuration of a bent $H_2X$ molecule is

$$(a_1)(b_2)(2a_1)(b_1)(3a_1)(2b_2)$$

**Table 2**

| Reactants | Product | Point Group Maintained |
|---|---|---|
| $BeH_2 + H^-$ | $BeH_3^-$ $D_{3h}$ | $C_{2v}$ |
| $CH_2 + H^+$ [a] | $CH_3^+$ $D_{3h}$ | $C_{2v}$ |
| $CH_2 + H^-$ [a] | $CH_3^-$ $C_{3v}$ | $C_s$ |
| $H_2O + H^+$ | $H_3O^+$ $C_{3v}$ | $C_s$ |
| $H_2O + H^-$ | $H_3O^-$ $C_{2v}$ | $C_s$ |
| $XeF_2 + F^+$ | $XeF_3^+$ $C_{2v}$ | $C_{2v}$ |
| $ML_2 + L$ [b] | $ML_3$ $D_{3h}$ | $C_{2v}$ |
| $ML_2 + H^+$ [b] | $MX_2H^+$ $C_{2v}$ | $C_{2v}$ |

[a] For the singlet state.
[b] M is transition-metal atom of $d^{10}$ configuration.

Thus a proton will approach the filled $2a_1$ orbital of singlet $CH_2$ in the molecular plane, and maintaining a $C_{2v}$ point group,

$$H_2C{:} + H^+ \longrightarrow H_2C{-}H^+ \qquad (114)$$

This gives maximum overlap, and leads directly to the planar methyl cation.

A hydride ion would approach from above to overlap with the empty $b_1$ orbital.

$$H_2C + H^- \longrightarrow H_2C{-}H^- \qquad (115)$$

Only a plane of symmetry is maintained. Because of the changes in bond angles that are required, the approach need not be directly perpendicular to the molecular plane. Note that in $C_s$ symmetry both the $a_1$ and $b_1$ orbitals become $a'$ and are mixed to become the C—H bond orbital and the lone pair.

A hydrogen atom could approach either from above, or in the plane of $CH_2$. However singlet methylene, with its filled $a_1$ orbital, would react in a repulsive manner with the half-filled $s$ orbital of H. The product must form from approach to the empty $b_1$ orbital. Unstable pyramidal $CH_3$ would result. Triplet methylene could react with a free atom in the favored manner, however. Conversely $CH_2$, in its ground state $^3B_1$, could not react with either $H^+$ or $H^-$. The product would necessarily be a high-energy excited state of either $CH_3{}^+$ or $CH_3{}^-$.

Approach of a proton to $H_2O$ would be expected to be above the plane to overlap with the filled $b_1$ orbital, which is the HOMO. The structure of $H_3O^+$ is nearly planar, and it is clear that approach is nearly in the molecular plane. Thus overlap with the filled $a_1$ orbital is more important. Again both the $a_1$ and $b_1$ mix in the $C_s$ point group.

Addition of a nucleophile to $H_2O$ should utilize the LUMO that is of $a_1$ species, an antibonding orbital for the OH bonds. The structure of $H_3O^-$ should be that of $ClF_3$, which is essentially isoelectronic.

$$H{-}O{-}H + H^- \longrightarrow H{-}O(H){-}H^- \qquad (116)$$

Thus the approach of $H^-$ is in the molecular plane, but off center because of repulsion by the filled $a_1$ orbital lying in the plane also.

While $H_3O^-$ is very high energy, and unlikely to form, (116) is important as a model for nucleophilic substitution on sulfenyl compounds, for instance,

$$RSCl + Br^- \longrightarrow RSBr + Cl^- \qquad (117)$$

The TS is predicted to be A, rather than B or C.[71]

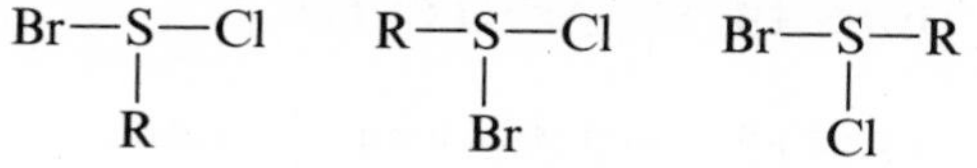

$$\underset{\text{A}}{\begin{array}{c}\text{Br—S—Cl}\\ |\\ \text{R}\end{array}} \qquad \underset{\text{B}}{\begin{array}{c}\text{R—S—Cl}\\ |\\ \text{Br}\end{array}} \qquad \underset{\text{C}}{\begin{array}{c}\text{Br—S—R}\\ |\\ \text{Cl}\end{array}}$$

Unfortunately no experimental test of this seems possible especially since A, B, and C can probably interconvert rather rapidly.

An electrophile, such as $F^+$, could add smoothly to the filled $2\pi_u$ orbital of linear $XeF_2$, even though it is not the HOMO, which is $3\sigma_g$. The product would be $XeF_3{}^+$, also with the T-shaped structure of $ClF_3$. This would have to rearrange before it could lead to a substitution reaction

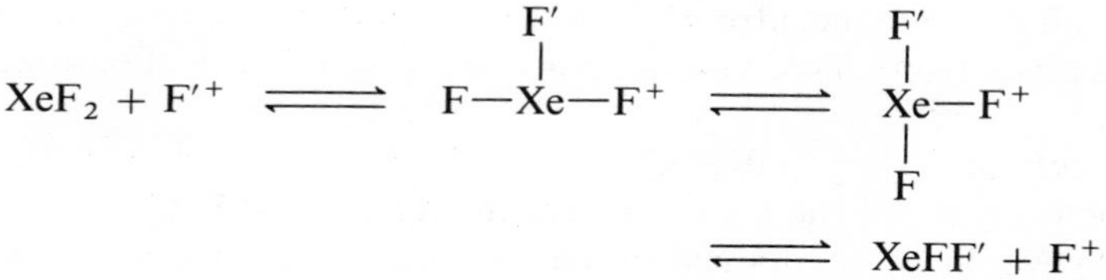

$$XeF_2 + F'^+ \rightleftharpoons \begin{array}{c}\text{F}'\\ |\\ \text{F—Xe—F}^+\end{array} \rightleftharpoons \begin{array}{c}\text{F}'\\ |\\ \text{Xe—F}^+\\ |\\ \text{F}\end{array}$$

$$\rightleftharpoons XeFF' + F^+ \qquad (118)$$

A nucleophile would add to the empty $2\pi_u$ orbital of carbon dioxide. This orbital is concentrated on the central atom for both $CO_2$ and $XeF_2$ (see p. 169). An electrophile would add to the filled $\pi_g$ orbital for $CO_2$. This is exclusively on the oxygen atoms.

A transition-metal molecule would have the *d* manifold between the bonding orbitals and the antibonding ones. A ligand, such as a halide ion, would normally add to form a trigonal planar $ML_2X^-$ complex. Any of these *d* orbitals, if empty, could help in adding $X^-$. More commonly these orbitals are filled, as in a $d^{10}$ molecule. In this case it becomes possible to add an electrophile to the central metal atom.

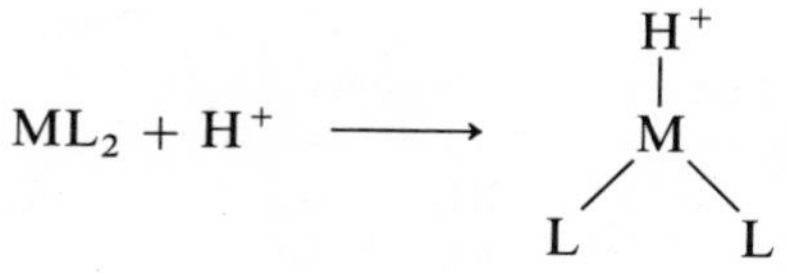

$$ML_2 + H^+ \longrightarrow \begin{array}{c}\text{H}^+\\ |\\ \text{M}\\ \diagup\quad\diagdown\\ \text{L}\qquad\text{L}\end{array} \qquad (119)$$

Assuming a $C_{2v}$ structure for the product, the $\sigma_g(d_{z^2})$ orbital can best supply the necessary bonding electrons. Since only a $d^{10}$ complex will have two electrons in the high energy $\sigma_g$ orbital, these are the ones that can react most readily in this way. Notice that reaction (119) is formally an oxidation of the metal atom by two units. Only $ML_2$ molecules where M is easily oxidized can bind an electrophile well. If this criterion is met, two electrons in a $\delta_g$ orbital $(d_{x^2-y^2})$ can also be used to bind a proton.

## REACTIONS OF $XY_3$ MOLECULES

Next in order of increasing complexity would be reactions such as

$$BH_3 + H^+ \longrightarrow BH_4^+ \qquad (120)$$

$$BH_3 + H^- \longrightarrow BH_4^- \qquad (121)$$

Table 3 shows the structures of the products and the presumed reaction paths for stable $XY_3$ molecules. The MO sequence for planar $XY_3$ is

$$(a_1')(e')(a_2'')(2a_1')(2e')$$

Therefore a proton adding to $BH_3$ would seek out the HOMO of $e'$ species. These orbitals are concentrated in the molecular plane, and the product, $BH_4^+$, is predicted to have a square planar structure. The LM path would be $C_{2v}$.

A nucleophile adding to $BH_3$ would use the LUMO, which is $a_2''$. Approach would be from above the plane to form the tetrahedral $BH_4^-$ product. $C_{3v}$ symmetry would be maintained in the approach. The same reaction path is predicted for the addition of a proton to $NH_3$. For pyramidal $NH_3$ the MO ordering is

$$(a_1)^2(e)^4(2a_1)^2(3a_1)^0(2e)^0$$

The lone pair is $(2a_1)$, which is also the HOMO.

Since $BH_4^+$, $BH_4^-$, and $NH_4^+$ have equivalent positions for all four hydrogen atoms, no rearrangement is necessary for their substitution reactions. Planar $BH_3$-like molecules do not exist in stereoisomeric forms,

**Table 3**

| Reactants | Product | | Point Group Maintained |
|---|---|---|---|
| $BH_3 + H^+$ | $BH_4^+$ | $D_{4h}$ | $C_{2v}$ |
| $BH_3 + H^-$ | $BH_4^-$ | $T_d$ | $C_{3v}$ |
| $NH_3 + H^+$ | $NH_4^+$ | $T_d$ | $C_{3v}$ |
| $NH_3 + H^-$ | $NH_4^-$ | $C_{2v}$ | $C_s$ |
| $ClF_3 + F^+$ | $ClF_4^+$ | $C_{2v}$ | $C_s$ |
| $ClF_3 + F^-$ | $ClF_4^-$ | $D_{4h}$ | $C_{2v}$ |
| $ML_3 + L$[a] | $ML_4$ | $T_d$ | $C_{3v}$ |
| $ML_3 + H^+$[a] | $ML_3H^+$ | $D_{4h}$ | $C_{2v}$ |

[a] M is a $d^{10}$ transition-metal atom.

but pyramidal $NH_3$-like molecules do. The addition of one electrophile and the removal of another would occur with retention of configuration at the chiral central atom for each step. Nevertheless, the entire process would occur with *inversion* of configuration. This is made clear in Fig. 14*a*. Examples are known in reactions of chiral sulfoxides, $R_1R_2SO$, which are electronically equivalent to $NH_3$.[72]

The stereochemistry of nucleophilic substitution at $NH_3$-like molecules is more complex. The adduct $NH_4^-$ would have the $C_{2v}$ structure of $SF_4$. It would be formed from the addition of $H^-$ to the LUMO, which is the $3a_1$ N—H antibonding orbital.

$$NH_3 + H^- \longrightarrow H{-}N{-}H^- \text{ (with two further H on N)} \qquad (122)$$

The $3a_1$ is largely an *s* orbital of N (see p. 177). The intermediate can be formed in two ways such that it is one of the apical atoms, or one of the equatorial atoms. This corresponds to the two possible modes of dissociation of $SF_4$.

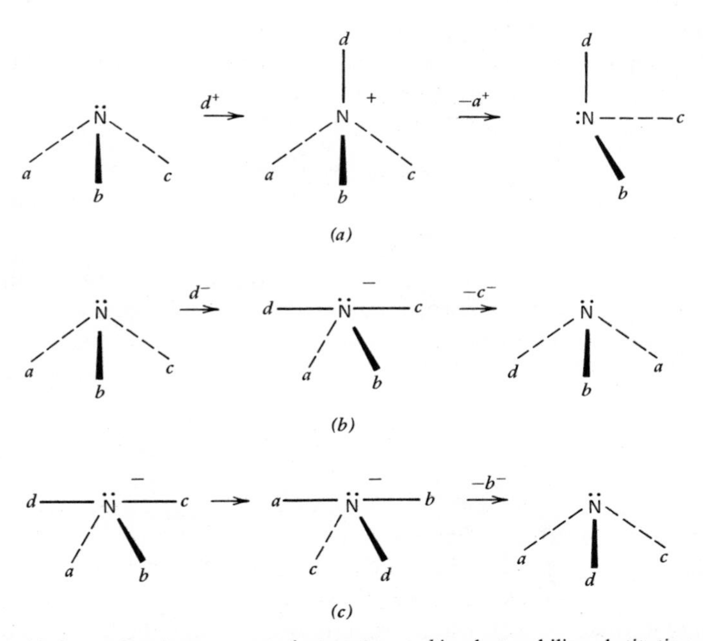


**Figure 14.** (*a*) Stereochemistry at central atom *inverted* in electrophilic substitution of $XY_3$ molecule; (*b*) stereochemistry *inverted* at central atom in nucleophilic substitution of $XY_3$; (*c*) stereochemistry *retained* at central atom during nucleophilic substitution of $XY_3$, due to a pseudorotation of intermediate.

If the original molecule were chiral, nucleophilic substitution would occur with inversion of configuration at the central atom. This assumes that the leaving group is either apical or equatorial, but equivalent to the entering group in either case. This is illustrated in Fig. 14*b*. However it is known that $SF_4$ rapidly exchanges its apical and equatorial positions (p. 187). If this occurs, and a group originally equatorial, becomes apical and leaves, the net result is substitution with *retention* of configuration. This is shown in Fig. 14*c*.

Much of the experimental evidence again deals with chiral sulfoxides.[73] Most nucleophilic substitution reactions occur with inversion of configuration, for instance,

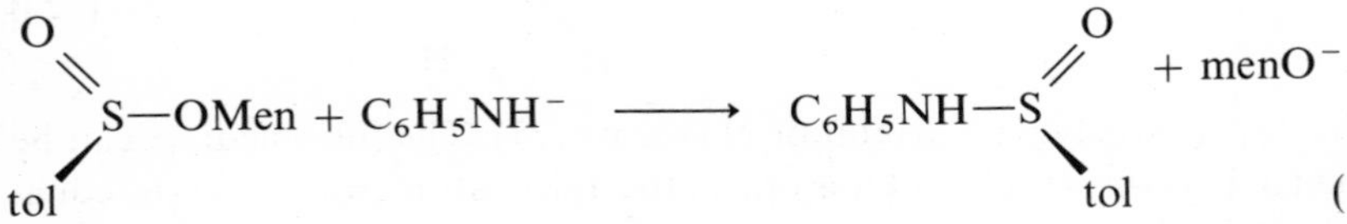

$$\text{tol}(\text{O}{=})\text{S—OMen} + C_6H_5NH^- \longrightarrow C_6H_5NH\text{—S}(\text{=O})\text{tol} + \text{menO}^- \qquad (123)$$

tol = $p$-$CH_3C_6H_4$ men = menthyl

However in a number of cases, there is retention of configuration, or racemization occurs.[73,74] It appears that normally the entering and leaving groups are axial, and that decomposition of the intermediate occurs before pseudorotation. Some stereochemical control is possible by utilizing two rules:[73]

1. Electronegative groups prefer to be axial.
2. Small rings (four and five membered) prefer to span one axial and one equatorial position.

Similar considerations apply to the electrophilic substitution of molecules such as $ClF_3$. In this case addition of $F^+$ must occur equatorially, and the leaving group would also depart from an equatorial position.

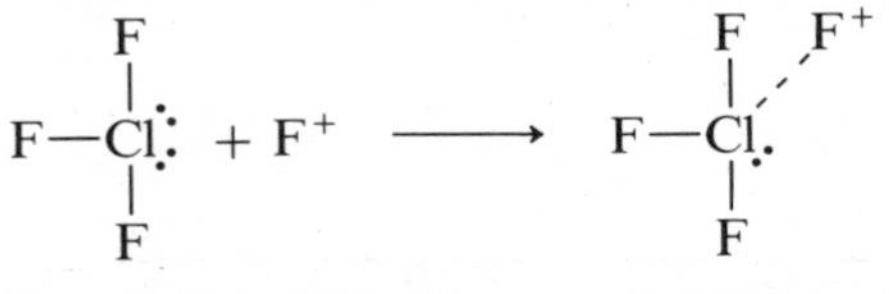

$$ClF_3 + F^+ \longrightarrow ClF_4^+ \qquad (124)$$

The reaction coordinate has only a plane of symmetry. Both lone-pair orbitals on chlorine are $a'$ in the $C_s$ point group, and hence are mixed to form the new Cl—F bond and the new lone pair in $ClF_4^+$. Configuration at the central atom is accordingly maintained in that the same two groups remain axial. Rapid pseudorotation in $ClF_3$ equilibrates the axial and equatorial fluorine atoms.[50] Fortunately trialkyliodonium compounds $R_3I$, isoelectronic with $ClF_3$, are known which pseudorotate much more slowly.[75]

Addition of a nucleophile, $F^-$, to $ClF_3$ should give square planar $ClF_4^-$. The approach should be to the HOMO, an $a_1$ orbital lying in the molecular plane, and antibonding to all Cl—F bonds.

$$\text{F—ClF}_2\text{(lone pair)} + \text{F}^- \longrightarrow \text{F—ClF}_2\text{—F}^- \qquad (125)$$

The point group maintained is $C_{2v}$. Since all four groups are equivalent, any one can be lost in a substitution reaction. Hence either axial group retention or axial–equatorial inversion is possible as a net stereochemical change.

Transition-metal complexes $ML_3$ normally would have a planar trigonal structure. Addition of a nucleophile would give either a tetrahedral adduct, as does $BH_3$, or a square planar one, depending on the number of $d$ electrons. There is also the possibility of adding an electrophile, corresponding to oxidation by two units. A probable example is

$$Pt[P(C_6H_5)_3]_3 + H^+ \longrightarrow HPt[P(C_6H_5)_3]_3^+ \qquad (126)$$

The product is a platinum(II) complex, expected to be planar. The LM approach of $H^+$ in the plane gives $C_{2v}$ symmetry. The Pt—H bond formed is $a_1$ and the electrons must come from the high-energy $d_{x^2-y^2}$ orbital of the metal (the $xy$ plane is that of the molecule). This orbital is only filled for a $d^{10}$ complex, such as those of Ni(0), Pd(0) and Pt(0).[75]

## REACTIONS OF $XY_4$ MOLECULES

The next group of examples include the most important reactions in chemistry since $CH_4$ is the prototype of tetrahedral carbon compounds.

$$CH_4 + H^- \longrightarrow CH_5^- \qquad (127)$$

$$CH_4 + H^+ \longrightarrow CH_5^+ \qquad (128)$$

Thus (127) and 128) can be considered as models for all bimolecular nucleophilic ($S_N2$) and electrophilic ($S_E2$) reactions at saturated carbon.

The MOs of methane are shown elsewhere (p. 186). The entire electron density has spherical symmetry, since it corresponds to filled shells, $(a_1)^2(t_2)^6$. Except to avoid the protons, there is no obvious reason to prefer any direction of attack of an electrophile on this molecule that has no unshared pairs of electrons. However the structure of the carbenium ion, $CH_5^+$, is known from accurate SCF calculations, including correlation energy.[77] The point group is $C_s$ and there is an arrangement of three short C—H bonds (1.08 Å) and two long C—H bonds (1.27 Å).

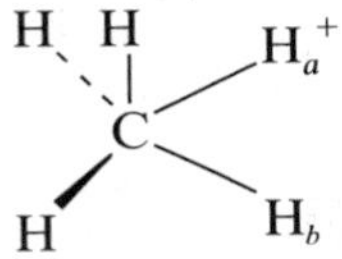

H H $H_a^+$
C
H $H_b$

The long bonds are also associated with a short $H_a$—$H_b$ distance.

This structure is reached directly by an approach of the proton along one edge of the $CH_4$ tetrahedron, but slightly off center. Calculations show that approach to an edge is favored over approach to a corner or to a face.[77] This is only true at short carbon—proton distances. At distances which are larger, approach to a corner is favored. This would correspond to attack of the proton on a bound hydrogen atom. Note that, because of the spherical symmetry of the $t_2$ orbitals of methane, a filled MO can be provided to point directly at the proton no matter which direction of approach is taken. That is, the $z$ axis can be selected arbitrarily.

There is ample evidence that $CH_5^+$ is formed when methane is dissolved in solutions of very strong acid, such as HF—$SbF_5$ or $HSO_3F$—$SbF_5$.[78] This is evidenced by deuterium exchange and by evolution of hydrogen,

$$CH_4 + DSbF_6 \longrightarrow CH_3D + HSbF_6 \tag{129}$$

$$CH_4 + HSbF_6 \longrightarrow H_2 + CH_3^+ + SbF_6^- \tag{130}$$

as well as by products formed from attack of the methyl cation on methane. Other alkanes behave similarly to methane, and the carbenium ions formed in these "superacid" solutions undergo reactions corresponding to (129) and (130).

The adduct $CH_5^+$ is admirably suited for electrophilic substitution, with $H_a$ as the entering group and $H_b$ as the leaving group. The stereochemical result would be *retention* of configuration at carbon.

Note that $CH_5^+$ is prone to lose $H_2$, as well as to undergo simple substitution. Some help from solvent or $SbF_6^-$ would be needed to overcome the gas-phase endothermicity of (130). The trigonal bipyramid structure for $CH_5^+$, $D_{3h}$, is not so much higher in energy than the $C_s$ structure, about 16.5 kcal by ab initio calculation.[77] If it were formed, there could be electrophilic substitution, $H_a$ replacing $H_b$, with *inversion* of configuration.

The experimental facts for electrophilic substitution on carbon are that retention of stereochemistry is most commonly observed.[79] However there are many examples of electrophilic substitution which go with inversion.[80]

Apparently structures related to $C_s$ are favored, but structures related to $D_{3h}$ are also common. In these cases it can be considered that the electrophile has attacked a filled C—H bonding orbital, but on the backside.

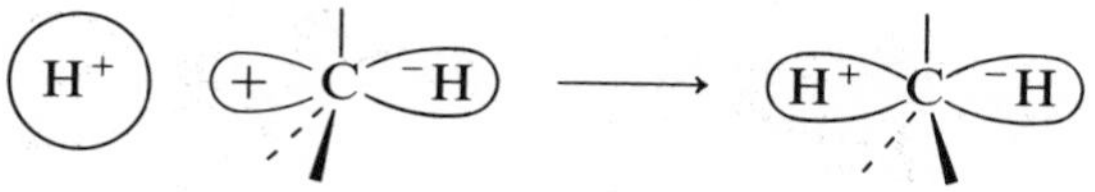

$$H^+ \quad (+\,C\,{}^-H) \longrightarrow (H^+\,C\,{}^-H)$$

Clearly depending on the electrophile, the leaving group, the other groups attached to carbon, and even the solvent, structures related to $C_s$ or $D_{3h}$ might be favored in various cases.

The most common examples of electrophilic substitution on carbon relate to cleavage of alkyl—metal bonds by halogen or halogen acid.

$$R\text{—}M + Br_2 \longrightarrow R\text{—}Br + Br\text{—}M \tag{131}$$

$$R\text{—}M + HBr \longrightarrow R\text{—}H + Br\text{—}M \tag{132}$$

Caution must be used in interpreting the stereochemical results of such reactions. The mechanisms are often complex. For example, the rates of reactions such as (132) depend on the concentration of HBr, not on $H^+$ only.[81] Hence the reaction is not a simple electrophilic substitution. Reactions (131) and (132) for transition metals often involve a mechanism of oxidative addition of HBr or $Br_2$ to the metal, followed by reductive elimination of RH or RBr.[80,82] In these cases the stereochemistry depends on factors discussed in Chapter 5.

## The $S_N2$ Mechanism

The nucleophilic substitution reaction, $S_N2$, for which (127) is a model, is probably the most important reaction of organic chemistry. An examination of the unoccupied $2a_1$ and $2t_2$ orbitals of $CH_4$ shows again overall spherical symmetry, so that selection of the direction of attack by a nucleophile is not obvious. Fortunately there have been several ab initio calculations of the energetics of simple $S_N2$ reactions.[83] The results show unambiguously that attack of the nucleophile is on the face of the tetrahedron which is furthest from the leaving group (backside attack). The elements of $C_{3v}$ symmetry are maintained. For identical entering and leaving groups, the TS has $D_{3h}$ symmetry.

Potential energy barriers have been calculated for the following reactions:[83d,e]

| | $E_0$ | $\Delta H$ | |
|---|---|---|---|
| $H^- + CH_4 \rightarrow CH_4 + H^-$ | 60 kcal | 0 | (133) |
| $F^- + CH_3F \rightarrow CH_3F + F^-$ | 7.3 | 0 | (134) |
| $H^- + CH_3F \rightarrow CH_4 + F^-$ | 3.8 | −68 | (135) |
| $CN^- + CH_3F \rightarrow CH_3CN + F^-$ | 22.6 | +5 | (136) |

These calculations are all for the gas phase. The overall heat of reaction is also given. Reactions (133) and (135) cannot be studied in a protic solvent, but both (134) and (136) would have experimental activation energies of about 25 kcal.[84] Clearly solvation energies are of great importance in determining observed rates of reaction. For example, $F^-$ is a better nucleophile than $CN^-$ in the gas phase, but the reverse is true in water or methanol.[85]

Unlike $CH_5^+$, whose formation according to (128) is exothermic by 129 kcal mole$^{-1}$, the formation of $CH_5^-$ is highly endothermic by 60 kcal mole$^{-1}$. The MO configuration of the $D_{3h}$ $CH_5^-$ molecule is [84b]

$$(a_1')^2(e')^4(a_2'')^2(2a_1')^2(2a_2'')^0(2e')^0$$

The lowest-energy transition is from $2a_1'$ to $2a_2''$. This gives a transition density ($\rho_{0k}$) of $A_2''$ symmetry. The $A_2''$ mode corresponds to decomposition into $CH_4$ and $H^-$ (p. 194). It may be mentioned that in all of the reactions (133)–(136), the TS equatorial C—H bonds are shortened compared to $CH_4$, and the axial bonds are lengthened compared to $CH_4$ or $CH_3F$. Also positive charge increases on the carbon atom in the TS, compared to $CH_4$ or $CH_3F$.

It is instructive to look at the orbital-correlation diagram for a typical $S_N2$ reaction. This is shown in Fig. 15 for a reaction such as

$$Cl^- + CH_3Cl \longrightarrow CH_3Cl + Cl^- \qquad (137)$$

The point group is $C_{3v}$, except at the TS, where it is $D_{3h}$. The critical MOs are the filled $2a_1$, essentially the C—Cl bonding orbital in $CH_3Cl$, the filled $3a_1$, which is a $p_z$ orbital on the chloride ion (the $z$ axis defines the threefold rotation), and the empty $4a_1$, which is the antibonding partner to $2a_1$. The energies shown are due to a semiempirical calculation by Fukui.[86]

These three orbitals are all of the same symmetry and mix during formation of the activated complex. They become, in order, $a_2''$, a filled orbital that is bonding for both C—Cl bonds, $2a_1'$, a filled orbital on both chlorine atoms that is essentially nonbonding, and an empty $2a_2''$ orbital, antibonding for both C—Cl bonds. The energy changes on forming the activated complex are only shown schematically. However they are not great since the nature of the

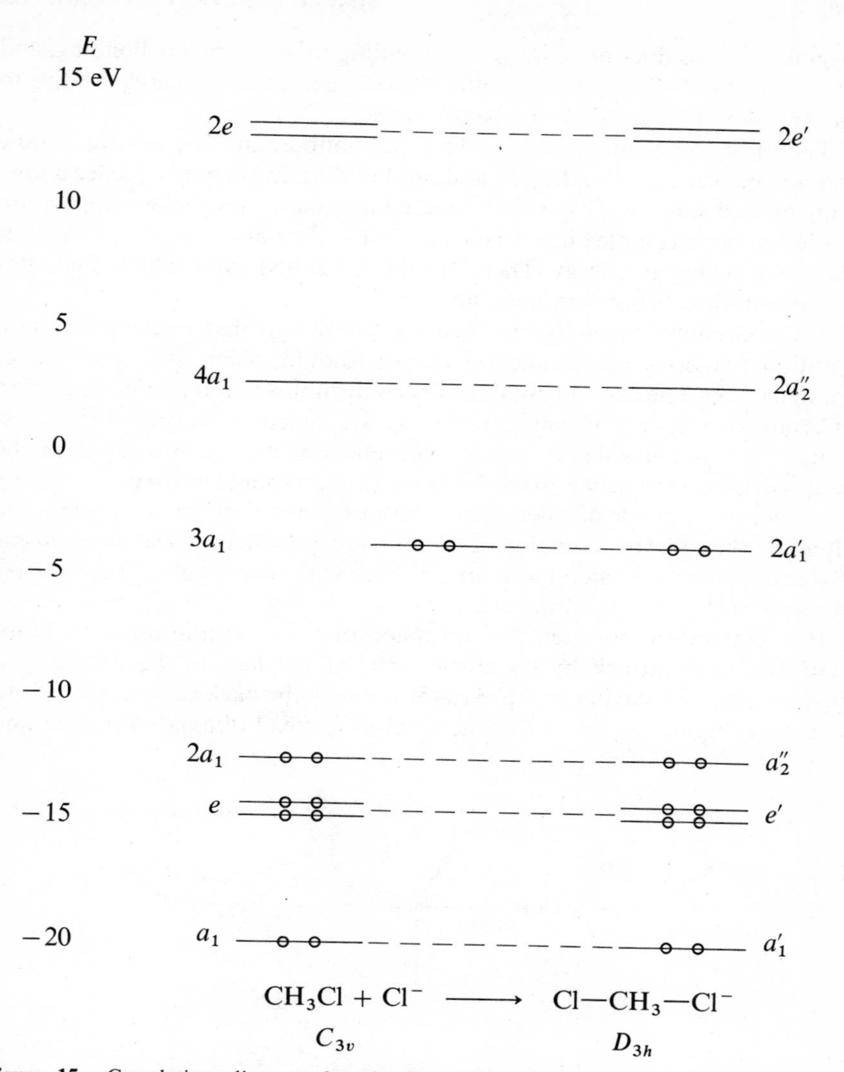


**Figure 15.** Correlation diagram for the formation of the TS $Cl—CH_3—Cl^-$ from $CH_3Cl + Cl^-$; only the $\sigma$ orbitals are shown (energies from reference 86).

original orbitals does not change, the bonding orbital remains bonding, and so on. It is likely that the $2a_2''$ orbital actually decreases in energy relative to $3a_1$ because of the long C—Cl bonds in the TS.[86]

The most important feature is that the nonbonding $2a_1'$ orbital is now inserted between the bonding $a_2''$ and nonbonding $2a_2''$ orbitals. Hence a low-lying excited state of $A_2''$ species is created that facilitates the decomposition of the activated complex into $CH_3Cl$ and $Cl^-$. Note also that the $2e'$ orbitals are much higher in energy. Therefore the $E'$ excited state, which facilitates pseudorotation, is not very accessible.

These circumstances allow a clear-cut prediction that nucleophilic substitution reactions on tetrahedral carbon should occur with *inversion* of configuration. The experimental results confirm this to a remarkable degree. "Certainly inversion of configuration in $S_N2$ reactions is one of the most unqualified and absolutely dependable phenomena ever observed in the field of organic stereochemistry."[83a] Some caution should be used, since many examples are known of nucleophilic substitutions that go with retention. However the belief in mandatory inversion is so great that special mechanisms ($S_Ni$ mechanism, two-step mechanisms, each with inversion, etc.) are always invoked to explain these anomalies.[87]

It is sometimes forgotten that inversion of configuration does not automatically mean attack by the nucleophile at the face of the tetrahedron furthest from the leaving group. Attack at one of the back edges also leads to inversion. Figure 16 shows this mechanism. Instead of axial approach and

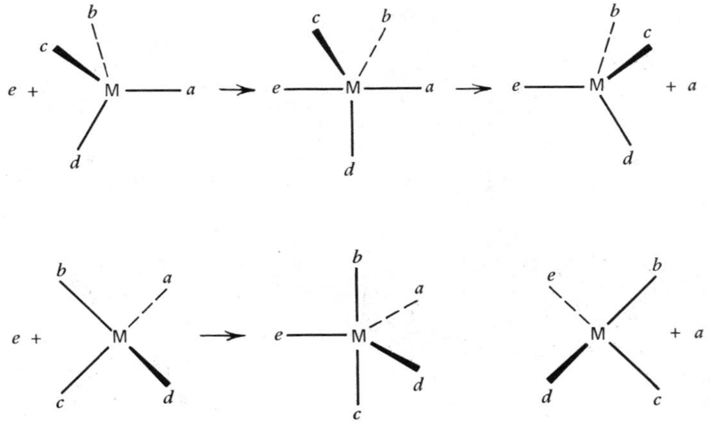

**Figure 16.** Both attack at a face (upper equation) and at an edge (lower equation) can lead to inversion in nucleophilic substitution at a tetrahedral molecule; *e* is the nucleophile and *a*, the leaving group.

axial departure with respect to the trigonal bipyramidal intermediate, there is now an equatorial approach of the nucleophile and equatorial departure of the leaving group.

It should be recalled that the $D_{3h}$ structure cannot be the activated complex for such a process. The reason is that it can decompose in three equivalent ways (see p. 11). The true activated complex would probably be a slightly distorted trigonal bipyramid.

Since the theoretical calculations of (133)–(136) are for the gas phase, and since solvent effects are a major perturbation, it is most interesting that recently a number of gas-phase nucleophilic substitution reactions of alkyl halides have been studied.[88] The methods used are ion cyclotron resonance and the flowing afterglow technique. The rates of reactions such as

$$X^- + RBr \longrightarrow RX + Br^- \tag{138}$$

are measured for different ionic nucleophiles $X^-$ and alkyl groups R. The results are somewhat surprising, since the rates are all very fast, with zero or small activation energy. The rate constants are generally of the order of $10^{11}$ $M^{-1}s^{-1}$, which is the expected figure for an encounter-controlled reaction.[89] Furthermore there is little variation in rates due to changes in R or $X^-$, compared to enormous variations in solution.

The reason for this behavior is not difficult to see. The collisions of an ion with a neutral molecule are not random, but are controlled by rather strong ion-dipole, and ion-induced dipole forces. The reactants are brought together with a net lowering of potential energy of some 10–20 kcal mole$^{-1}$. By conservation of energy, the kinetic, vibrational, and rotational energy of the ion–molecule adduct must increase by the same amount. Since deactivating collisions are rare in a dilute gas, there is usually adequate time for this energy to find its way into the proper vibrational mode to cause nucleophilic substitution. The needed activation energy must not exceed that available in the complex, however.

Accordingly all reactions of moderate activation energy occur at near the capture rate. Reactions such as (136), with a theoretical barrier of 22.6 kcal, simply do not occur.[88c] Those reactions that do occur show effects expected if they proceed by a trigonal bipyramidal activated complex. For example, inversion of configuration has been found in a gas-phase reaction of the type (138), where $X^-$ is the chloride ion and R is the 4-cyclohexanol ring.[89]

In the case of simple reactions such as (138) it is likely that the direction of approach favored by ion-dipole forces coincides with that favored by orbital overlap considerations. This has been shown to be so for (133) and (134).[90] Furthermore, at distances of approach of about 5 Å, energy lowering due to ion-induced dipole forces exceeds all other effects. At shorter distances covalency effects will take over and dominate the energetics.

It must be realized that ion-dipole, and similar forces, are somewhat extraneous to the arguments based on orbital symmetry. For example, in the reaction

$$Br^- + ClCH_2CH_2I \longrightarrow ClCH_2CH_2Br + I^- \quad (139)$$

the initial, energetically favored, direction of approach might be to the backside of the carbon—chlorine bond. However the lowest-energy path eventually would be for displacement of iodide ion. Based on the overall lowest energy path from infinite separation, the reaction coordinate would be quite complex, with no symmetry. Only the latter part of the reaction path could be discussed in symmetry terms.

Having seen how electrophiles and nucleophiles attack $CH_4$ and its derivatives, it is of interest to examine the free-radical reaction of a hydrogen atom with methane. The possible processes are abstraction and exchange.

$$D\cdot + CH_4 \longrightarrow DH + CH_3\cdot \quad (140)$$

$$D\cdot + CH_4 \longrightarrow DCH_3 + H\cdot \quad (141)$$

Calculations at the ab initio level have been made for the potential energy surfaces of both (140) and (141).[91] The dynamics of both reactions have also been considered.[91b] Reactions (140) and (141) are important models for many reactions of free radicals with organic molecules.

Compared to $H^+$, which approaches an edge, and $H^-$, which approaches a face, H atom prefers to approach a corner of the $CH_4$ tetrahedron. This reaction coordinate preserves $C_{3v}$ symmetry and leads to the abstraction reaction, (140). The calculated activation energy is 17 kcal, compared to the experimental value of 10–12 kcal. It is known that at lower temperatures only abstraction is found experimentally. Exchange reactions are found only in "hot-atom" chemistry, the threshold being about 35 kcal.[92]

The theoretical activation energies for the exchange reaction 141 are 42 kcal for an inversion mechanism (backside attack) and 64 kcal for a retention mechanism (frontside approach). Also an activation energy of about 70 kcal was calculated for the formation of a $CH_5$ molecule with $C_{4v}$ symmetry.[91a] This is the energy required for formation of a $D_{3h}$ species, followed by a Berry pseudorotation so that racemization would be the observed stereochemical result. In spite of the higher barrier, the retention mechanism is most probable when dynamics is taken into account.[91b] However this is only true for very hot atoms which are sufficiently energetic.

Again the $C_{4v}$ structure cannot be the activated complex for any displacement reaction, since it can decompose in four equivalent ways. It lies midway between two lower-energy $D_{3h}$ structures, which are the activated complexes for the inversion mechanisms. It also lies in the center of four lower energy

$C_s$ structures, which interconvert at a lower energy than that of $C_{4v}$, and give retention of configuration.

An approximate MO calculation has been made for the interaction of a $CH_3$ radical with $CH_4$.[93] This would be similar to (140) at medium distances of approach. The main interactions are between the singly occupied MO (SOMO) of $CH_3$ and both the LUMO and HOMO of $CH_4$. These correspond to the charge-transfer processes shown in Fig. 17. The HOMO is an $a_1$ orbital formed mainly from the $p_z$ orbital of carbon and the $s$ orbital of the hydrogen atom to be removed.

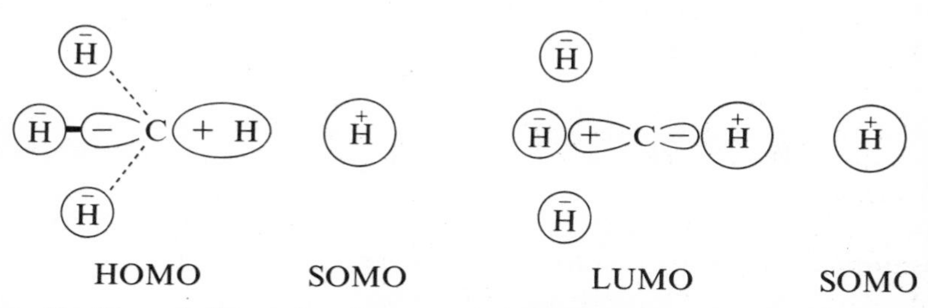


**Figure 17.** The main interactions between a hydrogen atom and a methane molecule are between the HOMO and LUMO of $CH_4$ and the SOMO of H. The HOMO and LUMO of $CH_4$ are linear combinations of the $t_2$ and $2t_2$ orbitals, respectively, of $CH_4$ (see p. 186). These combinations are equally valid descriptions, and are appropriate to the symmetry imposed by the approaching H atom.

### Phosphorus and Silicon Compounds

In addition to prototype organic molecules, the $XY_4$ class includes many examples of molecules that can be considered as derivatives of $SiH_4$, $GeH_4$, $SnH_4$, $PbCl_4$, $POCl_3$, $AsOCl_3$, $SO_2Cl_2$, and so on. In all cases the $\sigma$ valence electron count is the same as for $CH_4$. The only lone pairs of electrons are on the attached groups. The most important examples are the four-coordinated phosphorus compounds, $R_3PO$, $R_2PO(OR)$, $RPO(OR)_2$, and $PO(OR)_3$, the phosphine oxides, and phosphinic, phosphonic, and phosphoric esters. Next in importance would be silicon compounds.

In the case of all of the molecules mentioned above, important differences exist between them and tetrahedral carbon molecules. The first difference is in the low electronegativity of the central atom compared to carbon. This, plus the nature of the attached groups, puts a higher positive charge on the central atom. As a result, electrophilic attack at the central atom is rare. A reagent such as the proton almost invariably attaches itself to one of the four groups around the central atom, since that is the region of high electron density.

A second difference is that coordination numbers higher than four are common for Si(IV), Ge(IV), P(V), As(V), S(VI), and so on. Most common is coordination number six, as in $SiF_6^{2-}$, $PF_6^-$, and $SF_6$, but coordination number five is common, and in the case of P(V) most common, as in $PF_5$. The number of examples for silicon is small, but definite, with $SiF_5^-$ as an example.[94] This raises the interesting possibility that a nucleophile will add to these $XY_4$ molecules to form a rather stable adduct, rather than an activated complex. Surely the five-coordinated species will be longer lived than the corresponding carbon complexes. In this case, pseudorotation, interchanging axial and equatorial positions, can readily occur.

Theoretical calculations confirm these expectations.[95] An ab initio calculation of the reaction

$$SiH_4 + H^- \longrightarrow SiH_5^- \qquad (142)$$

shows that it is exothermic by 17 kcal, when a $D_{3h}$ structure is formed. This structure is only 3 kcal more stable than a $C_{4v}$ form. Approach to $SiH_4$ is favored at a face, corresponding to an *inversion* mechanism for nucleophilic substitution. Somewhat surprisingly there is an activation energy of 8 kcal, even though the overall process is exothermic.

Approach of a hydride ion to the center of an edge can also lead to a $D_{3h}$ structure.

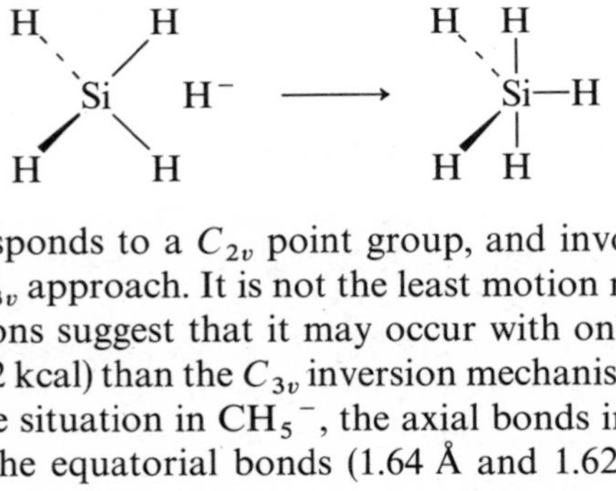

```
H      H                      H  H
 `   /                         ` |
  `Si     H⁻   ———→            `Si—H        (143)
  ◢   \                         ◢ |
H      H                      H  H
```

This motion corresponds to a $C_{2v}$ point group, and involves more nuclear motion than the $C_{3v}$ approach. It is not the least motion mechanism but less complete calculations suggest that it may occur with only a slightly higher activation energy (2 kcal) than the $C_{3v}$ inversion mechanism.[95]

Compared to the situation in $CH_5^-$, the axial bonds in $SiH_5^-$ are only a little longer than the equatorial bonds (1.64 Å and 1.62 Å). The MO configuration is [95,96]

$$(a_1')^2(a_2'')^2(e')^4(2a_1')^2(2e')^0(2a_2'')^0$$

The $2e'$ orbital is now below the $2a_2''$ orbital because strong bonding in $a_2''$ (axial bonds) implies strong antibonding in $2a_2''$. The lowest energy excited state comes from the $2a_1'$ to $2e'$ transition. A reaction mode of $E'$ symmetry is favored. Such a mode leads to a $C_{2v}$ point group. This can either be en route to the square pyramid of a pseudorotation process, or the result of a reversal of reaction (143), equatorial dissociation (see p. 270).

The stereochemical prediction is that an $S_N2$ reaction with *inversion* is slightly favored. However two alternatives can lead to *retention* of configuration. One is the pseudorotation process, and the other is gain or loss of groups in the equatorial position. It is necessary that a nucleophile adds axially and a group leaves equatorially, or the reverse process.[97]

The experimental results on chiral silicon compounds show that inversion is a common result but that retention is almost as frequent.[98] Good leaving groups, such as $Cl^-$, usually give inversion and poor leaving groups, such as $CH_3O^-$, usually give retention. The situation is complicated because when the leaving group is a poor one, there is often electrophilic assistance by the cation associated with the nucleophile. This has led to the postulation of four-center TS, for instance,

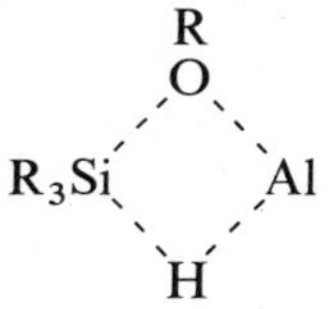

R
O
$R_3Si$ Al
H

The intermediate could be formed either by attack at a face of the silicon tetrahedron adjacent to the leaving group, or by edge attack by the nucleophile with assisted axial loss of the leaving group.[99]

There are also clear examples of nucleophilic substitution of good leaving groups, devoid of electrophilic assistance, which are replaced with retention. These are usually cases where small rings, including silicon, are involved, or where bridgehead silicon exists.[100] For example, 1,2-dimethyl-1-chloro-1-silacyclobutane is reduced by $LiAlH_4$ with retention.[101]

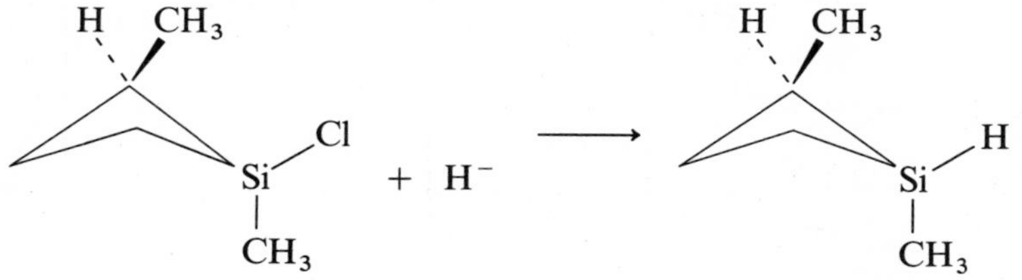

H $CH_3$ H $CH_3$

$$\text{Cl} \qquad\qquad\qquad\qquad\qquad\qquad \text{H} \qquad (144)$$

$$\text{Si} \quad + \quad H^- \quad \longrightarrow \quad \text{Si}$$

$CH_3$ $CH_3$

Backside attack would force the four-membered ring to span two equatorial positions, leading to strain. In this case it appears that edge attack is favorable. Thus there could be equatorial gain and axial loss of nucleophiles with retention. The assumption is that the chlorine atom prefers to be axial in the TS. In bridgehead examples, where backside approach and inversion are impossible, there could be simple addition of the nucleophile to an adjacent tetrahedral face. A slight rearrangement leads to a structure where entering and leaving groups are equivalent.

A theoretical analysis of the reactions of four-coordinate phosphorus(V) compounds would presumably give results similar to those for silicon(IV).[96] The five-coordinated intermediate in nucleophilic substitution reactions would be long-lived and have an electronic arrangement favorable to pseudorotation. Equatorial gain and loss of nucleophiles would become competitive with apical gain and loss. Indeed it was the classical work by Westheimer that first proved that pseudorotation occurred in hydrolysis reactions of phosphate esters.[102]

Figure 18 shows the pseudorotation mechanism applied to the acid hydrolysis of methyl ethylene phosphate. This cyclic ester hydrolyzes $10^6$ times faster than does trimethyl phosphate.[103] The explanation for this reactivity lies in the strain energy of the five-membered ring. This strain is alleviated in forming the trigonal–bipyramidal structure providing the ring spans one apical and one equatorial position, with its 90° bond angle. Pseudorotation exchanges the entering water molecule and the leaving methanol molecule, which enter and leave from apical positions. Alternatively, the water molecule

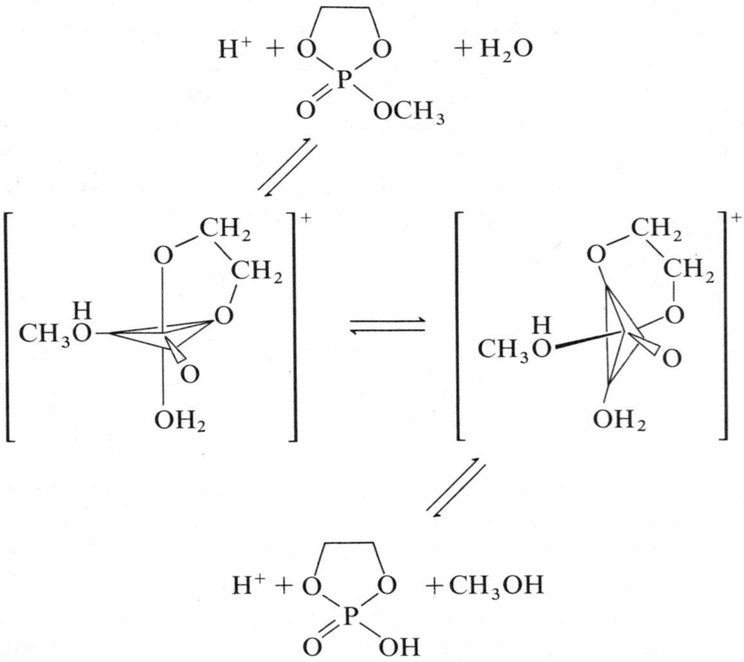

**Figure 18.** Pseudorotation mechanism in the acid hydrolysis of methyl ethylene phosphate; the process leads to retention of configuration at phosphorus (after reference 61a).

could enter in an equatorial position directly, avoiding the rotation step. Similarly, the methanol molecule could leave from an equatorial position.

The mechanism shown in Fig. 18 would lead to retention of configuration in a similar but asymmetric molecule. So would edge attack. Many stereochemical results are available for four-coordinated P(V) systems.[97,103] The reactions are usually highly stereospecific. Generally acyclic systems give inversion of configuration and cyclic systems give retention, during nucleophilic substitution. This is explicable in terms of the reluctance of small rings to span two equatorial positions, with the resulting 120° bond angle. Note that methyl ethylene phosphate would have a very strained ring if both water and methanol occupied apical positions (Fig. 18). The additional rule which explains the observed stereochemistry of phosphorus is that electronegative atoms prefer the apical positions (Muetterties's rule).

### Tetrahedral-Metal Complexes

Tetrahedral complexes of the transition metals offer several new features due to the presence of the *d* shell. These orbitals are chemically important since in a molecule like $TiCl_4$, or $CrO_4^{2-}$, the four $\sigma$ bonds are due chiefly to $sd^3$ bonding, rather than $sp^3$. That is, it is the $t_2$ subshell of *d* atomic orbitals rather than the *p* orbitals of the metal that enters into bonding.[104]

Figure 19 shows the crystal field splittings of the *d* manifold in fields of $T_d$, $C_{3v}$, and $D_{3h}$ symmetry. The $t_2^*$ orbitals are actually the antibonding partners of the bonding $t_2$ set. They are the ones that a nucleophile would seek to overlap with in an $S_N2$ process. Clearly it is desirable to have at least one of the $t_2^*$ orbitals empty in such a reaction. This is the one that becomes $a_1$ in $C_{3v}$ and finally $a_1'$ in a $D_{3h}$ TS. We can expect that $d^0$ to $d^4$ complexes would be most susceptible to nucleophilic attack. Complexes with $d^{10}$ configurations would resist nucleophilic attack.

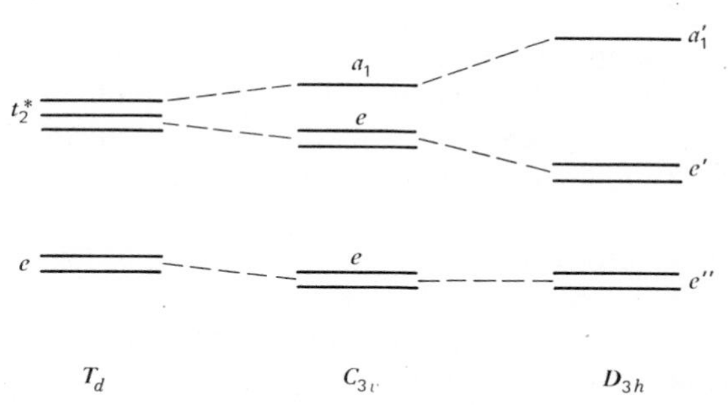


**Figure 19.** Correlation diagram for the *d* orbitals of a tetrahedral complex as a nucleophile approaches a face ($C_{3v}$) and forms a trigonal bipyramidal intermediate or TS ($D_{3h}$).

Not much evidence is available for the effect of different numbers of $d$ electrons on reactivity of tetrahedral complexes. However it is known that $d^{10}$ complexes such as $Ni(CO)_4$ or $Ni[P(OR)_3]_4$ react by a dissociation ($S_N1$) mechanism and not by an $S_N2$ mechanism.[28] Also $d^0$ complexes such as $CrO_4{}^{2-}$ and $MnO_4{}^-$ are quite susceptible to nucleophilic attack.[105] The chromate ion, in particular, is much more reactive to $S_N2$ processes than the sulfate ion, which must use a $t_2$ set based on atomic $p$ orbitals.

For $d^{10}$ metal ions with a positive charge on the central metal, nucleophilic attack is obviously possible because five-coordinated complexes of these metal ions exist. A most unusual and graphic illustration exists of the reaction coordinate and TS for the generalized reaction[106]

$$XCdS_3 + Y \longrightarrow X\text{---}CdS_3\text{---}Y \longrightarrow CdS_3Y + X \quad (145)$$

Here X and Y are a variety of electronegative donor atoms and $S_3$ refers to a trigonal arrangement of sulfur atoms (from various ligands) around cadmium. The data consist of a set of bond distances, determined by x-ray diffraction in the solid state, for a number of compounds in which cadmium is surrounded by $S_3$, X, and Y.

The relevant information consists of the values of $\Delta x$, $\Delta y$, and $\Delta z$ as shown in Fig. 20. To obtain $\Delta x$ and $\Delta y$ it is necessary to select standard Cd—X and Cd—Y bond distances from sums of covalent radii, and to calculate the deviations from these standard values. The values of $\Delta z$ are obtained directly from the deviation of Cd from the trigonal plane. Figure 20 shows the values of $\Delta x$ and $\Delta y$, plotted as a single function of $\Delta z$. To the observed points another one can be added with $\Delta x = 0$, $\Delta y = \infty$, and $\Delta z = 0.84$ Å. This corresponds to an idealized $CdS_4$ tetrahedron with an infinitely distant fifth ligand, Y.

It is clear from Fig. 20 that the approach of Y toward the Cd atom causes a flattening of the $CdS_3$ pyramid, with a corresponding increase in the Cd—X distance. When $\Delta z = 0$, $\Delta x = \Delta y = 0.32$ Å, the situation for a trigonal bipyramid with rather long and weak axial bonds. Thus $XCdS_3Y$ at this point is a model for the activated complex when X and Y are the same. Figure 20 is a representation of the major geometry changes for the backside attack of a nucleophile on a tetrahedral metal complex.

For tetrahedral $d^{10}$ complexes with the metal atom in a zero-valent or univalent state, another reaction is possible. Electrophiles, such as $H^+$, can add to the metal atom.[107] The structure of $Ni[P(OC_2H_5)_3]_4H^+$ is that of a trigonal pyramid with the hydrogen atom in an axial position.[108] The nickel(0) has become formally Ni(II) and the proton is considered to be hydride ion. The LM approach of the proton is to a tetrahedral face of the nickel tetrakis-phosphite. In methanol solvent there is an activation energy of 13 kcal for the formation of the adduct. Once formed, the adduct is quite stable, since the $a_1$ orbital, shown in Fig. 19 for $C_{3v}$ symmetry, is empty in a $d^8$ complex.

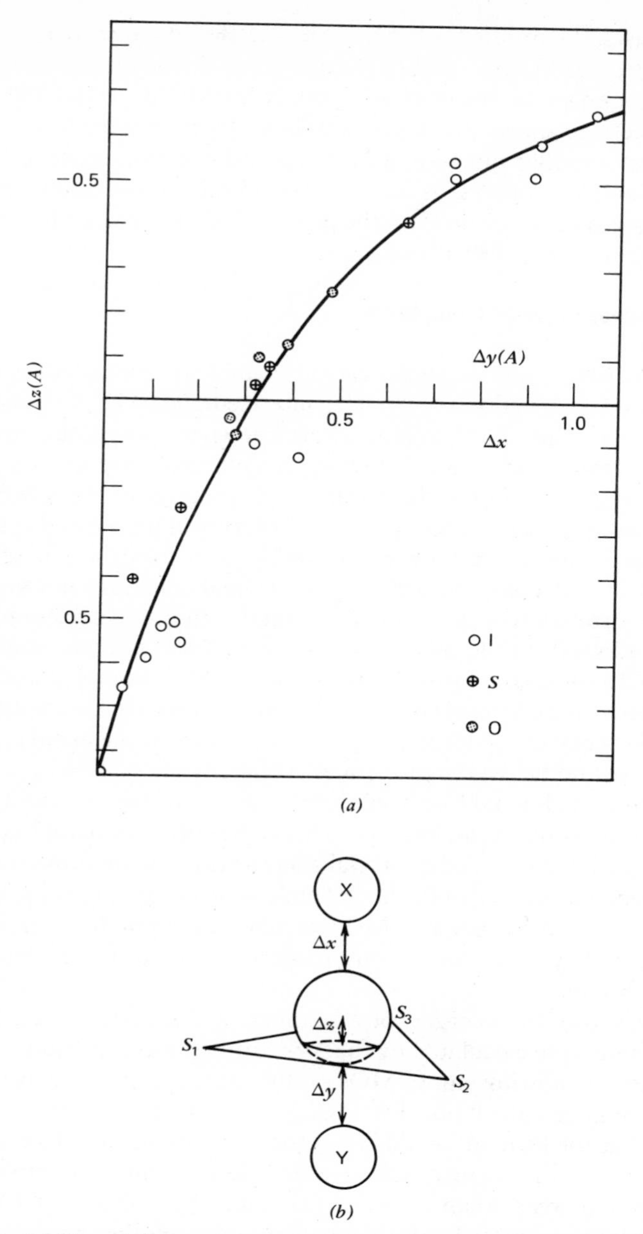


**Figure 20.** (*a*) Plot of distance increments $\Delta x$ and $\Delta y$ against $\Delta z$ for $CdS_3XY$ complexes; (*b*) definitions of $\Delta x$ $\Delta y$, and $\Delta z$; the diameters of the atoms indicate covalent radii. [Reprinted with permission from H. B. Burgi, *Inorg. Chem.*, **12**, 2321 (1973). Copyright by the American Chemical Society.]

In other cases, the product is labile. Thus $MP_4H^+$ dissociates to a greater or lesser extent into $MP_3H^+$ and P. Here M is Ni, Pd, or Pt and P refers to a phosphine or phosphite. An interesting result for $M[P(C_2H_5)_3]_4H^+$ is that the n.m.r. spectrum shows that equatorial loss of a phosphine ligand occurs rapidly, and at a similar rate to that for the pseudorotation process.[107] Note that the dissociation process is not an electrophilic substitution since the leaving group is a nucleophile. Also the product $MP_3H^+$ has a square planar structure and not a tetrahedral one.

## Square Planar Complexes

Square planar $ML_4$ complexes offer classic examples of bimolecular, nucleophilic substitution reactions.[27,105] The intermediates, $ML_4Y$, must be of moderate stability since stable analogues are known in many cases. Structures are usually trigonal bipyramidal, but square pyramids are known as well (see p. 190). Figure 21 shows the crystal field splittings of the $d$ orbitals in fields of $D_{4h}$ and $D_{3h}$ symmetries. The exact ordering of the $a_{1g}$ and $e_g$ levels in $D_{4h}$ is not important since they are normally both filled in square planar complexes. For a $d^8$ complex, which is the normal configuration for planar systems, the correlation of the empty $b_{1g}$ orbital with $a_1'$ is sufficient to let the reaction be allowed. While this orbital is the LUMO, it is not suitable for use by a nucleophile. It lies in the plane of the $ML_4$ complex and points directly at the ligands. It is, of course, antibonding. Instead the empty orbital that is readily accessible to form a $D_{3h}$ or $C_{4v}$ structure, is the metal $p_z$ orbital. This lies above and below the plane, and is of $a_{2u}$ species.

The LM approach would be from directly above or below, and at least a $C_{2v}$ point group would be maintained. This is the correlation used in Fig. 21. It may be argued that the filled $d_z^2$ orbital offers a barrier to this direct addition, and an off-center approach of the nucleophile with a $C_s$ point group has often been assumed.[109] Approximate MO calculations seem to bear out this assumption.[110] Experimental activation energies are not large, being of the order of 8–15 kcal.

Figure 21 shows the energies of a number of the valence-shell MOs of $PdCl_4^{2-}$. These were calculated by the X$\alpha$-scattered wave method, and have the advantage of showing empty MOs on the same energy basis as the filled MOs.[111] A similar calculation for $PtCl_4^{2-}$ shows that the energies of all MOs are lower for Pt than for Pd. Thus the nucleophile in $ML_4Y$ is bound more strongly in the former case. Nevertheless platinum complexes are usually slower to react than those of palladium by a factor of $10^5$ or so. Nickel complexes are more labile than palladium by another power of ten.[113]

This is partly due to the role played by the antibonding $a_{1g}$ orbital, shown as $12a_{1g}$ in Fig. 21. In $C_{4v}$, $C_{2v}$, or $C_s$ point groups this orbital has the same

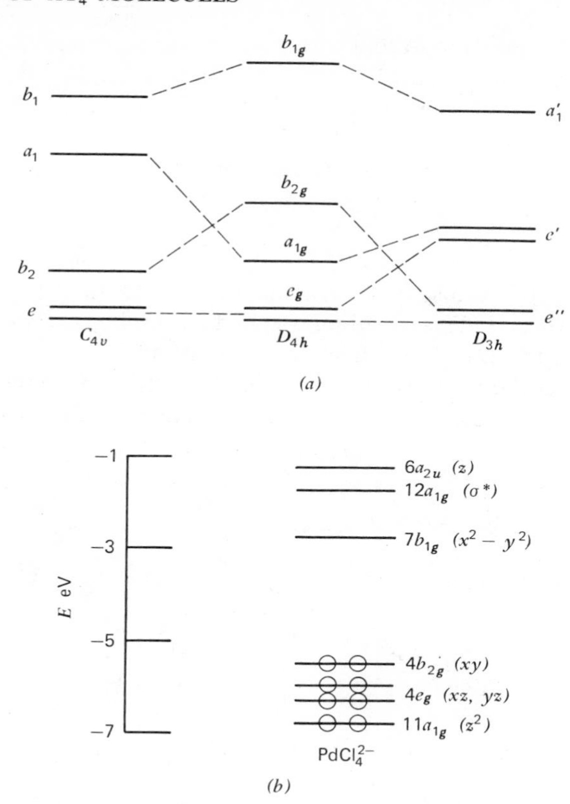


**Figure 21.** (*a*) Correlation diagram for *d* orbitals in fields of $D_{4h}$, $C_{4v}$ and $D_{3h}$ symmetry; crystal field levels are shown; (*b*) energies from an all-electron calculation for a number of higher MOs of $PdCl_4^{2-}$ (filled ligand $\pi$ orbitals of $e_u$, $a_{2g}$, and $b_{2u}$ symmetry also have energies of about $-6$ eV, but are not shown); a constant potential due to the cations of $K_2PdCl_4$ is included (energies from reference 111).

symmetry as $a_{2u}$, and also participates in the mixing with the HOMO of the nucleophile. One of its effects is to weaken the four original M—L bonds. This bond weakening is more difficult for Pt than for Pd or Ni. Also changes in the *d* manifold have a significant effect on the activation energies and rates (crystal-field activation energy). In this case the principal effect is the relative lowering of the empty orbital, $b_{1g}$ or $a'_1$, as shown in Fig. 21. This results in a loss of ligand-field stabilization, which falls off in the order Pt > Pd > Ni.[105]

One of the characteristic features of nucleophilic substitution of Pt(II) complexes is the failure to observe isomerization. Thus the entering group takes the place of the leaving group with *retention* of configuration. This is expected for a trigonal bipyramid with both the entering and leaving groups in equatorial positions

$$\begin{array}{c} A \\ | \\ T-Pt-X + Y \\ | \\ A \\ \textit{trans} \end{array} \longrightarrow \begin{array}{c} A \quad X \\ | \,/ \\ T-Pt \\ | \,\backslash \\ A \quad Y \end{array} \longrightarrow \begin{array}{c} A \\ | \\ T-Pt-Y + X \\ | \\ A \\ \textit{trans} \end{array} \qquad (146)$$

The group labeled T is also in the trigonal plane. It is the group *trans* to the leaving group, X, and plays a leading role in determining ease of replacement of X. The phenomenon is known as the *trans* effect.[27,105,110]

In order to get the isomeric product, the alternative choice of two ligands originally *cis* to X must go into the trigonal plane, followed by a pseudo-rotation,

$$\begin{array}{c} A \\ | \\ T-Pt-X + Y \\ | \\ A \end{array} \longrightarrow \begin{array}{c} Y \quad A \\ \backslash \,/ \\ T-Pt-X \\ | \\ A \end{array}$$

$$\swarrow$$

$$\begin{array}{c} T \quad X \\ \backslash \,/ \\ Y-Pt-A \\ | \\ A \end{array} \longrightarrow \begin{array}{c} T \\ | \\ Y-Pt-A + X \\ | \\ A \\ \textit{cis} \end{array} \qquad (147)$$

While this is a complicated process, it is not observed for energetic, rather than entropic, reasons. The nature of T, Y, and X is such that they prefer to be in the equatorial positions: Y because it is the entering group, X because it is the weakest held ligand, by definition, and T because of the nature of the *trans* effect.[27,110,112] If a number of the ligands are the same, as in $Fe(CO)_5$ or $HIr(PF_3)_4$, then pseudorotation, or an equivalent process, is rapid since the n.m.r. spectrum shows all the ligands of the same kind to be equivalent.[114]

In trigonal bipyramidal complexes of the transition metals, it is the axial positions where simple ligands are held most strongly (see p. 271). This is a consequence of the incomplete *d* shell. Figure 21 shows that crystal-field stabilization is greater for short axial bonds. That is the empty $a_1'$ orbital is raised in energy. This increases the gap between the $e'$ and $a_1'$ levels in Fig. 21, making the transition density of $E'$ species more difficult to attain, and stabilizing the structure. Thus the complexes $NiP_3X_2$, where P is a phosphine and $X^-$ is either a halide ion or cyanide ion, have axial cyanide groups, but equatorial halide ligands. This is predicted from the crystal-field sequence $CN^- > P >$ halide ion.[115a]

Ligands that are good $\pi$ acceptors, like CO and $PF_3$, prefer to be in the equatorial positions, even though they may have large crystal-field parameters.[115b] This results from the fact that $\pi$ bonding can occur with all four filled $d$ orbitals of the metal, both $e'$ and $e''$. In the axial positions, only the $e''$ orbitals can be used for $\pi$ bonding. This accounts also for the large *trans* effect of such ligands, since the intermediate of coordination number five is stabilized.

Electrophiles can also add to square planar complexes of $d^8$ metal ions. The most common result is that HX adds to give a *trans* adduct, for example,

$$Rh[P(OC_2H_5)_3]_4^+ + HBr \longrightarrow trans\text{-}Rh[P(OC_2H_5)_3]_4HBr^+ \quad (148)$$

In the case of a poorly coordinating anion such as $FSO_3^-$ or $CF_3SO_3^-$, the proton only may add to give a five-coordinated species. The best characterized examples are actually $CH_3^+$ adducts, formed from $CH_3SO_3F$ and similar strong alkylating reagents.[116]

$$(OC)(P)(P)(Cl)Ir + CH_3SO_3F \longrightarrow (OC)(P)(P)(Cl)Ir(CH_3)^+ + FSO_3^- \quad (149)$$

The structures in these cases are known to be square pyramidal, rather than trigonal bipyramidal. This is the expected structure for a $d^6$ metal ion in five coordination (p. 198).

Addition of $H^+$ or $CH_3^+$ to Ir(I) causes oxidation to Ir(III) and a change from $d^8$ to $d^6$. Considering the $d$ orbitals as the source of electrons, Fig. 21 shows that the $b_{2g}$ orbital (or $d_{xy}$) is the HOMO of a square planar complex. However orbital correlation shows that the electrons must come from the $a_{1g}$ (or $d_{z^2}$) orbital in order to prevent a cross over between the $a_1$ and $b_2$ orbitals in $C_{4v}$. The LM approach of the electrophile to the $a_{1g}$ orbital leads smoothly to the formation of a square pyramidal adduct. Other calculations put the $a_{1g}$ orbital highest in energy (p. 188).

Electrophilic substitution could result if the original planar complex had a hydride or some other easily oxidized ligand. However a rearrangement would be necessary to put this group in the apical position. The reaction would occur with retention of configuration.

$$(H)(D)(L)(L)(X)M \longrightarrow L{-}M(D)(H)(X){-}L \longrightarrow (X)(H)(L)(L)(D)M \quad (150)$$

Deuterium exchange is often found when transition-metal hydrides are put in deuterated, protic solvents.

### Other $XY_4$ Molecules

Table 4 summarizes the conclusions, to this point, for the mechanisms of addition of simple electrophiles and nucleophiles to $XY_4$ molecules. Also included in the table are some remarks concerning reactions of the remaining members of the $XY_4$ family of molecules, for example, $SF_4$. The MO sequence[29] for sulfur tetrafluoride, with its $C_{2v}$ structure, was given earlier (p. 269).

**Table 4**

| Reactants | Product | | Point Group Maintained |
|---|---|---|---|
| $CH_4 + H^+$ | $CH_5^+$ | $C_s$ | $C_s$ |
| $CH_4 + H^-$ | $CH_5^-$ | $D_{3h}$ | $C_{3v}$ |
| $ML_4 + L^{a,b}$ | $ML_5$ | $D_{3h}$ | $C_{3v}$ |
| $ML_4 + H^{+a}$ | $ML_4H^+$ | $C_{3v}$ | $C_{3v}$ |
| $ML_4 + L^c$ | $ML_5$ | $D_{3h}$ | $C_{2v}$ |
| $ML_4 + H^{+c}$ | $ML_4H^+$ | $C_{4v}$ | $C_{4v}$ |
| $SF_4 + F^+$ | $SF_5^+$ | $D_{3h}$ | $C_{2v}$ |
| $SF_4 + F^-$ | $SF_5^-$ | $C_{4v}$ | $C_s$ |
| $XeF_4 + F^+$ | $SeF_4^+$ | $C_{4v}$ | $C_{4v}$ |

[a] M is a transition metal of $d^{10}$ configuration.
[b] Also $d^0$, and so on, if $ML_4$ is tetrahedral.
[c] M is a transition metal of $d^8$ configuration; $ML_4$ is square planar.

The HOMO is $3a_1$ which is largely an $s$ orbital on sulfur, combined in an antibonding fashion with $\sigma$ orbitals on the two apical fluorine atoms.[29] The approach of an electrophile, such as $F^+$ would probably cause mixing of $2a_1$ and $3a_1$ to produce an orbital with more $p_z$ character on sulfur. Addition would be in a $C_{2v}$ point-group process to give $SF_5^+$, isoelectronic with $PF_5$ and thus with a $D_{3h}$ structure. The $SF_5^+$ molecule would be stereochemically nonrigid.

$$SF_4 \; (-)\!S\!(+) \quad (+)\!F\!(-) \longrightarrow S{-}F^+ \qquad (151)$$

The LUMO is the $2b_2$ orbital, which is largely a $p_y$ orbital on sulfur lying in the equatorial plane. The addition of $F^-$ would give $SF_5^-$, isoelectronic with $IF_5$, and with a $C_{4v}$ structure.

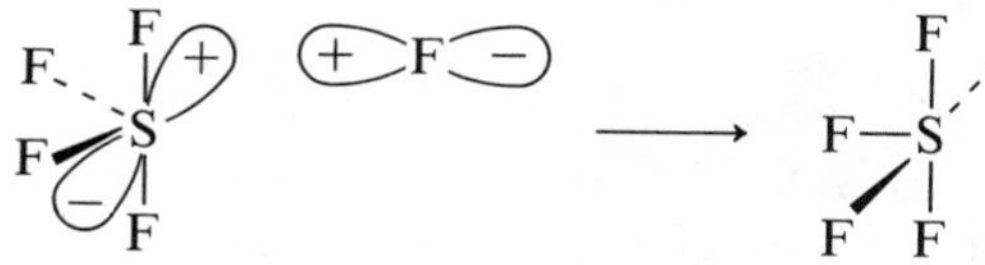

(152)

Only a $C_s$ point group would be preserved, the plane of symmetry being the equatorial one. The $4a_1$ would also mix in to some degree.

Unlike $PF_5$, the molecule $IF_5$ is stereochemically rigid.[118] The n.m.r. spectrum shows separate peaks for the apical and equatorial fluorine up to 115°C. Any of the four ligands in the plane of $SF_5^-$ could be the leaving group in a substitution reaction. One of these would correspond to a retention of configuration (original axial groups remaining axial) and two of the four would correspond to an isomerization of axial and equatorial groups.

Not much is known concerning substitution reactions of sulfuranes. $SF_4$ hydrolyzes quite rapidly in a

$$SF_4 + 2H_2O \longrightarrow SO_2 + 4HF \tag{153}$$

process that is markedly catalyzed by both hydrogen ion and hydroxide ion.[119] The latter is simply nucleophilic attack on sulfur by $OH^-$. The acid catalysis may be due to the formation of $SF_4H^+$, analogous to (151), followed by nucleophilic attack by $H_2O$.

$XeF_4$ has the MO configuration

$$(a_{1g})^2(e_u)^4(b_{1g})^2(a_{2u})^2(2a_{1g})^2(2e_u)^0$$

considering only the $\sigma$-bonding and antibonding orbitals. The addition of an electrophile, $F^+$, would occur easily to the $a_{2u}$ orbital ($p_z$ orbital of Xe) initially. The antibonding $2a_{1g}$ orbital would also mix in the $C_{4v}$ point group generated by direct approach from above. The

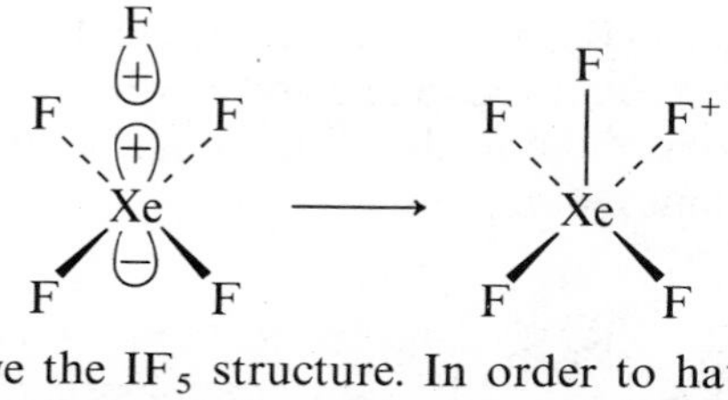

(154)

product would have the $IF_5$ structure. In order to have electrophilic substitution, a rather high energy isomerization of $XeF_5^+$ would be required, interchanging the apical position for an equatorial one.

The only empty orbitals of $XeF_4$ are the $2e_u$ orbitals. These are antibonding orbitals derived from the $p_x$ and $p_y$ orbitals of Xe and lying in the plane of the molecule. As in the case of $XeF_2$, it would be difficult for a nucleophile to use an orbital of this kind to bond to Xe. Instead attack

$$F{-}Xe{-}F + F^- \longrightarrow F^- + Xe + F_2 \qquad (155)$$

at fluorine is more favorable. The resulting electron transfer from $F^-$ would cause complete dissociation of $XeF_4$ or $XeF_2$ into its elements, since a $\sigma^*$ orbital is filled.

The reactions of these noble-gas fluorides with nucleophiles gives products quite in agreement with these expectations. For example,

$$2XeF_2 + 4OH^- \longrightarrow 2Xe + 4F^- + O_2 + 2H_2O \qquad (156)$$

$$XeF_4 + 4I^- \longrightarrow Xe + 4F^- + 2I_2 \qquad (157)$$

Presumably HOF is formed first in (156). This unstable molecule would decompose into molecular oxygen and HF.

## REACTIONS OF $XY_5$ MOLECULES

The HOMO of the hypothetical $PH_5$ molecule is $2a_1'$ (see p. 192). The orbital is concentrated entirely on the hydrogen atoms. Therefore attachment of an electrophile to the central atom would be difficult. The product, $PH_6^+$ for example, would be a molecule for which no known stable, or even quasi-stable, models exist. For real molecules like $PF_5$ it is even less likely that a proton would attack itself to phosphorus. Instead attack would be at a ligand.

$$PF_5 + H^+ \longrightarrow PF_4^+ + HF \qquad (158)$$

On the other hand, nucleophilic attack should be facile, since molecules like $PF_6^-$ are quite stable. The structure of the adduct would be octahedral. The LUMO of $PH_5$, or $PF_5$, would be the antibonding $2e'$ orbital (p. 192). These orbitals lie in the equatorial plane and are derived from the $p_x$ and $p_y$ orbitals of P. Approach of the nucleophile would be in the plane with the elements of the $C_{2v}$ point group present.

$$PF_5 + F^- \longrightarrow PF_6^- \qquad (159)$$

The original equatorial P—F bonds would all be weakened, and the axial bonds less affected. The final product would have six equivalent bonds. Any of the groups could leave in a substitution process. Three of the original groups would lead to retention of configuration, and two to an exchange of axial and equatorial positions.

Not much is known about nucleophilic substitution reactions of phosphoranes. They do occur, and with great facility, as predicted.[120] A kinetic study of the hydrolysis of pentaaryloxyphosphoranes has been carried out.[121]

$$P(OAr)_5 + H_2O \longrightarrow OP(OAr)_3 + 2ArOH \quad (160)$$

There is catalysis by both $H^+$ and $OH^-$. The steric effects of changing the aryl group in ArOH are strongly in favor of typical association mechanisms ($S_N2$). Acid catalysis is undoubtedly due to coordination at the oxygen atom of the aryloxyl ligand.

Five-coordinated transition-metal complexes may behave like $PF_5$, or the incomplete $d$ shell may cause substantial differences. Metal atoms of $d^0$ and $d^{10}$ configuration would behave like the nontransition elements. Sometimes the $ML_5$ systems, in these cases, have square pyramidal structures,[122] which makes it even easier to add a nucleophile to the vacant $a_1$ orbital on the $z$ axis.

Metal complexes with $d^6$ configuration also have the square pyramidal shape, making it quite easy to add a sixth ligand to form octahedral products.[115] However $d^8$ metal complexes will resist the addition of a nucleophile. Thus a square planar complex (low-spin $d^8$) will readily add a fifth ligand, but not a sixth.

The explanation lies largely in the crystal-field energies of the $d$ orbitals, shown in Fig. 22 for complexes of $D_{3h}$, $C_{4v}$, and $O_h$ point groups. If a trigonal

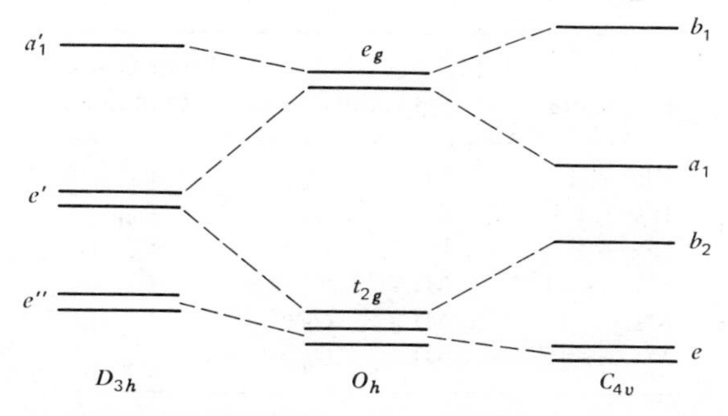


**Figure 22.** Correlation diagram for $d$ orbitals in fields of $O_h$, $C_{4v}$, and $D_{3h}$ symmetry; the effect of adding a ligand to an $ML_5$ complex that is either a square pyramid or a trigonal bipyramid is illustrated.

bipyramid is converted to an octahedron by adding a ligand, a $d^8$ configuration will be raised in energy. The ground state will be a triplet, rather than a singlet. The singlet state will be subject to a SOJT instability and distort tetragonally. In any event, nucleophilic substitution in five-coordinated $d^8$ complexes occurs almost exclusively by dissociation mechanisms and only rarely by association.[123]

The situation is quite different for addition of an electrophile to metal complexes, where the metal atom is in a low oxidation state. According to Fig. 22, it would be favorable to go from $D_{3h}$ to $O_h$ symmetry for a $d^8$ system. The electrophile would add in the equatorial plane and two electrons from the $e'$ orbital would be transferred to it. The addition of an electrophile to a $d^6$ square pyramid would not be a favorable process, since a $d^4$ configuration is not stable in $O_h$ symmetry. It is known that protons add readily to $d^8$ complexes such as $Fe(CO)_5$ and its derivatives.[124] The infrared spectrum of $Fe(CO)_5H^+$ shows it to have the expected $C_{4v}$ structure.[125] Electrophilic substitution in suitable cases would require an exchange of the axial group with a group in the basal plane.

Table 5 summarizes the stereochemical predictions for $XY_5$ or $ML_5$ molecules. Also included are the reactions of $IF_5$. Electrophiles should add to form octahedral $IF_6^+$, isoelectronic with the very stable $SF_6$. The HOMO of $IF_5$ is a lone-pair orbital of $a_1$ species, occupying an apical site.

The LUMOs of $IF_5$ are $e$ orbitals lying in the equatorial plane. However orbital correlation shows that a nucleophile must add to $IF_5$ to form $IF_6^-$ by using the high-energy vacant $a_1$ orbital (see p. 193). The analogy is to the isoelectronic, octahedral $XeF_6$ molecule. The configuration of this molecule is

$$(a_{1g})^2(t_{1u})^6(e_g)^4(2a_{1g})^2(2t_{1u})^0$$

**Table 5**

| Reactants | Product | | Point Group Maintained |
|---|---|---|---|
| $PF_5 + F^-$ | $PF_6^-$ | $O_h$ | $C_{2v}$ |
| $IF_5 + F^+$ | $IF_6^+$ | $O_h$ | $C_{4v}$ |
| $IF_5 + F^-$ | $IF_6^-$ | $O_h$ | $C_s$ |
| $ML_5 + H^{+a}$ | $ML_5H^+$ | $C_{4v}$ | $C_{2v}$ |
| $ML_5 + L^a$ | $ML_6$ | $O_h$ | $C_{2v}$ |
| $ML_5 + L^b$ | $ML_6$ | $O_h$ | $C_{4v}$ |

[a] M is $d^8$ transition metal, $ML_5$ is trigonal bipyramid.
[b] M is $d^6$ transition metal, $ML_5$ is square pyramid.

for the $\sigma$ electrons only. This shows that the antibonding orbitals derived from the $p$ orbitals of the central atom must be vacant.

$IF_5$ and $BrF_5$ are very reactive molecules. They hydrolyze very rapidly and react with other nucleophiles. Salts of $IF_6^-$ are known, however electrophiles apparently react with $IF_5$ to remove fluoride ion and form $IF_4^+$.[49] Thus the presence of the lone pair in $IF_5$ seems to neither hinder approach of a nucleophile nor help access of an electrophile.

## REACTIONS OF $XY_6$ MOLECULES

Only octahedral, six-coordinated molecules will be considered because of their overwhelming occurrence. The simplest example would be $SF_6$, a notoriously inert molecule. Referring back to the MO sequence for $XeF_6$, we see that the HOMO is $e_g$, a set of orbitals located entirely on the fluorine atoms. The LUMO is $2a_{1g}$, chiefly an $s$ orbital on the central atom and antibonding to all six attached ligands. This orbital is shown in Fig. 23.

We conclude that electrophiles are not likely to attach themselves to the central atom of $XY_6$ molecules when $d$ electrons are missing from the valence shell. Instead attack at the ligands is much more likely. This conclusion is reenforced by examining Fig. 23, which shows that the central atom is well shielded from all reagents by the six ligands. Attachment of a nucleophile would be equally difficult for steric reasons.

The $2a_{1g}$ orbital is nondirectional, which makes it difficult to predict where a seventh group would add to an octahedron. In addition, seven-coordinate

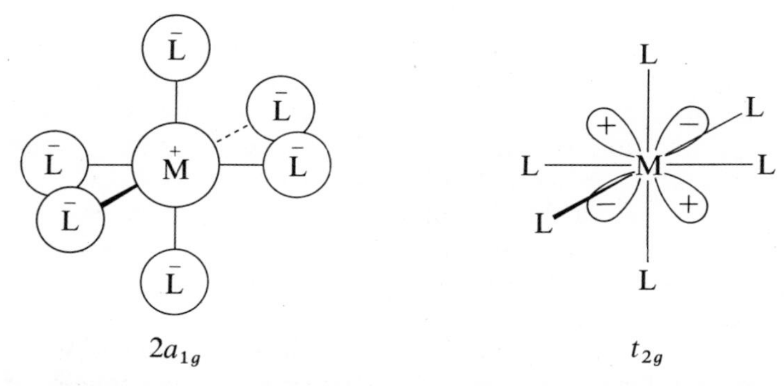


**Figure 23.** The antibonding orbital, $2a_{1g}$, of an $ML_6$ complex; normally, the $\sigma$ orbitals of the ligands will be largely $p_z$ orbitals of the donor atom, and not $s$ orbitals as shown; one of the $t_{2g}$ orbitals is also shown—the latter orbitals are only capable of $\pi$ bonding.

compounds of the nontransition elements are almost unknown, the only simple example being $IF_7$.[126] This has a pentagonal bipyramidal structure (p. 205). If metal ions of $d^0$ configuration are also included, such as $NbOF_6^{3-}$ and $TaF_7^{2-}$, two other structures are found. These are a 1,3,3 form, based on an octahedron with a ligand added to the face, and a 1,4,2 form, based on a trigonal prism with a seventh group added to a square face. Alternatively the 1,4,2 could come from an octahedron with two groups sharing a corner.

Kinetic studies show that nucleophilic substitution reactions of octahedral $XY_6$ molecules occur almost exclusively by dissociation mechanisms.[127] There are few, if any, clear cut cases of associative, or $S_N2$, reactions. Somewhat surprising is the observation that $XeF_6$ hydrolyzes to give typical substitution products.[25]

$$XeF_6 + 3H_2O \longrightarrow XeO_3 + 6HF \qquad (161)$$

This result may be compared to reactions 156 and 157. However the mechanism of (161) probably involves stepwise dissociation of fluoride ions.

Most kinetic studies of octahedral complexes have been concerned with those of the transition metals. The most thoroughly studied systems are those of the labile complexes of the divalent metal ions of the first transition series, and the inert complexes of cobalt(III).

The former react by the interchange mechanism first suggested by Eigen.[128]

$$M(H_2O)_6^{n+} + X^- \rightleftharpoons M(H_2O)_6, X^{(n-1)+} \qquad (162)$$

$$M(H_2O)_6, X^{(n-1)+} \longrightarrow M(H_2O)_5X, H_2O^{(n-1)+} \qquad (163)$$

In this mechanism, an ion pair is first formed, in a diffusion-controlled process. The ion pair then interchanges the ion (or neutral molecule) in the second coordination sphere with a water molecule (or other ligand) in the first coordination sphere. The rate of this last step is nearly independent of the nature of $X^-$. Therefore the process is a dissociative one primarily. The TS may have a 1,4,2 structure leading to *retention* of configuration. The incoming group simply takes the place of the leaving group.

The reactions of cobalt(III) complexes are very similar, except that in favorable cases a discrete intermediate of coordination number five is formed.[129] In the latter case, stereochemical change is possible, isomerization or racemization.[127] The slowness of reactions of cobalt(III) is due chiefly to the large crystal-field stabilization of the filled $t_{2g}$ subshell. It should be mentioned that, for this reason, cobalt(III) may be exceptionally prone to dissociation reactions, and other metal ions may be more susceptible to nucleophilic attack.

Consider the MO configuration for a typical octahedral complex of a transition-metal ion.[130]

$$(a_{1g})^2(e_g)^4(t_{1u})^6 \,\vdots\, (t_{2g})^x(2e_g)^y \,\vdots\, (2a_{1g})^0(2t_{1u})^0$$

Only $\sigma$ bonding is considered. The $d$ manifold $(t_{2g})^x(2e_g)^y$ is spherically symmetric, if filled. The $2e_g$ orbitals appear as pockets in the spherical shell pointing directly at the six ligands. The $(t_{2g})^6$ electrons, plus the ligands constitute a barrier for the addition of a nucleophile to the metal in any possible direction (see Fig. 23). In particular the octahedral faces and edges are equally blocked.

Removal of electrons from the $t_{2g}$ subshell makes it easier to add a seventh ligand. It was suggested by Taube many years ago that $d^0$–$d^4$ metal complexes might react by an $S_N2$ mechanism.[131] Semiempirical estimates of ligand-field activation energies bear out this suggestion,[132] but experimental verification from kinetics is scanty. An empty $t_{2g}$ orbital would accept electron density from the nucleophile. Certainly the seven-coordinate complexes known are almost all $d^0$–$d^4$ systems.[126]

Chromium(III) complexes are more labile than those of cobalt(III), related to the $(t_{2g})^3$ configuration of chromium. Reactions are still primarily of a dissociative type, but there is evidence that there is more bond-making by the nucleophile in the TS for chromium than for cobalt.[133] The reactions are highly stereoretentive, indicating frontside attack. The TS would resemble

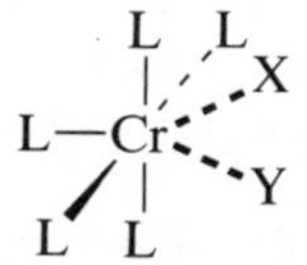

```
    L  L
    | ,' ..X
L—Cr:
   / |  `..Y
  L  L
```

where Y and X are the entering and leaving groups, both with some bonding to chromium. However Y cannot bond until the Cr—X bond has lengthened markedly.

Complexes with a filled $t_{2g}$ subshell, and with the metal in a low oxidation state should be able to bind electrophiles. Addition could be at a face, to form a 1,3,3 structure, or at an edge to form a pentagonal bipyramid. The proton is known to add to derivatives of $Cr(CO)_6$, but the structures of the adducts are not known.[124] Also several reactions of $Ru(NH_3)_6{}^{2+}$ have been reported in which apparently electrophiles such as $H^+$ or $NO^+$ add to the ruthenium atom.[134] The stability of ruthenium(IV) makes this plausible since oxidation of the metal accompanies addition of an electrophile.

Finally it may be noted that the ligands of an octahedral complex are much more open to attack by reagents than is the metal (Fig. 23). There are many examples known of both electrophiles, particularly $H^+$, and nucleophiles reacting with the ligands. Reaction with nucleophiles is particularly common when the ligand contains unsaturated groups, such as CO.[135]

## REACTIONS OF $X_2Y_n$ MOLECULES

The chemistry of molecules such as $C_2H_2$, $N_2H_2$, and $H_2O_2$ is dominated by the orbitals formed by combining the $p_x$ and $p_y$ AOs of the central atoms (the molecular axis is $z$). In linear acetylene, these orbitals generate the filled $\pi_u$ and empty $\pi_g$ orbitals (see p. 210). Addition of an electrophile is to the filled $\pi_u$ subshell and of a nucleophile, to the $\pi_g$. The molecule, of course, has axial symmetry.

A proton can approach acetylene either at the midpoint of the C—C bond to form a $\pi$-complex, or toward one carbon atom to form a $\sigma$ complex.

$$\underset{\pi \text{ complex}}{\text{H—C}\overset{\text{H}}{\overset{|}{\equiv}}\text{C—H}^+} \qquad \underset{\sigma \text{ complex}}{(\text{H})_2\text{C=C—H}^+}$$

The $\pi$ complex would remain symmetric along the C—C axis, with some slight bending away of both H atoms. The $\sigma$-complex is identical with the vinyl cation, the structure for which is shown. An empty $p_y$ orbital lies in the plane of the molecule. Both structures have $C_{2v}$ symmetry, but the $\pi$ complex maintains this symmetry during its formation, whereas the vinyl cation maintains only a mirror plane.

When the proton–molecule distance is relatively large, the $\pi$ complex is probably more stable. Eventually at shorter distances, the $\sigma$ complex becomes more stable. Electrophilic substitution requires the formation of the $\sigma$ complex, in any case. Normally the vinyl cation would capture a nucleophile before electrophilic substitution could occur. If the vinyl cation were formed in another way, by loss of a halide ion from a vinyl bromide, there would also be a loss of stereochemistry

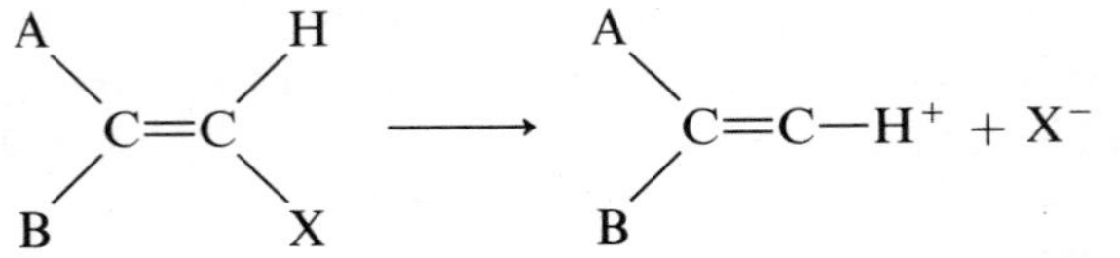

$$\underset{cis \text{ or } trans}{(\text{A})(\text{B})\text{C=C}(\text{H})(\text{X})} \longrightarrow \underset{\text{same cation}}{(\text{A})(\text{B})\text{C=C—H}^+ + \text{X}^-} \qquad (164)$$

Addition of a nucleophile, such as $H^-$, must necessarily be directed to one carbon atom only, because of the node at the midpoint of the empty $\pi_g$

orbital. In this case a vinyl carbanion is formed

$$H^- + H{-}C{\equiv}C{-}H \longrightarrow \begin{array}{c} H \\ \diagdown \\ C{=}\ddot{C}^- \\ \diagup \quad \diagdown \\ H \qquad H \end{array} \tag{165}$$

with a filled $p_y$ orbital lying in the plane. Nucleophilic substitution would result if (165) were reversed with the originally bonded hydrogen atom leaving as hydride ion. The addition–elimination mechanism is indeed common for acetylenic derivatives, for instance,[136]

$$RS^- + C_6H_5{-}C{\equiv}C{-}Br \longrightarrow RS{-}C{\equiv}C{-}C_6H_5 + Br^- \tag{166}$$

A competing reaction would be addition of a proton, or other electrophile, to the vinyl-carbanion intermediate.

The vinyl carbanion can also be formed in other ways; for example, by proton removal from a vinyl compound.[136]

$$\underset{cis}{\begin{array}{c} A \qquad H \\ \diagdown \quad \diagup \\ C{=}C \\ \diagup \quad \diagdown \\ B \qquad CN \end{array}} + CH_3O^- \longrightarrow \underset{cis\ \text{anion}}{\begin{array}{c} A \\ \diagdown \\ B{=}C^- \\ \diagup \quad \diagdown \\ B \qquad CN \end{array}} + CH_3OH \tag{167}$$

Unlike the vinyl cation, the vinyl anion is quite stereoretentive.[137] The lone pair maintains the stereochemistry. However isomerization would occur in rather short times, a few seconds or less. The transition state would undoubtedly be the linear form, isostructural with the vinyl cation.

There is also the addition of a free radical to the acetylenic bond. This leads to the eventual formation of a vinyl radical.

$$R\cdot + CH_3{-}C{\equiv}C{-}CH_3 \longrightarrow \begin{array}{c} R \\ \diagdown \\ C{=}C\cdot \\ \diagup \quad \diagdown \\ CH_3 \qquad CH_3 \end{array} \tag{168}$$

The $p_y$ orbital is now half-filled. Vinyl radicals are configurationally less stable than vinyl carbanions, but more stable than alkyl radicals. Depending on the attached groups, a wide range of lifetimes is possible for the isomerization process.[138]

Diimide, $N_2H_2$, has a pair of electrons in one of the $\pi_g$ orbitals. The two hydrogen atoms bend into the *trans* configuration. Addition of a proton would be to the HOMO, which is the former $\pi_g$ orbital, now of $a_g$ species, in the molecular plane

$$\mathrm{H^+ + H\ddot{N}{=}\underset{\cdot\cdot}{N}H \longrightarrow H_2N{=}\underset{\cdot\cdot}{N}H^+} \tag{169}$$

The resulting species is isoelectronic with the vinyl carbanion. The $p_y$ orbital is filled. Electrophilic substitution would be accompanied by *cis–trans* isomerization.

Addition of a nucleophile to $N_2H_2$ would be to the LUMO, which is the $\pi^*$ partner of the double bond of $b_g$ symmetry. The product, $NH_2NH^-$ for a hydride ion addition, would have a nonplanar structure. It would, in fact, be the anion of hydrazine (see p. 217). There would be rotation about the N—N bond with a very small barrier, since the bond is now single. Nucleophilic substitution would be accompanied by isomerization in a nonstereospecific manner.

In $H_2O_2$ another pair of electrons has been added to fill the remaining $\pi_g$ orbital of $C_2H_2$. The HOMO and the LUMO are shown in Fig. 24. The HOMO is one of the former $\pi_g$ orbitals. Addition of a proton would give $H_2OOH^+$, isoelectronic with $NH_2NH^-$. However a nucleophile must now use an orbital derived from a $\sigma_g^*$ orbital of acetylene. The lowest energy orbital is $4b$, which is antibonding for the O—O bond.

Thus nucleophilic attack on peroxides leads to breaking of the oxygen—oxygen bond.[139]

$$\mathrm{I^- + H_2O_2 \longrightarrow HOI + OH^-} \tag{170}$$

The best nucleophiles are soft bases, which are good reducing agents. The net reaction is one of oxidation–reduction. A hard base, like $OH^-$, is not capable of reacting with peroxides, except by proton removal, and a higher

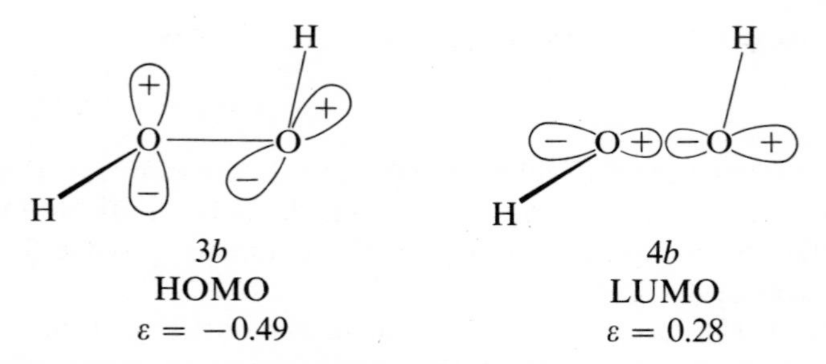


**Figure 24.** The HOMO and LUMO of hydrogen peroxide, $C_2$ point group. [energies from W. H. Fink and L. C. Allen, *J. Chem. Phys.*, **46**, 2261 (1967)].

energy orbital $6a$, must be used to break a oxygen—hydrogen bond. This is not serious since the interaction between a hard base and an electrophilic center is mainly electrostatic in nature.[140] There is little transfer of electron density to an orbital on the electrophile.

## Reactions of Olefins

The addition of nucleophiles, electrophiles, and free radicals to the double bond of olefins is of great importance in organic chemistry. The HOMO of ethylene is, of course, the $\pi$ orbital of $b_{2u}$ symmetry, and the LUMO is the $\pi^*$ orbital of $b_{3g}$ symmetry in $D_{2h}$. Addition of a proton to the $\pi$ orbital can occur symmetrically ($\pi$ bonding) or asymmetrically ($\sigma$ bonding).

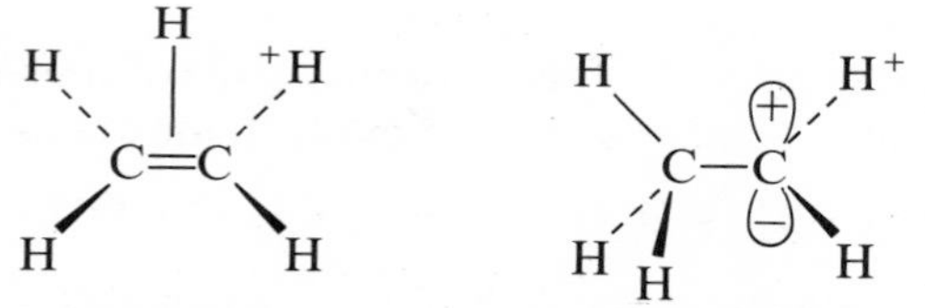

$\pi$ complex $\qquad$ $\sigma$ complex

The former has $C_{2v}$ symmetry, and the latter has only $C_s$. These point groups are maintained for the LM reaction paths. Note that the $\sigma$ complex is the ethyl carbenium ion. It is shown in the stable bisected form which is reached directly by a $C_s$ path.

At larger distances of separation, the $\pi$ complex is more stable; at shorter distances the carbenium ion is more stable, but only by 3–4 kcal.[141] A shift of the proton from one carbon to the other should be facile (see p. 95). Also the carbenium ion is the precursor to electrophilic substitution. Since the carbon—carbon bond is now single, rotation is easy and isomerization can occur. It is convenient to use Newman projections to visualize the stereochemical results.

We start with a *trans* olefin with an easily replaced group, for example,

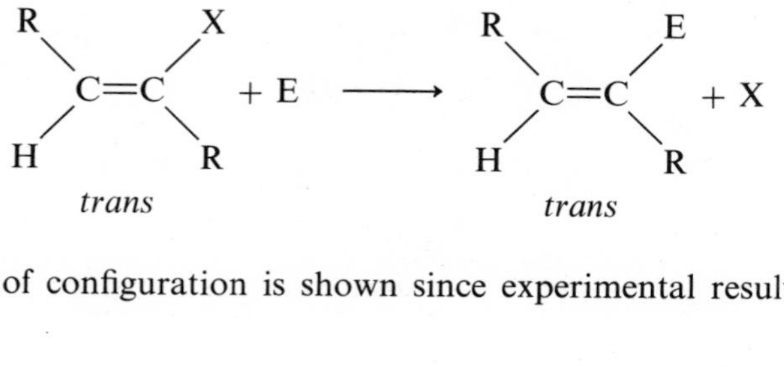

$$\underset{trans}{\text{R(H)C=C(X)R}} + \text{E} \longrightarrow \underset{trans}{\text{R(H)C=C(E)R}} + \text{X} \qquad (171)$$

Retention of configuration is shown since experimental results show that this is the most common result in electrophilic substitution on olefins.[142]

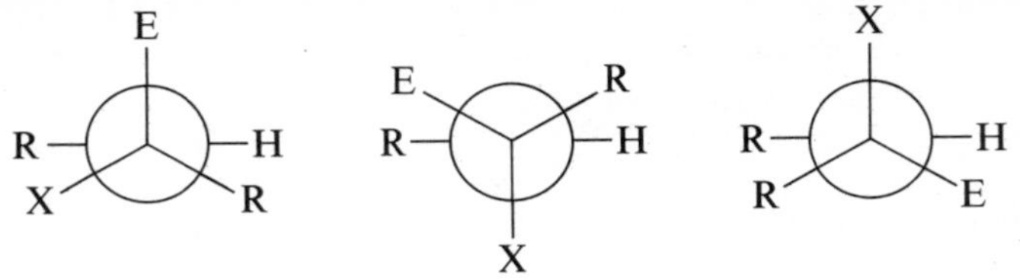

A B C

A is the projection formula for the initially formed $\sigma$ complex, B is the rotamer that gives retention of configuration, and C is the rotamer giving *trans–cis* isomerization. It can be seen that B requires a rotation of only 60° about the C—C bond axis, and C requires 120°. Accordingly retention is favored. However the relative stabilities of the rotamers and the isomeric olefins could also determine which product is formed.

Another possible reaction of the carbenium ion, of course, is addition of a nucleophile to form a stable, saturated product, for example,

$$CH_3CH_2^+ + H_2O \longrightarrow CH_3CH_2OH + H^+ \quad (172)$$

This kind of reaction could also occur for the $\pi$ complex. Especially with the halonium ions as electrophiles, the $\pi$ complex is known to be more stable than the $\sigma$ complex.[143] It may be considered as a three-membered ring compound, isoelectronic with ethylene oxide. The nucleophile opens the ring by backside attack on carbon.[144]

$$H_2C(Br^+)CH_2 + H_2O \longrightarrow BrCH_2CH_2OH + H^+ \quad (173)$$

In this case addition of the electrophile, $Br^+$, and the nucleophile, $OH^-$, occurs with a *trans* positioning of the two groups in the product. The reaction is stereospecific because the bromine ion holds the configuration in the intermediate. If the carbenium ion reacts rapidly enough with a nucleophile, there will also be *trans* addition. That is, rotamer A will add a group *trans* to the added proton to form the stable staggered ethane derivative. The opposite direction of additions forms the eclipsed conformation.

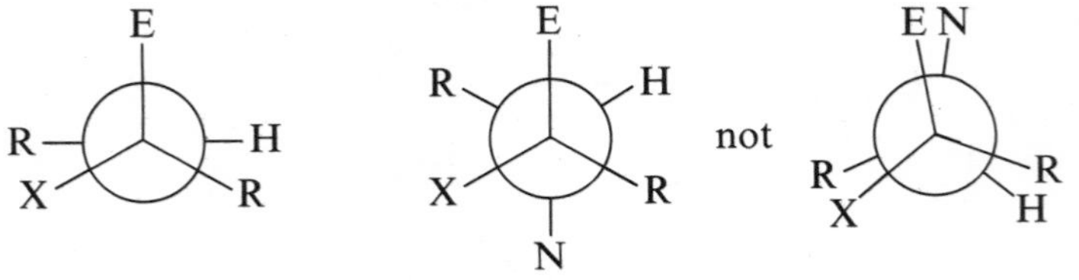

not (174)

*trans* *cis*

Addition of a nucleophile to the $\pi^*$ orbital of an olefin must occur to one carbon atom only. A carbanion is formed from an anionic nucleophile. The arrangement of groups around both carbon atoms is pyramidal, with an electron pair occupying one site.

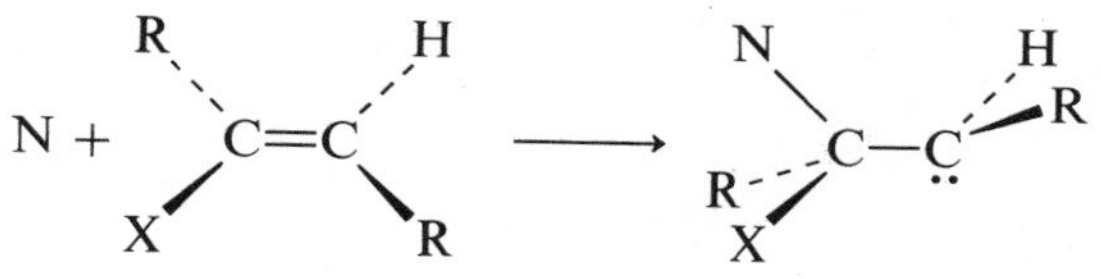

(175)

This electron pair must be *trans* to the added nucleophile, in order to obtain a staggered form. Now there are two ways in which structural changes can occur: (1) rotation about the C—C bond and (2) inversion at the anionic carbon.

Rotational barriers for simple molecules are in the range of 2–5 kcal, and the energy barriers for inversion of amines and carbanions are also in the 4–6 kcal range.[145] Actually the two processes are coupled together and described by a single potential energy surface.[146] For example, a 60° rotation plus an inversion converts D to E

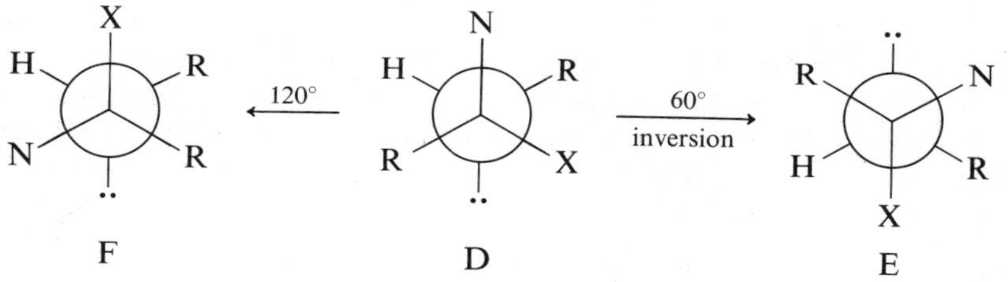

(176)

A rotation of 120°, without inversion, converts D to F, which are rotational isomers, or rotamers. Now both E and F are in a position to lose a group such as X, if it is weakly held. Loss of X from E will form an olefin with the R groups *trans*, in other words, with retention of the original olefin geometry. If X is lost from rotamer F, the olefin will have the R groups *cis*, and nucleophilic substitution with isomerization will have occurred. Other combinations of inversion and rotation can give isomerization without substitution.[142]

The experimental results on nucleophilic substitution at vinylic halides show that *retention* of configuration is by far the most common result.[147] Apparently the D → E conversion is easier than D → F, although it is by no means clear that this result could have been predicted.

Note that the inversion step is necessary to get E. No combination of rotations alone will convert D to E. However E could be formed from the *cis* isomer of the olefin. The experimental observations also rule out another mechanism that might be imagined. This is an approach of the nucleophile in the plane of the olefin molecule.

$$\mathrm{N} + \begin{matrix} \mathrm{R} & & \mathrm{X} \\ & \diagdown\ \diagup & \\ & \mathrm{C} & \\ & \| & \\ & \mathrm{C} & \\ & \diagup\ \diagdown & \\ \mathrm{H} & & \mathrm{R} \end{matrix} \longrightarrow \begin{matrix} & \mathrm{R} & \\ & | & \\ \mathrm{N}- & \mathrm{C} & -\mathrm{X} \\ & \| & \\ & \mathrm{C} & \\ & \diagup\ \diagdown & \\ \mathrm{H} & & \mathrm{R} \end{matrix} \longrightarrow \begin{matrix} \mathrm{N} & & \mathrm{R} \\ & \diagdown\ \diagup & \\ & \mathrm{C} & \\ & \| & \\ & \mathrm{C} & \\ & \diagup\ \diagdown & \\ \mathrm{H} & & \mathrm{R} \end{matrix} + \mathrm{X} \quad (177)$$

This mechanism, which must use a high $\sigma^*$ orbital ($b_{2u}$ on p. 215) is very unfavorable energetically.[148]

Besides nucleophilic substitution, the carbanion formed by addition of a nucleophile to an olefin can also react with an electrophile, for example, a proton from the solvent

$$RSCH_2CH_2^- + H_2O \longrightarrow RSCH_2CH_3 + OH^- \quad (178)$$

The stereochemistry of overall addition depends on which form reacts. The initially formed D would give *trans* addition. So would F and any other rotamer formed from D by rotation only. However E would give the diastereomeric product. This would correspond to *cis* addition of the electrophile and nucleophile. Other rotamers of E would also give *cis* addition.

The experimental results on addition reactions of olefins show that *trans* addition is overwhelmingly preferred. One can also look at the reverse reaction, which is the formation of an olefin by loss of two atoms or groups from a saturated molecule, in solution.

$$C_2H_4XY \longrightarrow C_2H_4 + XY \quad (179)$$

The results here also are that *anti* or *trans* loss of the two groups, Y and X in (179), almost invariably occurs. The elimination reaction, such as (179), has been subject to a great deal of study and much discussion as to mechanisms.[149]

The consensus is that a range of mechanisms exist from extremes where the electrophilic group is lost first (usually by base catalysis), followed by a loss of the nucleophilic group, to cases where the nucleophile is lost first, followed by loss of the electrophile. In between are reactions where both groups are lost in a synchronous, or near synchronous, fashion. By applying the PMR, the previous discussions show that nearly synchronous elimination reactions should be *anti*. Limiting mechanisms in which free carbanions or carbenium ions are formed must give products determined by rotamer or isomer stability.

The meaning of nearly synchronous implies only that both groups leave within a time too short for rotation or inversion. An additional feature can be deduced for a truly synchronous, or concerted reaction leading to *syn* elimination. Such reactions are partly forbidden by orbital symmetry (see p. 100), although they do occur in the gas phase.

$$\begin{array}{cc} CH_2-CH_2 \\ | \quad\quad | \\ H \quad\quad Cl \end{array} \longrightarrow \begin{array}{c} CH_2=CH_2 \\ \\ H-Cl \end{array} \quad (180)$$

If two groups are displaced in a *cis* fashion, it is not possible to prevent their interaction with each other. Accordingly they leave as a single molecule, rather than as two discrete particles.

The addition of free atoms and radicals to olefins is also an important reaction.

$$R\cdot + CH_2=CH_2 \rightleftharpoons RCH_2-CH_2\cdot \quad (181)$$

The radical product can readily invert at the free-radical center. Hence the addition of an atom or radical to an olefin, followed by its loss, can be a mechanism for isomerization of olefins.[150] The product of reaction 181 can also abstract H atom or Cl atom, or dimerize, so that overall two radicals have been added to an olefin. The lifetime of the intermediate radical and its stereostability determine the stereochemical results.

Alkyl radicals or hydrogen atoms can only act effectively as electron donors. Hence they add to the $\pi^*$ orbital, combining with a single carbon atom to form $\sigma$-bonded radicals. Electronegative radicals, such as Br· or RO·, can also act as electron acceptors. It is likely that they form $\pi$ complexes.[151] Such three-membered ring radicals would be stereoretentive, somewhat like (173).

The reactions of sulfur and oxygen atoms with olefins have received much attention. The ground state of these atoms is $^3P$, and a $^1D$ state lies about 1 eV higher. The singlet states can react directly with an olefin by a LM path with $C_{2v}$ symmetry to form ground-state ethylene oxide or sulfide.[152] The reactions would be stereospecific for substituted olefins. The orbital interactions and the symmetries of the bonds and electron pairs are shown in Fig. 25.

The $^1D$ state has the configuration $(p_x)^2(p_y)^2(p_z)^0$, so that the empty $p_z$ orbital can accept electron density from the filled $\pi$ orbital of the olefin. To a lesser degree, because of electronegativity differences, the $p_y$ orbital can donate electrons to the empty $\pi^*$ orbital. The $^3P$ state can have the configuration $(p_x)^1(p_y)^2(p_z)^1$, as well as $(p_x)^2(p_y)^1(p_z)^1$, with the latter more favorable for reaction. Figure 25 shows that the product would be an excited state of ethylene oxide or sulfide, the $^3B_1$. In familiar language, the reaction is forbidden by both spin and state symmetry. Figure 26 shows the state correlation diagram. The $S_0$–$T_1$ and $S_0$–$S_1$ separations for the product are estimated from the uv absorption spectra. They are, of course, energies at the equilibrium geometry of the ground state, because of the Franck–Condon effect.

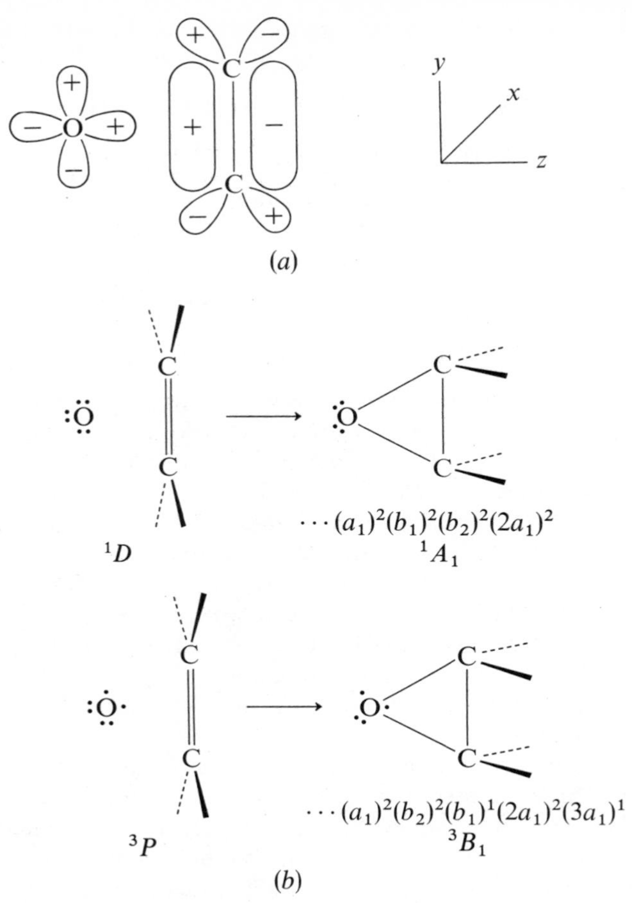


**Figure 25.** (*a*) Chief orbital interactions in reaction of an oxygen atom with ethylene, assuming a $C_{2v}$ point group; the $\pi$ and $p_z$ and $\pi^*$ and $p_y$ orbitals all have positive overlaps; (*b*) orbital occupancies in reactions of $^1D$ and $^3P$ oxygen atoms with ethylene to form $C_2H_4O$.

In spite of the forbiddenness, the addition of triplet O and S atoms to olefins does take place readily. The products are ground singlet state molecules and the reactions are highly stereospecific for sulfur and less so for oxygen. The explanation lies in the distorted equilibrium geometry of the $^3B_1$ state.[152] This may be represented as

$$H_2C\!-\!CH_2S \quad \text{or} \quad H_2\dot{C}\!-\!CH_2\!-\!\dot{S}$$

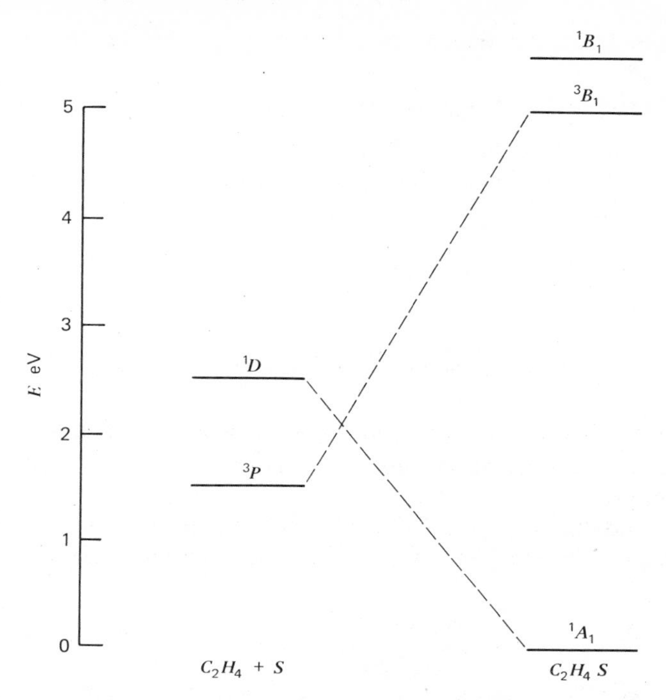


**Figure 26.** Correlation diagram for some low-lying states of $C_2H_4$ + S and $C_2H_4S$; energies estimated from thermochemical and spectral data, $C_2H_4S$ set at the zero of energy. [Reprinted with permission from O. P. Strausz, H. E. Gunning, A. S. Dener, and I. G. Csizmadia, *J. Am. Chem. Soc.* **94**, 8318 (1972). Copyright by the American Chemical Society.]

with a C—C—S bond angle of 110° and one very long C—S bond. In the limit this can also be considered as an open chain biradical. The long bond holds the configuration, accounting for stereoselectivity. Oxygen is not as effective as sulfur in maintaining this bond.

The sequence of events in that the $^3B_1$ state is formed by a $C_s$ reaction coordinate at its equilibrium geometry. This puts it near the crossover point shown in Fig. 26. An intersystem crossing occurs, induced by molecular collision. The $^1A_1$ state is formed in a vibrationally excited state, which then decays to the right geometry by internal conversion. The reason for the distorted shape of the $^3B_1$ state will be discussed later (p. 482).

The addition of carbenes and nitrenes to olefinic bonds is also important. Both $CH_2$ and NH are isoelectronic with the O or S atom, and have triplet ground states, plus a low-lying singlet state. Their addition reactions follow Skell's rules:[153]

1. Addition of the singlet state occurs in a concerted manner, and is stereospecific.
2. Addition of the triplet state occurs stepwise, and is nonstereospecific.

These rules follow from our previous analysis, with two amendments.[154] The triplet state must form a biradical, rather than a distorted, excited state of the product.

$$C_2H_4 + {}^3B_1\,CH_2 \longrightarrow \cdot CH_2CH_2CH_2\cdot \tag{182}$$

or

$$C_2H_4 + {}^3\Sigma\,NH \longrightarrow \cdot NHCH_2CH_2\cdot$$

This accounts for the loss of stereochemistry in the product. Reaction (182) is then followed by ring closure to form ground-state cyclopropane, or ethyleneimine.

The second change in mechanism has to do with the reaction coordinate for the singlet reaction. The LM $C_{2v}$ approach is forbidden by symmetry.[154]

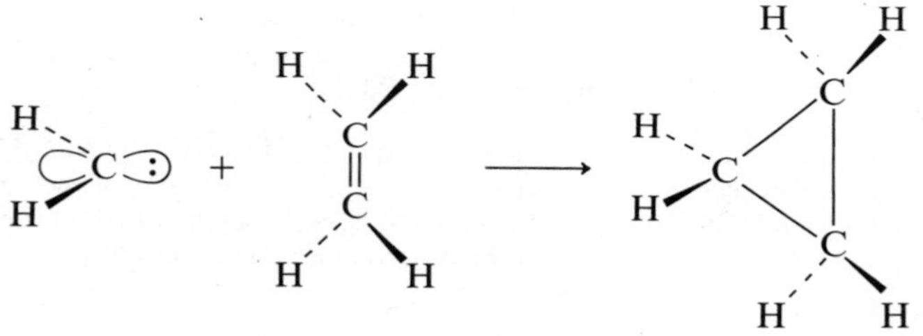

$$a_1 \quad + \quad a_1 \longrightarrow a_1 + b_1 \tag{183}$$

Consequently a $C_s$ reaction path is of lower energy, in which the carbene approaches first in a tipped orientation.

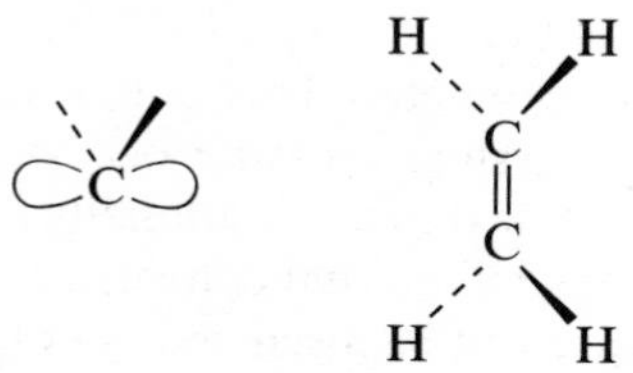

This allows overlap between the filled $\pi$ orbital of the olefin and the empty $b_1$ orbital of the carbene. At closer distances the $CH_2$ fragment tips back toward its final equilibrium position.

## Other $X_2Y_n$ Molecules

Replacing two carbon atoms in ethylene by nitrogen gives hydrazine, $N_2H_4$. The orbitals of this molecule that are the reactive ones for electrophiles and nucleophiles are very similar to those for $H_2O_2$. The HOMO is an anti-bonding combination of two *p* orbitals on the two nitrogen atoms (Fig. 24). A proton would add to give $NH_2NH_3^+$. Removal of electron density from the HOMO would strengthen the N—N bond. A nucleophile would add to a $\sigma^*$ orbital that is antibonding for the N—N bond.

The ensuing reaction for nucleophilic attack would be similar to reaction (170). The N—N bond would break and products corresponding to various redox reactions would form. Protonation prior to nucleophilic attack would be helpful. The N—N bond would be only slightly strengthened because the proton does not remove much electron density. However electron donation to the $\sigma^*$ orbital would be greatly facilitated. Thus $NH_3$ is a better leaving group than $NH_2^-$ in nucleophilic displacement.

The HOMO and LUMO of diborane, $B_2H_6$, are shown elsewhere (p. 222). The HOMO is strongly B—H bonding and weakly B—B anti-bonding. A proton would attack this orbital, perhaps at the midpoint of one B—H, or perhaps in a colinear fashion, B—H---$H^+$. Hydrogen evolution would result, but probably not until a nucleophile intervenes. The LUMO is more strongly anti-bonding in the B—B bond. It is essentially the $\pi^*$ orbital of ethylene.

A nucleophile adding to $B_2H_6$ would bond strongly to one boron atom, and promote a breaking of the boron—boron linkage. One of the bridging H atoms would also be repelled. For a hydride ion nucleophile, an intermediate $B_2H_7^-$ might well be formed,

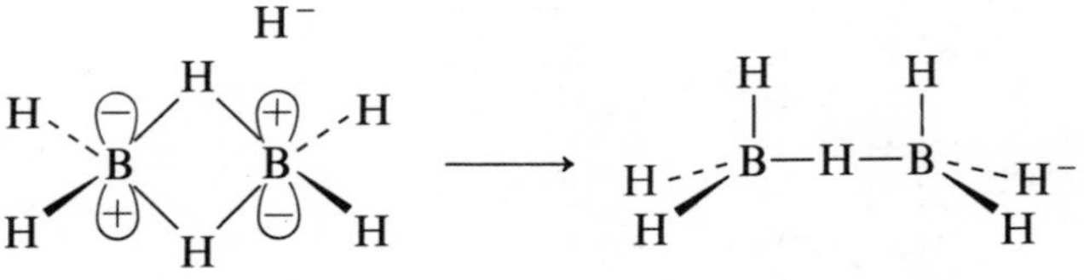

$$H^- + B_2H_6 \longrightarrow H_3B\text{—}H\text{—}BH_3^- \qquad (184)$$

Such an ion does exist and has the structure shown, or a similar one with staggered B—H bonds.[155] $B_2H_6$ is rapidly hydrolyzed to $B(OH)_3$ and $H_2$ and reacts with various HX molecules to give $BX_3$ plus $H_2$. With neutral donor molecules, such as ethers, $BH_3 \cdot$ ether adducts are formed.

The frontier orbitals of $C_2H_6$ are shown elsewhere (p. 220). The sequence is

$$\ldots(3a_{1g})^2(e_g)^4(a_{2u})^0(e_u)^0$$

The $3a_{1g}$ orbital is strongly C—C bonding, while the $e_g$ is C—H bonding.

Their energies are virtually the same. Actually ethane reacts with strong acid to give primarily cleavage of the carbon—carbon bond.[156]

$$C_2H_6 + H^+ \longrightarrow CH_3^+ + CH_4 \qquad (185)$$

The intermediate, $C_2H_7^+$, like $CH_5^+$, is stable in the gas phase.

The balance is close enough between the $3a_{1g}$ orbital and $e_g$ so that other paraffin hydrocarbons also react with the proton at both the C—C and C—H bonds. Also a variety of other strong electrophiles, such as $NO_2^+$ and $CH_3^+$, will react with saturated carbon—carbon linkages.[156] The reaction coordinate could either be one that leads to attack at the midpoint of the bond, effective $C_{2v}$ approach, or a backside approach to one carbon atom, $C_{3v}$ like symmetry. The former mechanism would lead to retention of configuration of the product and the latter, to inversion.

Nucleophilic attack on ethane should utilize the empty $a_{2u}$ orbital. This is strongly C—C antibonding and leads to breaking this bond. The reaction coordinate would have effective $C_{3v}$ symmetry, as in other reactions of $CH_4$ derivatives.

$$H^- + C_2H_6 \longrightarrow CH_4 + CH_3^- \qquad \Delta H = -37 \text{ kcal} \qquad (186)$$

This would be a classical inversion process. An alternative would be for the nucleophile to use the empty $e_u$ orbital, which is C—H antibonding

$$H^- + C_2H_6 \longrightarrow C_2H_6 + H^- \qquad \Delta H = 0 \qquad (187)$$

**Table 6**

| Reactants | Product | Point Group Maintained |
|---|---|---|
| $C_2H_2 + H^+$ | $C_2H_3^+$ $C_{2v}$ | $C_s$ or $C_{2v}$ |
| $C_2H_2 + H^-$ | $C_2H_3^-$ $C_s$ | $C_s$ |
| $N_2H_2 + H^+$ | $N_2H_3^+$ $C_s$ | $C_s$ |
| $N_2H_2 + H^-$ | $N_2H_3^-$ $C_1$ | $C_1$ |
| $H_2O_2 + H^+$ | $H_3O_2^+$ $C_1$ | $C_1$ |
| $H_2O_2 + H^-$ | $H_3O_2^-$ $C_1$ | $C_1$ |
| $C_2H_4 + H^+$ | $C_2H_5^+$ $C_s$ or $C_{2v}$ | $C_s$ or $C_{2v}$ |
| $C_2H_4 + H^-$ | $C_2H_5^-$ $C_s$ | $C_s$ |
| $N_2H_4 + H^+$ | $N_2H_5^+$ $C_s$ | $C_1$ |
| $N_2H_4 + H^-$ | $N_2H_5^-$ $C_s$ | $C_1$ |
| $B_2H_6 + H^+$ | $B_2H_7^+$ $C_s$ | $C_s$ |
| $B_2H_6 + H^-$ | $B_2H_7$ $D_{3h}$ or $D_{3d}$ | $C_s$ |
| $C_2H_6 + H^+$ | $C_2H_7^+$ $C_{2v}$ or $C_{3v}$? | $C_{2v}$ or $C_{3v}$? |
| $C_2H_6 + H^-$ | $C_2H_7^-$ $C_{3v}$ | $C_{3v}$ |

The fact that the $a_{2u}$ orbital is much more stable than the $e_u$ orbital is shown by the exothermicity of (174) (in the gas phase), compared to the thermoneutrality of (187). With a substituted ethane, such as $C_2H_5Br$, it is the reaction analogous to (187) that occurs. Cleavage of carbon—carbon bonds by nucleophiles, as in (186), is a rare event. Only strongly stabilized carbanions, such as $C(NO_2)_3^-$, could be displaced. Table 6 summarizes the conclusions on reactions of $X_2Y_n$ molecules.

## SOME MISCELLANEOUS ADDITION REACTIONS

Many other unsaturated organic molecules add electrophiles or nucleophiles. Several of these are important prototype reactions. The formaldehyde molecule serves as a model for reactions of the carbonyl group. The HOMO of $CH_2O$ is the so called $n$ orbital of $b_2$ symmetry. This is largely a lone pair on oxygen, hence the label $n$ for nonbonding. However the orbital is weakly antibonding for the C—O bond. The LUMO is the $\pi^*$ orbital of $b_1$ species (see p. 224).

Addition of an electrophile is to the $n$ orbital to give an oxycarbenium ion.[157]

$$H_2C{=}\ddot{O} \quad H^+ \longrightarrow H_2C{=}O{-}H^+ \qquad (188)$$

The reaction coordinate is $C_s$, and the product has a plane of symmetry. Many other Lewis acids besides the proton will add to oxygen in a similar fashion. The addition of weak nucleophiles to the carbonyl group is enhanced by such interactions.

Addition of a nucleophile will be to the $\pi^*$ orbital and specifically to the carbon atom. This is favored by two factors: the $\pi^*$ orbital has its maximum amplitude on carbon rather than oxygen, the reverse of the $\pi$ orbital case. The polarity of the C—O bond favors addition of a negative charge, or dipole, to the carbon end.

$$N^- + H_2C{-}O \longrightarrow N{-}CH_2{-}O^- \qquad (189)$$

The reaction coordinate will be in the plane bisecting the H—C—H bond angle. As N approaches, the two H atoms will drop below the original plane.

While maximum overlap would suggest an initial approach nearly perpendicular to the molecular plane, the polarity of the molecule favors a greater separation between N (the nucleophile) and O. An ab initio calculation of the reaction path for

$$H^- + CH_2O \longrightarrow CH_3O^- \tag{190}$$

is very informative.[158] Figure 27 shows the coordinates used, and Table 7 shows how they vary, together with the calculated energy.

**Table 7 Structural Parameters and Total Energies for the $H^-$ + $H_2CO$ Reaction**[a]

| | $d(H^- \cdots C)$ Å | $\alpha(H^- \cdots CO)$ deg | $\Delta$, Å | $r$(CO), Å | $E$, AU |
|---|---|---|---|---|---|
| A | 3.0 | 180 | 0.00 | 1.203 | −114.0334 |
| B | 2.5 | 126 | 0.00 | 1.22 | −144.0643 |
| C | 2.0 | 119 | 0.03 | 1.23 | −114.0672 |
| D | 1.5 | 118 | 0.13 | 1.25 | −114.0760 |
| E | 1.12 | 109.5 | 0.40 | 1.405 | −114.1106 |

[a] Reprinted with permission from H. B. Burgi, J. M. Lehn, and G. Wipff, **96**, 1956 (1974). Copyright by the American Chemical Society.

The initial direction of approach is in the molecular plane ($180° = \alpha$). This is dictated entirely by classical ion–dipole forces, since orbital overlap is zero. At about 2.5 Å, covalency effects begin to take over; the angle $\alpha$ approaches its final value, which is the tetrahedral one, the C—O distance changes from that of a double bond to that of a single bond, and the methylene hydrogens drop below the plane. The energy drops steadily, and no barrier exists.

The reversal of (189) would lead to nucleophilic substitution. This would be very marked, of course, for a substituted molecule which had a good leaving

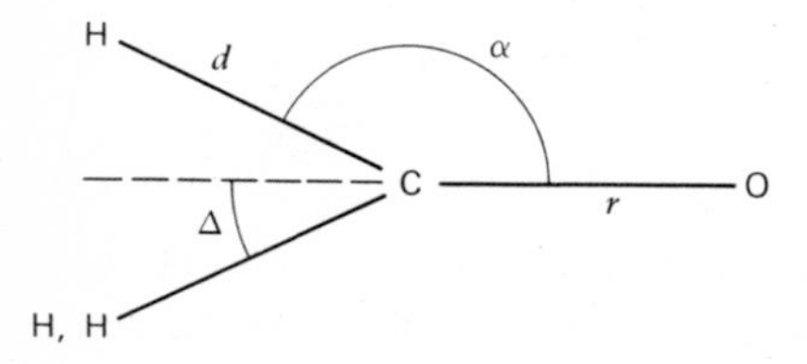


**Figure 27.** Coordinates used in reaction $CH_2O + H^- \rightarrow CH_3O^-$; $d$ is C—H distance, where H is the hydride ion; $s$ is C—O distance; $\alpha$ is angle of approach of $H^-$; $\Delta$ is deviation of two H atoms below the original plane (from reference 158).

group, such as HCOCl, or $HCOOC_2H_5$. Because of the polarity of the carbonyl group, hard bases such as $OH^-$, $F^-$, and $NH_3$ will be good nucleophiles. This is in contrast to nucleophilic displacement on saturated carbon compounds, or on vinylic halides. Here soft bases, such as $RS^-$, $I^-$, and $PR_3$, are the best nucleophiles, particularly in protic solvents.

### Aromatic Substitution

Benzene serves as a model for aromatic substitution reactions. The relevant orbitals are shown in Fig. 28. They are the doubly degenerate $e_{1g}$ pair, which are occupied, and the degenerate $e_{2u}$ pair, which are empty. The shape of $e_{1g}$ suggests addition of an electrophile to a single carbon atom, where the amplitude of the wavefunction is a maximum. The $e'_{1g}$ orbital suggests that a $\pi$ complex centered above two carbon atoms might be more stable.

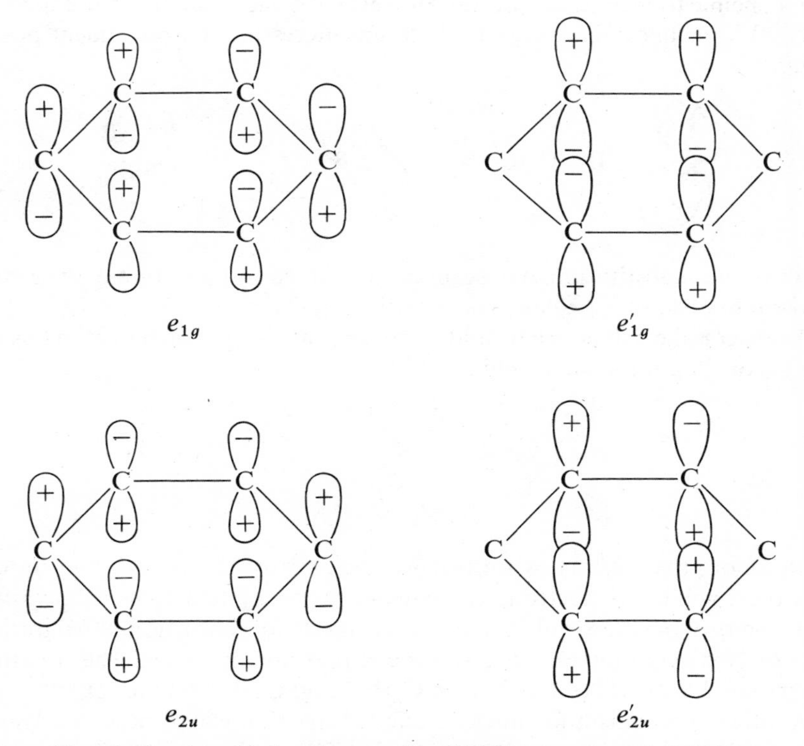


**Figure 28.** The frontier orbitals of benzene. The HOMOs are $e_{1g}$ and the LUMOs, $e_{2u}$.

Ab initio calculations show that the $\sigma$-bonded form is the most stable.[159]

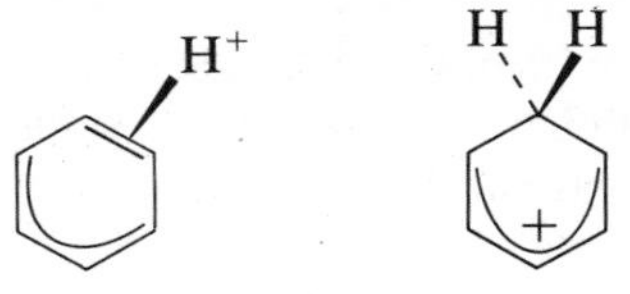

$\pi$ complex $\qquad$ $\sigma$ complex

This ion is called the "benzenium ion."[157b] The structure has $C_{2v}$ symmetry and the reaction coordinate would maintain $C_s$ symmetry during formation. The ion may be considered as a tetrahedral carbon bonded to a resonance stabilized five-carbon system with four $\pi$ electrons. Naively, one might expect that the benzenium ion would be hard to form, because of the loss of the resonance energy of benzene. In fact the proton affinity of benzene is very large, 183 kcal, which is somewhat greater than that of ethylene.[160]

Electrophilic substitution will result if the proton originally on benzene is lost from the benzenium ion. Because of the planar structure of benzene, no stereochemical change can occur. However the fact that the $\pi$ complex is only 20 kcal higher in energy leads to an interesting rearrangement possibility.[159]

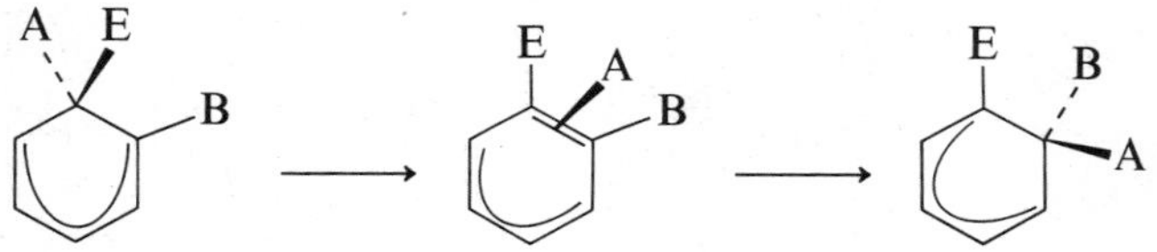

(191)

Electrophilic substitution may occur accompanied by migration of the group in the *ipso* position[161] to an *ortho* position.

A nucleophile will normally add to benzene at a single carbon atom, using the $e_{2u}$ orbital. If the nucleophile

is an anion, the $\sigma$ complex formed will be negatively charged, and in any case there will be six electrons in the five-carbon $\pi$ system. Loss of the group A at the *ipso* position will be the normal mode for substitution. Migration of A or N is very unlikely because a $\pi$ complex does not exist. The situation is very similar to that of 1,2-shifts in $C_2H_5^+$ and $C_2H_5^-$ (see p. 95).

Another mechanism for nucleophilic substitution exists, however, which does lead to anomalous positional results. This is the benzyne mechanism, discovered by Roberts in 1953.[162] A base, especially a strong, hard one such as $NH_2^-$, can attack an empty $\sigma^*$ orbital that is C—H antibonding. Proton-removal results, which may be followed by the loss of a weakly held group, X, in the *ortho* position.

$$C_6H_4(H)(X) \xrightarrow{-HX} C_6H_4 \text{ (benzyne)} \xrightarrow{+HY} C_6H_4(Y)(H) \qquad (192)$$

The highly reactive intermediate formed is called "benzyne." It has an additional $\pi$ bond lying in the molecular plane. Addition of a group, Y, followed by proton addition results in nucleophilic substitution, but at a switched position.

This unlikely looking mechanism takes place because ordinary nucleophilic substitution at a molecule like chlorobenzene is very slow. While electrophiles add readily even to benzene itself, nucleophiles will not add unless strong activating groups, such as $NO_2$ or CN, are present. This results from the fact that an unstable, antibonding orbital must be used, as well as from the low electronegativity of carbon and hydrogen. A similar situation applies to ethylene, which adds electrophiles much more readily than nucleophiles. In acetylene, however, the tendency to add bases is greater, because carbon has a higher electronegativity as a result of relatively more use of its 2*s* orbital in bonding.

If there are good activating groups such as $NO_2$, and no good leaving group, then stable ions or molecules may form as a result of addition of nucleophiles to unsaturated systems. The classic example are the Meisenheimer complexes[163]

$$C_6H_2(OCH_3)(NO_2)_3 + C_2H_5O^- \longrightarrow [C_6H_2(CH_3O)(OC_2H_5)(NO_2)_3]^- \qquad (193)$$

The stability in these cases arises from the nitro groups, which can further stabilize the already stabilized $\pi$ carbanion. Similar products should be formed from nitroolefins and nitroacetylenes.

Unsaturated benzene and ethylene are normally much better bases than they are Lewis acids. Exceptions can occur in reactions with molecules that can function as $\pi$ bases. The typical examples are transition-metal ions of low-oxidation state and $d^8$ or $d^{10}$ configuration (soft Lewis acids). Bonding to ethylene or benzene usually is a mixture of $\sigma$ bonding, in which the metal

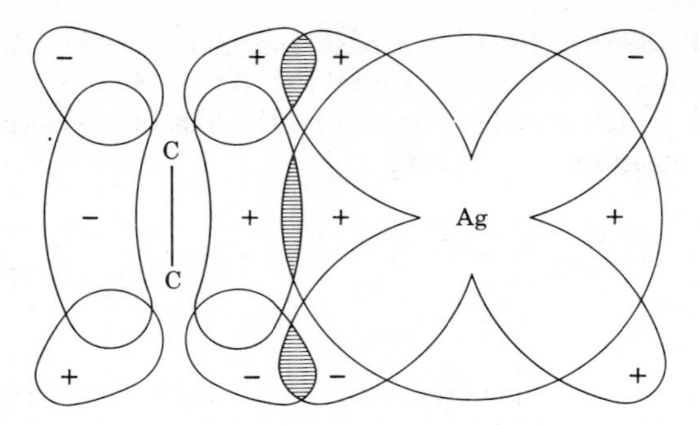


**Figure 29.** The Chatt–Dewar model for bonding of silver ion and an olefin molecule; there is donation from the $\pi$ orbital of ethylene to the $5s$ orbital of $Ag^+$, and back donation from the $4d$ orbital of $Ag^+$ to the $\pi^*$ orbital of the olefin.

atom acts as a Lewis acid, and $\pi$ bonding, in which the metal atom acts as a base. Figure 29 shows the classical Chatt–Dewar model for bonding in metal-olefin complexes.

For ethylene or benzene, the $\sigma$ interaction is more important than the $\pi$ interaction. Electronegative substituents on the organic molecule, for example, tetracyanoethylene, can reverse this situation. Even for the unsubstituted molecules the importance of $\pi$ bonding is shown by the structures of the complexes. Normally these have the $\pi$-complex structures, with the metal atom centered above two carbon atoms. Sometimes the $\pi$ complex is an intermediate en route to the formation of a more stable $\sigma$ complex.

$$H_2C{=}CH_2\text{—}Hg^{2+} + H_2O \longrightarrow {}^{+}Hg\text{—}CH_2\text{—}CH_2\text{—}OH + H^+ \qquad (194)$$

## Free-Radical Addition

Free radicals can react with benzene either by hydrogen atom abstraction in a direct process, or by addition to the benzene ring, followed by loss of a hydrogen atom.

$$C_6H_6 + R\cdot \longrightarrow C_6H_5\cdot + RH \qquad (195)$$

$$C_6H_6 + R\cdot \longrightarrow C_6H_6R\cdot \longrightarrow C_6H_5R + H\cdot \qquad (196)$$

This was first pointed out by Hey in 1934, who studied cases where R· is an aryl radical and where the addition mechanism is preferred. Rates of reactions such as (195) are less than for abstraction reactions of aliphatic hydrocarbons.

Orbital considerations suggest that addition to the benzene ring should be easier than abstraction. Interaction would be with both the $e_{1g}$ and $e_{2u}$ orbitals. The adduct would be $\sigma$-bonded, with five electrons in the five-carbon $\pi$ systems.

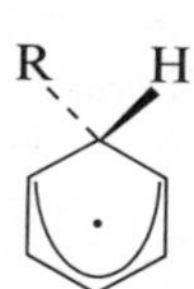


Note that with a reaction coordinate that maintains a plane of symmetry, there is no problem taking an electron from a $\sigma$ orbital and placing it in a $\pi$ orbital. The odd electron appears in an orbital of $a'$ species, which is the same symmetry as its initial orbital.

The species $C_6H_6R\cdot$ would be a stabilized free radical. It would not abstract atoms from stable molecules, but would probably lose a hydrogen atom to other free radicals in the system, rather than lose an H· directly as shown in (196). Another possibility is that (195) is not direct, but occurs by the decomposition of the intermediate.

$$C_6H_6R\cdot \longrightarrow C_6H_5\cdot + RH \qquad (197)$$

In this case there is a symmetry problem, since the electron is initially in an orbital of $b_1$ symmetry (when R = H) and must appear in an orbital of $a_1$ species. The assumption is that an LM path of $C_{2v}$ point group is followed.

As indicated earlier free radicals do not normally attack a saturated carbon atom, as in (141). However it is known that certain free atoms or radicals will readily attack the metal atoms of organometallic compounds.[165] The radicals are those where the odd electron is on an electronegative atom, though some examples of alkyl radicals reacting are known. The net reaction is a bimolecular homolytic substitution, or $S_H2$ reaction.

$$X\cdot + MR_n \longrightarrow XMR_n \longrightarrow XMR_{n-1} + R\cdot \qquad (198)$$

The metal M can be Li, Zn, Cd, Hg, B, Al, Ga, Sn, P, As, Sb, or Bi. Free-radical substitution on a P(III) system has been shown to occur with inversion of configuration.[166] This means the entering and leaving groups are both axial, or both equatorial.

## REFERENCES

1. F. O. Rice and E. Teller, *J. Chem. Phys.*, **6**, 489 (1938).
2. J. Franck and E. Rabinowitsch, *Z. Elektrochem.*, **36**, 794 (1930).
3. See S. I. Miller, *in* V. Gold (ed.), *Advances in Physical Organic Chemistry*, Vol. 6, Academic, New York, 1968, p. 185, for examples.
4. (a) J. Hine, *J. Org. Chem.*, **31**, 1236 (1966); *J. Am. Chem. Soc.*, **88**, 5525 (1966); (b) O. S. Tee, *J. Am. Chem. Soc.*, **91**, 7144 (1969); O. S. Tee and K. Yates, ibid., **94**, 3074 (1972); (c) S. Ehrenson, *J. Am. Chem. Soc.*, **96**, 3778, 3784 (1974).
5. H. F. Schaefer, III, *The Electronic Structure of Atoms and Molecules*, Addison-Wesley, Reading, Mass., 1972; J. W. McIver, Jr. and A. Komornicki, *J. Am. Chem. Soc.*, **94**, 2625 (1972).
6. R. Hoffmann, R. Gleiter, and F. B. Mallory, *J. Am. Chem. Soc.*, **92**, 1460 (1970).
7. J. W. McIver, Jr., *Acc. Chem. Res.*, **7**, 72 (1974).
8. J. R. Durig, B. M. Gimarc, and J. D. Odom, *Vibrational Spectra and Structures*, Vol. II, Dekker, New York, 1973.
9. H. Basch, *J. Chem. Phys.*, **55**, 1700 (1971).
10. J. P. Simons, *Nature*, **205**, 1308 (1965).
11. B. M. Gimarc, *J. Am. Chem. Soc.*, **92**, 266 (1970).
12. M. C. Lin and K. J. Laidler, *Can. J. Chem.*, **46**, 973 (1968).
13. J. L. Magee, W. Shand, Jr., and H. Eyring, *J. Am. Chem. Soc.*, **63**, 677 (1941).
14. R. B. Cundall, *Prog. React. Kinetics*, **2**, 167 (1964).
15. W. D. Stohrer and R. Hoffmann, *J. Am. Chem. Soc.*, **94**, 1661 (1972).
16. W. J. Hehre and P. v. R. Schleyer, *J. Am. Chem. Soc.*, **95**, 5837 (1973).
17. For a general discussion, see C. H. Bamford and C. F. H. Tipper (eds.), *Comprehensive Chemical Kinetics*, Elsevier, Amsterdam, 1972, Vols. 4 and 5.
18. D. F. Shriver and B. Swanson, *Inorg. Chem.*, **10**, 1354 (1971).
19. P. F. Alewood, P. M. Kazmeier, and A. Rauk, *J. Am. Chem. Soc.*, **95**, 5466 (1973).
20. R. A. Gangi and R. F. W. Bader, *Chem. Phys. Lett.*, **11**, 216 (1971).
21. C. F. Bender, H. F. Schaefer, III, D. R. Franceschetti, and L. C. Allen, *J. Am. Chem. Soc.*, **94**, 6888 (1972).
22. J. N. Murrell, J. B. Pedley, and S. Durmaz, *J. Chem. Soc.*, *Faraday Trans.*, **II**, 1370 (1973).
23. V. H. Dibeler and H. M. Rosenstock, *J. Chem. Phys.*, **39**, 1326 (1963).
24. J. P. Day, R. G. Pearson, and F. Basolo, *J. Am. Chem. Soc.*, **90**, 6933 (1968); M. Meier, F. Basolo, and R. G. Pearson, *Inorg. Chem.*, **8**, 795 (1969).
25. H. Selig, *in* V. Gutmann, (ed.), *Halogen Chemistry*, Academic, New York, 1967, Vol. 1, p. 407.
26. J. Jortner, E. G. Wilson, and S. A. Rice, *J. Am. Chem. Soc.*, **85**, 814, 815 (1963).
27. C. H. Langford and H. B. Gray, *Ligand Substitution Processes*, Benjamin, New York, 1965, Chapter 2.
28. F. Basolo and R. G. Pearson, *Mechanisms of Inorganic Reactions*, Wiley, New York, 1967, Chapter 2.

29. (a) R. D. Willett, *Theor. Chim. Acta*, **2**, 393 (1964). (b) R. D. Brown and J. B. Peel, *Austr. J. Chem.*, **21**, 2605, 2617 (1968); (c) T. Grillbro and F. Williams, *J. Am. Chem. Soc.*, **96**, 5032 (1974).

30. (a) J. B. Lambert, D. H. Johnson, R. G. Keske, and C. E. Mixan, *J. Am. Chem. Soc.*, **94**, 8172 (1972); (b) K. J. Synne, *Inorg. Chem.*, **10**, 1868 (1971).

31. K. Mislow, *Acc. Chem. Res.*, **3**, 321 (1970).

32. E. L. Muetterties, W. Mahler, and R. Schmutzler, *Inorg. Chem.*, **2**, 613 (1963).

33. R. G. Pearson and C. A. Grimes, *Inorg. Chem.*, **13**, 970 (1974).

34. N. Bartlett and F. P. Sladky, *J. Am. Chem. Soc.*, **90**, 5316 (1968); D. E. McKee, A. Zalkin, and N. Bartlett, *Inorg. Chem.*, **12**, 1713 (1973).

35. A. Dutta-Ahmed and E. A. Boudreaux, *Inorg. Chem.*, **12**, 1597 (1973).

36. P. George and D. S. McClure, *Progr. Inorg. Chem.*, **1**, 38 (1959).

37. J. Chatt and B. L. Shaw, *J. Chem. Soc.*, 705 (1959); D. P. M. Mingos, *Chem. Commun.*, 165 (1972).

38. R. G. Pearson and F. Basolo, *J. Am. Chem. Soc.*, **78**, 4878 (1956).

39. J. Fischer and R. K. Steunenberg, *J. Am. Chem. Soc.*, **79**, 1876 (1957).

40. R. A. Geanangel, *J. Inorg. Nucl. Chem.*, **34**, 1083 (1972).

41. W. N. Lipscomb, *Acc. Chem. Res.*, **6**, 257 (1973).

42. E. V. Waage and B. S. Rabinovitch, *Int. J. Chem. Kinet.*, **3**, 105 (1971).

43. W. L. Jorgensen and L. Salem, *The Organic Chemists' Book of Orbitals*, Academic, New York, 1973, p. 98.

44. K. Takagi and R. J. Crawford, *J. Am. Chem. Soc.*, **93**, 5910 (1971).

45. P. B. Shevlin and J. L. Greene, Jr., *J. Am. Chem. Soc.*, **94**, 8447 (1972).

46. J. P. Collman and W. R. Roper, *J. Am. Chem. Soc.*, **87**, 4008 (1965).

47. J. A. McGinnety, N. C. Payne, and J. A. Ibers, *J. Am. Chem. Soc.*, **91**, 6301 (1969).

48. For reviews see J. P. Collman, *Acc. Chem. Res.*, **1**, 136 (1968); L. Vaska, ibid., **1**, 335 (1968); J. Halpern, ibid., **3**, 386 (1970).

49. L. Stein, *in* V. Gutmann, (ed.), *Halogen Chemistry*, Academic, New York, 1967, Vol. 1, p. 133 ff.

50. E. Fluck, *Die Kernnagnetische Resonanz*, Springer–Verlag, Berlin, 1963, p. 136.

51. C. H. Bamford and C. F. H. Tipper, *Comprehensive Chemical Kinetics*, Elsevier, Amsterdam, 1972, Vol. 4, p. 91.

52. R. C. Dobson, D. M. Hayes, and R. Hoffmann, *J. Am. Chem. Soc.*, **93**, 6188 (1971).

53. W. Braun, A. M. Bass, and M. Pilling, *J. Chem. Phys.*, **52**, 5131 (1970).

54. D. Seyferth and Y. M. Chang, *J. Am. Chem. Soc.*, **95**, 6763 (1973).

55. C. K. Ingold, *Structures and Mechanisms in Organic Chemistry*, 2nd ed., Cornell U. P., Ithaca, N. Y., 1969, p. 518 ff.

56. H. Kwart and P. S. Strilko, *Chem. Commun.*, 767 (1967).

57. B. M. Trost, *Fortschr. Chem. Forsch.*, **41**, 1 (1973).

58. M. H. Ford-Smith, J. J. Habeck, and J. H. Rawthorne, *J. Chem. Soc. Dalton*, 2116 (1972).

59. A. Tamaki, S. A. Magennis, and J. K. Kochi, *J. Am. Chem. Soc.*, **95**, 6487 (1973).

60. H. E. Kluksdahl and G. H. Cady, *J. Am. Chem. Soc.*, **81**, 5285 (1959).

61. (a) F. H. Westheimer, *Acc. Chem. Res.*, **1**, 70 (1968); (b) R. F. Hudson and C. Brown, *Acc. Chem. Res.*, **5**, 204 (1972); (c) G. Aksnes, *Acta Chem. Scand.*, **20**, 2463 (1966).

62. D. B. Denney, D. Z. Denney, C. D. Hall, and K. L. Marsi, *J. Am. Chem. Soc.*, **94**, 245 (1972).

63. R. F. Hudson, *Structure and Mechanism in Organo-Phosphorus Chemistry*, Academic, New York, 1965, pp. 174, 205.

64. A. J. Kirby and S. G. Warren, *The Organic Chemistry of Phosphorus*, Elsevier, Amsterdam, 1967, pp. 38–44.

65. G. Zon, K. E. DeBruin, K. Naumann, and K. Mislow, *J. Am. Chem. Soc.*, **91**, 7023 (1969).

66. D. B. Denney and W. H. Hanifin, *Tetrahedron Lett.*, 2177 (1963); D. B. Denney and N. G. Adin, ibid., 2569 (1966).

67. R. D. Adamcik, L. L. Chang, and D. B. Denney, *J. Chem. Soc. Chem. Commun.*, **1**, 986 (1974).

68. R. G. Pearson and J. Rajaram, *Inorg. Chem.*, **13**, 246 (1974).

69. Transition-metal atoms in low oxidation states are well known to act as Lewis bases or nucleophiles; see D. E. Shriver, *Acc. Chem. Res.*, **3**, 231 (1970).

70. J. S. Bradley, D. E. Connor, D. Dolphin, J. A. Labinger, and J. A. Osborn, *J. Am. Chem. Soc.*, **94**, 4043 (1972); J. A. Labinger, A. V. Kramer, and J. A. Osborn, ibid., **95**, 7908 (1973).

71. W. A. Pryor and K. Smith, *J. Am. Chem. Soc.*, **92**, 2731 (1970).

72. D. J. Cram, et al., *J. Am. Chem. Soc.*, **92**, 7369 (1970).

73. For a review, see D. J. Cram and J. M. Cram, *Fortschr. Chem. Forsch.*, **31**, 1 (1972).

74. J. Day and D. J. Cram, *J. Am. Chem. Soc.*, **87**, 4398 (1965); R. Tang and K. Mislow, ibid., **91**, 5644 (1969).

75. H. J. Reich and C. S. Cooperman, *J. Am. Chem. Soc.*, **95**, 5078 (1973).

76. F. Cariati, R. Ugo, and F. Bonati, *Inorg. Chem.*, **5**, 1128 (1966).

77. V. Dyczmons and W. Kutzelnigg, *Theor. Chim. Acta*, **33**, 239 (1974).

78. G. A. Olah, G. Klopman, and R. H. Schlosberg, *J. Am. Chem. Soc.*, **91**, 3261 (1969); G. A. Olah, Y. Halpern, J. Shen, and Y. K. Mo, ibid., **95**, 4960 (1973).

79. D. S. Matteson, *Organometallic Reaction Mechanisms*, Academic, New York, 1974; F. R. Jensen and B. Rickborn, *Electrophilic Substitution*, McGraw-Hill, New York, 1968.

80. P. L. Bock, D. J. Bocchetto, J. R. Rasmussen, J. P. Demers, and G. M. Whitesides, *J. Am. Chem. Soc.*, **96**, 2814 (1974); H. L. Fritz, J. H. Espenson, D. A. Williams, and G. L. Molander, ibid., 2378 (1974).

81. R. E. Dessy and F. Paulik, *J. Chem. Ed.*, **40**, 185 (1963); R. W. Johnson and R. G. Pearson, *Inorg. Chem.*, **10**, 2091 (1971).

82. U. Belluco, M. Gustiniani, and M. Graziani, *J. Am. Chem. Soc.*, **89**, 6494 (1967).

83. (a) N. L. Allinger, J. C. Tai, and F. T. Wu, *J. Am. Chem. Soc.*, **92**, 579 (1970); (b) C. D. Ritchie and G. A. Chappell, ibid., **92**, 1819 (1970); (c) A. Dedieu and A. Viellard, *Chem. Phys. Lett.*, **5**, 328 (1970); (d) A. Dedieu and A. Viellard, *J. Am. Chem. Soc.*, **94**, 6730 (1972); (e) R. F. W. Bader, A. J. Duke, and R. R. Messer, ibid., **95**, 7715 (1973).

84. R. H. Bathgate and E. A. Moelwyn-Hughes, *J. Chem. Soc.*, 3642 (1959).

85. For a review of nucleophilic reactivity see R. G. Pearson, *in* N. B. Chapman and J. Shorter (eds.), *Advances in Linear Free Energy Relationships*, Plenum, London, 1972.

86. K. Fukui, H. Fujimoto, and S. Yamabe, *J. Phys. Chem.*, **76**, 232 (1972).

87. For a short review, see S. R. Hartshorn, *Aliphatic Nucleophilic Substitution*, Cambridge U. P., 1973, Chapters 3 and 4; also G. S. Koermer, M. L. Hall, and T. G. Traylor, *J. Am. Chem. Soc.*, **94**, 7204 (1973); H. G. Kuivala, J. L. Considine, and J. D. Kennedy, ibid., 7206.

88. (a) J. D. Baldeschweiler and S. S. Woodgate, *Acc. Chem. Res.*, **4**, 114 (1971); (b) L. B. Young, E. Lee-Ruff, and D. K. Bohme, *J. Chem. Soc. Chem. Commun.*, **35** (1973); (c) D. K. Bohme, G. J. Mackay, and J. D. Payzant, *J. Am. Chem. Soc.*, **96**, 4027 (1974); (d) J. J. Brauman, W. N. Olmstead, and C. A. Lieder, ibid., 4030.
89. C. A. Lieder and J. J. Brauman, *J. Am. Chem. Soc.*, **96**, 4028 (1974).
90. J. P. Lowe, *J. Am. Chem. Soc.*, **93**, 301 (1971).
91. (a) K. Morukama and R. E. Davis, *J. Am. Chem. Soc.*, **94**, 1060 (1972); (b) L. M. Raff, *J. Chem. Phys.*, **60**, 2220 (1974).
92. C. C. Chou and F. S. Rowland, *J. Chem. Phys.*, **50**, 2763, 5132 (1969).
93. H. Fujimoto, S. Yamabe, T. Minato, and K. Fukui, *J. Am. Chem. Soc.*, **94**, 9205 (1972).
94. F. Klanberg and E. L. Muetterties, *Inorg. Chem.*, **7**, 155 (1968).
95. D. L. Wilhite and L. Spialter, *J. Am. Chem. Soc.*, **95**, 2100 (1973).
96. R. Hoffmann, J. M. Howell, and E. L. Muetterties, *J. Am. Chem. Soc.*, **94**, 3047 (1972).
97. K. Mislow, *Acc. Chem. Res.*, **3**, 321 (1970).
98. (a) L. H. Sommer, *Stereochemistry, Mechanism and Silicon*, McGraw-Hill, New York, 1965; (b) R. Corriu and J. Masse, *Bull. Soc. Chim. France*, 3491 (1969).
99. R. Corriu and G. F. Lanneau, *J. Organomet. Chem.*, **67**, 243 (1974).
100. G. D. Horner and L. H. Sommer, *J. Am. Chem. Soc.*, **95**, 7700 (1973).
101. B. G. McKinnie, N. S. Bhacca, F. K. Cartledge, and J. Fayssoux, *J. Am. Chem. Soc.*, **96**, 2637 (1974).
102. See reference 6; also N. K. Hamer, *J. Chem. Soc.*, **B**, 404 (1966).
103. M. J. Gallagher and I. D. Jenkins, *Topics in Stereochemistry*, Wiley, New York, 1968, Vol. 3.
104. M. Wolfsberg and L. Helmholz, *J. Chem. Phys.*, **20**, 837 (1952).
105. See reference 28, Chapter 5.
106. H. B. Burgi, *Inorg. Chem.*, **12**, 2321 (1973).
107. C. A. Tolman, *J. Am. Chem. Soc.*, **92**, 4217 (1970).
108. P. Meakin, R. A. Schunn, and J. P. Jesson, *J. Am. Chem. Soc.*, **96**, 277 (1974).
109. J. Chatt, L. A. Duncanson, and L. M. Venanzi, *J. Chem. Soc.*, 4456 (1955).
110. D. R. Armstrong, R. Fortune, and P. G. Perkins, *Inorg. Chim. Acta*, **9**, 9 (1974).
111. R. P. Messmer, L. V. Interrante, and K. H. Johnson, *J. Am. Chem. Soc.*, **96**, 3847 (1974).
112. L. G. Vanquickenborne, J. Vranckx, and C. Görler-Walrand, *J. Am. Chem. Soc.*, **96**, 4121 (1974).
113. F. Basolo, J. Chatt, H. B. Gray, and B. L. Shaw, *J. Chem. Soc.*, 2207 (1961); R. G. Pearson and M. J. Hynes, *Kungl. Teknis. Högsk. Handl.*, **285**, 481 (1972).
114. P. Meakin, E. L. Muetterties, and J. P. Jesson, *J. Am. Chem. Soc.*, **94**, 5271 (1972).
115. (a) J. W. Dawson, T. J. McLennan, W. Robinson, A. Merle, M. Dartiguenave, Y. Dartiguenave, and H. B. Gray, *J. Am. Chem. Soc.*, **96**, 4428 (1974); (b) J. R. Shapley and J. A. Osborn, *Acc. Chem. Res.*, **6**, 305 (1973).
116. D. Strope and D. F. Shriver, *J. Am. Chem. Soc.*, **95**, 8197 (1973); *Inorg. Chem.*, **13**, 2652 (1974).
117. C. D. Falk and J. Halpern, *J. Am. Chem. Soc.*, **87**, 3003 (1965).
118. E. L. Muetterties and W. D. Phillips, *J. Am. Chem. Soc.*, **81**, 1084 (1959).
119. K. D. Asmus, W. Grünbein, and J. H. Fendler, *J. Am. Chem. Soc.*, **92**, 2625 (1970).
120. F. Ramirez, R. Tasaka, and R. Hershberg, *Phosphorus*, **2**, 41 (1072).

121. W. C. Archie, Jr. and F. H. Westheimer, *J. Am. Chem. Soc.*, **95**, 5955 (1973).

122. A. P. Gaughan, G. Joy, and D. F. Shriver, *Inorg. Chem.*, **14**, 1795 (1975).

123. C. A. Tolman, *Chem. Soc. Rev.*, **1**, 337 (1972); however see D. A. Sweigart, D. E. Cooper, and J. M. Millican, *Inorg. Chem.*, **13**, 1272 (1974), for exceptions.

124. A. Davison, W. McFarlane, L. Pratt, and G. Wilkinson, *J. Chem. Soc.*, 3653 (1962).

125. Z. Iqbal and T. C. Waddington, *J. Chem. Soc.*, **A**, 2958 (1968).

126. J. E. Ferguson, *Stereochemistry and Bonding in Inorganic Chemistry*, Prentice-Hall, Englewood Cliffs, N. J., 1974, Chapters 10 and 11.

127. See reference 27, Chapter 3 and reference 28, Chapters 3 and 4.

128. M. Eigen and K. Tamm, *Z. Elektrochem.*, **66**, 93, 107 (1962); M. Eigen, *Bunsenges. phys. Chem.*, **67**, 753 (1963).

129. H. Eyring (ed.), *Physical Chemistry, an Advanced Treatise*, Academic, New York, 1975, Vol. VII, Chapter 5.

130. For calculations that are near ab initio in accuracy, see A. Dutta-Ahmed and E. A. Boudreaux, *Inorg. Chem.*, **12**, 1597 (1973).

131. H. Taube, *Chem. Rev.*, **50**, 69 (1952).

132. H. Yamatera, *Bull. Chem. Soc. Jap.*, **41**, 2817 (1968).

133. G. Guastalla and T. W. Swaddle, *Can. J. Chem.*, **51**, 821 (1973).

134. P. C. Ford, J. R. Kuempel, and H. Taube, *Inorg. Chem.*, **7**, 1976 (1968).

135. For example, see T. Kruck and M. Noack, *Chem. Ber.*, **97**, 1693 (1964); W. F. Edgell and B. J. Bulkin, *J. Am. Chem. Soc.*, **88**, 4839 (1966).

136. A. Fujii and S. J. Miller, *J. Am. Chem. Soc.*, **93**, 3694 (1971).

137. H. M. Walborsky and L. M. Turner, *J. Am. Chem. Soc.*, **94**, 2273 (1972).

138. L. A. Singer, *in* B. S. Thyagarajan (ed.), *Selective Organic Transformations*, Vol. II, Wiley, New York, 1972, p. 239; R. C. Bingham and M. J. S. Dewar, *J. Am. Chem. Soc.*, **95**, 7180 (1973).

139. J. O. Edwards, *Peroxide Reaction Mechanisms*, Wiley, New York, 1962.

140. G. Klopman, *J. Am. Chem. Soc.*, **90**, 223 (1968).

141. (a) P. C. Hariharan, W. A. Latham, and J. A. Pople, *Chem. Phys. Lett.*, **14**, 385 (1972). (b) J. M. Lehn and G. Wipff, *J. Chem. Soc. Chem. Commun.*, 747 (1973).

142. S. J. Miller, *in* V. Gold (ed.), *Advances in Physical Organic Chemistry*, Vol. 6, Academic, New York, 1968, p. 185 ff. This has an extensive analysis of addition reactions to unsaturated systems.

143. G. A. Olah, J. M. Bollinger, and J. Brinish, *J. Am. Chem. Soc.*, **90**, 2587 (1968); W. J. Hehre and P. C. Hiberty, ibid., **96**, 2665 (1974).

144. H. Fujimoto, M. Katata, S. Yamabe, and K. Fukui, *Bull. Chem. Soc. Jap.*, **45**, 1320 (1972).

145. A. Rauk, J. D. Andose, W. G. Frick, R. Tang, and K. Mislow, *J. Am. Chem. Soc.*, **93**, 6507 (1971).

146. S. Wolfe, A. Rauk, and I. G. Csizmadia, *J. Am. Chem. Soc.*, **91**, 1567 (1969).

147. Z. Rappoport, *in* V. Gold (ed.), *Advances in Physical Organic Chemistry*, Vol. 7, 1969, p. 1 ff; G. Modena, *Acc. Chem. Res.*, **4**, 173 (1971).

148. D. R. Kelsey and R. G. Bergman, *J. Am. Chem. Soc.*, **93**, 1953 (1971).

149. W. H. Saunders, Jr. and A. F. Cockerill, *Mechanisms of Elimination Reactions*, Wiley-Interscience, New York, 1973.

150. H. Steinmetz and R. M. Noyes, *J. Am. Chem. Soc.*, **74**, 4141 (1952).

151. See R. G. Pearson, *J. Am. Chem. Soc.*, **85**, 3533 (1963) for a discussion.

152. R. Hoffmann, C. C. Wan, and V. Neagu, *Mol. Phys.*, **19**, 113 (1970); O. P. Strausz, H. S. Gunning, A. S. Denes, and I. G. Csizmadia, *J. Am. Chem. Soc.*, **94**, 8317 (1972).

153. P. S. Skell and R. C. Woodworth, *J. Am. Chem. Soc.*, **78**, 4496 (1956).

154. R. Hoffmann, *J. Am. Chem. Soc.*, **90**, 1475 (1968); N. Bodor, M. J. S. Dewar, and J. S. Wasson, ibid., **94**, 9095 (1972).

155. R. K. Hertz, H. D. Johnson, II, and S. G. Shore, *Inorg. Chem.*, **12**, 1875 (1973).

156. G. A. Olah, Y. Halpern, J. Shen, and Y. K. Mo, *J. Am. Chem. Soc.*, **93**, 1251 (1971); G. A. Olah, J. R. DeMember, and J. Shen, ibid., **95**, 4952 (1973).

157. (a) G. A. Olah, A. M. White, and D. H. O'Brien, *Chem. Rev.*, **70**, 561 (1970); (b) for a general discussion of carbocations and their naming see G. A. Olah, *J. Am. Chem. Soc.*, **94**, 808 (1972).

158. H. B. Burgi, J. M. Lehn, and G. Wipff, *J. Am. Chem. Soc.*, **96**, 1956 (1974); H. B. Burgi, J. D. Dunitz, and E. Shefter, ibid., **95**, 5065 (1973).

159. W. J. Hehre and J. A. Pople, *J. Am. Chem. Soc.*, **94**, 6901 (1972).

160. M. A. Haney and J. L. Franklin, *J. Phys. Chem.*, **73**, 4328 (1969).

161. For terminology and examples, see R. C. Hahn and D. L. Strack, *J. Am. Chem. Soc.*, **96**, 4335 (1974); R. C. Hahn and M. W. Galley, ibid., 4337 (1974).

162. J. D. Roberts, H. S. Simmons, L. A. Carlsmith, and C. W. Vaughan, *J. Am. Chem. Soc.*, **75**, 3290 (1953).

163. For a review, see M. J. Strauss, *Acc. Chem. Res.*, **7**, 181 (1974).

164. G. H. Williams, *Homolytic Aromatic Substitution*, Pergamon, New York, 1960.

165. A. G. Davies and B. P. Roberts, *Acc. Chem. Res.*, **5**, 387 (1972).

166. W. G. Bentrude, W. A. Khan, M. Murakami, and H. W. Tan, *J. Am. Chem. Soc.*, **96**, 5568 (1974).

# CHAPTER 5

# MECHANISMS OF MORE COMPLEX REACTIONS

Chapters 3 and 4 were largely concerned with key reactions of certain very simple molecules. Next to consider are the reactions of more complex molecules, both with simple molecules and with each other. The objective is to give examples of fundamental classes of reactions. Mechanisms that are predicted for these reactions by the LMP will be examined for allowedness.

The warning should be repeated that a reaction that is completely allowed, and thermodynamically favorable, may still not occur with a measurable rate at moderate temperatures, or in the absence of a catalyst. As an example, consider the isotope-exchange reaction

$$^{16}O{=}C{=}^{16}O + C{=}^{18}O \longrightarrow {}^{16}O{=}C + {}^{16}O{=}C{=}^{18}O \quad (1)$$

Obviously this reaction is completely allowed in a $C_{\infty h}$ point group, or in any less symmetric approach. Yet in fact it requires a heterogeneous catalyst, plus rather high temperatures, to be observable.

While (1) is thermoneutral, a more extreme example is the isoelectronic process

$$N{=}N{=}O + C{=}O \longrightarrow N{\equiv}N + O{=}C{=}O \quad (2)$$

Here $\Delta H$ is $-87$ kcal, but still this extremely favorable reaction has no measurable rate at room temperature. What is demonstrated in (1) and (2) is a rather general rule: "Multiple bonds are broken one bond at a time." This rule has no theoretical basis, but is based on empirical observations and the mechanisms deduced from such observations. That is, if one examines the mechanisms proposed for multiply bonded molecules, such as $O_2$, $N_2$, $(CH_3)_2CO$, and $CH_3CN$, one finds that stepwise breaking of the multiple bonds is always proposed. Some reactions of free atoms, such as oxygen, are exceptions.

The allowedness of (2) is most easily seen by considering the $\pi$ bonds only. By this is meant those bonds whose nodal plane is the plane of the page. $N_2O$ has four electrons in orbitals of this kind, just as $CO_2$ does. CO has two $\pi$ electrons, and $N_2$ also has two. Hence the $\pi$ electron count is satisfied. If the $\pi$ electrons are balanced, then the $\sigma$ electrons must also be.

## SOME CYCLOADDITION REACTIONS

The symmetry rules for cycloaddition reactions were discussed earlier (p. 136). It was pointed out that the condensation of ethylene to form cyclobutane was forbidden, but that symmetrical substituents could lower the activation energy by configuration interaction. The effect of unsymmetrical substitution may also be considered.[1] Take two propylene molecules, which may cyclize in either a head-to-head or tail-to-head manner

$$2CH_3CH{=}CH_2 \longrightarrow \underset{\text{head-to-head}}{\begin{array}{c}CH_3{-}CH{-}CH_2\\ \quad\;\; | \qquad\; | \\ CH_3{-}CH{-}CH_2\end{array}} \quad \text{or} \quad \underset{\text{tail-to-head}}{\begin{array}{c}CH_3{-}CH{-}CH_2\\ \quad\;\; | \qquad\; | \\ H_2C{-}HC{-}CH_3\end{array}} \tag{3}$$

We assume first that the reactions are concerted antarafacial and consider the mode of addition which maximizes the overlap of the HOMO and LUMO. The methyl group may be considered to lower the electronegativity of the olefinic carbon to which it is attached. This reduces the coefficient of its atomic orbital in the $\pi$ MO and increases it in the $\pi^*$ MO, as shown in Fig. 1. Simple analysis shows that maximum overlap occurs with the tail-to-head orientation of Fig. 1. We have union of the two atoms with highest frontier orbital density and union of the two atoms of lowest frontier orbital density.[1]

This rule holds for any substituents, whether electron donating or withdrawing. However vinyl or phenyl substituents are predicted to prefer head-to-head. The cases of acrylonitrile and butadiene are also shown in Fig. 1. Experimentally it is found that most olefin cyclizations occur in a head-to-head manner. This indicates that a concerted mechanism is not being followed. A two-step mechanism with a biradical intermediate is expected to give head-to-head products, because the radical centers are stabilized by the substituents.[2]

Another test of concertedness is the stereospecificity. A biradical intermediate can rotate about the carbon—carbon single bond to lose the configuration of the original olefins. Electron-withdrawing substituents on one olefin, coupled with electron-donating substituents on the other, usually increase stereoselectivity. However if the substitution is asymmetric, the stereoselective character is lost.[3] This is the expected result for a zwitterionic mechanism.[4]

$$\mathrm{(H)(H_3C)C{=}C(H)(OC_2H_5)} + \mathrm{(NC)_2C{=}C(CN)_2} \rightarrow \mathrm{(H)(H_3C)C{-}\overset{+}{C}(H)(OC_2H_5)}\ \text{with}\ \mathrm{C(CN)_2{-}\underline{C}(CN)_2} \longrightarrow \mathrm{H_3C{-}C(H){-}C(OC_2H_5)(H)},\ \mathrm{(NC)_2C{-}C(CN)_2} \tag{4}$$

The electron-rich propenyl ether donates a pair of electrons to tetracyanoethylene, a good electron acceptor, in an asymmetric fashion, one carbon atom of each olefin being favored. This permits rotation about the new bond, as shown. The formation of a charge-transfer complex is shown by an immediate color change on mixing the two reactants. However the zwitterion intermediate is impossible to distinguish from a polar biradical intermediate, providing the latter is postulated to be a singlet. Biradicals, whether polar or not, may ring close very quickly, if the electron spins are paired. This would lead to stereospecific suprafacial addition on each olefin.

The reactions of $^1\Delta_g$ oxygen are similar to those of an electron-poor olefin, in that the $O_2$ molecule acts as a Lewis acid. Cycloaddition occurs with retention of olefin configuration, even though suprafacial addition is forbidden (p. 136). A zwittrionic intermediate is strongly indicated because of the high electronegativity of oxygen. An estimate of initial interactions shows[5] that the preferred mode of approach is one with a $C_s$ point group. This maximizes the overlap between the HOMO of ethylene and the LUMO of $O_2$.

$$(-)\mathrm{C}(+),\ (-)\mathrm{C}(+) \cdots\rightarrow (+)\mathrm{O}(-),\ (-)\mathrm{O}(+) \longrightarrow \mathrm{H_2C{-}H_2C{>}O^+{-}O^-} \tag{5}$$

While it seems very reasonable that the perepoxide of (5) should form, it is by no means clear that it has a major role in the mechanism. It would still

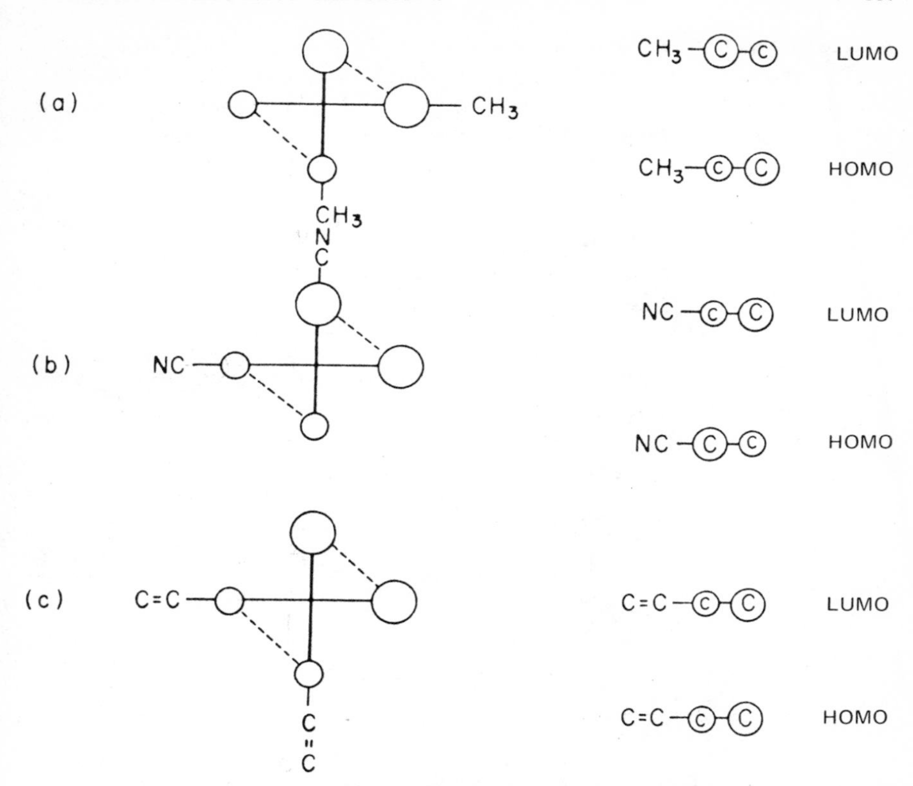


**Figure 1.** Orbital interaction in antarafacial cyclodimerization of: (*a*) propylene, (*b*) acrylonitrile, and (*c*) butadiene; the orientation is such as to give maximum overlap between the HOMO and LUMO (from reference 1).

have to rearrange to the 1,2 dioxetane product in a stereospecific manner. The entropy and energy of the TS in the rearrangement determines the overall rate of reaction, not the energy of the perepoxide.

The *ortho*-benzyne molecule, $C_6H_4$, has properties similar to that of an electron-poor olefin. The HOMO and the LUMO have similar symmetry properties to a normal olefin.[6]

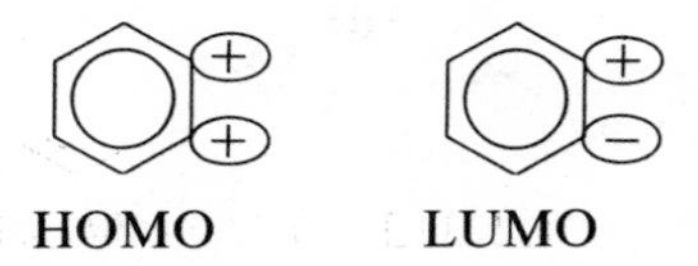


HOMO LUMO

The LUMO is more electronegative than that of a simple olefin because the orbital is really a $\sigma$ orbital with carbon $2s$ contributing. Olefins add in a stereospecific manner to benzyne. Suprafacial addition is forbidden, and antarafacial addition on benzyne is sterically impossible. Antarafacial addition on the olefin would lead to inversion of the configuration, which is not observed. It appears likely in this case that the analog to the perepoxide must be formed first and then rearrange.[7]

(6)

A carbonyl group is another example of an unsymmetrical double bond. Several possible reactions exist.

$$>C{=}C< \; + \; >C{=}O \longrightarrow \text{C—C / C—O four-membered ring} \qquad (7)$$

$2a'$ → $2a'$

$$>C{=}O \; + \; >C{=}O \longrightarrow \text{C—O / C—O four-membered ring} \qquad (8)$$

$a_1 + b_2$ → $2a_1$

$$>C{=}O \; + \; O{=}C< \longrightarrow \text{C—O / C—C four-membered ring} \qquad (9)$$

$a_g + b_u$ → $a_g + b_u$

Considering the product molecules as planar, which is not strictly true, (7) is allowed in the $C_s$ point group, (8) is forbidden in $C_{2v}$, and (9) is allowed in $C_{2h}$. The $\pi$ bonds lie in the plane of the paper in all cases and suprafacial addition is assumed.

The conclusions must be modified by the realization that in all cases there is the problem of poor overlap between the HOMO and LUMO, the overlap that is exactly zero for two olefins. The overlaps are shown schematically in Fig. 2. As expected, the overlaps are small, but not zero for (7) and (9), and exactly zero for the forbidden reaction (8). Actually none of the reactions occur thermally, as a rule, because of unfavorable thermodynamics. Reactions (7) and (9) occur photochemically, although the product is unstable and decomposes again, in the latter case.[8] Cyclic trimers, A, do form from aldehydes in solution,

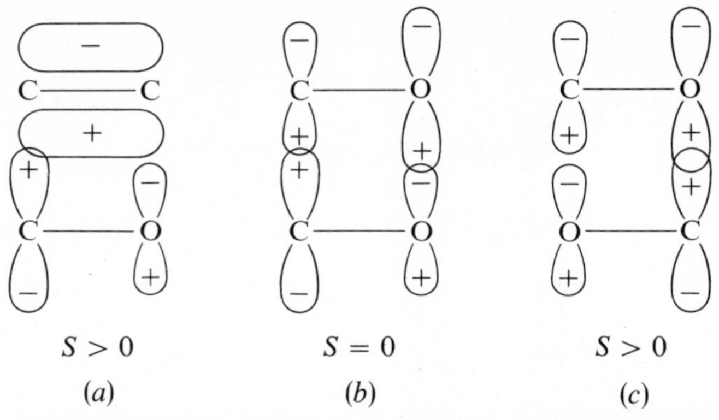


**Figure 2.** Overlaps of HOMO and LUMO in: (*a*) cycloaddition of olefin and carbonyl groups, (*b*) head-to-head addition of two carbonyl groups, and (*c*) head-to-tail addition of carbonyl groups.

$$3\,CH_2O \longrightarrow \underset{\text{A}}{\text{cyclo-}(CH_2O)_3} \tag{10}$$

but a likely intermediate is the dipolar, or zwitterionic, species.

$$^{-}O{-}\overset{|}{C}{-}O{-}C^{+}\!\!< \; + \; CH_2O \longrightarrow \text{A} \tag{11}$$

Reactions reverse to (7) and (8) are known.[9] Both are believed to involve bond rupture to form biradicals. The cleavage of trimethylene oxide to ethylene and formaldehyde is quite clean with no chain processes.

$$\text{cyclo-}(C{-}C{-}O{-}C) \longrightarrow \dot{C}{-}C{-}C{-}\dot{O} \longrightarrow {>}C{=}C{<} \; + \; {>}C{=}O \tag{12}$$

Note that the unbalanced overlap between the HOMO of ethylene and the LUMO of formaldehyde (Fig. 2) formally leads to a dipole structure which is,

$$-\overset{+}{C}{-}C{-}C{-}O^{-}$$

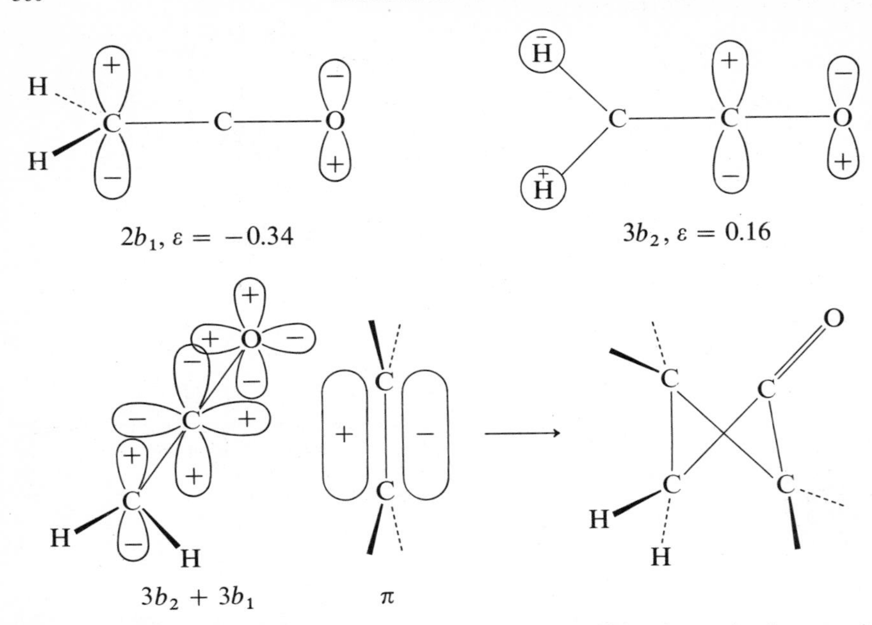


**Figure 3.** The HOMO and LUMO of ketene—also the stabilizing interaction between the filled $\pi$ orbital of ethylene, and the empty $3b_2$ orbital of ketene; the final product, however, involves antarafacial addition to the $3b_1$ orbital [energies from J. H. Letcher, M. L. Unland, and J. R. Van Wazer, *J. Chem. Phys.*, **50**, 2185 (1969)].

in fact, indistinguishable from the biradical structure of (12). Both are simply different canonical forms contributing to a single resonance hybrid, in valence bond language. The activation energy for (12) is 60 kcal, which is reasonable for breaking a strained carbon—oxygen bond.

Mildly activated olefins add readily to ketene, usually at the carbon—carbon double bond,

$$>C{=}C{=}O + >C{=}C< \longrightarrow \begin{array}{l} >C-C{=}O \\ \ \ \ | \quad\ | \\ >C-C< \end{array} \tag{13}$$

although some examples of addition to the carbonyl double bond are known. The reactions are stereospecific in the olefins and appear to be concerted. Two factors are important in making (13) a facile reaction, compared to cyclobutane formation.[10] One is the absence of two H atoms on the central carbon, which facilitates a $[_{\pi}2_s + {}_{\pi}2_a]$ process.† The other is the polarity of the carbonyl group that allows for a strong charge-transfer interaction.

† The expression $[\pi^2 s + \pi^2 a]$ is a shorthand for showing that two $\pi$ electrons which add suprafacially, and two $\pi$ electrons which add antrafacially, are involved.

Semiempirical MO calculations confirm these speculations and show that the preferred direction of approach of the olefin is perpendicular to the ketene molecule.[11] Figure 3 shows the geometry and indicates that the strongest interaction is with the HOMO of ethylene and the LUMO of ketene. The latter orbital is an antibonding $\pi^*$ orbital concentrated on the central carbon atom. It and the HOMO of ketene are also shown in Fig. 3.

Finally, consider the addition of $B_2Cl_4$ to an olefin or acetylene molecule, where $B_2Cl_4$ also is a Lewis acid and able to form a $\pi$ complex with the unsaturate. The addition is *cis* to an acetylene.[12]

$$B_2Cl_4 + C_2H_2 \longrightarrow \begin{matrix} H & & H \\ & C{=}C & \\ Cl_2B & & BCl_2 \end{matrix} \tag{14}$$

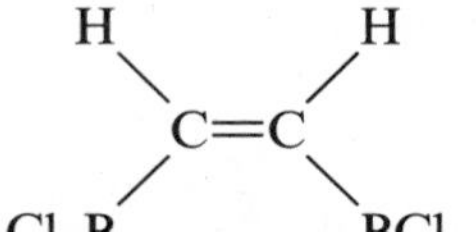

It is tempting to believe that the primary interaction is between the HOMO of acetylene and the LUMO of planar $B_2Cl_4$, a $\pi$ orbital similar to the filled orbital of $C_2Cl_4$. These orbitals have excellent overlap

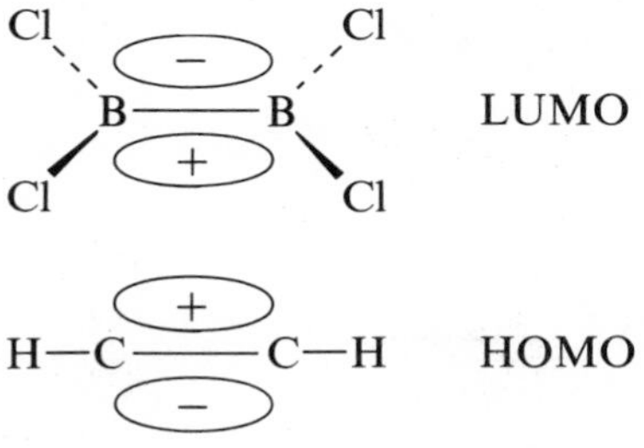

Cl Cl
B—B (−) above, (+) below
Cl Cl
LUMO

H—C—C—H (+) above, (−) below
HOMO

However it can be seen that such an interaction cannot possibly lead to the products of (14). The transfer of electrons from HOMO to LUMO will strengthen the B—B bond, rather than break it. It is easily seen that the least motion $C_{2v}$ approach of $B_2Cl_4$ to acetylene is a reaction coordinate forbidden by orbital symmetry. Thus (14) must occur by some stepwise process.

If butadiene rather than ethylene is the reactant, cycloaddition reactions become allowed in the sense of 1,4-addition. For example,[13]

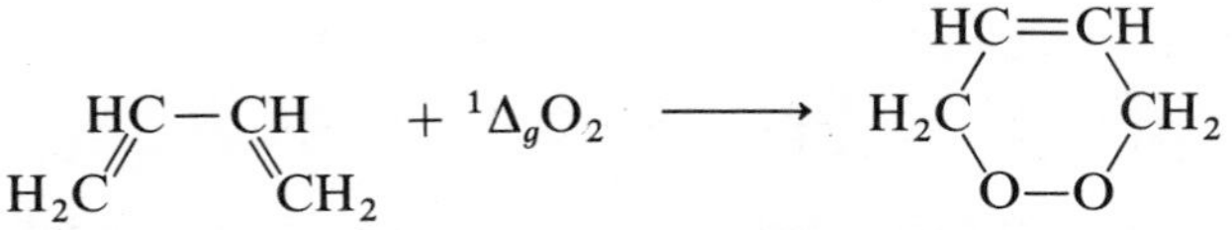

$$\begin{matrix} HC{-}CH \\ H_2C{\diagup\!\!\!\!\diagup} \quad {\diagdown\!\!\!\!\diagdown}CH_2 \end{matrix} + {}^1\Delta_gO_2 \longrightarrow \begin{matrix} HC{=}CH \\ H_2C \quad\quad CH_2 \\ O{-}O \end{matrix} \tag{15}$$

In the Diels–Alder reaction with substituted olefins or dienes, the stereoselection rules are the same as for (2 + 2) cycloadditions, assuming concertedness.[1] For unsymmetrical reactants, the stereochemistries actually observed are the opposite of those predicted for a concerted process. The results are consistent with a biradical intermediate, with the substituents stabilizing the radical centers. Accordingly a two-step mechanism with a biradical or dipolar intermediate is indicated. This does not contradict the prediction that the concerted process is allowed, since the two-step mechanism is also allowed.

One might also expect butadiene to react with itself in an allowed (4 + 2) reaction

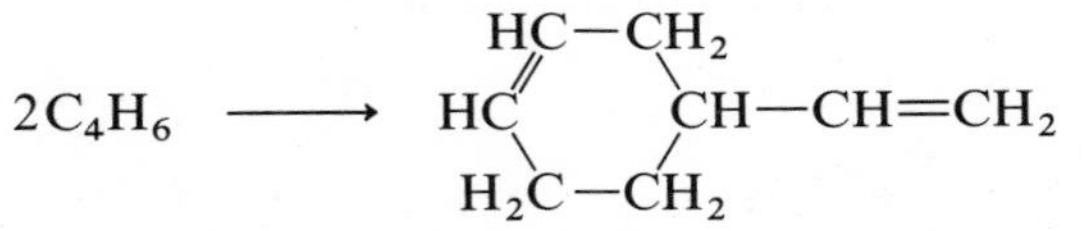

$$2C_4H_6 \longrightarrow \begin{array}{c} HC-CH_2 \\ HC \quad\quad CH-CH=CH_2 \\ H_2C-CH_2 \end{array} \qquad (16)$$

This does occur, but cyclobutane derivatives are also formed by (2 + 2) addition.

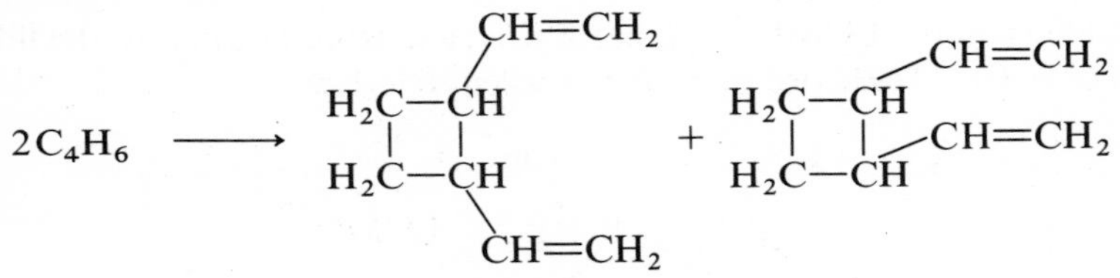

$$2C_4H_6 \longrightarrow \underset{trans}{\begin{array}{c} CH=CH_2 \\ H_2C-CH \\ | \quad\; | \\ H_2C-CH \\ CH=CH_2 \end{array}} + \underset{cis}{\begin{array}{l} \quad\quad\quad CH=CH_2 \\ H_2C-CH \\ | \quad\; | \quad CH=CH_2 \\ H_2C-CH \end{array}} \qquad (17)$$

The *cis*-1,2-divinylcyclobutane is not observed, but is converted rapidly to cycloocta-1,5-diene by a Cope rearrangement.[14]

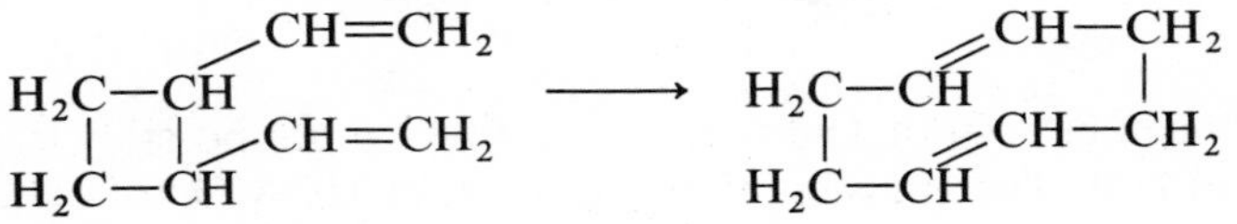

$$\begin{array}{l} \quad\quad\quad CH=CH_2 \\ H_2C-CH \\ | \quad\; | \quad CH=CH_2 \\ H_2C-CH \end{array} \longrightarrow \begin{array}{l} \quad\quad\quad CH-CH_2 \\ H_2C-CH \quad\quad | \\ | \quad\quad\quad\; CH-CH_2 \\ H_2C-CH \end{array} \qquad (18)$$

The direct formation of the latter product by 1,4-addition of two butadiene molecules is not allowed. It would require an antiaromatic eight-electron TS.

The head-to-head orientation of the two vinyl groups of (17) is consistent with a two-step mechanism. The biradical intermediate is unusually stable in the case of butadiene because of a double allylic resonance.

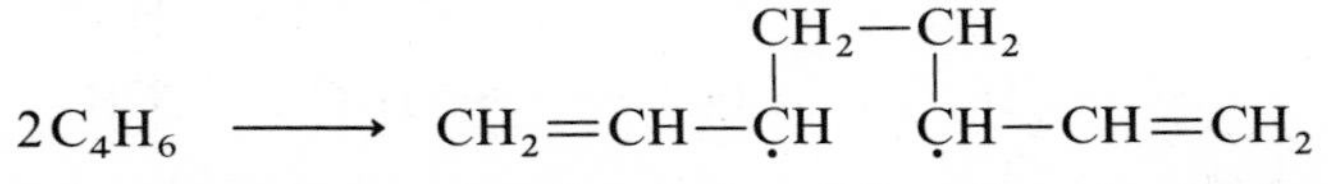

$$2C_4H_6 \longrightarrow \begin{array}{c} CH_2-CH_2 \\ | \quad\quad\; | \\ CH_2=CH-\dot{C}H \quad \dot{C}H-CH=CH_2 \end{array} \qquad (19)$$

It is therefore not surprising that the choice of (2 + 2) cycloaddition is made as well as the expected (2 + 4).

Benzene normally does not enter into cycloaddition reactions because of its stability. With suitable activating substituents, however, it will react. It can act both as a diene toward olefins or acetylenes,

$$+\ CF_3{-}C{\equiv}C{-}CF_3 \longrightarrow \quad \begin{matrix} C{-}CF_3 \\ \| \\ C{-}CF_3 \end{matrix} \qquad (20)$$

and as an olefin toward a diene.

$$+\ CH_2{=}CH{-}CH{=}CH_2 \longrightarrow \quad \begin{matrix} H_2 \\ C\ \ CH \\ \| \\ C\ \ CH \\ H_2 \end{matrix} \qquad (21)$$

The 1,2-addition to an olefin, or the 1,4-addition to a diene is forbidden. In the one case there would be a four-electron, and in the other case an eight-electron, TS.

It has been known for many years that cyclopropane behaves in some respects like an olefin. This shows up in many kinds of chemical reactivity, and in evidence for conjugation of a cyclopropyl ring just as with a vinyl group. The parallelism becomes more obvious if the MOs of cyclopropane are examined.[15] The highest occupied and the lowest vacant orbitals of $C_3H_6$ that are carbon—carbon bonding and antibonding, respectively, are shown in Fig. 4. The energy of the HOMO is high compared to that of a typical carbon—carbon bonding MO, $-0.419$ AU compared to $-0.49$ in ethane. The antibonding MO of cyclopropane, conversely, is low relative to the normal case, 0.28 AU compared to 0.57.

Also the symmetries of the HOMO and LUMO are similar to those of ethylene. It is not surprising that cyclopropane derivatives can enter in to

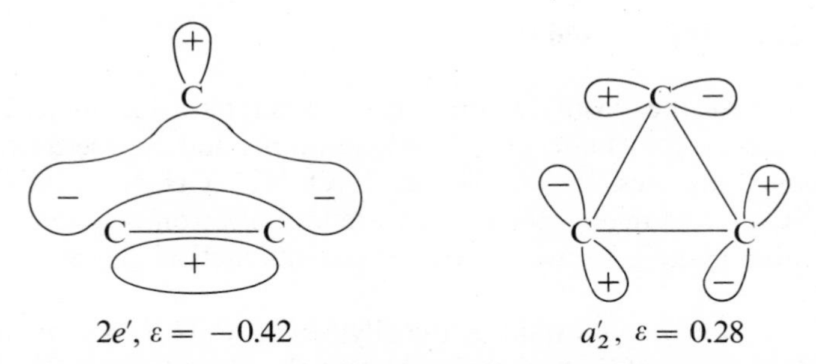


**Figure 4.** One of the degenerate pairs of $2e'$ HOMOs of $C_3H_6$, and the $a'_2$ orbital—the latter is the LUMO that is purely C—C antibonding; the $3e'$ and $3a'_1$ are slightly lower in energy but are also C—H antibonding (energies from reference 15).

cycloaddition reactions just as ethylene does. An example would be the 1,4-addition to butadiene of cyclopropanone.[16]

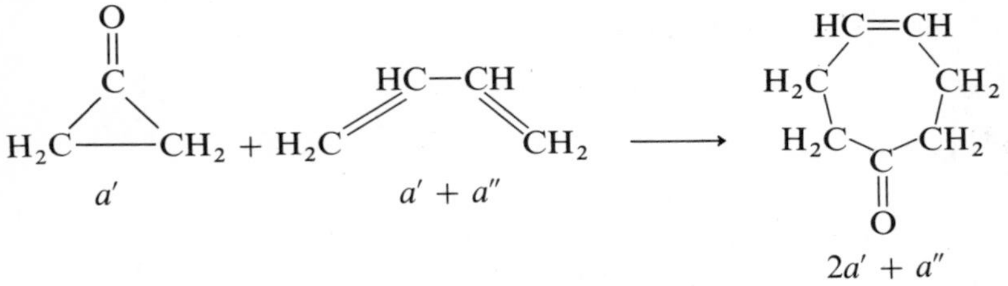

$$\text{cyclopropanone } (a') + H_2C{=}CH{-}CH{=}CH_2 \ (a' + a'') \longrightarrow \text{cyclohept-4-enone } (2a' + a'') \tag{22}$$

Cyclopropanone is very prone to ring opening to give the oxyallyl zwitterion, B,[17]

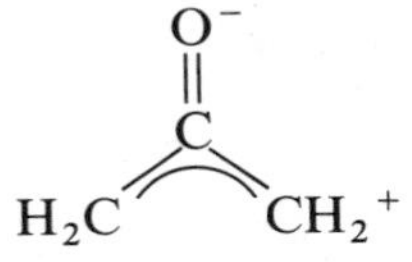

$$H_2C{\overset{\displaystyle O^-}{\overset{\|}{C}}}CH_2^{\,+}$$

B

This species can act as either a two-electron $\pi$ system, adding as in (22), or as a four-electron $\pi$ system, adding at carbon and oxygen. Addition across polar double bonds can now occur.

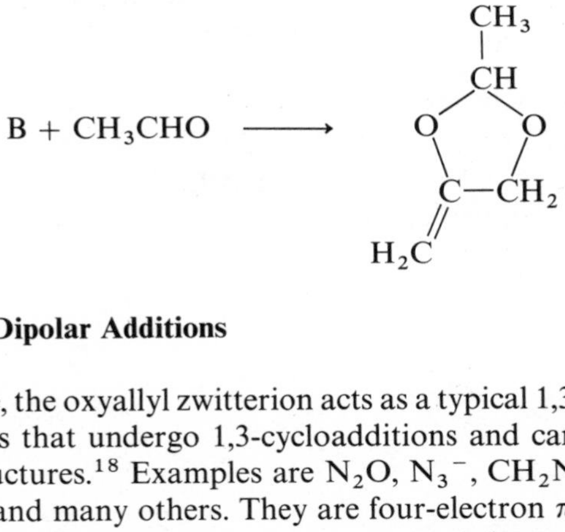

$$B + CH_3CHO \longrightarrow \text{2-methyl-4-methylene-1,3-dioxolane} \tag{23}$$

### 1,3-Dipolar Additions

In the latter case, the oxyallyl zwitterion acts as a typical 1,3 dipole. These are triatomic groups that undergo 1,3-cycloadditions and can be described by zwitterionic structures.[18] Examples are $N_2O$, $N_3^-$, $CH_2N_2$, RNCO, RNC, $RN_2O$, $NO_2^-$, and many others. They are four-electron $\pi$ systems and add in an allowed manner to two-electron $\pi$ systems such as olefins or carbonyl groups.

The simplest example would be the allyl anion, $C_3H_5^-$. The $\pi$ orbitals of this and other three-atom systems, are shown elsewhere (p. 172). The reaction with $C_2H_4$ maintains only a plane of symmetry, since the product is non-planar.

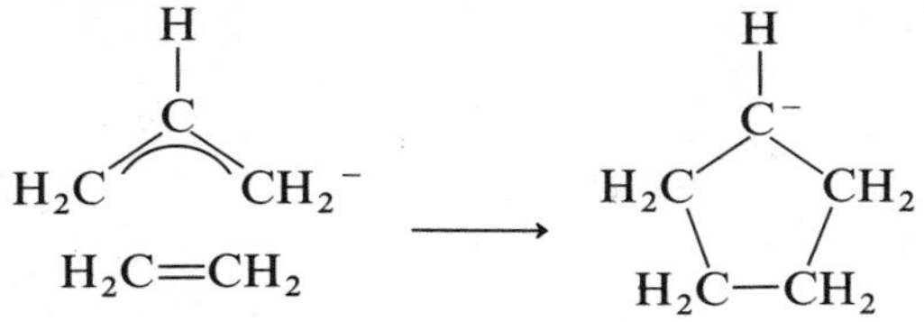

$$\text{H}_2\text{C}{\overset{\text{H}}{\text{C}}}\text{CH}_2^- + \text{H}_2\text{C}{=}\text{CH}_2 \longrightarrow \text{cyclo-}[\text{HC}^-\text{-CH}_2\text{-CH}_2\text{-CH}_2\text{-CH}_2] \tag{24}$$

$$a' + a'' + a' \qquad\qquad 2a' + a''$$

The new carbon—carbon bonds are $a' + a''$, and the lone pair of electrons is in an orbital of $a'$ symmetry. Note that an allyl cation would not be able to add to an olefin, since the $a''$ orbital is empty.

The additions are *cis*, or suprafacial, on the olefin. The regioselectivity in the case of unsymmetric 1,3-dipoles and olefins depends upon the coefficients of the various atomic orbitals in the HOMOs and LUMOs of each reactant. The orbital energies also play a role. A thorough discussion of the various factors has been given.[19] In another section of the present text are diagrams of a number of frontier orbitals of the reactants in question (p. 138).

An important example of a 1,3-dipole is ozone. The reaction of $O_3$ with an olefin is complex, since a primary adduct is formed first, which then rearranges to a moderately stable ozonide. The classical Criegee mechanism is as follows:[20]

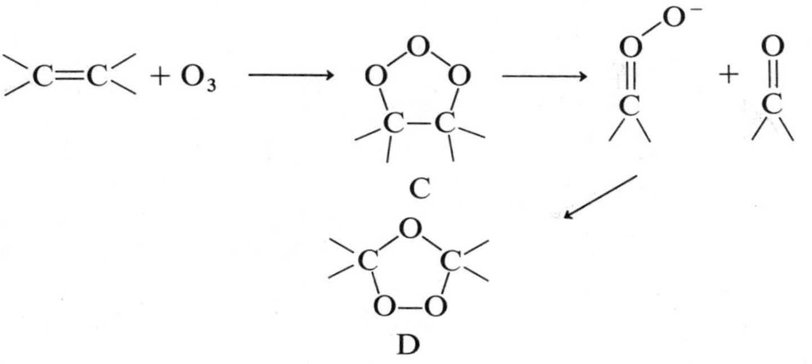

$$>\text{C}{=}\text{C}< + \text{O}_3 \longrightarrow \underset{\text{C}}{\text{(primary ozonide)}} \longrightarrow {>}\text{C}{=}\text{O}^+{-}\text{O}^- + {>}\text{C}{=}\text{O} \longrightarrow \underset{\text{D}}{\text{(ozonide)}} \tag{25}$$

The ozonide, D, may then be cleaved with water in the presence of zinc to produce two carbonyl compounds. The primary ozonide, C, is the product of 1,3-dipolar addition.

The structure is unknown but calculations indicate it has a carbon—carbon half-chair conformation.[21]

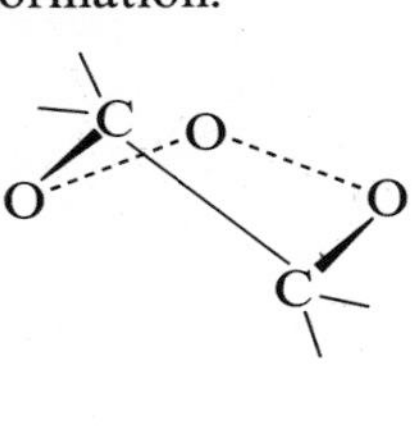

The point group is now $C_2$, in which the new carbon—oxygen bonds are *a* and *b*, and a lone pair on oxygen is also *b*. This matches the original $\pi$ orbitals of $O_3$, which are *a* and *b*, and the *b* orbital of ethylene. The addition is suprafacial.

One common 1,3-dipole that has not been reported to react with dipolarophiles is $CO_2$. However $CS_2$ does react with acetylene at 100°C to form a carbene.[22]

$$\begin{array}{c} S{=}C{=}S \\ H{-}C{\equiv}C{-}H \end{array} \longrightarrow \text{cyclo-}[\ddot{C}(S)_2 \text{—} HC{=}CH] \tag{26}$$

The observed products are those due to the reaction of the carbene with itself or added reagents.

### Addition of $N_3$ Radicals

The condensation of two 1,3-dipoles with each other to give a six-membered ring is forbidden (eight-electron TS). However removal of two electrons from the $\pi$ system would suggest that cyclization could occur. An example would be the reaction of two $N_3$ radicals.

$$2N_3 \longrightarrow N_6\,(\text{cyclic}) \longrightarrow 3N_2 \tag{27}$$

Since cyclic $N_6$ is isoelectronic with benzene, it would appear that (27) is favorable, even though the eventual products must be molecular nitrogen. Cyclic $N_6$ is calculated to lie some 11 eV above $3N_2$ in energy.[23]

Pulse radiolysis, or flash photolysis, of aqueous azide ion solutions produces $N_3$ radicals. These decay to an intermediate X with a second-order rate constant $k_1 = 9 \times 10^9\ M^{-1}\ s^{-1}$ at room temperature.[24] The species X decays at a much slower first order rate, $k_2 = 3.6 \times 10^3\ s^{-1}$, into molecular $N_2$.

$$\begin{aligned} 2N_3 &\xrightarrow{k_1} X \\ X &\xrightarrow{k_2} 3N_2 \end{aligned} \tag{28}$$

The absorption spectrum of X suggests that it is a conjugated system such as $N_6$ (cyclic), or $N_6$ (chain).

The radical $N_3$ is linear with a configuration

$$(\sigma_g)^2(\sigma_u)^2(\sigma_g)^2(\sigma_u)^2(\pi_u)^4(\pi_g)^3$$

If these orbitals are correlated with those benzene in the $D_{2h}$ point group, it is found that the cyclization step of (27) is forbidden. The $\pi$ electrons give no problem, since there are six of them in orbitals of $b_{1u}$, $b_{2g}$, and $b_{3g}$ species. However the $\sigma$ electrons give trouble.

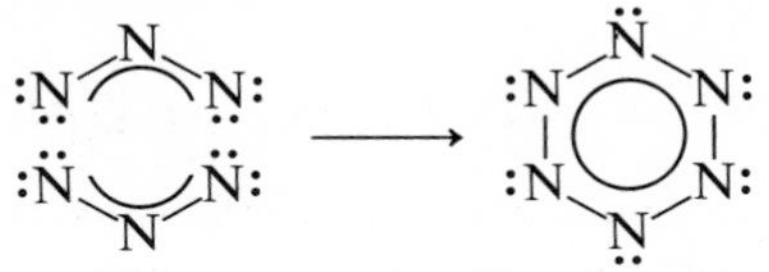

$$\text{:N=N}\cdots \longrightarrow \text{cyclic } N_6 \qquad (29)$$

$$a_g + b_{2u} + b_{3u} + b_{1g} \qquad 2a_g + b_{2u} + b_{3u}$$

We need focus only on four equivalent lone pairs, one on each terminal nitrogen atom. These must be converted to the two new N—N $\sigma$ bonds, and to two lone pairs on the central N atoms. This cannot be done in the least motion approach, which is $D_{2h}$.

The formation of a chain $N_6$ is allowed by symmetry.

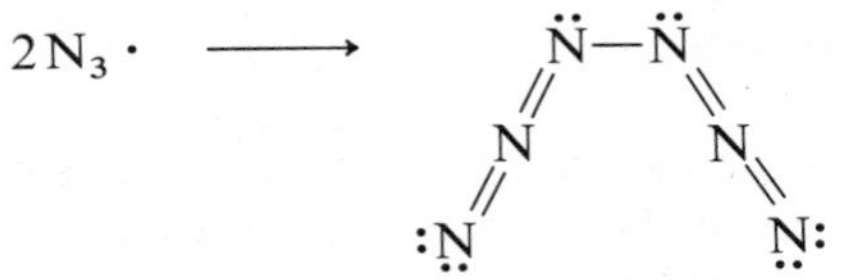

$$2N_3\cdot \longrightarrow \text{:N=N=}\ddot{N}\text{—}\ddot{N}\text{=N=N:} \qquad (30)$$

However the decomposition into $3N_2$ is now forbidden by symmetry, either as a concerted process, or by steps. There are eight $\pi$ electrons in chain $N_6$, as indicated by four double bonds. These are in orbitals with a node in the plane of the paper. Three $N_2$ molecules have only six electrons in such $\pi$ orbitals. There would be no difficulty with $N_6$ (cyclic) decomposing into $3N_2$, since it corresponds to the allowed (2 + 2 + 2) cycloaddition.

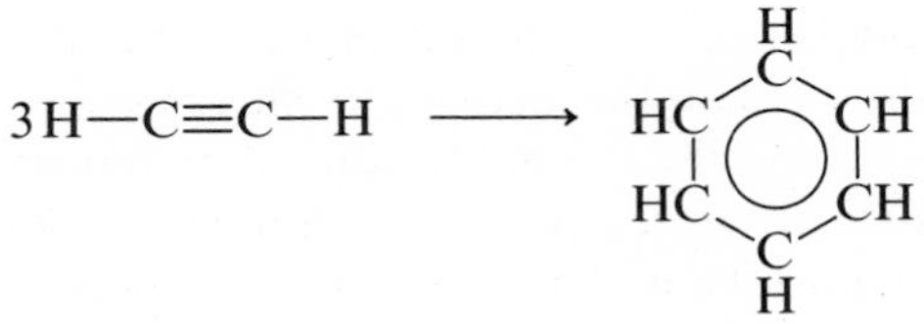

$$3\text{H—C}\equiv\text{C—H} \longrightarrow C_6H_6 \text{ (benzene)} \qquad (31)$$

However a rapid decomposition of $N_6$ is not compatible with the small value of $k_2$.

In view of the orbital-symmetry barrier, it seems unlikely that cyclic $N_6$ is formed at essentially a diffusion controlled rate ($k_1$). Instead either a chain $N_6$ is formed, or another possibility, ($N_4 + N_2$).

$$2N_3\cdot \xrightarrow{k_1} N_4 + N_2 \qquad (32)$$

The structure of $N_4$ is not known but it could have either a linear shape with a $^3\Sigma_g$ ground state, or a square shape with a $^3A_{2g}$ ground state. It lies some 10 eV above $2N_2$ in energy.[23] Its decomposition into molecular nitrogen is forbidden by spin and symmetry. Since the $^3\Pi_u$ state of $N_2$ lies only 7.5 eV above the ground state, decomposition into excited state products is a possibility.

$$N_4 \xrightarrow{k_2} (^3\Pi_u)N_2 + (^1\Sigma_g)N_2 \tag{33}$$

Formation of atomic products is also possible, since the dissociation energy of $N_2$ is 9.8 eV. However isotope labeling experiments rule this out.[25]

The undoubted fact that formation of $N_6$ in the cyclic structure is forbidden, should not be taken as a failure of the rule concerning aromatic and non-aromatic TS. Clearly it is the eight $\sigma$ electrons that must be considered, and not the six $\pi$ electrons. The accidental result that $8 + 6 = 14$, which satisfies the $4n + 2$ rule for aromaticity has no meaning. The two sets of electrons, $\pi$ and $\sigma$, are independent.

### Ester Pyrolysis

On the other hand, a six-electron TS does not necessarily make activation energies shrink to the vanishing point. A good example lies in the pyrolysis of organic esters, compared to the similar reactions of alkyl halides.

$$C_2H_5O{-}\overset{\overset{\displaystyle O}{\|}}{C}{-}CH_3 \xrightarrow{\Delta} C_2H_4 + CH_3{-}\overset{\overset{\displaystyle O}{\|}}{C}{-}OH$$

$$\Delta H = 9.5 \text{ kcal}, E_a = 48 \text{ kcal} \tag{34}$$

$$C_2H_5Cl \xrightarrow{\Delta} C_2H_4 + HCl \qquad \Delta H = 14.5 \text{ kcal}, E_a = 60 \text{ kcal} \tag{35}$$

In both cases pyrolysis occurs in the gas phase at 300–400°C. Examination of kinetic data shows that the activation energies are very substantial, being some 40 kcal above the thermodynamic requirement.

The ester reactions are stereospecific *cis* eliminations. Other information, including a maximum kinetic deuterium isotope effect, strongly suggest a concerted mechanism.[26] The classical mechanism involves a cyclic, six-membered, TS.[27]

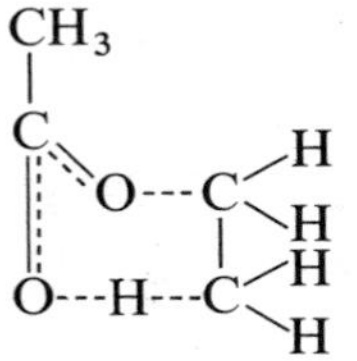

```
 CH3
 |
 C          H
 ⁞ ⸌O---C<
 ⁞       | H
 ⁞       | H
 O---H---C<
            H
```

An electron count gives the number six, two on each oxygen atom and two for the C—H bond. The TS is aromatic, and the reaction is favored. Reaction

(35), however, has a cyclic four-membered TS with only four electrons, and is disfavored. After correcting for the extra thermodynamic requirement of 5 kcal, the favored ester pyrolysis has only a modest 7-kcal advantage over the alkyl halide reaction.

## SOME RING-OPENING REACTIONS

Many heterocyclic molecules can be pyrolyzed to yield a smaller organic molecule plus an inorganic product, such as $N_2$, $CO_2$, CO, S, SO, $SO_2$, or HCN. Such reactions are usually called "extrusion" reactions.[28] In the special case where a ring of carbon atoms is held together by a single atom, the process and its reverse are called "chelotropic" reactions.[10] A simple example would be the pyrolysis of ethylene sulfide.

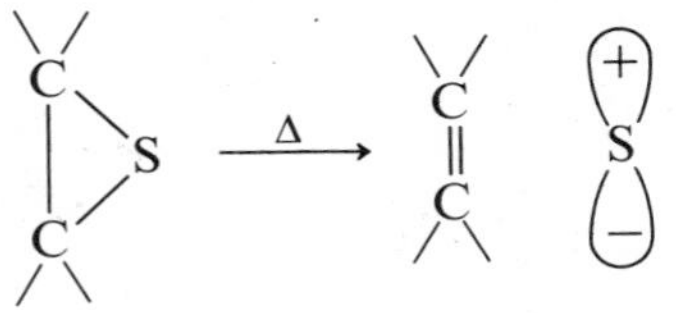

$$a_1 + b_2 \xrightarrow{\Delta} a_1 + b_2 \qquad (36)$$

To be allowed it is necessary that the $^1D$ sulfur atom be formed with a pair of electrons in the $p$ orbital of $b_2$ species. The other $p$ orbital of $a_1$ symmetry must be empty.

The reverse of the allowed (36) has already been discussed (p. 336). Because of the simplicity of a monatomic reactant, the 1,4 addition reaction of a sulfur atom with butadiene is also allowed.

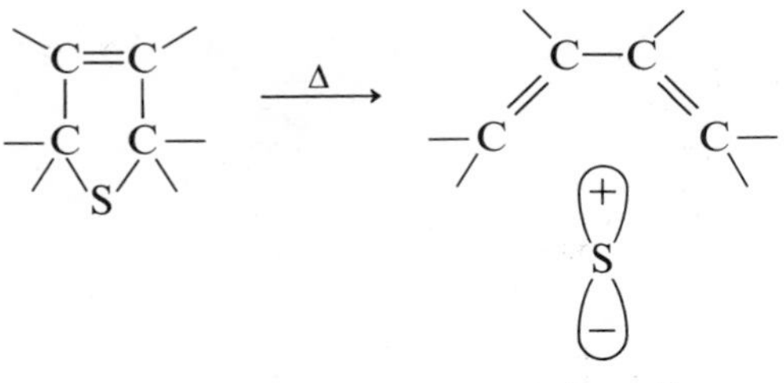

$$3a' + a'' \xrightarrow{\Delta} 3a' + a'' \qquad (37)$$

In this case, however, the sulfur $^1D$ atom must have the $p$ orbital of $a'$ species doubly occupied, and the $a''$ orbital empty. A more complex example is the decomposition of a 1,4-dithiadiene ring[29]

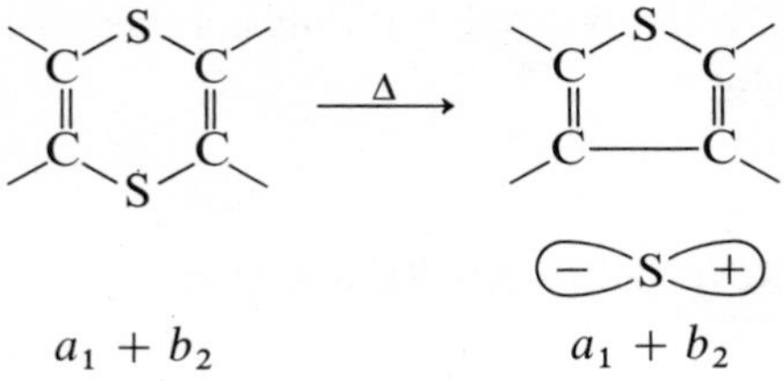

(38)

$a_1 + b_2$ $\qquad$ $a_1 + b_2$

If the molecule eliminated in an extrusion reaction is diatomic, the mechanisms are more restrictive. Typical examples are loss of CO or $N_2$. The loss of carbon monoxide from cyclopropanone is a forbidden reaction.

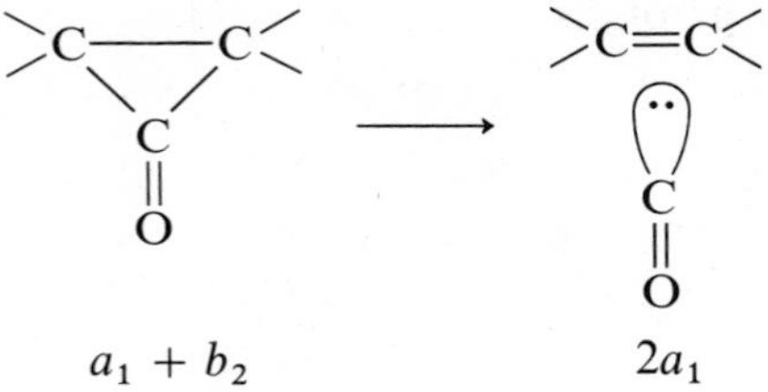

(39)

$a_1 + b_2$ $\qquad$ $2a_1$

Because of the axial symmetry of CO, a pair of electrons must appear in an orbital of $a_1$ species, a lone pair on carbon. Actually both cyclopropanone and cyclopropenone lose CO rather easily. This is explained by assuming that a non-LM reaction path is followed, similar to the reaction of a carbene with an olefin.[17] The loss of CO from cyclopentenone is allowed. A plane of

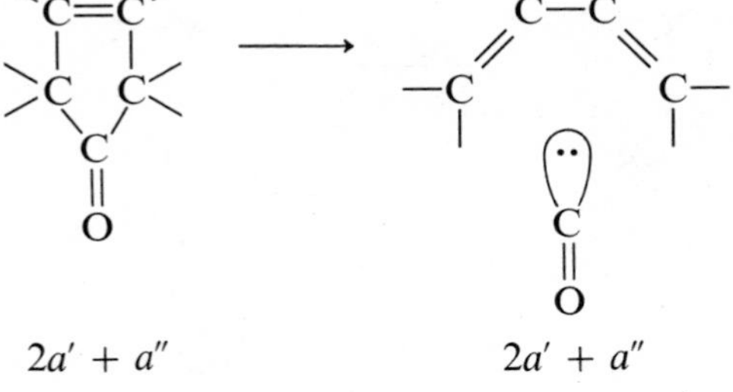

(40)

$2a' + a''$ $\qquad$ $2a' + a''$

symmetry is maintained and hence the rotation of the methylene groups is disrotatory.

A cycloheptadieneone should lose carbon monoxide by a conrotatory process.[10]

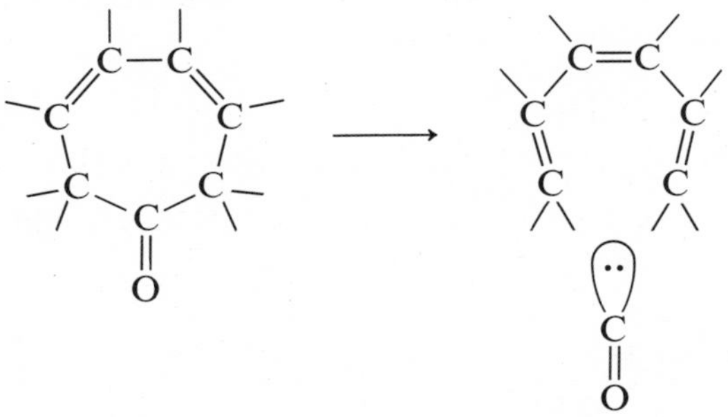

(41)

$2a' + 2a''$ $\quad$ $3a' + a''$ forbidden

$2a + 2b$ $\quad$ $2a + 2b$ allowed

This preserves a two fold axis. The bonds made and broken can be made to balance in this point group. The decomposition of a cyclopentadieneone in a concerted reaction is allowed.

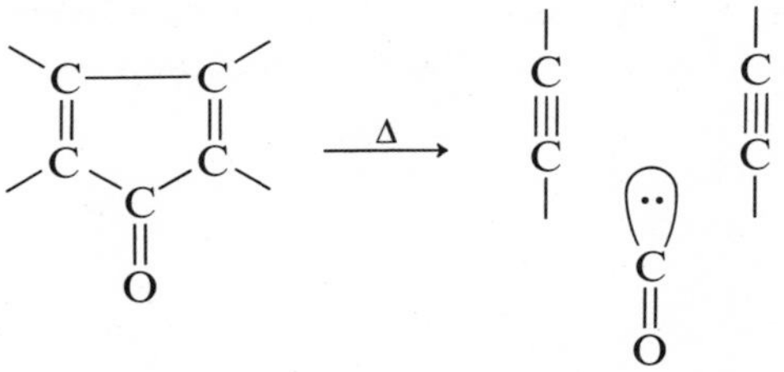

(42)

$2a_1 + b_2$ $\quad$ $2a_1 + b_2$

Note that cyclobutadiene cannot be an intermediate, since its decomposition into acetylene is forbidden.

Similar reactions to the above in which $N_2$ is lost are known. The

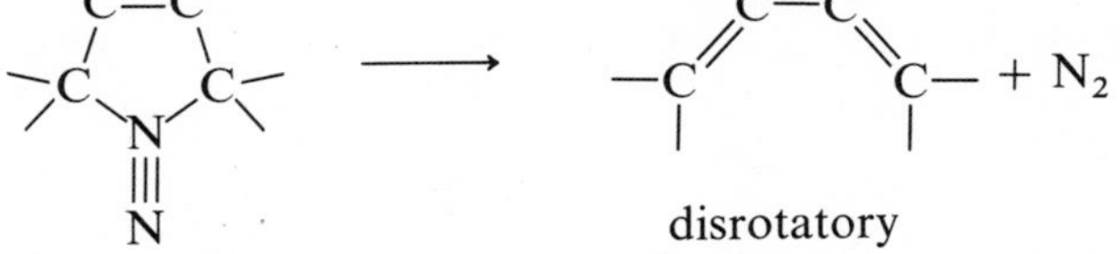

(43) $\longrightarrow$ $+ N_2$

disrotatory (44)

$\xrightarrow{\Delta}$ $>C{=}C< + N_2$

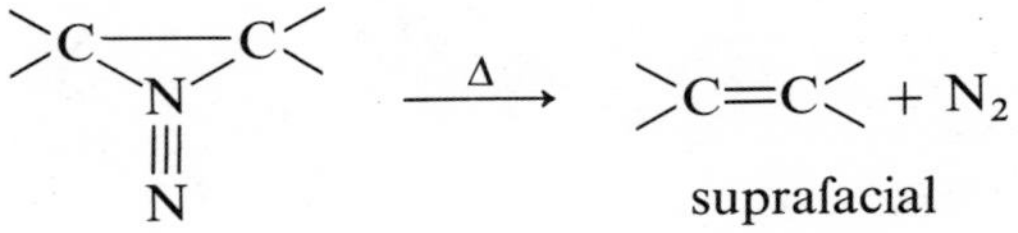

suprafacial

fact that loss of $N_2$ from an olefin corresponds to *cis* addition, suggests that a non-linear cheletropic path is followed.

Loss of $N_2$ from molecules where both N atoms are originally in the ring, would not be cheletropic reactions.

$$\text{cyclo-}[CH_2CH_2CH_2N{=}N] \;(a' + a'') \xrightarrow{\Delta} \text{cyclo-}C_3H_6 + N{\equiv}N \;(2a') \quad \text{forbidden} \tag{45}$$

$$\text{cyclo-}[HC{=}CH\,CH_2N{=}NCH_2] \;(2a' + a'') \xrightarrow{\Delta} H_2C{=}CH{-}CH{=}CH_2 + N{\equiv}N \;(2a' + a'') \quad \text{allowed disrotatory} \tag{46}$$

These are simple retrocycloaddition reactions.

Nitrous oxide is very easily lost from N-nitrosoaziridines. This is an allowed reaction if the structure has $C_s$ symmetry as shown.[30]

$$\text{N-nitrosoaziridine} \;(a' + a'') \longrightarrow {>}C{=}C{<} \;(a') + N{=}N{=}O \;(a'') \tag{47}$$

The antisymmetric combination of C—N bonds correlates with a nonbonding $\pi$ orbital of linear $N_2O$. Oddly enough neither the corresponding five- or seven-membered ring nitrosamines loses nitrous oxide at all readily.[30]

Loss of $SO_2$ from an episulfone is a forbidden reaction in the suprafacial mode.[31]

$$\text{episulfone} \;(a_1 + b_2) \longrightarrow {>}C{=}C{<} + SO_2 \;(2a_1) \tag{48}$$

Nevertheless it occurs readily and stereospecifically. The reaction path again seems to be a nonlinear one, at least in nonpolar solvents. In polar solvents a zwitterionic diradical mechanism seems to operate.[32] The following examples obey the normal symmetry rules.[33]

```
HC—CH2   .O                HC═CH2
||    \S·                  |              + SO2        (49)
HC—CH2 /  ◥O    ——Δ——→     HC═CH2

   2a' + a''               2a' + a''  allowed
```

```
     HC—CH2   .O                 CH═CH2
H2C<  |    \S·          ——→   H2C             + SO2    (50)
     HC—CH2 /  ◥O                CH═CH2

   2a' + a''                2a' + a''  allowed
```

```
   HC—CH                       HC═CH
 HC     CH                   HC     CH  + SO2
  \     /         ——→        ||     ||             (51)
H2C—S—CH2                   H2C     CH2
    O2

   2a' + 2a''               3a' + a''  forbidden
```

### Sulfur-dioxide Orbitals

Sulfur dioxide is a very versatile molecule that can form adducts with a large number of reagents. It is instructive to examine the frontier orbitals of $SO_2$ to understand the formation of these adducts, and their structures. Figure 5 shows the orbitals. The molecule can act as a Lewis acid, utilizing the LUMO of $b_1$ symmetry, or as a base, using either the $a_2$ orbital concentrated on the two oxygen atoms, or the $a_1$ orbital concentrated on sulfur. The geometries of the adducts formed in all three cases are also shown in Fig. 5.

Hard Lewis acids, where electrostatic bonding is important will bind to an oxygen atom. Examples include $AlCl_3$, $BF_3$, $SbF_5$, and $TiCl_4$. Soft Lewis acids, where covalency is more important will bind to the sulfur atom. The product should be coplanar, as shown in Fig. 5. The simplest example would be the reaction,

$$({}^3P)O + SO_2 \longrightarrow ({}^1A_1')SO_3 \qquad (52)$$

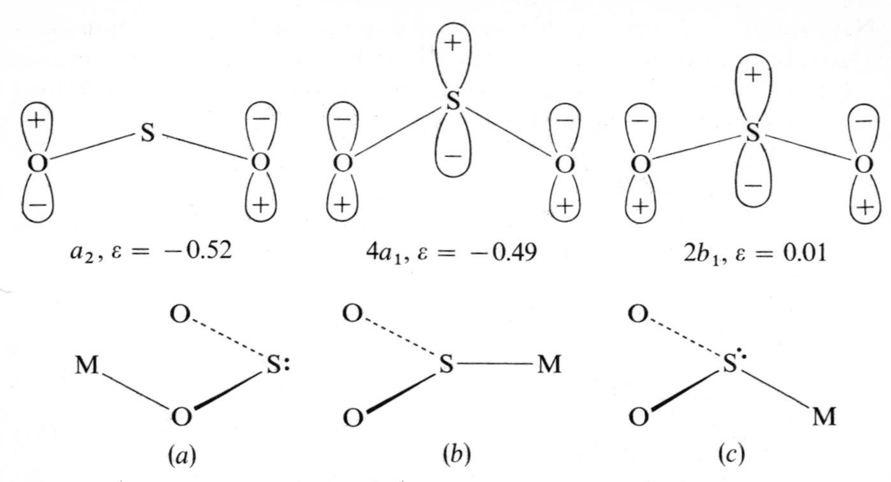


**Figure 5.** The two HOMOs of $SO_2$, and the LUMO, $2b_1$: (*a*) structure of adduct when M acts as a Lewis acid on oxygen; (*b*) structure when M acts as Lewis acid on sulfur (planar); (*c*) structure when M acts as a Lewis base (pyramidal) [energies from S. Rothenberg and H. F. Schaefer, III, *J. Chem. Phys.*, **53**, 3014 (1970)].

which is very fast, even though spin-forbidden.[34] The product, $SO_3$, is of course planar, as predicted. Also the metal complexes $Ru(NH_3)_4ClSO_2^+$ and $Pt[P(C_6H_5)_3]_3SO_2$ have structures with the metal atom coplanar with the $SO_2$ moiety.[35]

In other cases the sulfur-dioxide–metal-atom grouping is pyramidal. Examples are $Ir(CO)Cl[P(C_6H_5)_3]_2SO_2$ and the corresponding rhodium complex. The nonplanar structures show clearly that in these cases the metal atom is acting as a base. That is, two electrons from a *d* orbital are transferred to $SO_2$ and an oxidation to Ir(III) has occurred.[36] The adduct $(CH_3)_3N \cdot SO_2$, where the amine surely is a base, has the predicted pyramidal structure about sulfur. So also does the complex $Pt[P(C_6H_5)_3]_2CH_3I \cdot SO_2$, where iodine bonds to sulfur.[37]

Sulfur dioxide forms complexes with the halide ions in a variety of solvents. The order of stability in water is $I^- > Br^- > Cl^- > F^-$, but in acetonitrile the order is reversed.[38] These complexes are of some importance, since they increase the reducing ability of $SO_2$. It has already been pointed out that the direct reaction (p. 115)

$$SO_2 + Cl_2 \longrightarrow SO_2Cl_2 \qquad (53)$$

is forbidden. The reducing electrons of $SO_2$ are in an $a_1$ orbital (Fig. 5) and cannot lead to a concerted oxidative–addition reaction. The pyramidal complex $SO_2Cl^-$ still has its reducing electrons in a symmetric $a'$ orbital on sulfur, but they are now in an orbital of much higher energy. The nucleophilic substitution reaction

$$ClSO_2^- + Cl_2 \longrightarrow ClSO_2Cl + Cl^- \quad (54)$$

can occur readily. Similarly $HSO_3^-$ is a much better reducing agent than $SO_2$.

Cyclization reactions in which a nucleophilic atom adds to O and an electrophilic atom adds to S are also possible. For example, cyclopropanone will add to form a cyclic sulfite ester[16]

$$\text{cyclopropanone } (H_2C{-}CH_2 \text{ ring with } C{=}O) + SO_2 \longrightarrow \text{cyclic sulfite ester } (H_2C{-}C({=}CH_2){-}O{-}S({=}O){-}O{-}) \quad (55)$$

Sulfur dioxide undoubtedly forms a four-membered cyclic dimer with itself, since $SO_2^{16}$ and $SO_2^{18}$ exchange rapidly in $CCl_4$ solution or in the vapor phase.[39] However solid $SO_2$ does not appear to be associated, unlike $SeO_2$, which crystallizes in infinite chains.

## SIMPLE RING OPENINGS

Returning to the subject of ring-opening reactions, we consider examples of simple cyclic hydrocarbons. If a carbon—carbon bond is broken, a diradical must eventually result.

$$(CH_2)_{n+2} \xrightarrow{\Delta} \cdot CH_2(CH_2)_nCH_2\cdot \quad (56)$$

A diradical has two isolated electrons in degenerate, or nearly degenerate, orbitals.[40] The two electrons can be paired, leading to a singlet state, or unpaired, leading to a triplet. Of course two electrons sufficiently far apart would correspond to two independent doublets. However our interest is not in such non-interacting systems, but in cases of varying degrees of interaction.

The thermal reactions of cyclopropane, and its derivatives, have been the subject of a great deal of experimental and theoretical study.[41] There is a clean unimolecular decomposition into propylene. The classical mechanism postulates a trimethylene diradical as an intermediate.[42]

$$\text{cyclo-}(CH_2)_3 \underset{k_{-1}}{\overset{k_1}{\rightleftharpoons}} H_2C\cdot \; CH_2 \; \cdot CH_2 \text{ (}\cdot CH_2CH_2CH_2\cdot\text{)} \xrightarrow{k_2} CH_2{=}CH{-}CH_3 \quad (57)$$

Proof for the intermediate was found in the geometrical (*cis–trans*) isomerization which accompanied the reaction, for example,

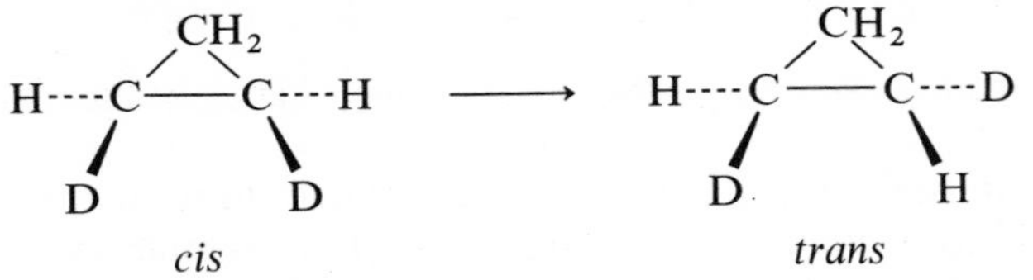

CH$_2$ / H····C—C····H / D D (*cis*) ⟶ CH$_2$ / H····C—C····D / D H (*trans*) (58)

Racemization of chiral cyclopropanes also occurs. These products are explicable by twisting of the methylene groups in the diradical, followed by ring closure. The overall activation energy is about 65 kcal, which agrees with estimates of the carbon—carbon bond energy in cyclopropane.

Several ab initio calculations have been made of the potential energy surface for singlet trimethylene.[43] The major conclusions can be described in terms of three conformations of trimethylene, "called face-to-face," "edge-to-face," and "edge-to-edge," as well as the interior bond angle, $\alpha$.

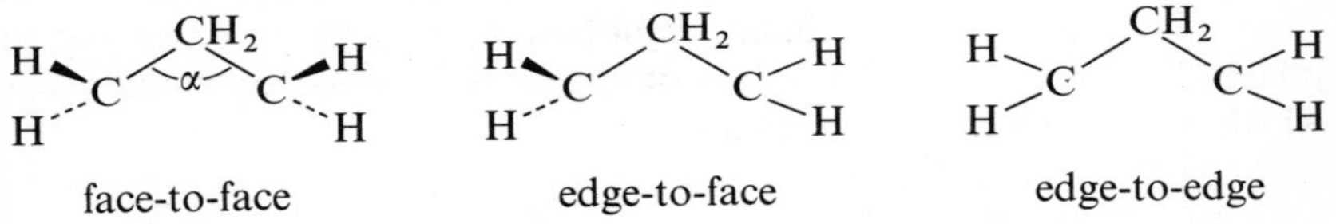

face-to-face edge-to-face edge-to-edge

Initially in cyclopropane, $\alpha$ is 60° and the conformation is face-to-face. Upon ring opening, $\alpha$ increases while the methylene groups remain face-to-face. At about $\alpha = 100°$, rotation of the methylene groups requires little energy, or even lowers the energy. Thus the edge-to-face and edge-to-edge conformers can be formed. For larger values of $\alpha$, the edge-to-edge conformer becomes slightly more stable than either face-to-face or edge-to-face. There are no deep local minima, corresponding to trimethylene as a long-lived intermediate. Also there are no preferred directions of ring closing, corresponding to stereospecific disrotatory or conrotatory processes.

The rearrangement to propylene requires the edge-to-edge conformer, with a value for $\alpha$ near 120°. A 1,2 hydrogen shift must occur with little activation energy, since it must compete with ring closure, which has zero or small activation energy.

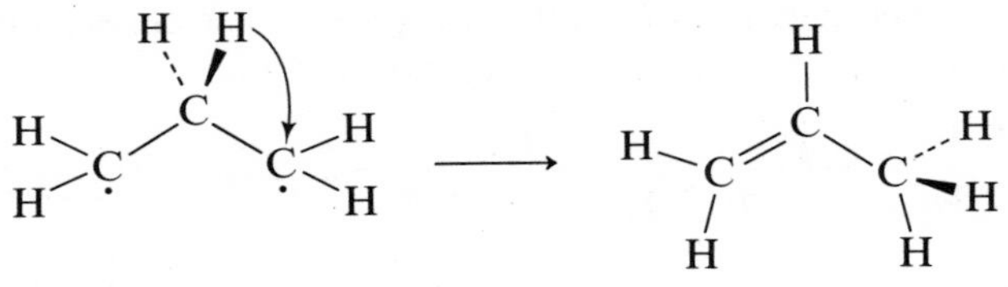

$\dot{C}H_2$—CH$_2$—$\dot{C}H_2$ ⟶ H$_2$C=CH—CH$_3$ (59)

While normally 1,2 hydrogen shifts in free radicals do not occur (p. 95), (59) is very favorable because of the concomitant creation of the double bond.

The MO configuration of cyclopropane is

$$(a_1')^2(e')^4(a_2'')^2(2a_1')^2(e'')^4(2e')^4(3e')^0(3a_1')^0(3a_2')^0$$

The frontier orbitals were shown in Fig. 4. The lowest energy excited state comes from $(2e') \rightarrow (3e')$, and leads to a reaction coordinate of $E'$ type. This agrees with the theoretical calculation since the face-to-face conformer has $C_{2v}$ symmetry, as generated by an $E'$ mode. Twisting of the methylene groups requires a reaction coordinate of $E''$ species. The small energy gap between $2e'$ and $3e'$ favors breaking the carbon—carbon bond, compared to acyclic hydrocarbons.

A similar situation exists in ethylene oxide, which has a strained three membered ring. The frontier orbitals of $C_2H_4O$ are shown in Fig. 6. The lowest energy excitation is $(6a_1) \rightarrow (4b_2)$, which gives a reaction coordinate of $B_2$ symmetry. This corresponds to the nonsymmetric ring-opening mode.

$$H_2C\text{—}CH_2 \text{ (with bridging O)} \xrightarrow{\Delta} \cdot CH_2\text{—}CH_2\text{—}O\cdot \qquad (60)$$

Admittedly other orbitals are so close to both the HOMO and LUMO in energy that this prediction cannot be taken seriously, even though it agrees with the experimental results.

The pyrolysis of ethylene oxide gives CO and $CH_4$ as products. These are presumably produced by the sequence of reactions given by (60), followed by a hydrogen atom transfer as in (59).[44]

$$\cdot CH_2\text{—}CH_2\text{—}O\cdot \longrightarrow CH_3CHO \longrightarrow CH_4 + CO \qquad (61)$$

Acetaldehyde is known to decompose into the observed products by a free-radical chain process. The rate is considerably faster than the overall rate for

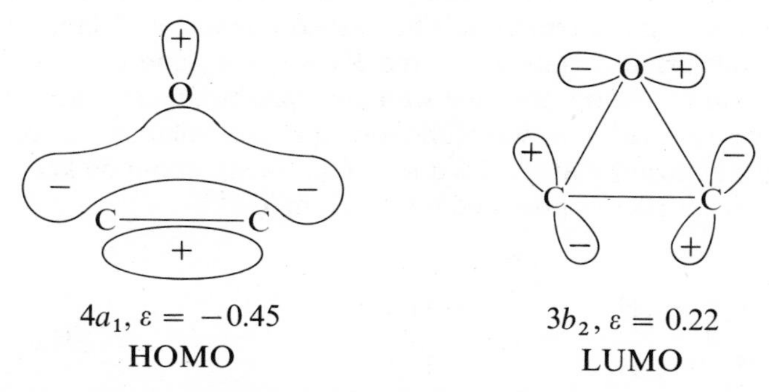


**Figure 6.** The frontier orbitals of ethylene oxide (energies from reference 15).

$C_2H_4O$ decomposition. The latter activation energy, which corresponds to (60), is 52 kcal.

## Cyclobutane Ring Opening and its Reverse

The ring-opening reaction of cyclobutane has also been studied theoretically. The MO sequence is

$$\cdots (b_{1g})^2(b_{1u})^2(2e_u)^4(2a_{2u})^0(3e_u)^0(a_{2g})^0$$

for cyclobutane, assumed to be planar $D_{4h}$.[46] All the lowest energy transitions predict a transition density of *gerade* type. This is definitely not correct, the preferred mode being an $E_u$ vibration producing the tetramethylene diradical. Figure 7 shows the potential surface for an assumed $B_{1g}$ reaction coordinate. This corresponds to the forbidden suprafacial dissociation into two ethylene molecules.

$$\text{cyclobutane} \xrightarrow{B_{1g}} \text{(C—C ring with } R_1, R_2) \xrightarrow{A_{1g}} 2\ \mathrm{C{=}C} \qquad (62)$$

The intermediate point group is $D_{2h}$.

The energy in Fig. 7 is plotted as a function of the two carbon—carbon distances $R_1$ and $R_2$. At $R_1 = 1.42$ Å and $R_2 = 2.21$ Å, a saddle point is reached. It lies 156 kcal above the energy of cyclobutane, and is the TS for the forbidden process. Figure 8 shows the orbital correlation diagram. There is a crossing of the $b_{2u}$ and $b_{3u}$ orbitals in going from cyclobutane to two ethylene molecules. Thus to the right of the diagram we have (for the $\pi$ electrons of ethylene only) a configuration $\psi_1 = (a_g)^2(b_{3u})^2(b_{2u})^0$, and to the left a configuration $\psi_2 = (a_g)^2(b_{2u})^2(b_{3u})^0$.

At the TS the energies of $\psi_1$ and $\psi_2$ become equal, or would if they did not mix by configuration interaction. The dashed line in Fig. 7 labeled $\Delta E = 0$ is the line for various values of $R_1$ and $R_2$ where $\psi_1$ and $\psi_2$ have the same calculated energy. Below this line we have cyclobutane and above the line two ethylene molecules, or wavefunctions that are chiefly $\psi_1$ or $\psi_2$, respectively. The stabilization at the TS due to CI is large, about 56 kcal.[46]

A much easier path is provided by the $E_u$ mode.[46a]

$$\begin{matrix} H_2C\text{—}CH_2 \\ | \quad\ \ | \\ H_2C\text{—}CH_2 \end{matrix} \xrightarrow{E_u} \begin{matrix} H_2C^{\alpha}\text{—}CH_2 \\ | \\ H_2C^{\alpha}\text{—}CH_2 \end{matrix} \longrightarrow 2C_2H_4 \qquad (63)$$

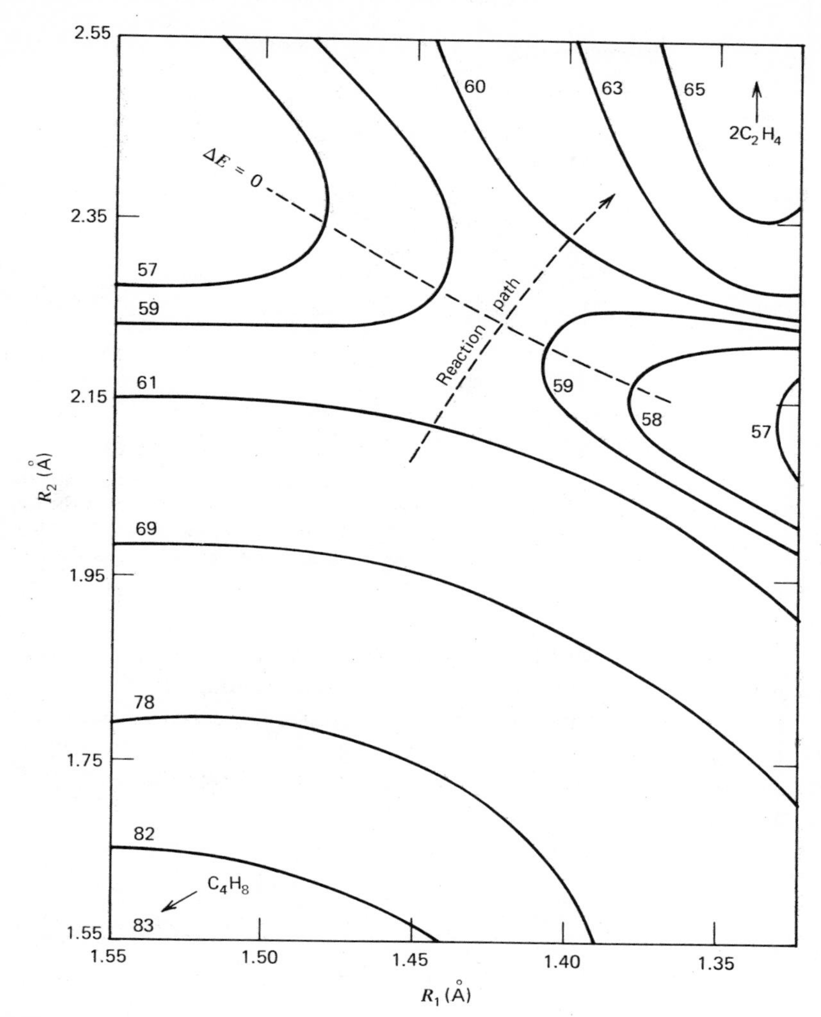


**Figure 7.** *Ab Initio* potential energy surface calculated for the decomposition of $C_4H_8$ into $2C_2H_4$ by a $B_{1g}$ reaction path, with $R_1$ and $R_2$ as the two carbon—carbon bond distances. (for the line $\Delta E = 0$, see text); the contour-line energies are the two digits after the decimal point in $-155.00$ AU [Reprinted with permission from J. S. Wright and L. Salem, *J. Am. Chem. Soc.*, **94**, 322 (1972). Copyright by the American Chemical Society.]

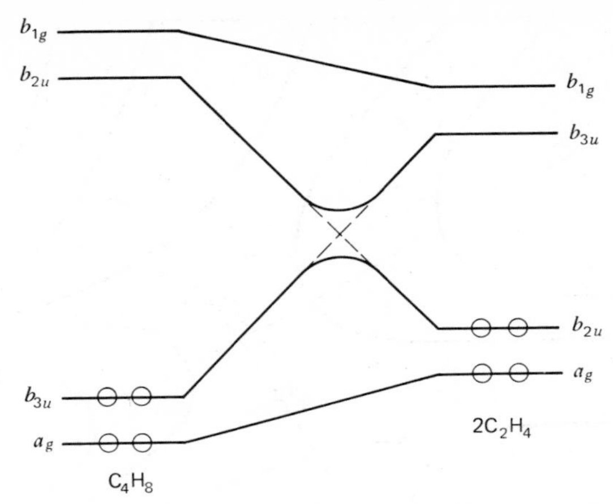


**Figure 8.** Correlation diagram for $C_4H_8 \rightarrow 2C_2H_4$ with the elements of $D_{2h}$ preserved; the sum and difference of the carbon—carbon $\sigma$ bonds of $C_4H_8$, are shown, along with the sum and difference of the $\sigma^*$ antibonding partners, and for ethylene, the sum and difference of the $\pi$ and $\pi^*$ orbitals is shown; the dashed line shows the intended correlation before configuration interaction.

Initially the two terminal $CH_2$ groups are face-to-face, but unlike the cyclopropane case, variation of the interior angle $\alpha$ is not enough to describe the lowest energy path. Twisting about the central carbon—carbon bond is also required. Also twisting of the terminal $CH_2$ produces small changes in the energy. In short, a number of conformations exist of which the most stable is the *trans* form, E,

H H
H C—C
H
C—C H
H
H H
E

While there are local minima, due to conformers, there are no deep minima due to a stabilized diradical. The path of two ethylene molecules is some 62 kcal downhill for all conformers, with no barrier in the way. However the existence of an energy plateau that is large in many-dimensional space does suggest that a tetramethylene free radical might be rather long-lived. That is, it would spend considerable time exploring this plateau until it found the right configuration for reaction. This gives time for isomerization, interception, or diversion.[46a] Such a diradical mechanism would not be stereospecific, in agreement with the facts.[46b]

If the triplet state were formed, by intersystem crossing, then a long-lived intermediate would truly exist. Normally the triplet state lies below the singlet state of the diradical in energy. While it might give a product in an excited state, a more normal behavior would be another intersystem crossing to return to the single diradical or a distorted form of a product.

By reversing the reactants and products, we now have an energy-economic path for the forbidden dimerization of ethylene. First two ethylene molecules are condensed to a tetramethylene diradical. This costs at least 43 kcal, since two ethylene molecules lie 19 kcal above cyclobutane in energy. It makes little difference which conformer is generated initially, since all lie on the same plateau. However it is likely that a higher-energy eclipsed conformer must be formed, for reasons of the original geometry. Once formed, the diradical will either dissociate again, or eventually close to cyclobutane with little or no additional energy required.

This still leaves it somewhat mysterious as to how the enormous energy barrier of the forbidden reaction (128 kcal for the dimerization direction) has been circumvented. Why does adding the second ethylene one carbon at a time produce such a favorable result? It is possible that simply lowering the symmetry from $D_{2h}$ to $C_{2v}$, at most, can have such an effect? This seems unreasonable since even a minute distortion of the high-energy $D_{2h}$ TS will lower the point group to $C_{2v}$ in a mathematical sense.

$$\begin{array}{c} H_2C-CH_2 \\ | \quad\; | \\ H_2C-CH_2 \\ D_{2h} \\ \text{high energy} \end{array} \xrightarrow{B_{3u}} \begin{array}{c} {}^{2}H_2C-\overset{3}{C}H_2 \\ / \qquad \backslash \\ \underset{1}{H_2C\cdot} \qquad \underset{4}{\cdot CH_2} \\ C_{2v} \\ \text{low energy?} \end{array} \tag{64}$$

The answer is that symmetry in a mathematical sense has little to do with it, except as a convenient tool. It is the improved nature of the bonding in the lower symmetry intermediate that is significant. This can be seen most easily by examining the interaction of the $\pi$ orbitals of two ethylene molecules as they approach each other in a $C_{2v}$ fashion. This is shown in Fig. 9. The corresponding analysis for a $D_{2h}$ approach is given in Fig. 8.

Initially the sum and difference of both the $\pi$ and $\pi^*$ orbitals of the two olefin molecules are taken. These generate MOs of $a_1$ and $b_2$ symmetry in $C_{2v}$ for the $\pi$, and of $a_1$ and $b_2$ for the $\pi^*$, as shown on the left side of Fig. 9. When the interaction between the two molecules is small, because of distance, the orbitals generated from the $\pi$ combinations will be most stable and will be occupied. The $b_2$ orbital has a node between the two remote carbon atoms, those labeled 2 and 3 in (64). It is antibonding between those atoms.

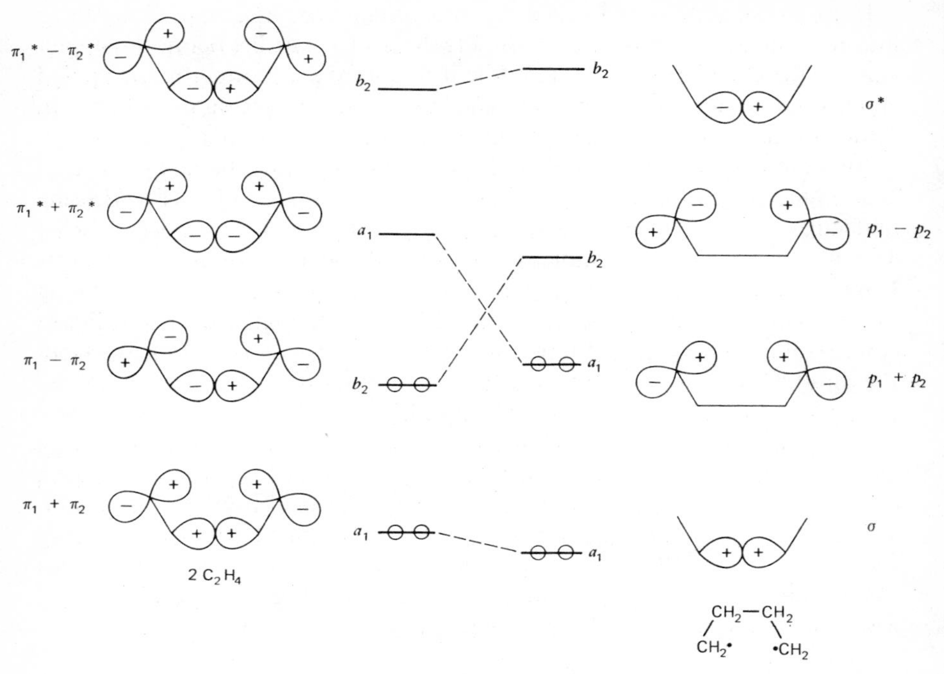


**Figure 9.** Interaction diagram for two ethylene molecules in a $C_{2v}$ mode of approach; left side, weak interaction and right side, strong interaction to form tetramethylene diradical (after reference 46).

Now bring the two olefin molecules close together so the two $p$ orbitals on atoms 2 and 3 form a $\sigma$ bond and a diradical is formed. The MOs will become as shown on the left side of Fig. 9. The $\sigma$ bond is $a_1$, and its empty antibonding partner is $b_2$. The two remaining MOs are the sum and difference of the two $p$ orbitals on atoms 1 and 4. They form a symmetric, or $a_1$, and an antisymmetric, or $b_2$, pair. Since the two atoms, 1 and 4, are now quite close, the $a_1$ orbital is lower in energy than the $b_2$. There are two electrons in each of the $a_1$ orbitals; each is a strong bonding orbital, and ring closure to cyclobutane can occur spontaneously.

However there is an orbital crossing in Fig. 9, just as there was in Fig. 8. In fact we still have a forbidden reaction, judged by this criterion.[47] The actual crossing is avoided, as usual, by a configuration interaction, the wavefunction being a mixture of $\psi_1 = (a_1)^2(b_1)^2$, and $\psi_2 = (a_1)^2(2a_1)^2$. The energetic advantage of the diradical mechanism is that $b_1$ has been raised in energy and $2a_1$ has been lowered in energy by the nonsymmetrical

nature of the bonding. This greatly reduces the additiona
bring the system to the crossing point.

On the right of Fig 9 there is a cyclobutane
carbon—carbon bond greatly stretched. The diradic
region just to the left of the crossing point. Further to
in energy between $b_1$ and $2a_1$ is too large to correspond t
diradical. The lower energy of $b_1$ and $2a_1$ in the diradical is due to in
through the $\sigma$-bond system. In the general case, either the symmetric or antisymmetric combination can lie lower.[6]

The existence of diradicals where the antisymmetric state lies lower can have chemical consequences. For example, the diradical of tricyclo[2.2.2.0] octane, F, might be generated in some way to serve as a precursor to the parent hydrocarbon

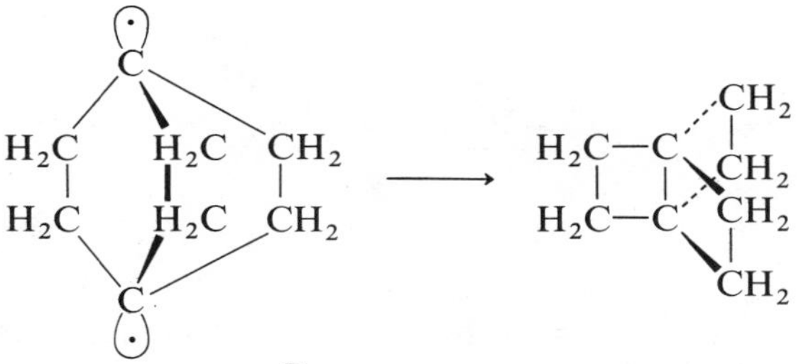

(65)

F

However this ring closure is forbidden by symmetry, since the product has a new carbon—carbon bond that is symmetric.[48] Instead F would probably rearrange in an allowed reaction to dimethylenecyclohexane.

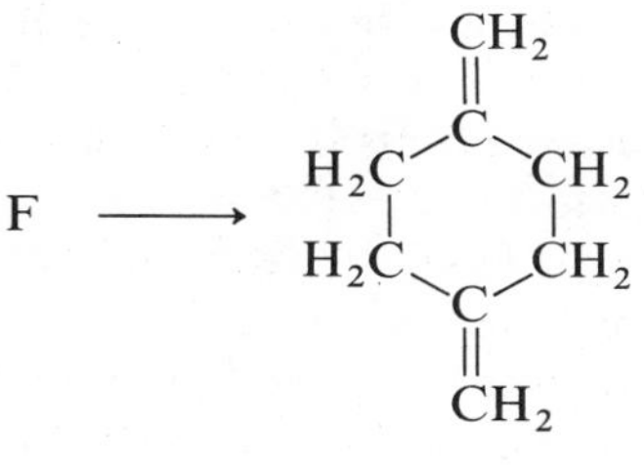

F ⟶ (66)

$a_1 + b_1$ $\quad$ $a_1 + b_1$

In view of the sometimes advantage of a diradical mechanism, as evidenced above, it might be that some reactions, allowed in a concerted fashion, still prefer a stepwise mechanism. This might be considered in the Diels–Alder condensation, for example. Fortunately there is rather convincing evidence the concerted nature of this $[_{\pi}2_s + _{\pi}4_s]$ cycloaddition.[49] This consists of the determination of the volume of activation, $\Delta V^{\ddagger}$, by measuring the rate as a function of very high pressures. The experimental values of $-30$ to $-40$ $cm^3/mol^{-1}$ are only consistent with two bonds forming at once. Of course this does not prove absolute concertedness, in the sense of a symmetrical TS, since one bond can still lead the other. Also the remarks only apply to symmetrical reactants (see p. 362).

## Electrocyclic Reactions

Detailed ab initio calculations of the ring opening of cyclobutene to *cis*-butadiene have been made, including configuration interaction.[50]

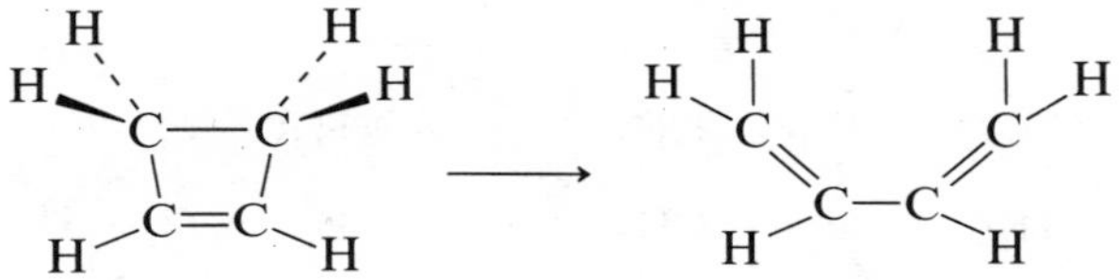

(67)

The important variables are $R$, the carbon—carbon bond distance, and $\theta$, the angle of twist of both methylenes, assumed to move synchronously. In the allowed conrotatory mode, it is found that $R$ increases markedly from 1.54 Å to 2.28 Å, before $\theta$ changes abruptly from its initial value of 0° to its final value of 90°. This seems to contradict the prediction (p. 73) that the reaction coordinate is of $A_2$ species or conrotatory.

Instead the initial reaction coordinate is $A_1$, followed eventually by $A_2$. This is not really a contradiction since totally symmetric reaction coordinates are always special in second-order PT. The effect of the initial C—C stretch on the critical MOs of cyclobutene is shown in Fig. 10. The $\sigma$ orbital, corresponding to the C—C bond, is increased in energy, while its antibonding $\sigma^*$ partner is lowered. The $\pi$ and $\pi^*$ orbitals are relatively unaffected. The key transitions $\sigma \rightarrow \pi^*$, and $\pi$ to $\sigma^*$, which promote the $A_2$ twisting mode, are now much easier. The forbidden disrotatory mode is calculated to lie some 14 kcal higher than the allowed mode.

Similar calculations have been made for the ring-opening reactions of cyclopropyl $\rightarrow$ allyl, cation, anion, and radical.[51] These reactions are usually described in the $C_{2v}$ point group, assuming planarity. In this case the cation reaction is predicted to be disrotatory (plane of symmetry saved), the anion reaction is predicted to be conrotatory ($C_2$ axis saved), and the free-radical reaction is forbidden in both modes.

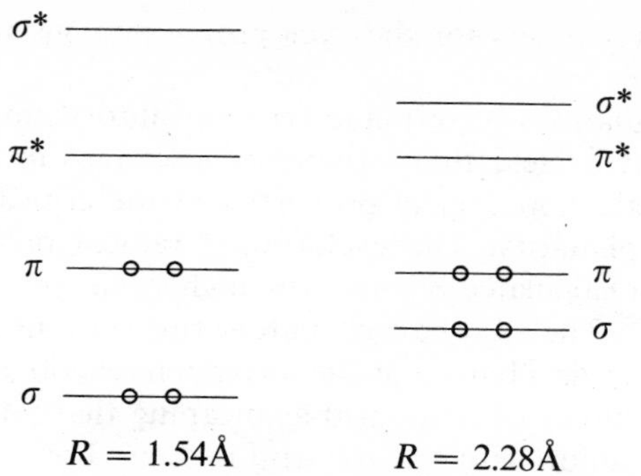

$\sigma^*$ ——— 

$\pi^*$ ——— ——— $\sigma^*$ ——— $\pi^*$

$\pi$ —⊖⊖— —⊖⊖— $\pi$

—⊖⊖— $\sigma$

$\sigma$ —⊖⊖—

$R = 1.54\text{Å}$ $R = 2.28\text{Å}$

**Figure 10.** Effect of stretching the carbon—carbon single bond in cyclobutene on critical orbital energies.

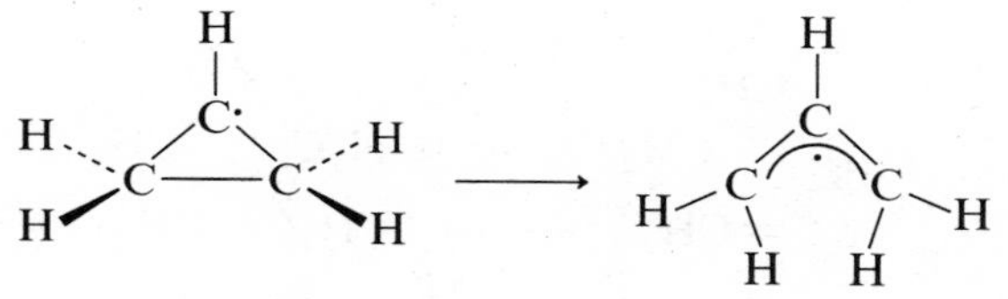

$$a' + \tfrac{1}{2}a' \qquad\longrightarrow\qquad a' + \tfrac{1}{2}a'' \tag{68}$$

$$a + \tfrac{1}{2}b \qquad\qquad b + \tfrac{1}{2}a$$

Actually in all cases the central C—H bond is moved down from the plane at some stage of the process. This reduces the point group to $C_s$, which is still consistent with an allowed disrotatory motion for the cation, and forbidden for the anion and radical. A conrotatory motion destroys all symmetry, and in principle allows all three reactions. Table 1 shows the

**Table 1 Energy Barriers (in kcal/mole$^{-1}$) for Various Cyclopropyl–Allyl Conversions**[a]

| | Disrotatory | | Conrotatory | |
|---|---|---|---|---|
| $C_3H_5^+$ | 1.6 | $(C_{2v})$ | 78.5 | $(C_{2v})$ |
| | 1.3 | $(C_s)$ | | |
| $C_3H_5\cdot$ | 43 | $(C_{2v})$ | 55 | $(C_{2v})$ |
| | 35 | $(C_s)$ | — | |
| $C_3H_5^-$ | 107 | $(C_{2v})$ | 43 | $(C_{2v})$ |
| | 97 | $(C_s)$ | | |

[a] From reference 51.

calculated energy barriers for the cyclopropyl systems in disrotatory and conrotatory modes.

While no calculations were made for conrotatory ring openings in the nonplanar case, it is clear that only small lowerings in the barrier would result. The essential topological properties of the critical orbitals are not changed by nonplanarity. The cyclopropyl radical prefers a disrotatory mode. The barrier calculated is somewhat higher than an experimental value of about 25 kcal.[52] The allowed conrotatory ring opening of the cyclopropyl anion has just as large a barrier as the forbidden reactions of the radical.

These results cannot be construed as meaning that a free radical can be considered as an anion with one electron missing. For example, (24) shows the concerted addition of an allyl anion to an olefin. The corresponding reaction of the radical is forbidden.[53]

$$H_2C{\cdot}CH{\cdot}CH_2 + H_2C{=}CH_2 \longrightarrow \text{cyclo-}(\dot{C}H{-}CH_2{-}CH_2{-}CH_2{-}CH_2) \qquad (69)$$

$$a' + \tfrac{1}{2}a'' + a' \qquad\qquad a' + a'' + \tfrac{1}{2}a'$$

Many polycyclic molecules exist even though subject to severe strain energies. The reason is often, but not always, that the logical isomerization path is forbidden. A classic example is Dewar benzene (p. 90). Another is the molecule prismane, G, which is some 90 kcal of energy higher than its isomer benzene. An examination of the transformation shows that it is forbidden.[10]

$$\text{G (prismane, } C_6H_6\text{, carbons 1–6)} \longrightarrow \text{benzene (carbons 1–6)} \qquad (70)$$

$$2a_1 + b_1 \qquad\qquad a_1 + b_1 + b_2$$

Prismane is stable at room temperature, the activation energy for decomposition being 34 kcal (hexamethylprismane).[54] While some Dewar benzene, H, is found as prismane decomposes, and may be the precursor to benzene, this is also a forbidden reaction.

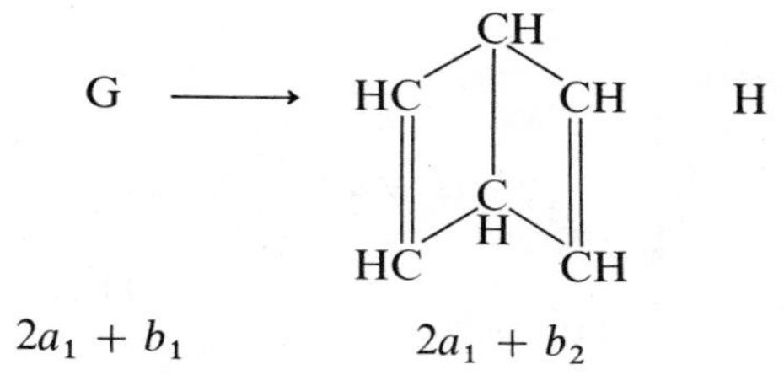

G ⟶ HC CH CH H HC C H CH (71)

$2a_1 + b_1$ $\qquad$ $2a_1 + b_2$

The molecular orbital sequence for prismane is $\cdots(e')^4(e'')^4(a_1')^0(a_2')^0(a_2'')^0$ with an unusually small gap between the HOMO and three nearly equal LUMOs, even less than for cyclopropane.[55] The predicted reaction coordinate is either $E''$ or $E'$. The latter generates a $C_{2v}$ point group, and would correspond to either an attempt at (70), or formation of a diradical by breaking the 2—6 bond. The $E''$ mode would correspond to breaking the 1—3 bond. This would be the diradical route to Dewar benzene.

It is likely that diradicals are often intermediates in the forbidden isomerizations. Dewar benzene could form a diradical with the symmetric $(a_1)$ orbital occupied:[56]

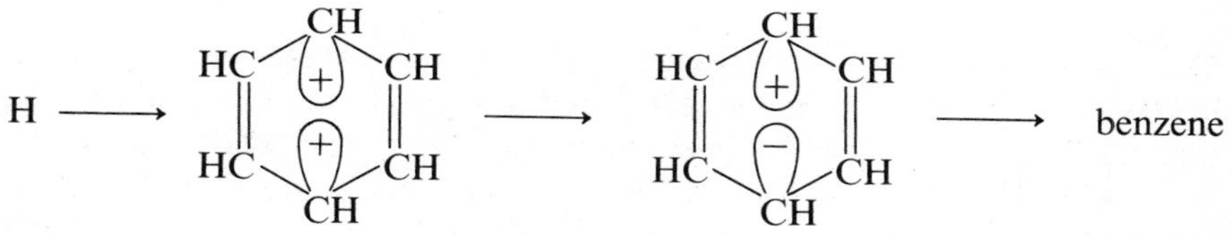

H ⟶ (HC CH CH HC CH CH, + / +) ⟶ (HC CH CH HC CH CH, + / −) ⟶ benzene

(72)

By configuration interaction this could pass over into the antisymmetric $b_1$ orbital, concomitant with the flattening out of the puckered structure to planar benzene.

Not all strained cyclic molecules owe their existence to symmetry considerations. Bicyclobutane has 70 kcal of strain energy, and yet an additional 40 kcal of activation energy must be supplied before isomerization to butadiene results.[10]

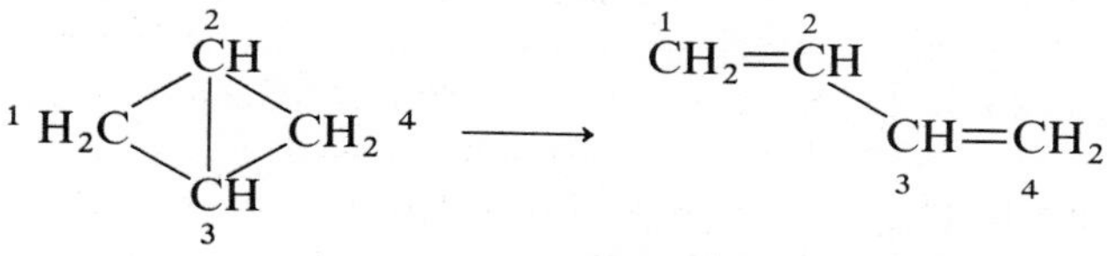

$^1H_2C(\overset{2}{C}H)(\underset{3}{C}H)CH_2{}^4 \longrightarrow \overset{1}{C}H_2{=}\overset{2}{C}H{-}\underset{3}{C}H{=}\underset{4}{C}H_2$ (73)

Offhand this looks like a typical forbidden reaction, but analysis shows that the mechanism can be an allowed $[_\sigma 2_s + {}_\sigma 2_a]$ process.[57] This is best shown by deuterium, or other, labeling. The wavy lines show which bonds break.

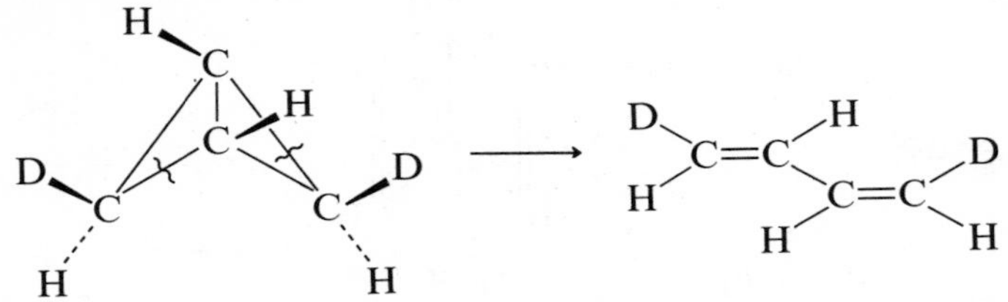

(74)

The *exo*-$d_2$ cyclobutane gives 1,4-*cis*, *trans*-$d_2$ butadiene. If the diolefin is now imagined to be condensed in an antarafacial manner back into cyclobutane, the original reactant is formed.

No elements of symmetry are conserved in the LM path for (74). The twofold axis that exists at the beginning and end is destroyed by the rotation of the methylene groups, both in the same direction. The activation energy is a result of the strain in the antarafacial TS. It is also possible that the bonds break stepwise, to give an intermediate diradical.[57c] The second bond must break rapidly, if the stereospecificity is to be saved.

## SOME SIGMATROPIC REACTIONS

Organic chemistry is rich in reaction mechanisms that involve a 1,2 shift; that is, a group moves from one carbon to the adjacent one. Examples are the pinacol–pinacolone and Wagner–Meerwein rearrangements, the Wolff, Beckmann, Hoffman, Curtius, and Schmidt reactions, and so on. When it is also demonstrated that a [1,2] sigmatropic shift is allowed in a linear polyene system (p. 95), it is a great temptation to believe that all 1,2 shifts are allowed, and conversely, all 1,3 shifts are forbidden. This is by no means the case and each system must be carefully examined to see if it meets the criteria of either: (1) an aromatic transition state for a Hückel system, or an antiaromatic one for a Möbius system or (2) positive overlap between the orbital of the migrating group and the terminal orbitals of the rest of the system. Of these, the first is usually more reliable, or less ambiguous.[58]

For example, the experimental results of the Claisen rearrangement are nicely explained by orbital-overlap considerations. Figure 11 summarizes both the results and the orbital explanation.[59] The migrating group is an allyl radical, and the residue is a phenoxy radical. The critical orbitals are $\psi_4$ in the phenoxy radical, and $\psi'_2$ in the allyl radical. The successive shifts are from oxygen to the *ortho* carbon, and from there to the *para* carbon, each step occurring with end-for-end inversion of the allyl group. Each step is considered a [3,3] sigmatropic shift, although we could count the other way around the ring and call it [3,7]. The smaller numbers seem simpler.

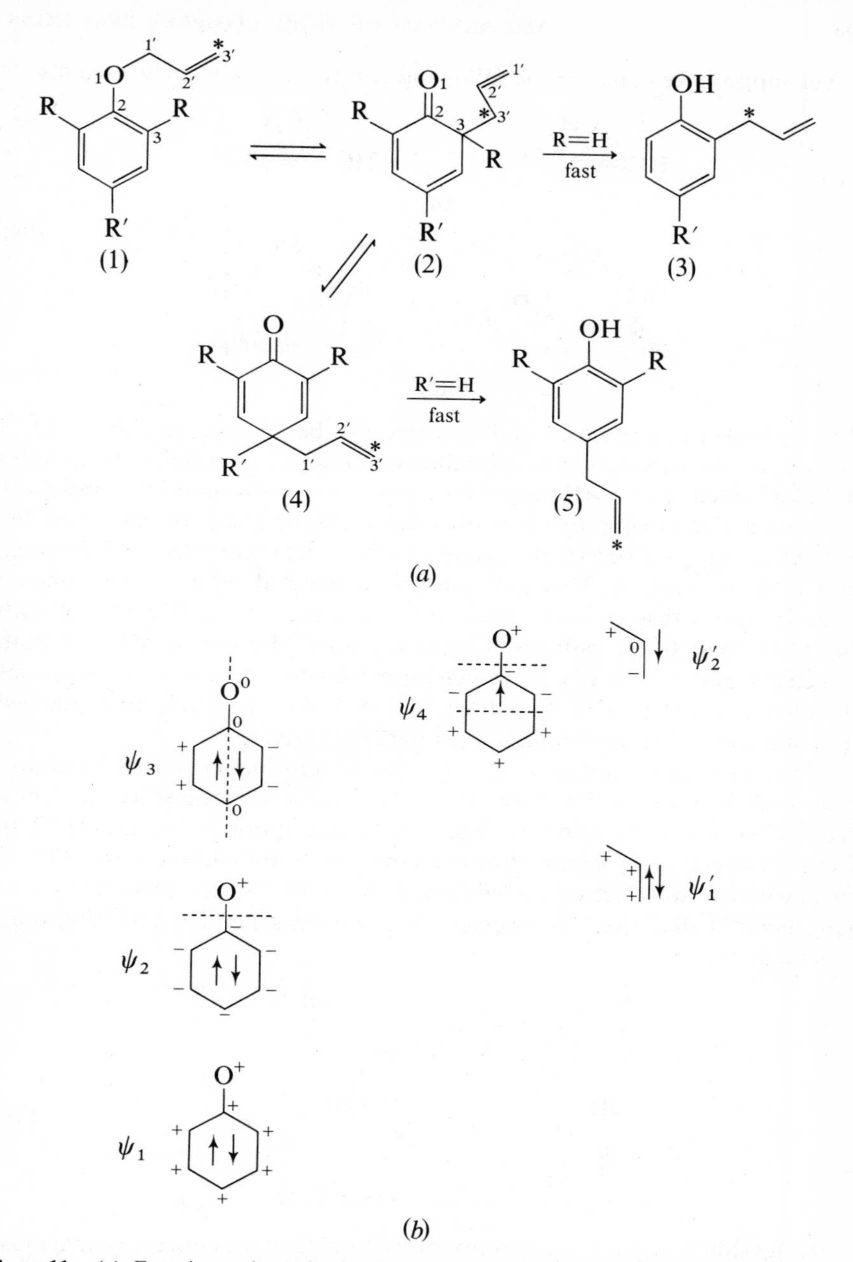


**Figure 11.** (*a*) Experimental results in the Claisen rearrangement summarized; the *para* migration occurs by a sequence of two [3,3] sigmatropic shifts; (*b*) orbital explanation for the Claisen reaction—orbital phases in $\psi_4$ and in $\psi'_2$ must match; dashed lines are the nodal planes (after reference 59).

Yet suppose we consider the following two reactions in cyclic systems:

$$\text{HC}{=}\text{CH}\text{ (ring with } CH_2) \longrightarrow \text{HC}{=}\text{CH}\text{ (ring with } CH_2) \quad (a)$$

$$\text{cyclopentadiene} \longrightarrow \text{cyclopentadiene (H shifted)} \quad (b) \qquad (75)$$

The hydrogen migration in cyclopropene can be counted as either a [1,2] process, in which case it seems to be allowed, or as a [1,3] example, whereupon it is forbidden. For cyclopentadiene, there is a corresponding ambiguity between a sigmatropic reaction of order [1,2], or [1,5]. In each case the correct answer is given by the larger numbers. Reaction (75*a*) is forbidden, and (75*b*) is allowed. This ambiguity is connected with another one in selecting the critical orbitals. There is a degeneracy in the HOMOs of both the $C_3H_3$ and $C_5H_5$ radicals. Figure 12 shows the lowest MOs of both radicals. Combination of a hydrogen atom with two adjacent carbon atoms is either allowed ($\psi_3$) or forbidden ($\psi_2$) in $C_3H_3$. In $C_5H_5$ it is allowed ($\psi_3$) and somewhat uncertain, but probably allowed ($\psi_2$).

The important feature of (75*a*), which makes it forbidden, is that it requires a four-electron cyclic TS. Similarly (75*b*) is allowed because it involves a six-electron cyclic intermediate. The correct description is that it is [1,5]. In fact suprafacial [1,2] sigmatropic rearrangements are allowed only in those cases where a two-electron cycle is adequate to describe the reacting system. This includes all of the "open-sextet" mechanisms generalized by Whitmore years ago.[60]

$$-\overset{R}{\underset{OH}{C}}-\overset{+}{C}< \longrightarrow -\underset{\overset{\|}{OH^+}}{C}-\overset{R}{C}<$$

$$O{=}\overset{R}{C}-\ddot{N}: \longrightarrow O{=}C{=}\overset{R}{N}: \quad \text{(etc.)} \qquad (76)$$

If a 1,2 shift brings a $\pi$ system into interaction, then the entire $\pi$ system must be included. In a cyclic molecule such as cyclopentadiene, an allowed [1,5] sigmatropic change incidentally creates a 1,2 shift. However this would not be true for rings of other sizes. In the cycloheptatriene ring, the allowed

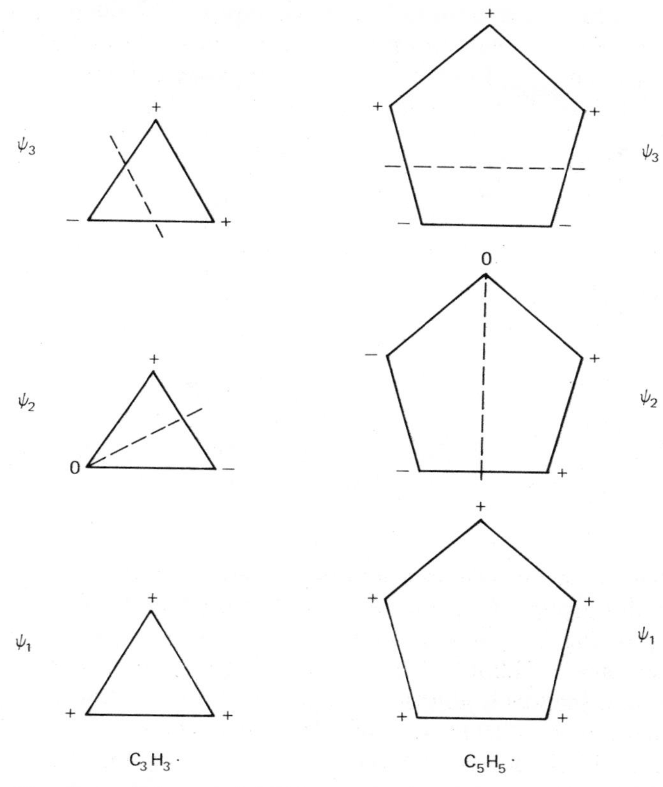


**Figure 12.** The phases of the occupied MOs of $C_3H_3\cdot$ and $C_5H_5\cdot$, showing the degeneracy of the HOMO; there are three electrons in the $\pi$ orbitals of cyclopropenyl radical and five electrons in cyclopentadienyl radical.

[1,5] process would occur, but the forbidden [1,7] sigmatropic reaction would not.

(77)

forbidden [1,7] allowed [1,5]

The latter, of course, would be a 1,2 shift. The actual reaction could be described as a 1,4 shift. The 1,2 shift does not take place because it creates an unstable eight-electron TS.

It is illuminating to examine an "allowed" and a "forbidden" sigmatropic change from the PT viewpoint. In order to have useful symmetry elements, take the forbidden 1,2 shift of a $\sigma$-bonded protonated cyclobutadiene and the allowed 1,2 shift of the benzenium ion.

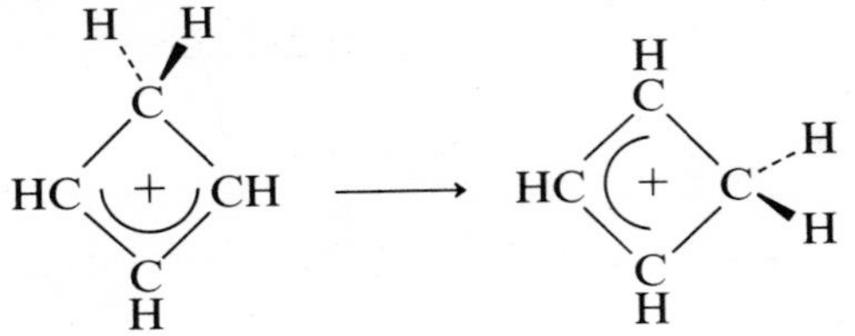

forbidden (78)

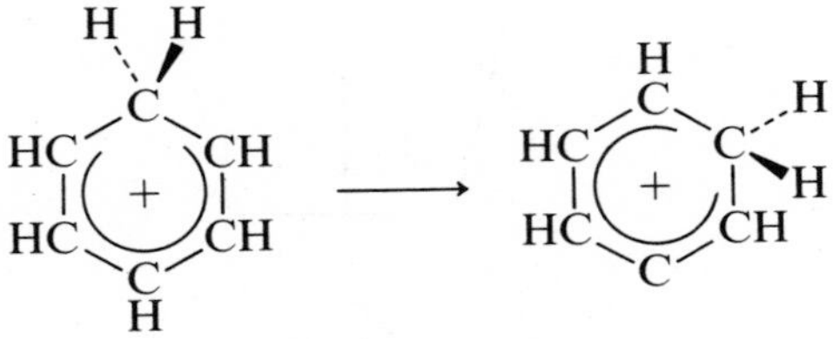

allowed (79)

The former has a four-electron TS and the latter, a six-electron one. The original point group is $C_{2v}$, and the key MOs for both $C_4H_5^+$ and $C_6H_7^+$ are shown in Fig. 13. These are the orbitals of the $CH_2$ group which corresponds to the C—H bonds. They are of $a_1$ and $b_1$ species. Also the occupied $\pi$ orbitals and the lowest empty $\pi$ orbitals are shown. These alternate from $b_1$ to $a_2$ in order, with the most stable always being $b_1$, in the $C_{2v}$ point group. In both cases it will be assumed that the $\pi$-bonded complex is the activated complex for the shift.

The electronic movement required must be from the C—H bond to be broken, into the electrophilic $\pi$ system. This single C—H bond is a mixture

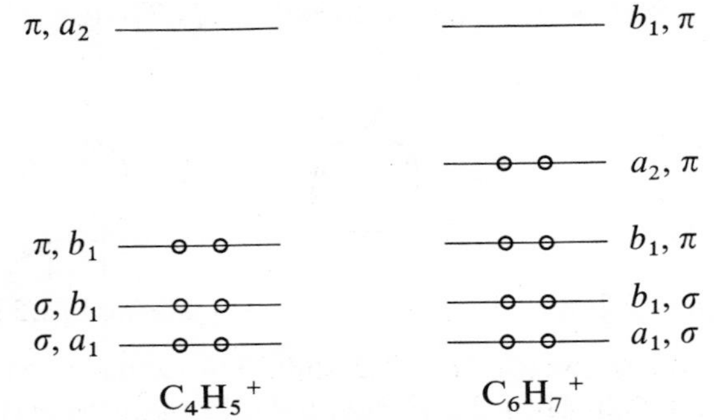

| $C_4H_5^+$ | $C_6H_7^+$ |
| --- | --- |
| $\pi, a_2$ | $b_1, \pi$ |
| | $a_2, \pi$ |
| $\pi, b_1$ | $b_1, \pi$ |
| $\sigma, b_1$ | $b_1, \sigma$ |
| $\sigma, a_1$ | $a_1, \sigma$ |

**Figure 13.** The critical molecular orbitals for a 1,2 shift in $C_4H_5^+$ and $C_6H_7^+$.

of $a_1$ and $b_1$ components. The accepting $\pi$ orbital is $a_2$ in $C_4H_5^+$ and $b_1$ in $C_6H_7^+$. We can now predict the initial reaction coordinates. In $C_4H_5^+$, it is a mixture of motions of $B_2$ and $A_2$ symmetry. In $C_6H_7^+$, it is a mixture of $A_1$ and $B_1$ motions. An $A_2$ mode is a twist of the $CH_2$ group, and a $B_2$ mode is a left-to-right wagging motion. An $A_1$ mode is simply a change in all bond lengths and bond angles of the $CH_2$ group, corresponding to a weakened bonding of one hydrogen atom, while the $B_1$ mode is an up-and-down rocking motion of the $CH_2$ group.

If these motions were continued, they would produce the following changes

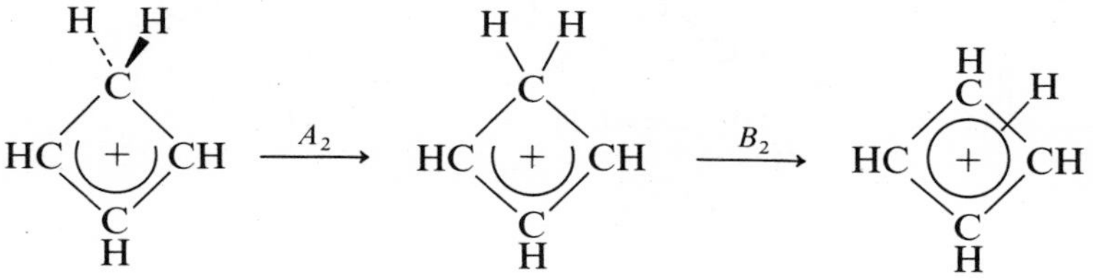

$$\xrightarrow{A_2} \qquad \xrightarrow{B_2} \qquad (80)$$

$$\xrightarrow{A_1 + B_1} \qquad (81)$$

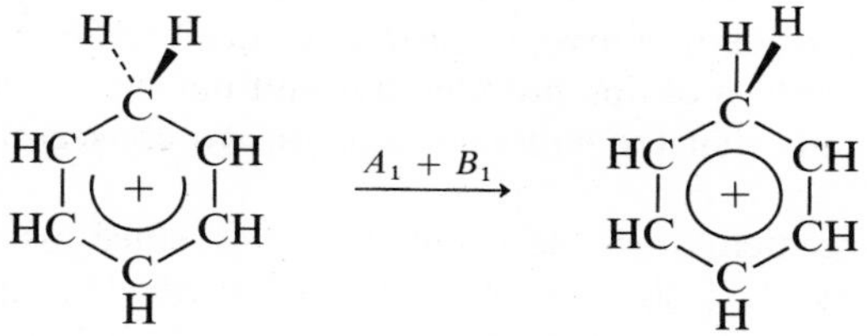

That is, the $\sigma$ complex of cyclobutane would convert into a $\pi$ complex with the proton lying in the plane of the ring. This is a very high energy structure, since the $\pi$ electron density has a node in this plane. Thus an activated complex suitable for the 1,2 proton jump has been reached, but it is an unfavorable one.

The $\sigma$-bonded benzenium ion would convert to a $\pi$ complex with the proton centered on one carbon, but correctly above the plane of the ring. This is not the stable structure where the proton bisects the carbon—carbon bond, but it is not much higher in energy. At any rate, the remainder of the reaction path would be downhill energetically. The actual reaction coordinate for 1,2 shifts in $C_6H_7^+$ would bring in mixing of $a'$ and $a''$ orbitals early on to give the left-to-right movement needed. The characteristic feature of the "allowed" reaction is that the symmetry of the chemically important orbitals favors nuclear motions that lead to the lowest-energy $\pi$ complex. The overall activation energy is small. In the "forbidden" reaction, the favored nuclear motions lead to a $\pi$ complex of very much higher energy. While it is not forbidden in the sense of an orbital crossing, the activation energy is large.

## Fluxional Metal Complexes

Sigmatropic reactions are important as a mechanism for the behavior of fluxional molecules. These molecules have several configurations of the nuclei that are equivalent in structure and in energy.[61] Interconversions between these structures takes place at rates ranging from very fast to very slow. An example would be a monohapto cyclopentadienide of a metal.[62]

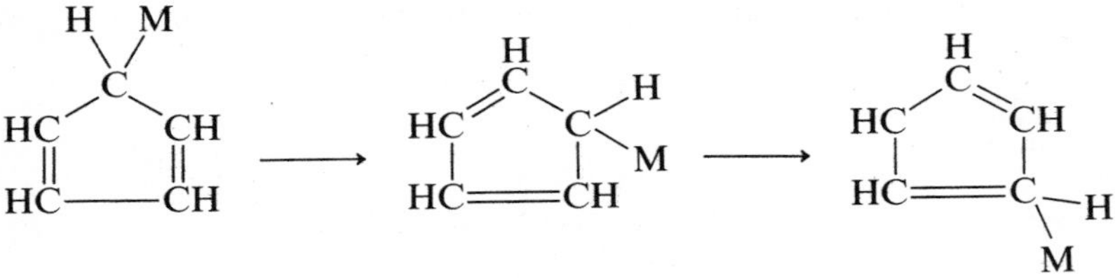

(etc.)

(82)

In these $\sigma$-bonded compounds, n.m.r. evidence shows that all five-ring protons are equivalent, above a certain critical temperature. At lower temperatures a pattern of line broadenings and line splittings occurs, which shows convincingly that the equivalence is due to consecutive 1,2 shifts, as shown in (82).[63]

This is in agreement with the symmetry rules, since just as in (75*b*), we have an allowed [1,5] sigmatropic rearrangement. This has been neatly proved by the preparation of the fluxional molecules trialkyl-7-cyclohepta-1,3,5-trienyl tin.[64]

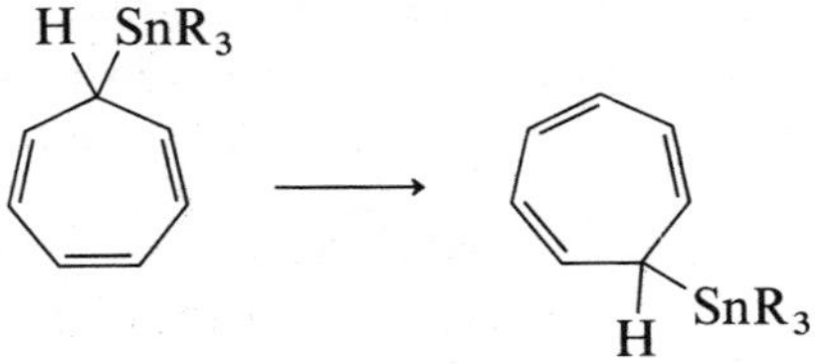

(etc.) (83)

The n.m.r. spectra of these molecules shows that only 1,4 shifts occur, following from allowed [1,5] processes. There are no 1,2 shifts, which would correspond to forbidden [1,7] rearrangements. The situation is the same as for (77).

The behavior of fluxional *monophapto* indenyl complexes has caused some confusion. The n.m.r. spectra are consistent with a series of 1,3 shifts, which corresponds to a forbidden [1,3] or [1,4] sigmatropic process. In the case of indene, the 1-isomer is much more stable than the 2-isomer, because of the loss of aromatic resonance energy in the latter. Therefore it is possible that the mechanism for the fluxional behavior is that of a rate determining 1,2 shift, followed by a very rapid 2,3 shift.[64]

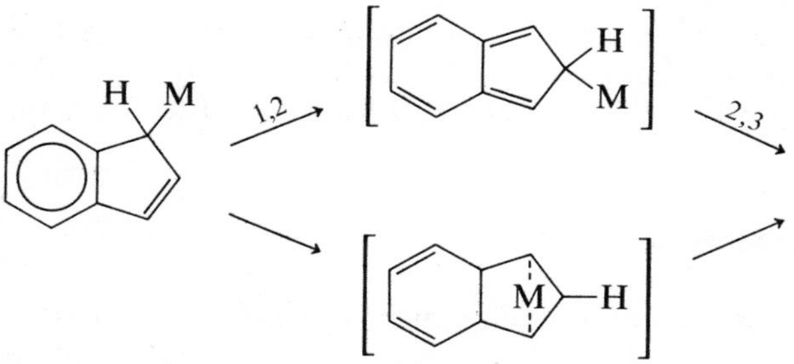

(84)

Support for this proposal comes from the observation that in 1,2(bis)-trimethyl silyl indene, exchange of the trimethylsilyl groups is going on.[65]

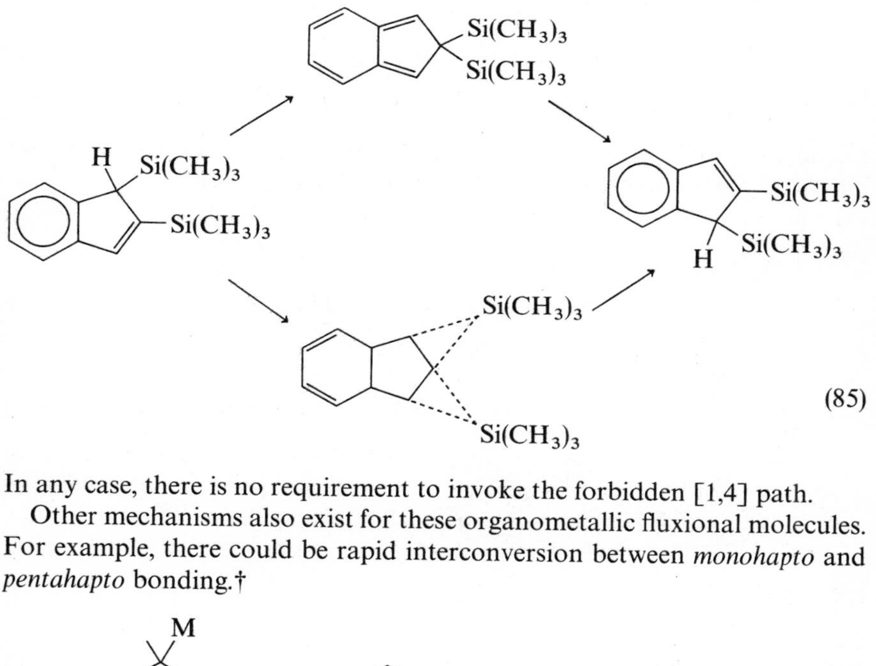

(85)

In any case, there is no requirement to invoke the forbidden [1,4] path.

Other mechanisms also exist for these organometallic fluxional molecules. For example, there could be rapid interconversion between *monohapto* and *pentahapto* bonding.†

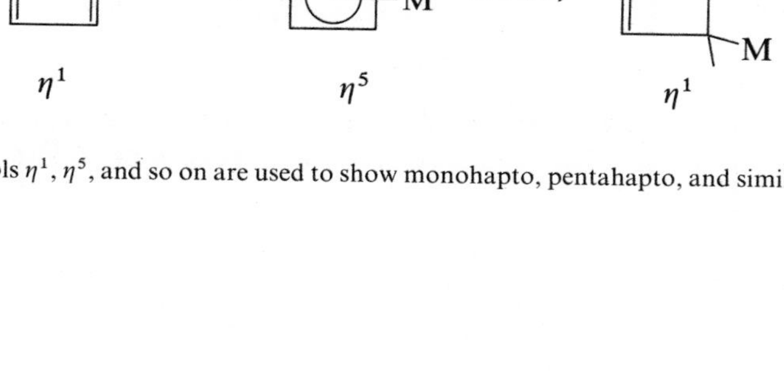

(86)

† The symbols $\eta^1$, $\eta^5$, and so on are used to show monohapto, pentahapto, and similar bonding.

This kind of structural change has been discussed earlier (p. 235). Because of the equivalence of all five positions in the $\eta^5$ structure, the net result can be a 1,3 shift, as shown, or a 1,2 shift. Indeed such molecules are known, an example being $(\eta^5\text{-}C_5H_5)_2Ti(\eta^1\text{-}C_5H_5)_2$. The n.m.r. spectrum shows that the two kinds of rings become equivalent at higher temperatures.[66] In this case, the equalization of all protons within the $\eta^1$ form is faster than the $\eta^1$–$\eta^5$ interchange.

### Some 1,3 and other Shifts

Although not as common as 1,2 shifts, organic chemistry has many examples of 1,3 shifts. A 1,3 shift means, in this connection, simply a transfer of an attached group to an atom two positions removed. Very common are 1,3 prototropic shifts.

$$R-C(=O)-CH_3 \longrightarrow R-C(OH)=CH_2 \tag{87}$$

$$C_6H_5CH_2-CH=CH_2 \longrightarrow C_6H_5CH=CH-CH_3 \tag{88}$$

Examination of these reactions in detail shows that invariably they are not concerted, but occur as acid, or base, catalyzed proton transfers,[67] for example,

$$C_6H_5CH_2-CH=CH_2 + CH_3O^- \rightleftharpoons C_6H_5\underline{CH-CH-CH_2}^- + CH_3OH$$

$$C_6H_5\underline{CH-CH-CH_2}^- + CH_3OH \longrightarrow C_6H_5CH=CH-CH_3 + CH_3O^- \tag{89}$$

There are no problems with symmetry in these proton-transfer reactions.

A number of 1,3 shifts take place with a change in the bonding atom of the migrating group. Examples would be the Claisen and Cope rearrangements, which are [3,3] processes. Another large class comprises of the [2,3] sigmatropic rearrangements of ylids,[68]

$$H_2C(-CH=CH_2)-\overset{+}{S}(R)-\overset{-}{C}R_2 \longrightarrow [H_2C\cdots CH\cdots CH_2,\ RS\cdots CR_2] \longrightarrow H_2C=CH-CH_2-CR_2-SR \tag{90}$$

The sulfur atom can be replaced by many other heteroatoms. The Wittig rearrangement of benzyl ethers is in the same category of [2,3] processes.

$$H_2C(-CH{=}CH_2)(-O-\bar{C}HC_6H_5) \longrightarrow H_2C{=}CH-CH_2-CH(C_6H_5)-O^- \qquad (91)$$

Reactions (90) and (91) have been shown to be both concerted and stereospecifically suprafacial, by the use of suitable substituents.[68] As indicated in (90), the TS is a six-electron system, similar to that for 1,3 dipolar additions.

It is likely that a number of rearrangements of aromatic compounds are also [2,3] processes, even though the observed products suggest that they are [1,3].[69] To illustrate the argument, consider the acid catalyzed rearrangement of nitramines.

$$C_6H_5NHNO_2 \xrightarrow{H^+} o\text{-}NH_2C_6H_4NO_2 \qquad (92)$$

Nitramines are inherently unstable and in acid the tautomeric nitritoamine could easily form. This would then rearrange, by way of an allowed [2,3] reaction.

$$C_6H_5NH_2NO_2^+ \longrightarrow C_6H_5NH_2{-}O{-}N{=}O^+ \longrightarrow \text{(cyclohexadiene: } {=}\overset{+}{N}H_2,\ H/NO_2) \longrightarrow o\text{-}NH_2C_6H_4NO_2 + H^+ \qquad (93)$$

The TS would hold six electrons.

Suprafacial [1,3] rearrangements are allowed, if the migrating group utilizes a *p* orbital for bonding to the two terminal atoms of the residue (p. 97). The rearrangement is accompanied by inversion of configuration at the bonding atom of the migrating group, a result that has been verified.[70] If the bonding atom is not tetravalent, the inversion would not be detectable and also migration should be relatively easy.[71]

$$CH_3-\underset{}{CH}(SR)-CH{=}CH_2 \xrightarrow{\Delta} CH_3CH{=}CH-CH_2SR \qquad (94)$$

allowed

An exception to the predictions of the symmetry rules lies in the Stevens rearrangement. This is an intramolecular migration of an alkyl group from a quaternary ammonium ion to an adjacent carbanionic center.

$$R_2\overset{\overset{R^*}{|}}{N^+}-\bar{C}H-CH_2C_6H_5 \xrightarrow{\Delta} R_2N-\overset{\overset{R^*}{|}}{C}H-CH_2C_6H_5 \quad (95)$$

The discrepancy in this four-electron 1,2 shift is that R* retains its configuration, if chiral.[72] Ad hoc explanations for this result are that the reaction goes by way of ion pairs, or radical pairs.[10] It is probably more sensible to postulate such mechanisms than to believe that the symmetry rules have been violated (however see p. 128).

As an example of the use of the rules for sigmatropic reactions in devising reaction mechanisms, the pyrolysis of vinyl ethyl ether is useful. At temperatures of 500° this compound decomposes quite cleanly into acetaldehyde and ethylene.[73] No free-radical

$$CH_2{=}CHOC_2H_5 \xrightarrow{\Delta} CH_2{=}CH_2 + CH_3CHO \quad (96)$$

chains are involved. Two possible cyclic TSs are worth considering; a four-membered ring, I, and a six-membered ring, J.

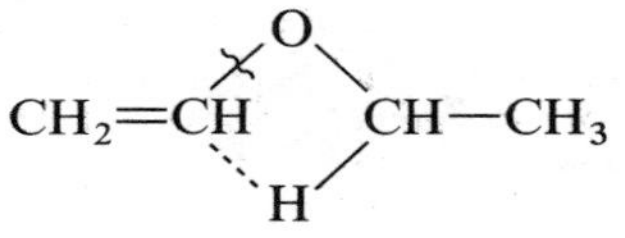

I: $CH_2{=}CH \cdots O \cdots CH{-}CH_3$ with bridging H (four-membered ring: CH–O–CH–H)

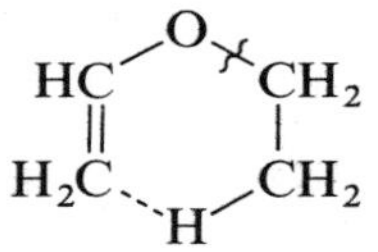

J: $HC(=H_2C)$–O–$CH_2$–$CH_2$–H (six-membered ring: HC, O, $CH_2$, $CH_2$, H, $H_2C$)

The carbon—oxygen bond to be broken is indicated by wavy lines. In 1952 the choice of J as the probable activated complex was made on the basis of the entropy of activation, which was $-10.2$ eu.[73] The conclusion was wisely tempered with a caution about the uncertainty of interpretation. At the present time we can select J as the activated complex with much more confidence, since it is a six-electron Hückel cycle, and I is a four-electron cycle.

## SOME REACTIONS OF ORGANOMETALLIC COMPOUNDS

It is well known that the transition metals often are excellent heterogeneous catalysts for a variety of reactions of small organic molecules. In recent years, with the development of the organometallic chemistry of the transition metals, many interesting and important homogeneous examples of catalysis have been found as well.[74] These include the *cis* addition of hydrogen to unsaturated molecules, olefin polymerization, olefin isomerization, olefin disproportionation, condensation of unsaturated molecules, isomerization of valence isomers, and so on. Most, but not all, of these catalyzed reactions would be forbidden as concerted processes in the absence of the metal catalyst. A fascinating problem is the attempt to deduce the mechanisms of these catalyzed reactions.

Certainly one of the functions of the metal atom is to bring the reactants together in a specified geometry. This greatly increases the rate of a complex reaction from the viewpoint of entropy. A reaction facilitated in this way is called a "template reaction,"[75] but this is only part of the story. It still must be explained how the forbiddenness of these reactions has been overcome. Just as in the case of heterogeneous catalysis, the *d* orbitals of the transition metals can be expected to play a major role. Nontransition metals, as a rule, are not catalytically active, nor are their complexes.

We start with an example where a simple explanation offers itself. The complex *syn*-tricyclooctadiene—iron tetracarbonyl in refluxing hexane is readily converted into bicyclooctatriene—iron tricarbonyl.[76]

Fe $(CO)_4$ $\underset{+CO}{\overset{-CO}{\rightleftharpoons}}$ Fe $(CO)_3$ $\longrightarrow$ Fe $(CO)_3$ (97)

The conversion rate is strongly inhibited by carbon monoxide, and it is assumed that prior loss of CO occurs as shown. The key reaction consists of the opening of a cyclobutene ring into a butadiene. Because of the geometric constraint of the other rings, this must occur in a disrotatory manner, forbidden in the absence of the metal.

A constant problem in studying these reactions is that the structure, and often the composition, of the complex undergoing the key step is not known. In this case we must assume that the three carbonyl ligands are arranged so that the complex has a constant plane of symmetry, passing through the iron atom and the center of the coordinated double bond. Figure 14 shows the correlation diagram for the disrotatory ring opening of free cyclobutene. A plane of symmetry is conserved and the point group is $C_s$. It can be seen that the reaction is forbidden because an orbital crossing occurs between a filled and an empty orbital.

Figure 15 shows the correlation diagram for a disrotatory ring opening of complexed cyclobutene. Only two *d* orbitals are considered. One is the $d_{z^2}$ orbital, pointing initially at the coordinated double bond, and the other is the $d_{xz}$ orbital, pointing finally at the two new double bonds of the products.

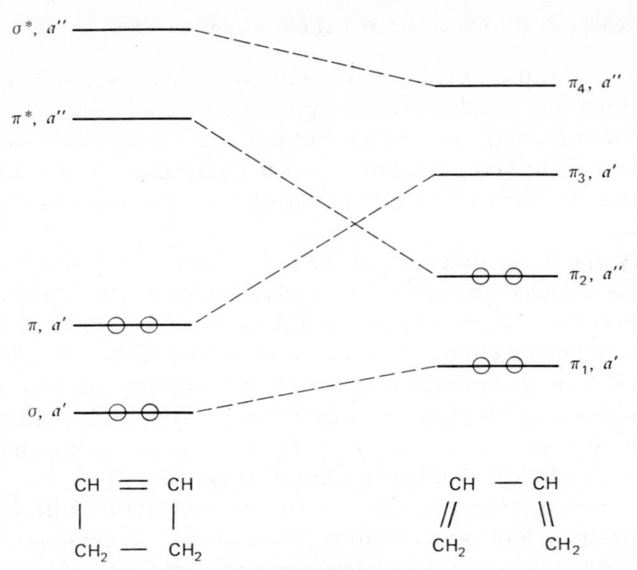


**Figure 14.** Correlation diagram for the forbidden disrotatory ring opening of cyclobutene; the point group maintained is $C_s$.

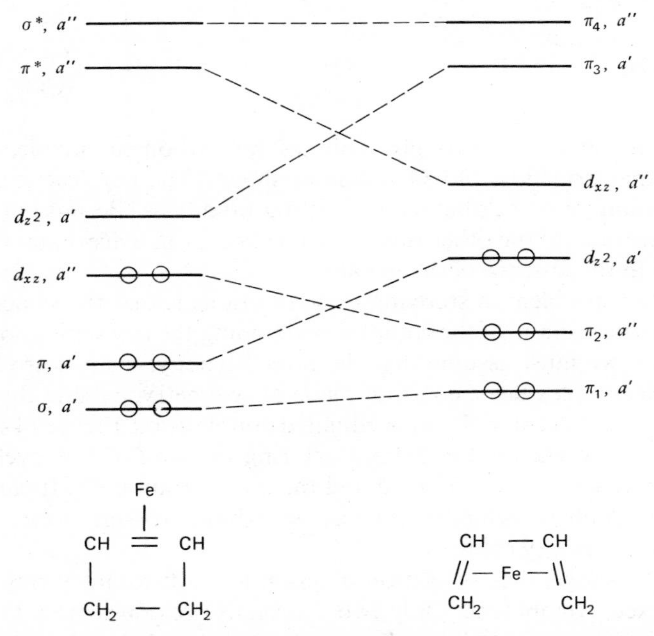


**Figure 15.** Correlation diagram for disrotatory ring opening of cyclobutene when complexed to a $d^8$ metal atom; reaction is allowed.

The other $d$ orbitals correlate with themselves and need not be considered. Only the necessary $\sigma$- and $\pi$-bond orbitals of cyclobutene and butadiene are shown. The metal complex is a $d^8$ system with iron having an oxidation state of zero. Six electrons are in the $d$ orbitals not shown. In the cyclobutene complex, the $d_{xz}$ orbital will be lower in energy than the $d_{z^2}$ orbital. Of the two, the $d_{xz}$ will be filled and the $d_{z^2}$ empty, as shown.

After ring opening, the situation is reversed. The $d_{xz}$ orbital is now strongly interacting with the filled $\pi_2$ orbital. Consequently it must become antibonding. The $d_{z^2}$ orbital has an overlap with the $\pi_1$ orbital, and contributes to the $\sigma$ bonding of the diolefin to the metal. However the overlap is small for geometric reasons. Thus the $d_{xz}$ orbital is above the $d_{z^2}$ orbital in energy and is unfilled. Correlation of the filled orbitals can now be made without crossing. The reaction is allowed.

Unfortunately for this neat explanation, it is found that $Ag^+$ and CuCl, which are $d^{10}$ metal ions, also are efficient catalysts for similar ring openings of strained cyclobutene molecules.[77] The diagram in Fig. 15 must now show both the $d_{z^2}$ and $d_{xz}$ orbitals as filled. The orbital crossing occurs again and the reaction is forbidden. Alternatively, we can invoke an empty 5$s$ orbital of $Ag^+$, which then must become filled in the products. Since the configuration $(4d)^8(5s)^2$ for $Ag^+$ lies 11 eV above $(4d)^{10}$, this cannot be considered as reasonable.[78a]

It appears that we must seek some other, mechanism for the action of $Ag^+$ in catalyzing this forbidden reaction. It may be mentioned that $Ag^+$ does not react with simple cyclobutenes, so we cannot say if the allowed conrotatory ring opening is also accelerated. Further, $Ag^+$ does not form a complex with, or react with, simple cyclobutanes or cyclopropanes. It does cause ring-opening reactions, otherwise forbidden, in strained, polycyclic molecules. Examples will be given later.

## The Mango–Schachtschneider Approach

Returning to the transition metals, if the orbital crossing can be converted entirely to the $d$ orbital manifold, this would be effective. The energy difference between the $d$ levels is usually small compared, for example, to $\pi$–$\pi^*$ or $\sigma$–$\sigma^*$ differences in organic molecules. This is the essential idea behind the arguments of Mango and Schachtschneider, who gave the first explanation for metal ion catalysis of symmetry-forbidden reactions.[79] The kind of reactions they considered in detail were the forbidden ethylene–cyclobutane interconversions. These are catalyzed by a number of metal complexes of Ni(0), Fe(0), Ru(I), and Pd(II), in other words, in $d^8$ and $d^{10}$ compounds. Simple olefins, or cyclobutanes, will not react, but more complex systems do.

For example, quadricyclene, K, is converted by $d^8$ metal ions to norbornadiene, L.[80] This, in turn, can be converted by $d^8$ or $d^{10}$ metal complexes to bisnorbornadiene, M.[81]

K ⟶ L (98)

L ⟶ M (99)

Figure 8 shows the orbital correlation for the uncatalyzed, forbidden conversions of this kind. In Fig. 16 is given the orbital correlation scheme when a $d^8$ metal complex is present.[79] It is assumed that two olefin molecules are first coordinated to a single metal center. Two planes of symmetry are also assumed, giving a $C_{2v}$ point group. This would be the case for a planar

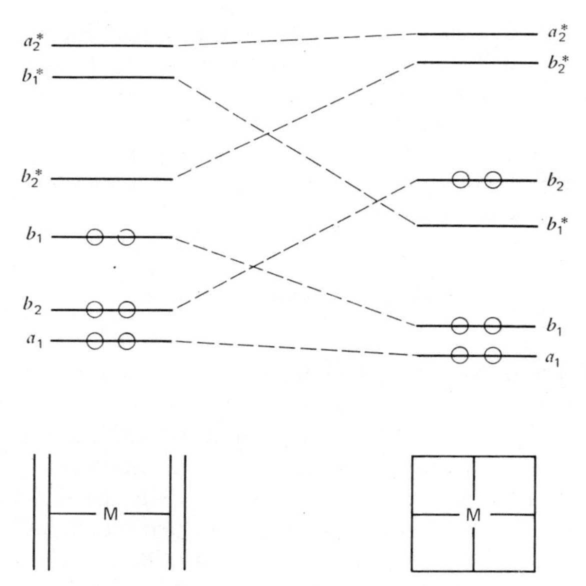


**Figure 16.** Orbital correlation in conversion of two-coordinated ethylene molecules to a coordinated cyclobutane molecule—the metal atom is $d^8$ (for a description of the orbitals, see Table 2).

complex, which is not unreasonable. The reaction coordinate is considered to be $A_1$, so that the two symmetry planes are maintained.

The essential orbitals are the $\pi$ and $\pi^*$ orbitals of the olefin molecules. The two $\pi$ orbitals combine to give $a_1$ and $b_2$ combinations. The two $\pi^*$ orbitals combine to give $b_1$ and $a_2$ pairs. The $d$ orbitals span $(2a_1 + b_1 + b_2 + a_2)$ representations in $C_{2v}$. The $x$ and $y$ axes are selected to point at the two olefin ligands. This gives the classification of Table 2, with the symmetry behavior on reflection in the two planes indicated. Metal and ligand orbitals of the same symmetry will mix to give a bonding orbital and an antibonding orbital.

**Table 2 Symmetries and Interactions of the $\pi$, $\pi^*$ and Metal $d$ Orbitals in $C_{2v}$[a]**

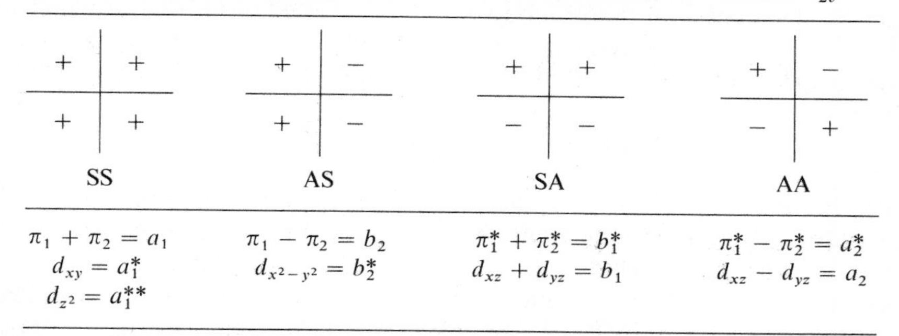

| SS | AS | SA | AA |
|---|---|---|---|
| $\pi_1 + \pi_2 = a_1$ | $\pi_1 - \pi_2 = b_2$ | $\pi_1^* + \pi_2^* = b_1^*$ | $\pi_1^* - \pi_2^* = a_2^*$ |
| $d_{xy} = a_1^*$ | $d_{x^2-y^2} = b_2^*$ | $d_{xz} + d_{yz} = b_1$ | $d_{xz} - d_{yz} = a_2$ |
| $d_{z^2} = a_1^{**}$ | | | |

[a] One lobe of the $p$ orbital on each carbon atom initially points at the metal atom, which is just above the intersection of the two mirror planes. As reaction proceeds, the lobes turn inward in a disrotatory manner.

Table 2 shows the major component of both the bonding and antibonding orbitals, in each case. The $a_1$ orbital, for example, is chiefly concentrated on the carbon atoms of the olefin molecules, but has some metal $d$ character. The reverse is true for the two $a_1^*$ orbitals. Of course the metal $s$ and $p$ orbitals also participate in the bonding, but for clarity they are omitted. Also for clarity, Fig. 16 does not show the two $a_1^*$ orbitals, nor the $a_2$ orbital. These metal $d$ orbitals are all filled, and all correlate with themselves.

The reaction we wish to carry out is ring closure of two ethylene molecules to form cyclobutane, all complexed to a metal center.

$$\|\!-\!M\!-\!\| \xrightarrow{A_1} \boxtimes M \tag{100}$$

Clearly this requires electron transfer from the $b_2$ orbital to the $b_1^*$ orbital.

As seen in Table 2, this will destroy the double bonding in the two olefin molecules, and create two new carbon—carbon $\sigma$ bonds between the two molecules. The direct transfer is forbidden, but the metal atom can act as a broker, electrons flowing from $b_1$ to $b_2^*$.

As shown in Fig. 16, this leads to an orbital crossing, in the usual case. However the diexcited configuration produced by the crossing lies within the $d$ manifold. Hence the energy difference is not large between it and the ground state. Alternatively, the configuration interaction that prevents the actual crossing will occur early since $b_1$ and $b_2^*$ are close in energy to begin with.

A requirement of this explanation is that the metal atom have at least two electrons in the $d$ shell, but no more than eight, but as already indicated, $d^{10}$ metal atoms such as Ni(0) are effective catalysts. From Fig. 16, filling the $b_2^*$ orbital in a $d^{10}$ metal would create a much larger energy gap between the excited state and the ground state than for a $d^8$ metal. It is difficult to believe that the two systems could be comparable in efficiency. Another mechanism must be found for $d^{10}$, and if found, it may well be valid for $d^8$.

## Tolman's Rule

At this point it is necessary to survey the reactions of transition-metal organometallic compounds that have been well studied, and where reaction mechanisms are thought to be well understood. Presumably these will be the usual steps that occur on a metal center after one or more of the eventual organic reactants have been assembled. It will be helpful to remember a well-known rule of organometallic chemistry: "Diamagnetic compounds of IV to VIII almost always have sixteen or eighteen electrons in their valence shell." This rule has been extended by Tolman to include the reactive intermediates of this group of molecules.[82]

The latter amendment is illustrated by these mechanisms for nucleophilic substitution,

$$\underset{18\text{ electrons}}{Ni(CO)_4} \longrightarrow \underset{16\text{ electrons}}{Ni(CO)_3} + CO \qquad S_N1,\ \text{no } S_N2 \tag{101}$$

$$\underset{16\text{ electrons}}{PtCl_4{}^{2-} + Cl^-} \longrightarrow \underset{18\text{ electrons}}{PtCl_5{}^{3-}} \qquad S_N2,\ \text{no } S_N1 \tag{102}$$

$$\underset{18\text{ electrons}}{Co(NH_6)_6{}^{3+}} \longrightarrow \underset{16\text{ electrons}}{Co(NH_3)_5{}^{3+}} + NH_3 \qquad S_N1,\ \text{no } S_N2 \tag{103}$$

In each case one mechanism is predicted by the rule, and the other would violate the rule. The rule must be used with caution since exceptions are known, both for stable molecules and for reactive intermediates. Included in the elementary reactions of organometallic compounds will be the gain or loss of simple nucleophiles, and the gain or loss of simple electrophiles. These reactions have already been discussed in Chapter 4.

## THE OXIDATIVE-ADDITION REACTION

This reaction and its reverse, reductive elimination, were discussed in part in Chapter 4 for transition-metal complexes (pp. 293). This earlier material serves as a background against which more detail can be presented. The goal is to understand catalytic processes in which oxidative-addition steps occur. Write the reaction in generalized form,

$$\mathrm{M + X{-}Y \rightleftharpoons X{-}M{-}Y} \tag{104}$$

where the stereochemistry of X and Y is not specified, and where other ligands attached to M are not shown. The most important point for catalysis is that X and Y are any two atoms or radicals held together by a single bond.

This means that carbon—carbon, carbon—hydrogen, and hydrogen—hydrogen bonds can all be cleaved, or formed, by either oxidative addition or its reverse. For example, the cleavage of carbon—carbon bonds occurs in the reaction[83]

$$\mathrm{PtL_3 + C_6H_5CN \longrightarrow PtL_2(C_6H_5)(CN) + L} \tag{105}$$

where L is triethylphosphine. The phenyl and cyano groups are bonded to platinum in the product in a *trans* arrangement. The cleavage of C—H bonds is also known, and many metal complexes are capable of adding molecular hydrogen. However it is far more common to observe carbon—carbon or carbon—hydrogen bond formation by reductive elimination.

$$\mathrm{R\overset{\overset{O}{\|}}{C}Fe(CO)_4R' \longrightarrow R{-}\overset{\overset{O}{\|}}{C}{-}R' + Fe(CO)_4} \tag{106}$$

As (106) shows, such reactions have considerable utility in organic synthesis.[84]

Three mechanisms for oxidative addition to square planar $d^8$ metal complexes were proposed earlier (p. 297). One was the concerted addition of both X and Y to the metal M, in (104). A second was a two-step mechanism, which may be written as

$$\mathrm{M + X{-}Y \longrightarrow M{-}X^+ + Y^-} \tag{107}$$

$$\mathrm{Y^- + M{-}X^+ \longrightarrow Y{-}M{-}X} \tag{108}$$

The rate-determining step is (107), which is an $S_N2$ reaction. A third mechanism is a free-radical reaction, not yet described in detail. It is expected to be important in the reaction of alkyl bromides and iodides with complexes of the third transition series. The primary interaction is between the metal and the halogen, and the heavy noble metals have a high affinity for the heavy halides.

Figure 17 shows the orbital picture of all three mechanisms. In the free-radical mechanism, a chain reaction would probably result.[85] A plausible sequence of events for an alkyl bromide reacting with an iridium(I) complex would be as follows:

$$\text{Ir(I)} + \text{RBr} \xrightarrow{k_1} \text{Ir(II)—Br} + \text{R}\cdot \tag{109}$$

$$\text{R}\cdot + \text{Ir(I)} \xrightarrow{k_2} \text{R—Ir(II)}\cdot \tag{110}$$

$$\text{R—Ir(II)}\cdot + \text{RBr} \xrightarrow{k_3} \text{R—Ir(III)—Br} + \text{R}\cdot \tag{111}$$

$$\text{R—Ir(II)}\cdot + \text{R}\cdot \xrightarrow{k_4} \text{R—Ir(III)—R} \tag{112}$$

Reaction (109) is chain-initiating, reactions (110) and (111) are chain-carrying, and (112) is chain-breaking. Applying the steady-state approximation to the reactive species R· and R—Ir(II)·, the simple rate law is found

$$\text{Rate} = \left(\frac{k_1 k_2 k_3}{k_4}\right)^{1/2} [\text{Ir(I)}][\text{RBr}] \tag{113}$$

The second-order rate law has been found experimentally, whenever the kinetics have been studied.[86] However both the concerted mechanism and the $S_N2$, two-step mechanism would also give second-order rate laws. The best test for the free-radical chain is the effect of inhibitors.[85] Small amounts (5–10 mole %) of free-radical scavengers can decrease the rate by factors ranging from 5 to 1000 fold. The maximum inhibition found is a measure of the chain length in (109)–(112). The reactions of $CH_3I$ and $C_6H_5CH_2Br$ are not inhibited. This does not necessarily mean a nonfree-radical mechanism. If $k_2$ in (110) is large enough, an inhibitor will not be effective since it must compete with a large excess of Ir(I).

A second kind of trapping experiment is the addition of a large excess of some anion other than the halogen of the alkyl halide. In the $S_N2$ mechanism, the second step (108) allows for the incorporation of an external anion. In the reaction of $CH_3I$ with $Ir(CO)Cl[P(C_6H_5)_3]_2$, for example, no trapping of external anions is found.[87] This is consistent either with a concerted mechanism, or with the free-radical mechanism, which allows no role for anions in solution. Conversely, a $RhL_4{}^+$ complex, in which $L_4$ is a macrocyclic ligand, reacts with $CH_3Br$ in the presence of $Cl^-$ to give only the $CH_3Cl$

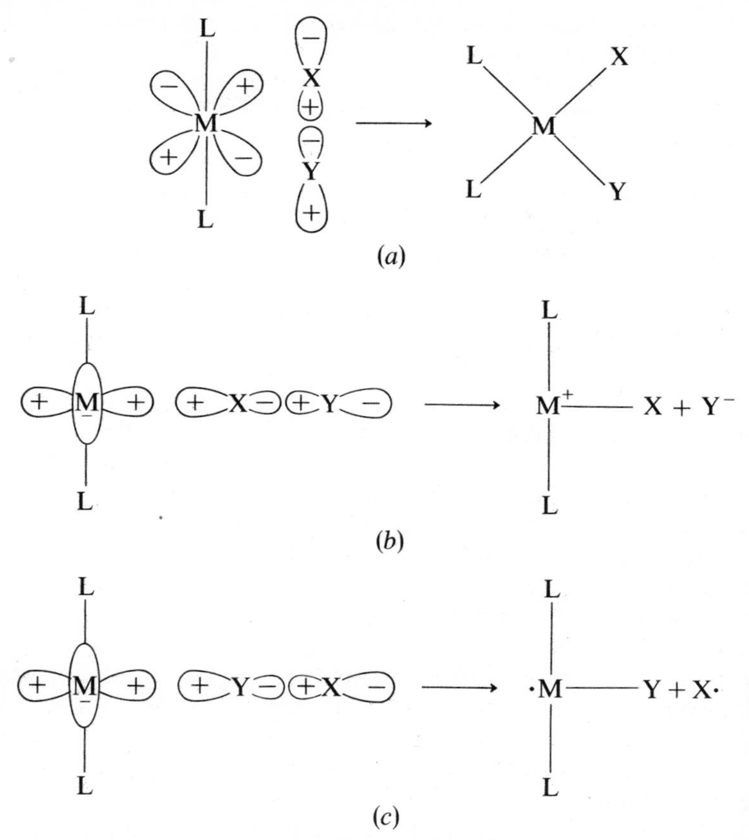


**Figure 17.** Three mechanisms for oxidative addition to $ML_4$ planar molecule. (*a*) concerted *cis* addition; (*b*) two-step $S_N2$ displacement (*trans* addition of $Y^-$ follows); (*c*) initial step in free-radical chain mechanism (*trans* product expected); two ligands are above and below the plane of the page.

adduct.[88] In this case $L_4$ has a structure which makes the concerted addition impossible. The $S_N2$ mechanism is strongly indicated.

The most definitive experiments which differentiate between the three mechanisms are the stereochemical results. Based on Fig. 17 and arguments given in Chapter 4, a set of predictions can be made (see Table 3). An examination of results already reported in the literature shows that all three mechanisms must be occurring. The factors selecting each mechanism are discernible in most of the known cases.

For example, in oxidative addition or reductive elimination reactions in which C—C, C—H, Si—H, or H—H bonds are involved, only the concerted

**Table 3 Stereochemical Predictions for Three Mechanisms of Oxidative Addition**

| Mechanism | Stereochemistry at Metal | Stereochemistry at X or R |
|---|---|---|
| 1. Concerted | *cis* | retention |
| 2. $S_N2$, Two-step | *trans*[a] | inversion |
| 3. Free radical | *trans*[a] | racemization |

[a] *Trans* products are most probable, but rearrangements leading to *cis* products cannot be ruled out.

mechanism has been found.[89,90] The evidence for mechanism is based on retention of configuration at a chiral carbon or silicon atom, as well as on observation of *cis* addition or elimination. These results are eminently reasonable, since it is evident that the two alternative mechanisms are unreasonable. It would be unrealistic to have $H^-$ or $CH_3^-$ as leaving groups in $S_N2$ reactions. Also free-radical mechanisms are unlikely because of the high bond energies and the rather low affinity of the metals for hydrogen or carbon.

Free-radical mechanisms are found, as expected for organic bromides and iodides reacting with iridium(I) and plantinum(0) complexes.[91] Rhodium and palladium are less likely to give free radical reactions.[92] Following Table 3, the evidence consists of formation of *trans* addition products and racemization of optically active alkyl groups. Presumably the formation of *trans* dihalides by oxidative addition of halogen to gold(I) and Pt(II) can also be explained by free-radical mechanisms.[93]

The nucleophilic substitution mechanism given by reactions (107) and (108) requires a good leaving group, such as a halide ion or an arylsulfonate ion. For example, the optically active alkyl chloride shown in (114) reacts with tris(triphenylphosphine)palladium(0), to give a product in which the alkyl group and chlorine are *trans* to each other, and the alkyl group has an inverted configuration.[92b]

$$(F_3C)(H_5C_6)(H)C{-}Cl + PdL_3 \longrightarrow (F_3C)(H)(C_6H_5)C{-}PdL_2{-}Cl + L \qquad (114)$$

(R) (S)

An $S_N2$ mechanism, with inversion, is also followed in complexes with a planar, macrocyclic ligand system, where a *cis* addition is impossible.[88,94]

There are also many reactions of alkyl halides or arylsulfonates with metal complexes where only the alkyl group adds to the metal. For example,

$$RBr + Mn(CO)_5^- \longrightarrow RMn(CO)_5 + Br^- \tag{115}$$

these reactions are not strictly to be called oxidative addition, but their close relationship is obvious. They correspond to (107), not followed by (108). As expected, they are simple nucleophilic substitutions, always occurring with inversion of configuration at the alkyl group.[84,95]

Changing the solvent has a predictable effect. Polar solvents favor the two-step $S_N2$ mechanisms. Nonpolar solvents, or the absence of solvents, favors a free-radical mechanism or a concerted process. For example, it is found that the hydrogen halides add *cis* to Vaska's compound in benzene, but add *trans* in methanol.[96] Allyl halides add *cis* in benzene but *trans* in methanol.[97] The *cis* addition found in nonpolar solvents suggests a concerted reaction mechanism.

With several mechanisms possible, and with substituents playing an important role, it is dangerous to predict relative reactivities. In general, reactivity parallels the tendency to increase the oxidation state by two units. For $d^8$ complexes the order is

$$\text{Os(0)} > \text{Ru(0)} > \text{Fe(0)} > \text{Co(I)} > \text{Ir(I)} > \text{Rh(I)} > \text{Pt(II)} > \text{Pd(II)} \gg \text{Ni(II)}.$$

For $d^{10}$ complexes the order is

$$\text{Pt(0)} \sim \text{Pd(0)} \sim \text{Ni(0)} > \text{Au(I)} > \text{Cu(I)} \gg \text{Ag(I)}.$$

Methyl halides, allyl, and benzyl halides are more reactive than ethyl halides and higher primary homologes. Secondary halides are considerably less reactive.

Aromatic halides and vinylic halides often react quite rapidly with $d^{10}$ systems, and less rapidly with $d^8$ systems.[98] This is unexpected in view of the low reactivity of such molecules to $S_N2$ mechanisms, and to the difficulty of forming vinyl or aryl free radicals. For example, styryl bromide, both *cis* and *trans*, reacts with $Pt[P(C_6H_5)_3]_3$ more rapidly than does methyl iodide.[99] The product has a *trans* arrangement of the added groups, and the olefin retains its stereochemistry.

$$PtL_3 + (H_5C_6)(H)C{=}C(H)(Br) \longrightarrow L + L{-}Pt(Br)(L){-}C(H){=}C(H)(C_6H_5) \tag{116}$$

The fact that a *trans* product is formed, here and in (114), is not indicative of mechanism since a five-coordinated intermediate could readily rearrange. The *trans* isomer would be the thermodynamically stable one.[100]

It seems almost inevitable that the initial stage of (116) is the formation of a $\pi$ complex, N. Just how N reacts further is not clear,

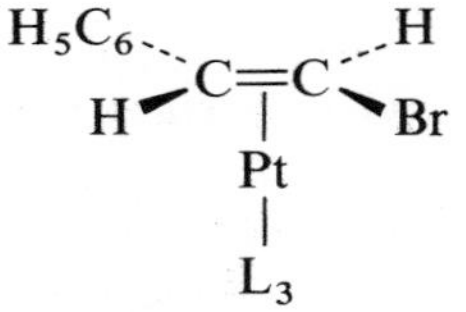

$$\begin{array}{c} H_5C_6 \quad\quad H \\ \;\; C{=}C \\ H \quad | \quad Br \\ Pt \\ | \\ L_3 \end{array} \qquad N$$

but it may be noted that adding an ordinary nucleophile to a vinyl halide is an endothermic process. The formation of N is exothermic and is equivalent to adding a nucleophile to the $\pi^*$ orbital of the olefin. The $\pi$ complex, N, must presumably rearrange to a $\sigma$ complex, since ordinary nucleophiles only form $\sigma$ complexes (p. 333).

### Reactions of Metal Dimers

The reactivity sequence $Au(I) > Cu(I) \gg Ag(I)$ requires special comment. The stability of Au(III) and the instability of both Cu(III) and Ag(III) oxidation states is a major factor. However copper has a stable oxidation state of plus two, which silver does not. This results in a special mechanism for copper(I) complexes in oxidative-addition reactions. An example is the Corey–Posner reaction, useful for forming carbon—carbon bonds.[100] The reagent is a lithium alkyl cuprate complex, with the stoichiometry $LiCuR_2$. This reacts with alkyl halides or *p*-toluenesulfonates to give coupling products, for instance,

$$LiCu(CH_3)_2 + C_2H_5Br \longrightarrow C_3H_8 + LiCuBrCH_3 \qquad (117)$$

The reaction can be made catalytic by continuous addition of $LiCH_3$. Use of an optically active halide or *p*-toluenesulfonate, gives inversion of configuration in the product, $R^*CH_3$.[101]

The catalyst has been shown to be a dimer, $Li_2Cu_2R_4$, with the cyclic structure[100]

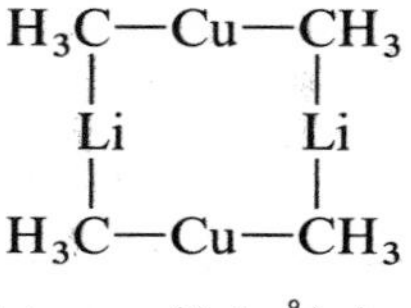

$$\begin{array}{ccc} H_3C & —Cu— & CH_3 \\ | & & | \\ Li & & Li \\ | & & | \\ H_3C & —Cu— & CH_3 \end{array}$$

While the copper–copper distance (4.4 Å) is too long for metal—metal bonding, the two atoms are electronically coupled via the cyclic system.

The rate law is[101,102]

$$\text{rate} = k[\text{dimer}][\text{alkyl halide}] \tag{118}$$

which suggests an $S_N2$ mechanism in which each copper atom provides one electron, being oxidized to Cu(II) in the process. This circumvents the unstable Cu(III) oxidation state. Reductive elimination of $RCH_3$ returns two electrons, one to each copper atom. Cuprous complexes are usually associated, which probably accounts for their usefulness as catalysts for many organic reactions.

Cuprous ion is not unique in its behavior. Other transition-metal ions, which will increase their oxidation state by one unit readily, should behave in the same way. An example is Co(II), which will easily form Co(III) but not Co(IV). The complex $Co(CN)_5^{3-}$ will cleave molecular hydrogen to form a monohydride.

$$2Co(CN)_5^{3-} + H_2 \longrightarrow 2Co(CN)_5H^{3-} \tag{119}$$

The rate law for (119) is first order in hydrogen and second order in $Co(CN)_5^{3-}$.[103] Two cobalt(II) ions are required to furnish the two electrons needed to split the $H_2$ molecules. With alkyl halides, $Co(CN)_5^{3-}$ reacts by the free-radical path.[104]

$$\begin{aligned} Co(CN)_5^{3-} + RBr &\longrightarrow Co(CN)_5Br^{3-} + R\cdot \\ Co(CN)_5^{3-} + R\cdot &\longrightarrow RCo(CN)_5^{3-} \end{aligned} \tag{120}$$

Now only one cobalt(II) ion is needed, and this is what is found.

## Nontransition Metals

Ag(I), on the other hand, does not react in well-defined oxidative-addition processes. The characteristic reaction with an alkyl halide is as an electrophilic catalyst, producing a carbonium ion.[105]

$$RBr + Ag^+ \longrightarrow R^+ + AgBr \tag{121}$$

Hg(II) behaves similarly, since both Hg(III) and Hg(IV) are quite unstable. With Tl(I), In(I), Sn(II), and Pb(II), oxidative addition again becomes possible, and such reactions with alkyl halides are well known, as well as the reverse, or reductive elimination.[106]

These examples are interesting because the reactive electrons of Tl(I), Pb(II), and so on, are in *s* orbitals. The concerted addition becomes forbidden (see p. 284) and the only mechanisms available are the $S_N2$ and free-radical ones. The orbital details would be similar to Figs. 17*b* and 17*c*, with an *s* orbital on the metal in place of a *d*. Evidence for the $S_N2$ mechanism in the case of Tl(I) comes in an indirect way.[107] Neopentyl thallium(III) dibromide is stable in solution, whereas methyl thallium(III) is not. The explanation is that the mechanism of decomposition is the reverse of that for formation. This requires the rate step to be the attack of $Br^-$ on the alkyl carbon, displacing Tl(I)

$$\begin{aligned} Br^- + CH_3TlBr^+ &\longrightarrow CH_3Br + TlBr \\ Br^- + (CH_3)_3CCH_2TlBr^+ &\longrightarrow \text{no reaction} \end{aligned} \tag{122}$$

An $S_N2$ attack on the neopentyl system would be extremely slow, accounting for the stability.

The simplest reaction of an alkyl halide with a metal would be the gas-phase reactions with alkali metal atoms studied long ago by M. Polanyi, for instance,

$$CH_3I + K \longrightarrow KI + CH_3\cdot \tag{123}$$

Molecular beam studies give details of the dynamics of such reactions.[108] It makes little difference whether the metal atom approaches the carbon end or the halogen end of the alkyl halide (factor of 1.0 : 1.5), the added electron still appears as iodide ion. If the metal atom has two reducing electrons in an *s* orbital, as in ground-state Mg, free radicals are not formed.[109a]

$$CH_3Br + Mg \longrightarrow CH_3MgBr \tag{124}$$

This remark applies to single Mg atoms isolated in a matrix. The important reaction of alkyl halides with magnesium metal to form the Grignard reagent is still not understood. Some experimental observations suggest a free-radical mechanism, and others do not.[109b] Reaction rates depend markedly on the magnesium surface and on the impurities present. Activation energies for different organic halides correlate with half-wave potentials for reduction of the halides, $E_{1/2}$. This suggests that $RX^{\cdot-}$ species are intermediates.

The alkali metals in the solid state apparently form organometallic compounds directly, without prior formation of free radicals. The free radicals that eventually occur in the Wurtz reaction appear to be produced in a second stage, by electron transfer from LiR, or NaR, to the alkyl halide.[110]

$$\begin{aligned} LiR + R'X &\longrightarrow Li^+ + R\cdot + R'\cdot + X^- \\ R\cdot + R'\cdot &\longrightarrow R{-}R' \end{aligned} \tag{125}$$

In addition to electron transfer (125), a metal–halogen interchange can also occur.

$$LiR + R^*X \longrightarrow LiR^* + RX \tag{126}$$

This can occur stereospecifically, with retention of configuration at R*.[111] A four-center mechanism is indicated from the stereochemistry, even though this is a disfavored process.

Inversion of configuration has been observed for the reduction by zinc metal of an alkyl halide, followed by hydrolysis.[112] It is

$$R^*Cl + Zn \longrightarrow R^*ZnCl \xrightarrow{H^+} R^*H + ZnCl^+ \qquad (127)$$

likely, but not proven, that the inversion step is at the reduction stage, and not the electrophilic cleavage stage. Reduction of an alkyl chloride electrolytically, at a mercury cathode, can also lead to inversion of configuration.[113] Electrolytic reduction is a one electron process at low potentials (less than the half-wave potential), and a two-electron (2e) process at higher potentials.[114]

$$\begin{aligned} RX + e &\longrightarrow R\cdot + X^- &\quad E < E_{1/2} \\ RX + 2e &\longrightarrow R^- + X^- &\quad E > E_{1/2} \end{aligned} \qquad (128)$$

It must be remembered that carbanions, if formed, can easily racemize and easily form free radicals by electron transfer, as in (125).

## THE LIGAND-MIGRATION REACTION

A large number of reactions consist of the addition of a metal compound, M—L, to an unsaturated molecule, X=Y, to form a new compound with X=Y inserted between M and L.

$$M{-}L + X{=}Y \longrightarrow M{-}(X{-}Y)L \qquad (129)$$

The molecule X=Y may be CO, olefin, diene, RNC, RCHO, RCN, $SO_2$, $O_2$ or other unsaturated systems. The ligand L may be $H^-$, $R^-$, $OR^-$, $NR_2^-$, $H_2O$, halide ion, or another metal atom. Reaction (129) is sometimes called the "insertion" reaction, but this is misleading. Mechanistic studies[115] show that X=Y must be coordinated to M before reaction, and that L moves or migrates onto X=Y.

$$\underset{\displaystyle \overset{|}{L}}{M}{-}(X{=}Y) \longrightarrow M{-}(X{-}Y)L \qquad (130)$$

For this reason a better name is the ligand-migration reaction.

A typical example is the methyl migration to CO in methylmanganese pentacarbonyl.

$$\begin{aligned} &\underset{18e}{CH_3Mn(CO)_5} \xrightarrow{\text{slow}} \underset{16e}{CH_3COMn(CO)_4} \\ &\underset{16e}{CH_3COMn(CO)_4} + L \xrightarrow{\text{fast}} \underset{18e}{CH_3COMn(CO)_4L} \end{aligned} \qquad (131)$$

L is a phosphine, amine, CO, and so on. If the concentration of L is large enough, the rate does not depend on it, or the nature of L. The stereochemical position of L is *cis* to the acetyl group. If an additional group, Z, is present as a stereochemical label, it is possible to prove that it is the methyl group which moves.[116] This is shown in Fig. 18. Though the reactive intermediate $CH_3COMn(CO)_4$ is written as coordinatively unsaturated, there may be a solvent molecule attached to hold the stereochemistry. In other cases, no solvent is involved and the intermediate either rearranges, or else the CO group moves toward the methyl in a cooperative fashion.[117]

Particularly important examples of ligand migration are the migration of $H^-$ and $R^-$ to an olefin. The former is a step in the homogeneous hydrogenation of olefins, with a transition metal catalyst, and the latter is a step in olefin dimerization or polymerization.[74]

$$IrH(CO)[P(C_6H_5)_3]_2C_2H_4 \longrightarrow Ir(CO)[P(C_6H_5)_3]_2C_2H_5 \quad (132)$$

Figure 19 shows the orbitals that must dominate the ligand-migration reaction of a coordinated olefin.[118] For simplicity a symmetric planar complex, $ML_3(C_2H_4)$, is taken. The initial point group is $C_{2v}$. There is a filled orbital corresponding to metal—ligand $\sigma$ bonding, and an empty $\pi^*$ orbital, which is the antibonding partner to the usual $\pi$-bonding orbital of metal—olefin complexes.

Now any one metal—ligand bond is usually a mixture of symmetries, corresponding to the extent to which *s*, *p*, and *d* orbitals on the metal contribute to bonding. We will assume that *d* orbitals dominate the bonding in the case at hand. In any event, the metal—ligand bond orbital is of $a_1$ species. It is also part of the $\sigma$ bonding of the olefin to the metal. The empty $\pi^*$ orbital is of $b_2$ species, and the initial reaction coordinate is accordingly $(A_1 \times B_2) = B_2$. This is an asymmetric motion in the plane, the ethylene moving from left to right.

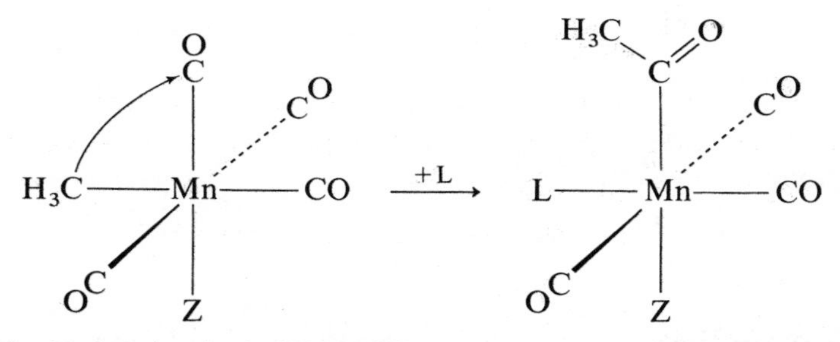


**Figure 18.** Methyl migration in $CH_3Mn(CO)_5$; a group Z, *trans* to the acetyl group, shows that $CH_3$ has migrated to CO.

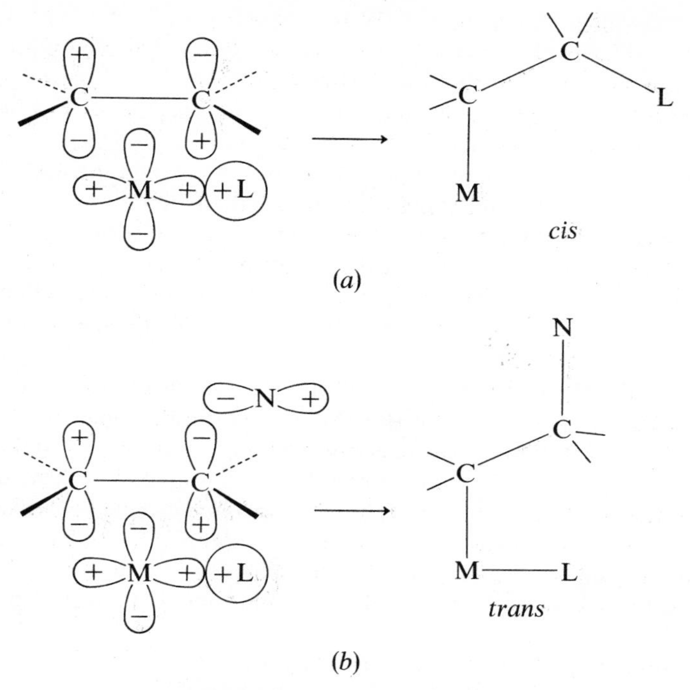


**Figure 19.** (*a*) Critical orbitals in the ligand migration reaction—the *d* orbital is filled and the $\pi^*$ orbital is empty; (*b*) attack by external nucleophile to give *trans* addition.

Electron flow from $a_1$ to $b_2$ accomplishes all of the following:

1. The metal—ligand $\sigma$ bond is broken.
2. The metal—olefin $\pi$ bond is broken.
3. The carbon—carbon double bond is broken.
4. A new bond is formed between the migrating ligand, L, and the olefin carbon nearest it. This bond is strengthened by the relative motions of the ethylene molecule and ligand.
5. A new metal—carbon $\sigma$ bond is also formed by the sidewise movement of the ethylene.

The final product is an alkyl—metal complex. Trial and error show that no other choice or orbitals will accomplish all the necessary functions. For example, use of the filled metal—olefin $\pi$ orbital, and the empty M—L $\sigma^*$ orbital will strengthen the carbon—carbon double bond, instead of destroying it. The direction of electron flow shows clearly that the ligand migrates as an anion, converting the neutral olefin to an anionic alkyl group.

The mechanism shown in Fig. 19 is stereochemically detailed. It shows that the olefin and the migrating ligand must lie in the same plane and in a *cis* arrangement to each other. This requirement has been confirmed by observing the behavior of model systems.[119] The ligand L adds to the side of the olefin closest to it. That is, the metal and L add *cis* to each other. This has been confirmed in many examples, for both olefins and acetylenes.[120] Also if L is an alkyl group having an asymmetric carbon atom bonded to the metal, migration will occur with a retention of configuration at that carbon. This has been proven the case in several examples in which migration is to a carbonyl group, rather than an olefin.[84, 121] The orbital analysis of carbonyl migrations would be very similar to Fig. 19, except for the orientation of the CO ligand.

However there are also many cases where *trans* addition of a nucleophile and a metal atom to an olefin or acetylene is found. These are normally explained as the result of the nucleophile adding from the external environment, not being coordinated to the metal first.[122] Figure 19*b* shows that this is a symmetry-allowed process. Thus a hydride ion can be added either *exo* (*trans* addition) or *endo* (*cis* addition) to an olefin.[123] To get *exo* addition, $BH_4^-$ is used as an external nucleophile. To get *endo* addition, $H^+$ is first added to the metal to form an M—H bond, and then ligand migration follows.

Insertion reactions for $SO_2$ are also well known.[124]

$$RFe(CO)_2C_5H_5 + SO_2 \longrightarrow RSO_2Fe(CO)_2C_5H_5 \qquad (133)$$

If R is chiral, then (133) takes place with inversion of configuration.[125] This is not surprising since the mechanism of Fig. 19*a* is not possible for sulfur dioxide. Binding to an electron rich metal atom would give the pyramidal structure of Fig. 5*c* for coordinated $SO_2$. There is a lone pair of electrons on the ligand, instead of an empty $\pi^*$ orbital. A mechanism in which uncoordinated $SO_2$ acts as an electrophile has been suggested.[126]

The previous discussion of mechanism has been based on an assumption that metal—ligand $\sigma$ bonding was due primarily to a *d* orbital. The case where an *s* orbital of the metal is dominant should now be examined. This would be the case for $Ag^+$ and $Hg^{2+}$, for example, where olefin $\pi$ complexes are well known.[127] For mercury, at least, addition of the metal and a nucleophile across the double bond is a common reaction

$$C_2H_4 + Hg^{2+} + X^- \longrightarrow HgC_2H_4X^+ \qquad (134)$$

The normal mode of addition of $Hg^{2+}$ and $X^-$ is *trans*,[128] suggesting that the nucleophile acts externally and is not coordinated to the metal.

Figure 20 shows that ligand migration is, in fact, a forbidden reaction when *s* orbital bonding is dominant. It can be seen that the orbital phase

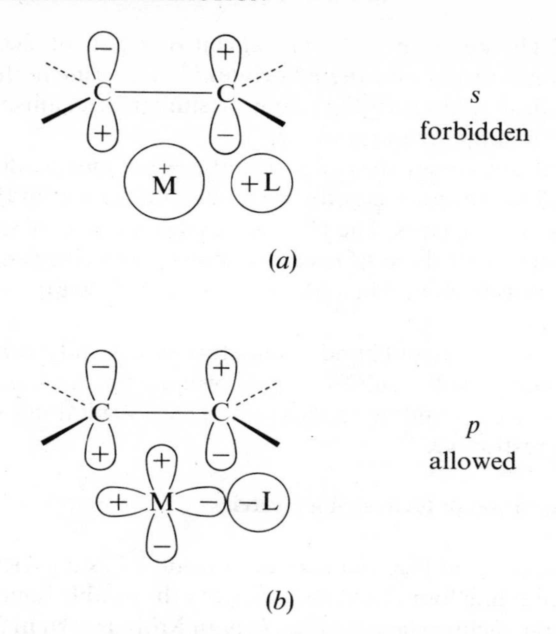


**Figure 20.** (*a*) Ligand migration when metal—ligand bond involves *s* orbital on metal; (*b*) ligand migration when metal—ligand bond involves *p* orbital on metal.

requirement cannot be met. The negative lobe of the carbon—carbon $\pi^*$ orbital cannot bond to the ligand, which necessarily has a positive sign for its orbital to match the *s* orbital of the metal. To complete the picture, Fig. 20*b* shows the case for predominant *p*-orbital binding. Two *p* orbitals must be used to bind two *cis* ligands. They are independent and therefore their phases can be assigned to fit the requirements. The reaction is allowed for *p*-orbital bonding.

An example of this is the hydroboration reaction,[129] which probably proceeds by the following sequence:

$$C_2H_4 + BH_3 \longrightarrow H_2C{=}CH_2 \cdot BH_3 \longrightarrow H_2BCH_2CH_2H \tag{135}$$

There are no filled *d* orbitals on boron to enter in to metal—ligand $\pi$ bonding, but there is no requirement for this since the empty $\pi^*$ orbital of ethylene is still available. The key step is the movement of a pair of electrons from a filled B—H bond orbital into the $\pi^*$ orbital.[130] Experimentally, boron and the hydrogen atom add *cis* to the olefin. A similar mechanism appears for the addition of aluminum alkyls to olefins.[131]

In these tetrahedral cases, the two *p* orbitals are not independent. The point group is $C_s$ and the reaction coordinate is $A'$, a movement in the plane, and both *p* orbitals are $a'$ species. The H-atom orbital is also of $a'$ species. Fortunately the phases of all three AOs can be matched to give positive overlap. This is shown in Fig. 20*b* by imagining L to move down to overlap both *p* orbitals at once.

In general a metal—ligand bond is a mixture of *s*, *p*, and *d* orbital bonding. An activation energy will result from the *s* orbital contribution to a reaction allowed by *d* orbital bonding, and so on. The *p* orbital contribution will usually not be restrictive.

## Migration in Extended $\pi$ Systems

The allowed reaction of Fig. 19*a* may be considered as a pericyclic reaction with a ring containing four electrons, two from the double bond of the olefin, and two from the metal—ligand bond. It is an Möbius system because of the nodal properties of the *d* orbital. The forbidden reaction of Fig. 20*a* is a four-electron Hückel system. We can now consider the migration reactions of extended $\pi$ ligands, such as the ally anion, and extended $\pi$ unsaturated molecules such as butadiene.

The allyl anion is a four-electron group. Its addition to ethylene would give a six-electron ring that would be Möbius if a transition metal served as the center. The migration would be forbidden. The basis for this is given in Fig. 21*a*, where the failure to match orbital phases is shown. The HOMO of the allyl anion and the LUMO of ethylene are the critical orbitals. This prediction is for a [3,3] sigmatropic shift as indicated in Fig. 21*a*. Of course the allyl group, like any ligand, could migrate in a [1,3] shift. Note that, because of the *d* orbital, a [1,3] shift is allowed, and the [3,3] shift forbidden, just the opposite from the usual organic case.

Similarly, Fig. 21*b* shows that a simple ligand, such as $H^-$, cannot migrate to complexed butadiene in a [1,5] sigmatropic process, corresponding to 1,4-addition of the metal and ligand to the diene. The six-electron ring is antiaromatic. However the concerted migration of an allyl group to butadiene in a [3,5] shift is allowed because of the eight-electron transition state. Figure 21*c* shows that the orbital phases match up. Some interesting examples of this last reaction have been found.[132]

There are in fact many examples of ligand migrations to butadiene.[115b] Presumably, these are usually allowed [1,3] shifts, in which only one double

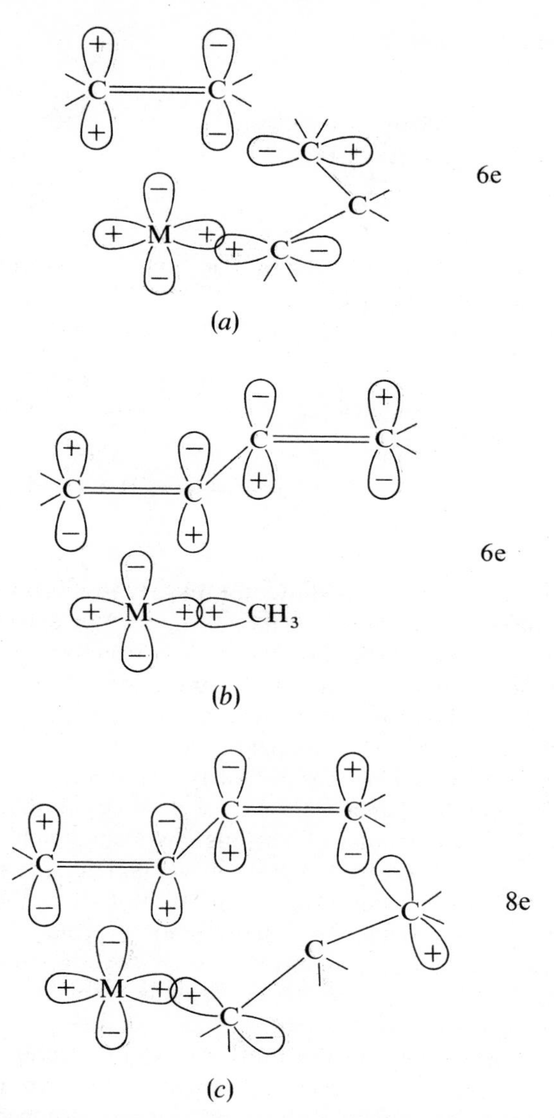


**Figure 21.** (*a*) Suprafacial migration of terminal carbon of an allyl ligand to an olefin is forbidden; (*b*) migration of simple ligand to terminal carbon of butadiene is forbidden; (*c*) migration of allyl ligand to butadiene is allowed.

bond of butadiene partakes.

$$\begin{array}{l} H_2C{=}CH \\ \quad\;\;\diagdown \\ \qquad CH{=}CH_2 \\ \qquad\quad | \\ \qquad\;\; M{-}L \end{array} \longrightarrow \begin{array}{l} H_2C{=}CH \\ \quad\;\;\diagdown \\ \qquad CH{-}CH_2L \\ \qquad | \\ \qquad M \end{array} \tag{136}$$

The initially formed $\sigma$-allyl complex is rapidly converted into the $\pi$-allyl complex, in the normal case.

$$\begin{array}{l} H_2C{=}CH \\ \quad\;\;\diagdown \\ \qquad CHCH_2L \\ \qquad | \\ \qquad M \end{array} \longrightarrow H_2C\overset{CH}{\underset{M}{\frown}}CHCH_2L \longrightarrow \begin{array}{l} CH_2CH{=}CHCH_2L \\ | \\ M \end{array} \tag{137}$$

The $\pi$-allyl complex in turn allows for an easy isomerization to the product of a forbidden 1,4-addition.[62] The addition of an external nucleophile to coordinated butadiene will also lead to the formation of a $\pi$-allyl complex. Allyl complexes are known to give ligand-migration reactions to olefins.[133] However it is difficult to tell if these are allowed [1,3] shifts or forbidden [3,3] shifts because of the isomerization reaction (137).

If the metal—ligand bond is primarily due to an *s* orbital, the situation is essentially reversed. Table 4 shows the predictions that are made for the ligand-migration reactions of various $\pi$-system ligands and various unsaturated molecules to which migration occurs. The rules are that $4n$ electrons give aromatic TSs and allowed reactions, and $4n + 2$ electrons give antiaromatic TSs and forbidden reactions, for *d*-orbital bonding. When the metal—ligand bonding is mainly due to an *s* orbital on the metal, the rules are opposite, since these are Hückel systems. Table 4 is a good example of the Dewar–Zimmerman procedure.

It has been assumed in the table that the addition to the unsaturated molecule is suprafacial on both components, as shown in Fig. 21. Certainly the rules are in agreement with a vast amount of experimental data, with few, if any, exceptions. However some of the predictions have not yet been tested. An example would be 1,4-addition to butadiene in the case of *s*-orbital bonding.

The $\beta$-elimination reaction is the reverse of a hydride-ion ligand migration. It requires a hydrogen atom, or some group with a reasonable affinity for metal atoms, to be in the $\beta$-position of a coordinated alkyl group.

**Table 4 Symmetry Predictions for Ligand Migrations**

| Migrating Group | Unsaturated Group | Predominant Metal Bonding | |
|---|---|---|---|
| | | *d* | *s* |
| $CH_3^-$ | Ethylene | a[b] | f[c] |
| | 1,3-Butadiene | f | a |
| | 1,3,5-Hexatriene | a | f |
| $C_3H_5^-$ | Ethylene | f | a |
| | Butadiene | a | f |
| | Hexatriene | f | a |
| $C_5H_7^{-a}$ | Ethylene | a | f |
| | Butadiene | f | a |
| | Hexatriene | a | f |

[a] $C_5H_7^-$ is $CH_2{=}CH{-}CH{=}CH{-}CH_2^-$. The predictions for it and the allyl anion are made assuming that the terminal free carbon of the migrating group attaches to the terminal carbon of the unsaturated group.
[b] Allowed.
[c] Forbidden.

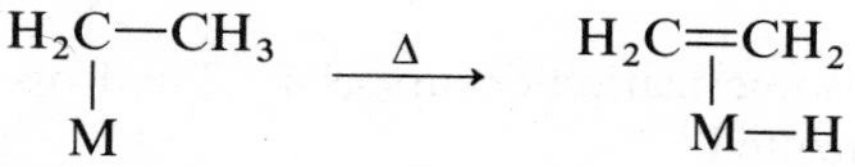

$$\underset{\displaystyle M}{\overset{\displaystyle H_2C{-}CH_3}{|}} \xrightarrow{\Delta} \underset{\displaystyle M{-}H}{\overset{\displaystyle H_2C{=}CH_2}{|}} \tag{138}$$

It is important because it plays the major role in determining thermal stabilities of metal—alkyl bonds of the transition metals.[134] The orbital requirements for (138) are just those of ligand migration, run in reverse. This suggests various ways in which a metal—alkyl bond can be stabilized. For instance, some cyclic organometallic compounds, such as P, contain very stable carbon—metal bonds

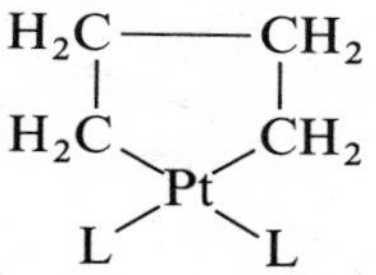

$$\begin{array}{ccc} H_2C & {-\!\!-\!\!-} & CH_2 \\ | & & | \\ H_2C & & CH_2 \\ & \diagdown\ \ \diagup & \\ & Pt & \\ & \diagup\ \ \diagdown & \\ L & & L \end{array} \qquad P$$

where L is a phosphine.[135] The stability results from the difficulty of attaining a dihedral Pt—C—C—H angle of 0°, as required by the mechanism of Fig. 19*a* for $\beta$ elimination.

From our symmetry analysis, we see that $\beta$-hydride elimination, (138), is forbidden for the alkyl derivatives of $d^{10}$ metal ions. Thus mercury(II) alkyls, thallium(III) alkyls, and silver(I) alkyl, all decompose thermally by the free-radical route.

$$C_2H_5\text{—}Hg\text{—}Cl \longrightarrow C_2H_5\cdot + \cdot HgCl \qquad (139)$$

Copper(I) alkyls are anomalous in that $\beta$-hydride elimination occurs,[134] even though the $d^{10}$ cuprous ion has no empty $d$ orbitals available for bonding. This may be a result of the polymeric structures invariably found for organometallic cuprous compounds.

In a similar way to the $\beta$-hydride elimination, the migration of an alkyl group from a $d^{10}$ metal to a coordinated carbonyl group is forbidden. There are no known migrations of this kind for Cu(I), Au(I), or Hg(II). Metals in which $p$ orbital bonding is predominant can give both $\beta$-hydride elimination and alkyl group migration. This is the case for aluminum alkyls, for example.

A combination of ligand migration and $\beta$ elimination gives an efficient mechanism for olefin isomerization. A transition-metal hydride, formed in some way, is normally the catalyst. For example, 1-butene can be isomerized to 2-butene by coordination to such a catalyst.

$$\underset{\text{H—M}}{H_2C{=}CHCH_2CH_3} \longrightarrow \underset{\text{M}}{CH_3\text{—}CHCH_2CH_3} \longrightarrow \underset{\text{M—H}}{CH_3CH{=}CHCH_3} \qquad (140)$$

It can be seen that isomerization occurs as a 1,3 hydrogen shift. This mechanism is well documented.[136]

Another mechanism for olefin isomerization relies on the monohapto–trihapto allylic rearrangement of (137). Oxidative addition of a C—H bond is a key step. This normally difficult reaction is presumably facilitated by the adjacent double bond.[137]

$$CH_3CH_2\text{—}CH{=}CH_2 + M \longrightarrow \underset{\text{M—H}}{CH_3\text{—}CH\text{—}CH{=}CH_2}$$

$$\longrightarrow CH_3\text{—}\overset{CH}{CH \quad CH} \;(\text{M—H, trihapto}) \longrightarrow CH_3\text{—}CH{=}CH\text{—}\underset{\text{M—H}}{CH_2}$$

$$\longrightarrow CH_3\text{—}CH{=}CH\text{—}CH_3 + M \qquad (141)$$

This also gives a 1,3 hydrogen shift, but it can be distinguished from (139) by the overall pattern of isotope positions when deuterium labeling is used. The mechanism of (140) allows for 1,2 shifts by hydride migration to the internal carbon of the olefin. The $\pi$ allyl mechanism has been demonstrated in several cases.[137]

## OXIDATIVE CYCLIZATION AND OTHER CYCLIZATION REACTIONS

In this section some cyclization reactions are considered where the mechanisms are not well understood. Nevertheless they represent a definite class of reactions of certain organometallic compounds, and simple concerted mechanisms can be postulated. The reactions involve two or more olefin or acetylene molecules cyclized under the influence of a transition-metal complex. Unfortunately it is not known whether the unsaturated molecules are both (or all) complexed to the metal atom when the cycloaddition step occurs. The initial assumption will be that they are.

The first class of reactions is conveniently called "oxidative cyclization."[118] Like oxidative addition, the reaction causes the oxidation state of the metal to increase by two units. Only $d^8$ and $d^{10}$ complexes where the metal is a good reducing agent have been found to react. Some examples are[139]

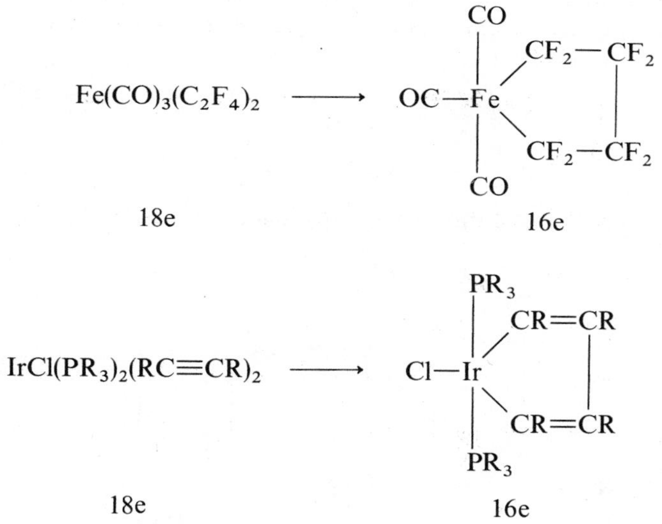

$$Fe(CO)_3(C_2F_4)_2 \longrightarrow (CO)_2(OC)Fe[CF_2-CF_2-CF_2-CF_2]$$

18e → 16e

$$IrCl(PR_3)_2(RC{\equiv}CR)_2 \longrightarrow (PR_3)_2ClIr[CR{=}CR-CR{=}CR] \qquad (142)$$

18e → 16e

The products are cyclic compounds with the metal atom incorporated in the ring. Such molecules are called "metallocycles." Compound P above is also a metallocycle. It was prepared in quite a different way, by the reaction of 1,4-dilithiobutane on *cis*-$PtL_2Cl_2$.

When a molecule similar to P is formed from $WCl_6$, it is spontaneously unstable, giving up ethylene.[140]

$$Cl_4W\begin{matrix} \diagup CH_2-CH_2 \\ \diagdown CH_2-CH_2 \end{matrix} \longrightarrow WCl_4 + 2C_2H_4 \qquad (143)$$

This is not surprising since tungsten in a high oxidation state should not form stable organometallic compounds. Therefore (143) shows that the oxidative-cyclization reaction is reversible. What is surprising, are the results when 1,4-dilithio-2,3-dideuteriobutane is used. The ethylene evolved is a mixture of $d_0$, $d_1$, and $d_2$.[140] Some unusual scrambling of carbon atoms has occurred prior to (142). We will return to this point later.

First we take up again the possible reactions of two olefin molecules coordinated to a low valent metal atom in a *cis* arrangement. In (100) it was shown that one possibility was the formation of a complexed cyclobutane. This was accomplished by an $A_1$ reaction coordinate, with electron transfer from $b_1$ to $b_1^*$, and from $b_2$ to $b_2^*$. Another possibility, not considered previously, is that electron transfer occurs directly from $b_2$ to $b_1^*$. The reaction coordinate now becomes $(B_1 \times B_2) = A_2$. As Fig. 22*a* shows, an $A_2$ motion corresponds to a conrotatory twisting of both ethylene molecules. The result is a complexed cyclobutane molecule, but one rotated by 45°, compared to (100). The other ligands,

$$\Vert\!-M-\!\Vert \xrightarrow{A_2} \Diamond\!-M-\!\Diamond \qquad (144)$$

not shown, could make this product distinguishable from that of (100).

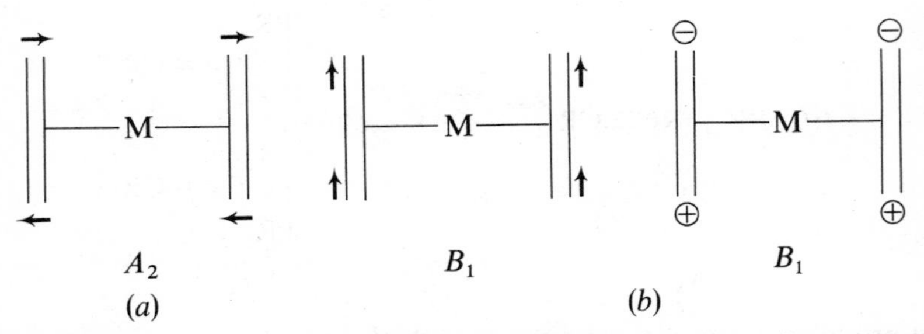


**Figure 22.** Nuclear motion for two coordinated olefin molecules corresponding to $A_2$ and $B_1$ reaction coordinates; the initial point group is $C_{2v}$.

In Fig. 16, three $d$ orbitals, all filled, were not shown because they did not change. These orbitals contain high energy, reducing electrons and it is possible that they are active, rather than the filled $b_2$ orbital. The $d_{z^2}$ and $d_{xy}$ orbitals are of $a_1$ species, as shown in Table 2. Electron flow from these orbitals into the empty $b_1^*$ orbital would create a reaction coordinate of $B_1$ species. In Fig. 22$b$ it is shown that this corresponds to two sets of nuclear motion for the coordinated olefin molecules. These are the required motions for the formation of a metallocycle.[118] The point group becomes $C_s$.

M $\xrightarrow{B_1}$ M (145$a$)

There would be no correlation problems. The filled $a_1$ and $b_2$ orbitals become two new metal—carbon $\sigma$ bonds and the $b_1^*$ orbital, a new carbon—carbon $\sigma$ bond. The empty $a_1^*$ orbital would be high energy in the product, especially if a fifth ligand is coordinated. As (142) show, the fifth ligand is normally present in the product, and probably in the reactant.

Structural studies show that two independent olefin molecules coordinated to a metal atom normally take up coplanar positions, and not perpendicular as shown in (145).[141]

M $\longrightarrow$ M (145$b$)

However this does not change the orbital analysis, as shown in Fig. 23$a$. If the point group is $C_{2v}$, the critical LUMO becomes $A_1$. The reducing electrons still come from an orbital of $A_1$ species, and the reaction coordinate is therefore also $A_1$. A $d^8$ metal atom might be reactive, or an atom with any number of $d$ electrons, including zero. In this case the reducing electrons could come from an $s$ or $p$ orbital of the metal of $a_1$ species.

However fewer than $d^4$ seems unlikely because it is necessary to coordinate two olefin molecules to have the concerted reaction (145). A nontransition element, lacking $d$ electrons, is not likely to accomplish this. Corresponding reactions of oxidative cyclization are known, but they probably go stepwise. For example, triphenylphosphine will couple acetylenes, yielding phosphoranes as initial products.[142]

$$R_3P + X{-}C{\equiv}C{-}X \longrightarrow R_3P^+{-}C(X){=}C^-{-}X$$

$$\xrightarrow{X-C\equiv C-X} R_3P^+{-}C(X){=}C(X){-}C(X){=}C^-{-}X \longrightarrow R_3P\begin{matrix} CX{=}CX \\ | \\ CX{=}CX \end{matrix} \quad (146)$$

It is quite possible that a similar stepwise process occurs with some transition-metal complexes.[143] In the gas phase, the dipolar intermediates would be replaced by equivalent diradical species. Reaction (146) requires that X be an electron-withdrawing substituent, to facilitate formation of the intermediates. Olefins are not coupled in this way, since they are less likely to add nucleophiles. However the reverse of (146) is known for olefins.[144] The two carbon—phosphorus bonds that are broken are axial and equatorial, which is forbidden for a concerted process (p. 290).

If four-coordinated initially, a $d^{10}$ metal atom encounters a special problem in oxidative cyclization. The initial geometry would undoubtedly be tetrahedral, whereas the $d^8$ metal complex formed would greatly prefer to be square planar.

(147)

unstable stable

For this reason it may be significant that only very light coordinated Ni(0) and Pd(0) complexes will give metallocycles. An example is the so-called "naked nickel," which is Ni(0) formed under conditions where only the olefin may be coordinated.[145] The cyclization of butadiene is particularly important. A variety of products are formed, but the initial reaction is oxidative cyclization.

(148)

$2a' + 2a'' + a'(d)$ $\qquad$ $3a' + 2a''$

It can be seen that this concerted reaction requires that the reducing electrons come from a *d* orbital of $a'$ symmetry. The $\sigma$-allylic complex may form first and be converted to the more stable $\pi$ allyl, or the *bis*-$\pi$ allyl may form directly.

Figure 23*a* shows the orbital-phase relationships between a filled metal MO of $a_1$ symmetry and the empty $(\pi^* + \pi^*)$ MO of two coordinated olefin molecules. Figure 23*b* shows the phases for the $(\pi^* + \pi^*)$ combination of two complexed butadiene molecules. The relevant orbital of the isolated butadiene molecule is the LUMO. We can see that in these cases, a six-electron and a 10-electron pericyclic reaction are allowed. The systems are Hückel systems since only one lobe of the *d* orbital is used. The nickel atom becomes $d^8$, with the high-energy empty *d* orbital pointing at the two $\pi$-allyl groupings.

Figure 23*c* shows that the case of one ethylene molecule and one butadiene molecule does not lead to a proper matching of the orbital phases. The metal orbital cannot overlap both $\pi$ orbitals simultaneously. This is an eight-electron forbidden reaction. Table 5 shows a compilation of predictions for oxidative-cyclization reactions, using the rule $4n + 2$ electrons, allowed, and $4n$ electrons forbidden. The predictions are in general agreement with the experimental facts. Olefins and dienes normally do not undergo cyclization reactions with each other,[145] although a few exceptions are known.[146a] The reaction could occur if the $d_{x^2-y^2}$ orbital were used, instead of $d_{xy}$. The predicted ethylene plus hexatriene cyclization has recently been found.

While there are many examples of olefin cyclization, there are few for diene cyclization, similar to (148). Even in the case of naked nickel, it is quite possible that a series of simple, stepwise reactions occurs to give the same product that the concerted reaction would yield. A reaction which requires a large number of atoms to be in specified locations is improbable, even if energetically favored.

A more characteristic reaction of acetylenes with $d^8$ and $d^{10}$ metal complexes is a series of cycloaddition reactions forming cyclobutadienes

**Table 5 Symmetry Predictions for Oxidative-Cyclization Reactions by Transition Metals**[a]

| | | |
|---|---|---|
| Ethylene + ethylene | allowed | 6e |
| Ethylene + butadiene | forbidden | 8e |
| Butadiene + butadiene | allowed | 10e |
| Ethylene + hexatriene | allowed | 10e |
| Butadiene + hexatriene | forbidden | 12e |
| Hexatriene + hexatriene | allowed | 14e |

[a] $4n + 2$ electrons allowed; $4n$ electrons forbidden.

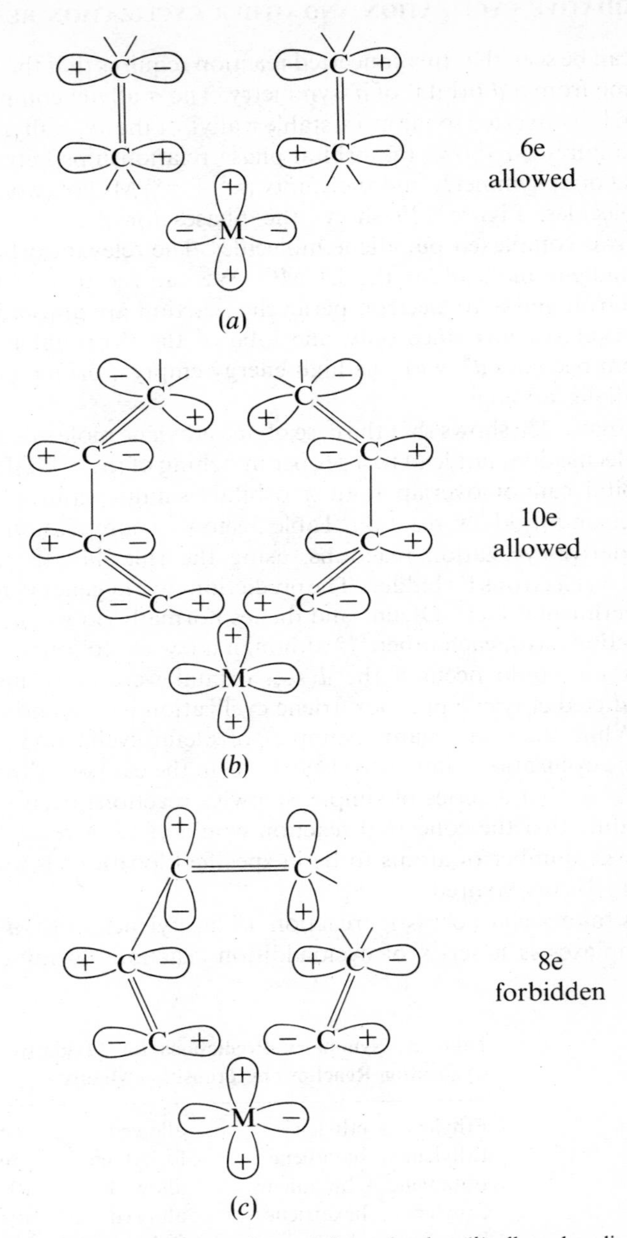


**Figure 23.** (*a*) Allowed oxidative cyclization of two olefin molecules; (*b*) allowed cyclization of two conjugated diene molecules; (*c*) forbidden cyclization of one olefin and one diene molecule. (The reducing electrons of the metal are assumed to come from an orbital of $a_1$ or $a'$ species.)

(complexed), substituted benzenes, or cyclooctatetraene.[115] The pioneering work in this area was done by Reppe. It is attractive to imagine that two, three or four acetylene molecules are coordinated to a single metal atom, and cyclize in a concerted manner. In the absence of the metal complex, we have the following

$$2C_2H_2 \longrightarrow C_4H_4 \quad 4e \quad \text{forbidden} \tag{149}$$

$$3C_2H_2 \longrightarrow C_6H_6 \quad 6e \quad \text{allowed} \tag{150}$$

$$4C_2H_2 \longrightarrow C_8H_8 \quad 8e \quad \text{forbidden} \tag{151}$$

Complexing would not change the nature of the allowed reaction, since no net change in the metal ion state needs to occur. In an all *cis* arrangement of three-complexed acetylenes, the point group would be $C_{3v}$.

M ⟶ M (152)

$2a_1 + 2e$ $2a_1 + 2e$

Localization of the double bonds is not implied in the coordinated benzene shown as a product.

In spite of this apparently favorable template reaction, detailed studies of cyclotrimerization for both Ni(0) and Pd(II) show complex mechanisms.[143,147] There is a series of ligand-migration steps, followed by a ring-closing step, which is not well defined. A similar stepwise mechanism also applies to the formation of complexed cyclobutadiene,[148] which is not an intermediate en route to benzene. The ring-closing step in this case appears to be[143,147b]

Pd L L X ⟶ Pd L L X

Pd L L X ⇌ Pd L L X (153)

The ring closure is a simple conrotatory electrocyclic reaction, which is allowed. The role of X is that of a substituent that is easily removed, such as $Cl^-$

Since (149) is forbidden as a concerted process in the absence of the metal ion, let us examine the effect of the latter in an assumed concerted reaction.[149] A square planar complex, reasonable for a $d^8$ metal ion, will be assumed, and an $A_1$ reaction coordinate.

M ⟶ M (154)

Part of the reaction can be accomplished by the same electron transfers required for the conversion of two complexed ethylene molecules to cyclobutane (Fig. 16). The two extra $\pi$ bonds of the acetylene molecules cannot be accommodated in this way. They form filled orbitals of $a_1$ and $b_2$ species, just as for the first two $\pi$ bonds. In the product they form a strongly bonding $a_1$ orbital, moderately bonding orbitals of $b_1$ and $b_2$ species, and an antibonding MO of $a_2$ species. The orbital correlations are shown in Fig. 24. The reaction is seen to be still strongly forbidden by the incipient formation

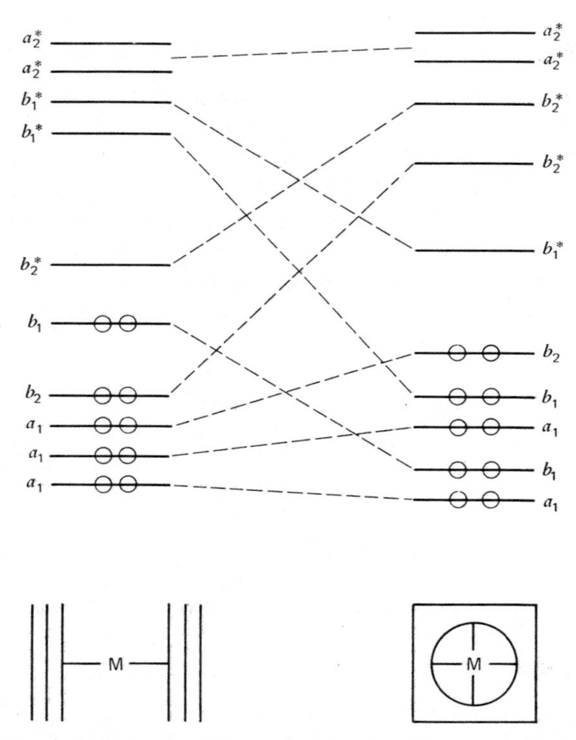


**Figure 24.** Orbital-correlation diagram for formation of complexed cyclobutadiene from two molecules of complexed acetylene.

of an excited state. The $d$ orbitals not shown are of $a_1$ and $a_2$ species and cannot help.

The situation is different if the $A_2$ reaction coordinate of (144) is selected.

$$\Vert\!-\!M\!-\!\Vert \longrightarrow \langle(M)\rangle \tag{155}$$

This coordinate changes the point group to $C_2$. The $b_1$ and $b_2$ species both become $b$, and the orbital crossings of Fig. 22 disappear. The reaction is now allowed. Inasmuch as stable cyclobutadiene metal complexes exist and those of cyclobutane do not, it is by no means clear that the concerted cyclization of acetylene should be any more difficult than that of ethylene.

## MECHANISMS OF SOME CATALYTIC REACTIONS

After this digression into the elementary, or possible elementary, reactions of transition-metal organometallic compounds, we can return to the subject of catalysis by these compounds. Homogeneous hydrogenation of unsaturated molecules has been the subject of many reviews, [82,82,150] and only a brief summary will be given. The general mechanism is well agreed on.

$$M + H_2 \longrightarrow MH_2 \xrightarrow{>C=C<} \begin{matrix} >C=C< \\ | \\ M-H \\ | \\ H \end{matrix}$$

$$\swarrow$$

$$\begin{matrix} & & C & \\ & C & & \backslash \\ M-H & & & H \end{matrix} \longrightarrow M + \begin{matrix} C-C \\ | \quad | \\ H \quad H \end{matrix} \tag{156}$$

There is oxidative addition of molecular hydrogen to the metal catalyst and coordination of the unsaturated substrate. This is followed by ligand migration of hydride ion and a reductive elimination of saturated product. From the discussion of the last two elementary steps given previously, it follows that the hydrogen is added stereospecifically *cis* to the olefin. This prediction has been amply verified.[90,151]

Variations of (156) are possible, since monohydrides can also be catalysts and can be generated in several ways. Halpern has pointed out that there are three fundamentally different ways in which metal complexes activate molecular hydrogen.[152] These can be illustrated by the behavior of Cu(I), Ag(I), and Rh(I).

(a) Heterolytic splitting:

$$[Cu(I)]_2 + H_2 \longrightarrow 2[Cu(II)H]$$

(b) Homolytic splitting:

$$Ag^+ + H_2 \longrightarrow Ag(I)H + H^+$$

(c) Oxidative addition:

$$Rh(I) + H_2 \longrightarrow Rh(III)H_2$$

Actually (a) may be regarded as oxidative-addition, with two atoms of copper(I) being required, because of the instability of Cu(III) (cf. p. 410). Also silver(I) reacts by an electrophilic substitution mechanism because of the instability of both Ag(II) and Ag(III).

Butadiene is hydrogenated both 1,2 and 1,4. In the latter case it has been suggested that simultaneous addition of both hydrogen atoms occurs.[153] This is an allowed concerted reaction in the absence of the metal catalyst, although it is found only as its reverse (p. 101).

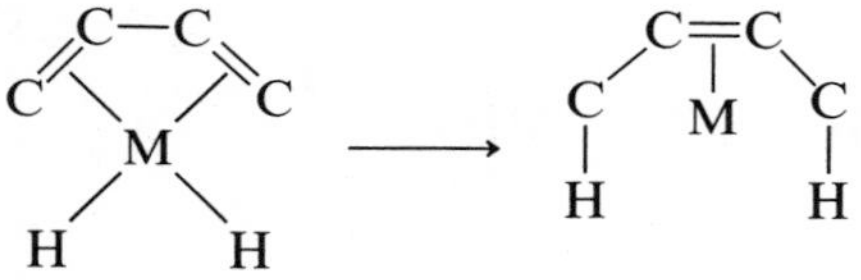

```
   C—C                  C=C
 C//  \\C             C/  |  \C
   \  /                   M
    M        ——→      |       |
   / \                H       H
  H   H                                   (157)
```

Examination of the orbital phases, as illustrated in Fig. 21, shows that (157) is a forbidden process, if both hydrogen atoms are bound by the same *d* orbital. This would be the case if they were *cis* to each other, as required on steric grounds. In any event, a stepwise mechanism is more realistic.

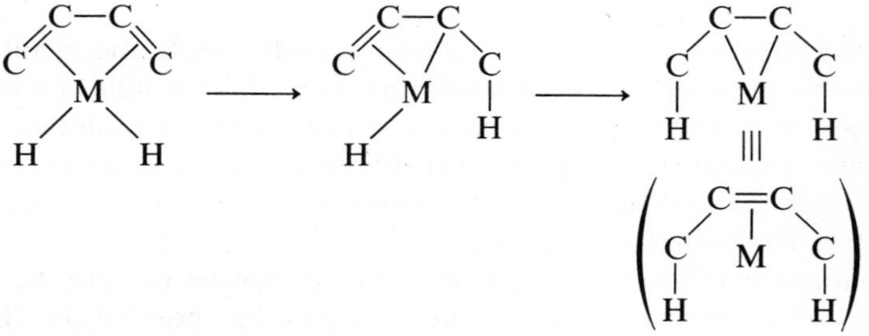

```
   C—C              C—C                C—C
 C//  \\C         C//  \  C          C/ \ / \C
   \  /             \ /   |          |   M   |
    M      ——→       M    |   ——→    H       H
   / \              /     H              |||
  H   H            H                                 (158)
                                     (   C=C    )
                                     ( C/ | \C  )
                                     ( |  M  |  )
                                     ( H     H  )
```

## Metal-catalyzed Ring-opening Reactions

The next catalytic reactions to consider are the ring-opening reactions of certain strained cyclic molecules, catalyzed by transition-metal complexes, notably $Ag^+$ and Rh(I).[154] These reactions often give products that are forbidden in the absence of the catalyst. Some examples are

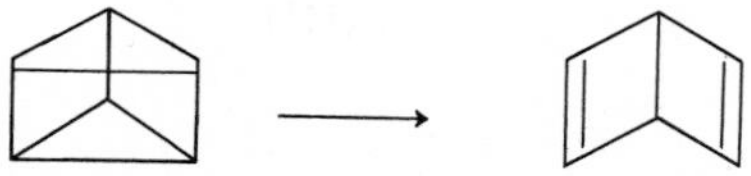

Prismane → Dewar benzene (159)

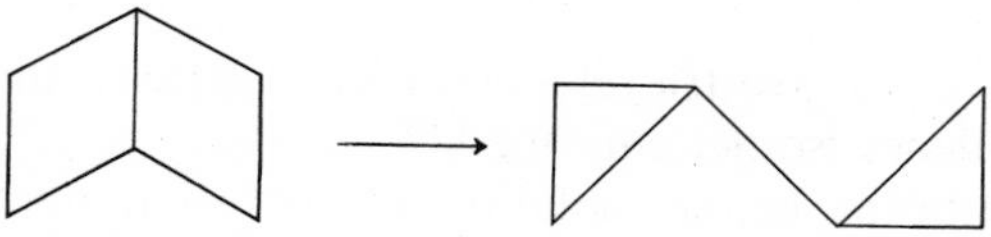

Bicyclohexane → Dicyclopropyl (160)

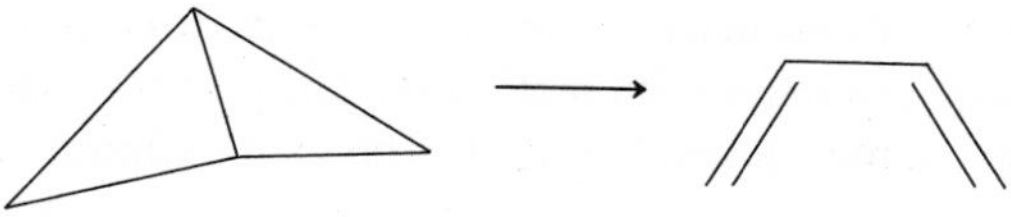

Bicyclobutane → *cis*-Butadiene (161)

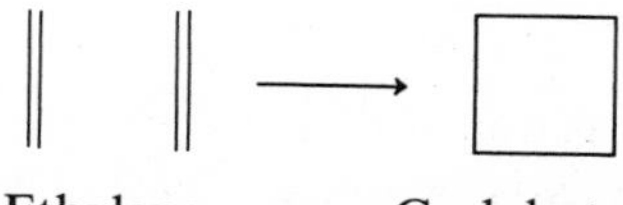

Ethylene → Cyclobutane (162)

Reaction (160) maintains a $C_2$ axis and the old and new bonds match up as $a$ and $b$. However the axis does not intersect any of these bonds and is misleading. The reaction is actually a forbidden $[_\sigma 2_a + {}_\sigma 2_a]$ process.[155] Similarly (161) is not the allowed $[_\sigma 2_s + {}_\sigma 2_a]$, but the forbidden $[_\sigma 2_s + {}_\sigma 2_s]$. A variety of other products can also be formed by ring opening of bicyclobutane under various conditions.

Generally speaking, metal ions that readily undergo oxidative addition give different products from $Ag^+$ and similar catalysts such as $Hg^{2+}$. An example is shown in the differing behavior of cubane, $C_8H_8$, toward Rh(I)

and $Ag^+$.

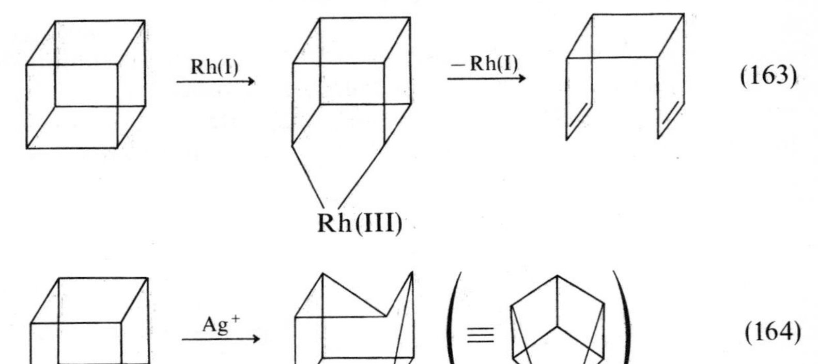


$Ag^+$ (164)

The former gives *syn*-tricyclooctadiene, whereas the latter yields the interesting valence isomer cuneane.[156]

More significant is the isolation of the rhodium(III) metallocycle in (163).[158] This is fairly convincing proof that the mechanism consists of an oxidative addition of a carbon—carbon $\sigma$ bond tothe metal, followed by the reverse of the oxidative-cyclization reaction to yield the diolefin. The same mechanism has been demonstrated in the reverse direction.[159] Norbornadiene has been shown to react with iridium(I) complexes to yield the metallocycle. On heating the metallocycle goes to the cyclic dimer.

—Ir— ⟶ Ir ⟶ + Ir (165)

As for the $Ag^+$ reaction, as well as that of $Hg^{2+}$, the evidence is strong that simple electrophilic catalysis is involved.[155,159] The $Ag^+$ reacts with the strained carbon—carbon bond by electrophilic substitution, for instance,

$Ag^+$ + ⟶ Ag + (166)

Ring opening occurs to give a silver—carbon bond at one end and a carbenium ion at the other. The subsequent reactions of the carbenium ion in various circumstances determine the products. With simple cyclopropane rings, reaction (166) does not occur readily, but weak complexation with $Ag^+$ or $Hg^{2+}$ allows for easy nucleophilic attack, which opens the ring.[160]

The behavior of $Ag^+$ in these cases is exactly analogous to its behavior in reactions with molecular hydrogen and with alkyl halides. Other Lewis acids will also behave as catalysts for (159)–(162).[156,159] Generally, however, the nontransition-metal catalysts, including $H^+$, are much less effective than transition-metal ions. As for those mechanisms requiring coordinated cyclobutane, (100) or (143), there seems to be no need to invoke them and little convincing evidence to support them.

There is another catalytic reaction to the transition metals where complexed cyclobutane has been postulated to play the major role.[162] This is the olefin disproportionation, or metathesis reaction.[163]

$$1{=}2 \quad + \quad 4{=}3 \xrightarrow{\text{cat}} 2{=}3 \quad + \quad 1{=}4 \qquad (167)$$

In this truly astonishing reaction, four carbon—carbon bonds must be broken, and four new ones made. In the absence of a catalyst, the reaction is, of course, strongly forbidden. Yet with certain heterogeneous and homogenerous catalysts containing transition metals, the reaction proceeds rapidly even at 0°C. The activation energy is only 6–7 kcal.

Unfortunately little is known about the exact nature of the catalyst in most cases, but molybdenum and tungsten complexes, in an oxidation state between 0 and +4, seem to be most effective.[163] These are $d^2$–$d^6$ systems. We will assume that two olefin molecules are coordinated to a metal atom in *cis* position. This brings us back to the prediction of the three possible reaction paths given by (100), (143), and (144). Obviously (100) and (143) can readily lead to olefin disproportions if they occur, since complexed cyclobutane could fall apart in two ways, one of which constitutes the metathesis.

However a new problem arises with the $A_1$ reaction coordinate. The geometry of the final product is different from that of the initial reactant. For example, a square planar complex would become tetrahedral.[164]

M, L, L → M, L, L → M, L, L (168)

$D_{4h}$ $T_d$

Since the activation energy for interconverting planar and tetrahedral forms is 10 kcal, even for the most favorable cases, this is a very serious drawback. If other common coordination numbers and geometries are assumed for the catalyst, similar difficulties will be found. The $A_2$ reaction mode has the advantage of avoiding the "tetrahedral dilemma" and related structural difficulties.

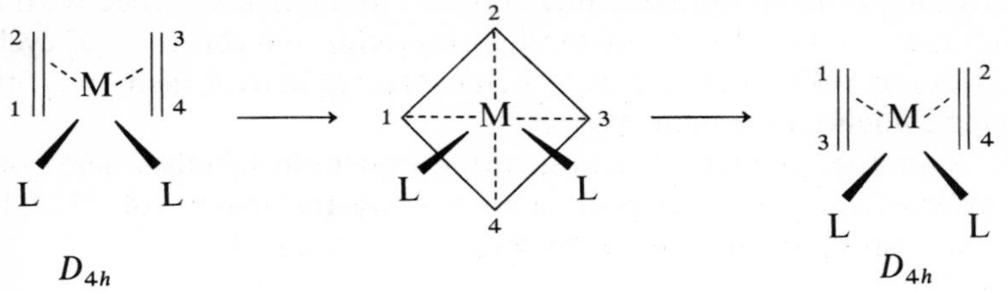

$D_{4h}$ $\longrightarrow$ $\longrightarrow$ $D_{4h}$ (169)

However if rotation of the complexed olefin molecules is reasonable, there is no reason why acetylenes should not also undergo disproportionation reactions, perhaps somewhat less readily than olefins. Experimentally it is found that acetylenes do not react in this way, except under extreme conditions (200–300°C).[165] It would appear that a mechanism is required that is strongly excluded for acetylenes. The oxidative-cyclization reaction (144) offers such a mechanism, although it appears at first to be an unlikely candidate. However it does utilize a known intermediate, whereas, complexed cyclobutane is unknown.

The trick is to find a method for scrambling the carbon atoms of the metallocycle, and to accomplish this with well accepted elementary processes. This can be done by the ligand-migration reaction.[140]

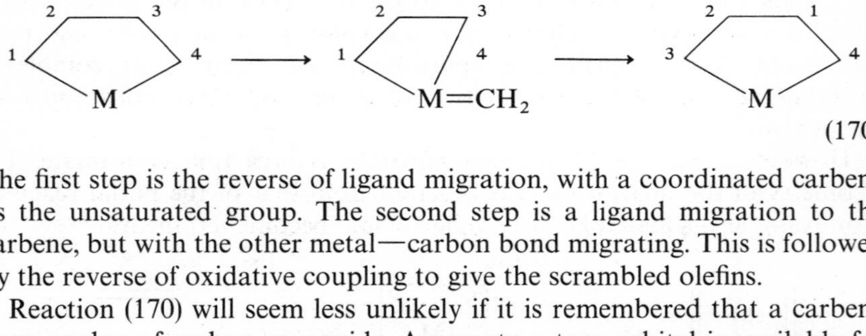

$\longrightarrow$ M=$CH_2$ $\longrightarrow$ (170)

The first step is the reverse of ligand migration, with a coordinated carbene as the unsaturated group. The second step is a ligand migration to the carbene, but with the other metal—carbon bond migrating. This is followed by the reverse of oxidative coupling to give the scrambled olefins.

Reaction (170) will seem less unlikely if it is remembered that a carbene is an analog of carbon monoxide. An empty $\pi$-type orbital is available for migration.

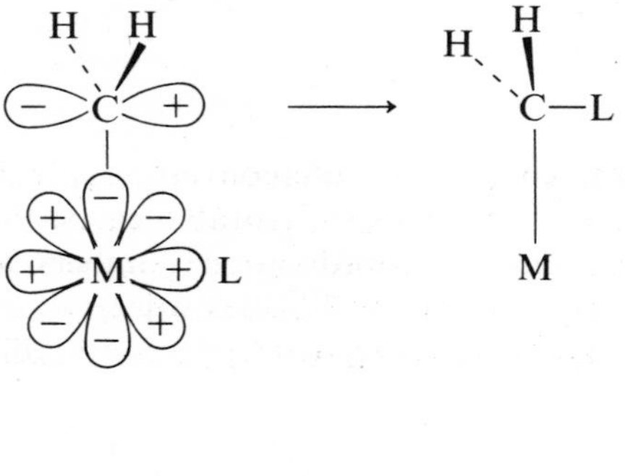

$\longrightarrow$ (171)

Actually carbene complexes of the transition metals are well known. Amino-olefins are disproportionated in the presence of $RhCl[P(C_6H_5)_3]_3$, and carbene complexes, $RhCl[P(C_6H_5)_3]_2CR_2$, which may be intermediates, can be isolated.[166b] This is a somewhat biased example since aminocarbenes are strongly stabilized by resonance.

$$:C(NH_2)_2 \longleftrightarrow \bar{:}C(=\overset{+}{N}H_2)(NH_2) \quad \text{(etc.)}$$

An even simpler mechanism for olefin disproportionation would be complete dissociation of olefins into carbenes, coordinated to the metal catalyst. This is an allowed process, but is not reasonable for simple olefins on energetic grounds. The bond energy to be broken is some 160 kcal (carbon—carbon double bond) and the sum of two metal carbene bonds probably does not exceed 130 kcal.

Other indications for the existence of carbenes exist. For example, in olefin-metathesis reactions, products containing an odd number of carbon atoms are often formed from ethylene. These can be explained by the formation of carbenes by any route. The reverse of (171) offers a way to generate carbenes. Almost any alkyl derivative of a metal could give a carbene by a simple hydride shift.

Many of the procedures used in preparing catalysts could form small quantities of carbene complexes. In this case a chain mechanism can be visualized in which the kep step is a cycloaddition of the carbene and the olefin to form a four-membered metallocycle.[167] This is an allowed reaction, using the orbital-overlap arguments shown in Fig. 23. The reverse of cycloaddition can then occur to give a different olefin and carbene. The latter can continue the chain.

$$M{=}C(R)(H) + R'(H)C{=}C(R')(H) \longrightarrow \begin{array}{l} R'CH{-}CHR' \\ \;|\qquad\;\; | \\ M{-}{-}CHR \end{array}$$

$$\longrightarrow \underset{Q}{M{=}C(R')(H)} + R'(H)C{=}C(R)(H)$$

$$Q + R''(H)C{=}C(R'')(H) \longrightarrow \begin{array}{l} R''CH{-}CHR'' \\ \;|\qquad\;\;\; | \\ M{-}{-}CHR' \end{array} \quad \text{(etc.)} \qquad (172)$$

It is possible to distinguish between the carbene mechanism and mechanisms involving complexed cyclobutane, or the five-membered metallocycle of (170). The carbene path exchanges only one group of an olefin at a time, whereas the other two mechanisms exchange two groups pairwise. By suitable labeling experiments, it has been shown that exchange occurs one group at a time.[168]

Another advantage of the carbene mechanism is that it explains the low reactivity of acetylenes in metathesis reactions. A carbyne, CH, complex would be needed and it would be more difficult to form these than carbene, $CH_2$, complexes. Also the four-membered metallocycle ring would probably be quite strained and unstable.

The above example shows that there is nothing wrong with a complex mechanism, requiring many steps, providing each step is a well-recognized elementary process. In fact this is the way that Nature commonly seems to work. We will take up one more catalytic reaction, and propose a highly speculative mechanism. Nevertheless, it seems to provide a good example of the way in which transition metal catalysts may operate. Both template properties and orbital symmetry loopholes are used.

The uncatalyzed reaction of nitric oxide and carbon monoxide does not occur to any measurable extent at room temperature, even though thermodynamically favorable.

$$2NO + CO \longrightarrow N_2O + CO_2 \qquad \Delta G^0 = -78.2 \text{ kcal} \tag{173}$$

However the reaction occurs in a time of minutes with certain iridium and rhodium complexes. Furthermore, the process can be made catalytic. The evidence, while incomplete, suggests that a species such as $IrL_2(NO)_2CO^+$ is the reactant. The infrared spectrum further shows that the two nitrosyl ligands are bonded together in some way, $N_2O_2$.

The uncatalyzed reaction (173) is forbidden by orbital symmetry, assuming that the unstable $N_2O_2$ dimer is formed first. This is easily seen by considering only those electrons in $\pi$ orbitals with respect to the plane of the paper.

$$\underset{4\pi e}{\text{O=N—N=O}} + \underset{2\pi e}{:\text{C=O}} \longrightarrow \underset{4\pi e}{\text{O=N=N}} + \underset{4\pi e}{\text{O=C=O}} \tag{174}$$

The HOMO of the reducing agent is in a $\sigma$ orbital (lone pair on carbon). The LUMO of $N_2O_2$ is a $\pi$ orbital, for instance,

$$\underset{4\pi e}{\text{O=N—N=O}} + 2e \longrightarrow \underset{6\pi e}{{}^-\text{O—N=N—O}^-} \tag{175}$$

In the hyponitrite ion, there are three lone pairs on each oxygen atom. One pair on each atom is in a $\pi$-type orbital.

Suppose that the reactants are all coordinated to the metal and lie in a plane, the $xy$ plane.

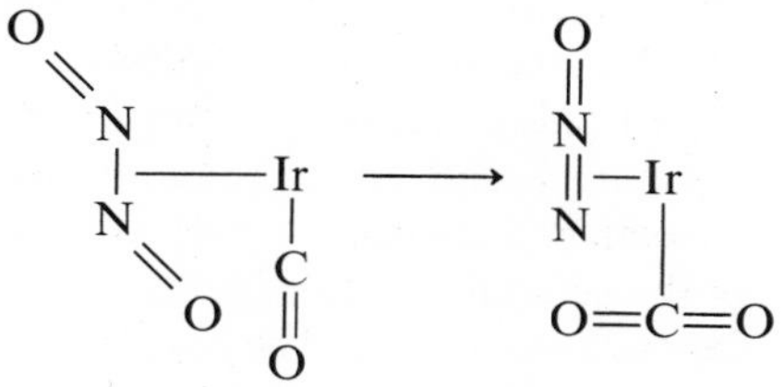

$$\text{O=N(–Ir)–N=O, Ir–C≡O} \longrightarrow \text{O=N=N–Ir, Ir–C(=O)=O} \quad (176)$$

Electrons from the $d_{xz}$ or $d_{yz}$ orbital are of $\pi$ type and can enter the LUMO of $N_2O_2$,[170] The resulting negative charge on oxygen makes it a nucleophile that can migrate to the adjacent carbonyl group. This can be considered as a kind of ligand migration, or as a simple nucleophilic attack on the carbonyl group. Such reactions are well known.[171] The poorly coordinating ligands $N_2O$ and $CO_2$ will leave, and in so doing the $CO_2$ will leave a pair of electrons in the $d_{xy}$ or $d_{x^2-y^2}$ orbital. Recoordination of NO and CO will force these electrons back into $d_{xz}$ or $d_{yz}$, and the process can be repeated indefinitely.

## SOME OXIDATION-REDUCTION REACTIONS

Many of the examples already discussed have been oxidation-reduction reactions. Some further comments on this broad class of reactions is in order. Redox reactions can be divided into categories in several ways[172]

(a) one-electron versus two-electron processes,
(b) electron transfer versus group transfer,
(c) inner-sphere versus outer-sphere reactions.

These are not mutually exclusive; a one-electron process can be either an electron transfer or an atom, or other free-radical, transfer. For example,

$$Cr(H_2O)_6{}^{2+} + Co(NH_3)_6{}^{3+} \longrightarrow Cr(H_2O)_6{}^{3+} + Co(NH_3)_6{}^{2+} \quad (177)$$

$$Cr(H_2O)_6{}^{2+} + Br_2 \longrightarrow Cr(H_2O)_5Br^{2+} + Br\cdot \quad (178)$$

Reaction (177) is an electron-transfer and (178), an atom-transfer reaction. Furthermore (177) is an outer-sphere process and (178), inner-sphere. This terminology relates to whether or not the inner coordination sphere of a complex ion remains intact in the TS. In (178) it does not, a bromine atom entering a coordination site vacated by a water molecule.

Two electron processes are rarely outer-sphere electron transfers. This is an energetic effect, based on the need for solvent reorganization in solution reactions. In the gas phase, two-electron transfers occur, but with low probability. If two reactants form a complex with intimate sharing of at least one atom, there can be a flow of two electrons from one part of the molecule to another, or transfer of the atom or a group of atoms.

In any of these kinds of redox reactions, orbital symmetry plays a major role, as it must in all chemical reactions. The requirement for a net positive overlap between the electron-donating orbital and the electron-accepting orbital is a key part of the general theory of electron transfer. In fact the rate of electron transfer depends strongly on the overlap in the gas phase.[173] In solution the role of the solvent reorganization tends to dominate, but positive overlap is still necessary.[174]

Previously the concept of electrons flowing from HOMO to LUMO was used to help visualize the mixing of the MOs of the reactants to form the new MOs of the products. In electron-transfer reactions there is no mixing, except briefly at the TS. An electron simply jumps from one orbital into another. This can perhaps be understood best by considering an electron in an orbital in a single molecule. The situation to examine would be one where the electron is in the wrong orbital for a certain reaction to occur, and must get to the right orbital.

For example, in (125) an electron is put, by any method, into the LUMO of an alkyl halide. This would be the $\sigma^*$ antibonding orbital of the carbon—halogen bond. The natural consequence would be dissociation into halide ion and alkyl radical, as shown. However if the molecule was an aromatic halide, the situation would be quite different. The LUMO would now be a $\pi^*$ orbital of the aromatic ring, but this orbital has little to do with the carbon halogen bond and will not cause bond breaking

$$C_6H_5Cl + e \longrightarrow C_6H_5Cl^{\overset{\cdot}{-}} \not\longrightarrow C_6H_5\cdot + Cl^- \qquad (179)$$

Another way of expressing it is that the odd electron in the phenyl radical must be in a $\sigma$ orbital in the ground state. The $\sigma$ and $\pi$ orbitals are, of course, orthogonal.

The radical ion $C_6H_5Cl^{\overset{\cdot}{-}}$ is actually quite stable and will last for a long time, as free radicals go. For example, the *p*-nitrobenzyl chloride radical ion has a lifetime of some 30 ms before decomposition.[175]

$$O_2NC_6H_5CH_2Cl^{\overset{\cdot}{-}} \longrightarrow O_2NC_6H_5CH_2\cdot + Cl^- \qquad (180)$$

What eventually happens is that, by internal conversion, the odd electron jumps from the $\pi^*$ orbital into the $\sigma^*$ orbital. Vibrations of the radical ion provide the coupling mechanism between these orthogonal orbitals.

That is, vibration lowers the symmetry momentarily and allows the states to mix.

Figure 25 shows the energetics of the process. The abscissa is the carbon—chlorine bond distance. At the equilibrium value, the $\pi^*$ state is lowest in energy. At large values the $\sigma^*$ state is lowest. At some intermediate value the two states cross. The $\sigma^*$ state is repulsive and leads always to dissociation. Different substituents can shift the $\pi^*$ and $\sigma^*$ states and cause large variations in the lifetime.[176] The $\pi^*$ state correlates in dissociation with an aromatic anion and chlorine atom.

The same trapping of an electron in the wrong orbital can occur in inorganic complexes. The reduction process for a cobalt(III) complex must eventually put the electron in a $\sigma^*$ orbital (the $d_{x^2-y^2}$ or $d_{z^2}$).

$$Co(NH_3)_5X^{2+} + e \longrightarrow Co(NH_3)_5X^{\dot{+}} \longrightarrow Co^{2+} + 5NH_3 + X^- \quad (181)$$

However if X is a complex ligand with a $\pi$ system, the electron normally appears there first, and may be trapped for times of the order of $\mu s$ to ms.[177]

The symmetry of the orbital into which a reducing electron must be placed can have a large effect on the rate of reduction. For example, Ru(III) complexes are reduced by $Cr^{2+}$ at rates $10^4$–$10^6$ times greater than Co(III).[178] The ruthenium ion is a $d^5$ system and the reducing electron accordingly goes into an orbital of the $t_{2g}$ set in octahedral complexes. In cobalt(III), as

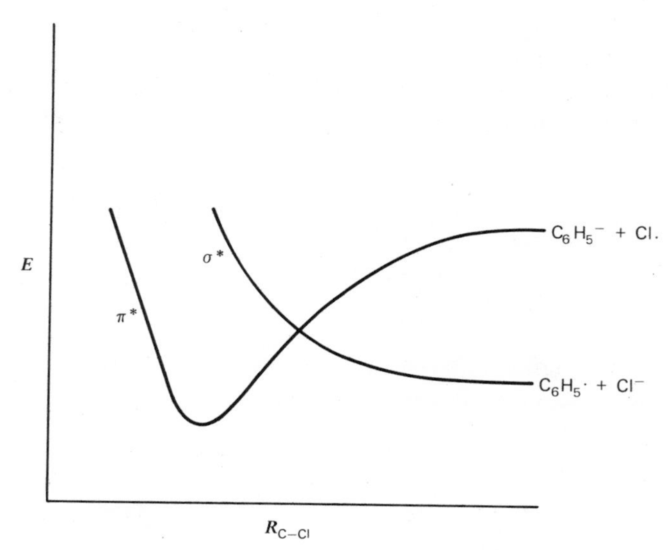


**Figure 25.** Potential energy diagram for internal conversion in $C_6H_5Cl^{\dot{-}}$.

mentioned, the electron goes into an $e_g^*$ orbital. The former orbitals are much more accessible to reducing agents than the latter, because of the shielding effect of the ligands. Also, as expected, $\pi$-electron donors are much better reducing agents toward Ru(III), in a relative sense, than toward Co(III).[179] The $t_{2g}$ orbitals can match the $\pi$-type symmetry.

### Some Atom-transfer Reactions

We conclude this chapter with some examples of redox reactions that are atom-transfer in nature. The analysis will be by the bond-symmetry rule, but a simple counting of electrons in the TS will usually work as well.[180] The forbidden nature of the concerted addition of halogen to a double bond has already been discussed. Halogenating agents such as $PbCl_4$, $SnCl_4$, and $SbCl_5$ do give an allowed concerted addition.

$$PbCl_4 + C{=}C \longrightarrow PbCl_2^{+} + ClC{-}CCl \qquad (182)$$

$$a_1 + b_1 \quad a_1 \qquad a_1 \quad a_1 + b_1$$

Evidence for this is available in the case of $SbCl_5$.[181] From previous arguments (p. 290), the chlorine atoms must come one from an axial and one from an equatorial position in $SbCl_5$.

Halogen molecule can simultaneously remove two hydrogen atoms from a hydrocarbon in an allowed process.

$$\begin{matrix} >C{-}H \\ | \\ >C{-}H \end{matrix} + \begin{matrix} Br \\ | \\ Br \end{matrix} \longrightarrow \begin{matrix} C \\ \| \\ C \end{matrix} + \begin{matrix} HBr \\ HBr \end{matrix} \qquad (183)$$

$$a_1 + b_1 \quad a_1 \qquad a_1 \quad a_1 + b_1$$

At the moment there is no direct evidence that this occurs. Another possible oxidation by molecular halogen is forbidden.

$$\begin{matrix} >C{-}OH \\ | \\ >C{-}OH \end{matrix} + \begin{matrix} Br \\ | \\ Br \end{matrix} \longrightarrow \begin{matrix} >C{=}O \\ >C{=}O \end{matrix} + \begin{matrix} HBr \\ HBr \end{matrix} \qquad (184)$$

$$2a_1 + b_1 \quad a_1 \qquad a_1 + b_1 \quad a_1 + b_1$$

However the corresponding reaction with $MnO_4^-$, or $OsO_4$, is allowed

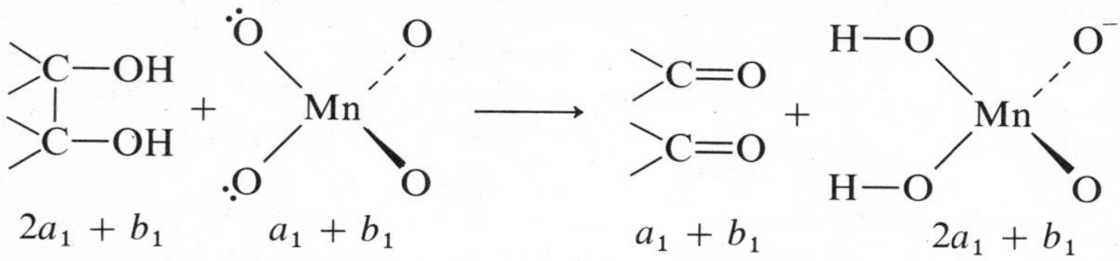

$$\begin{array}{c} \text{>C-OH} \\ | \\ \text{>C-OH} \end{array} + \text{MnO}_4 \longrightarrow \begin{array}{c} \text{>C=O} \\ \text{>C=O} \end{array} + (\text{HO})_2\text{MnO}_2^- \tag{185}$$

$2a_1 + b_1$ $\quad a_1 + b_1$ $\quad a_1 + b_1$ $\quad 2a_1 + b_1$

The manganese goes from +7 to +5 in oxidation state. This behavior is better understood if the well-understood hydroxylation reaction of olefins is analyzed. The initial reaction is the formation of a cyclic ester of Mn(V), if permanganate is the oxidant, or of Os(VI), if osmium tetroxide is used.[181]

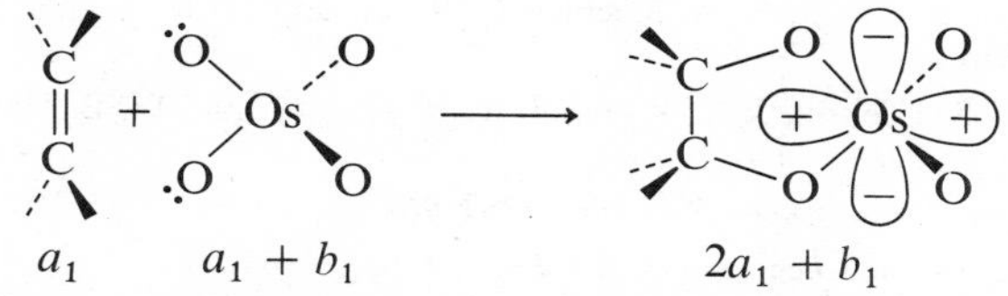

$$\text{C}{=}\text{C} + \text{OsO}_4 \longrightarrow \text{cyclic osmate ester} \tag{186}$$

$a_1$ $\quad a_1 + b_1$ $\quad 2a_1 + b_1$

Essentially the pair of electrons in the olefin $\pi$ bond are transferred to the metal atom. The tetracoordinated Os(VI) that results is in the $C_{2v}$ point group, but it is a pseudotehahedral complex. This means that the $d_{x^2-y^2}$ and $d_{z^2}$ orbitals (the $e$ set) lie lower than the $t_2$ set. Hence the electrons should appear in one of the $e$ orbitals. But in $C_{2v}$, both of these are $a_1$, which then matches the symmetry of the $\pi$ bond.

Permanganate will oxidize saturated hydrocarbons, but much more slowly. The major reason for the difference in rate is that a concerted process is not allowed.

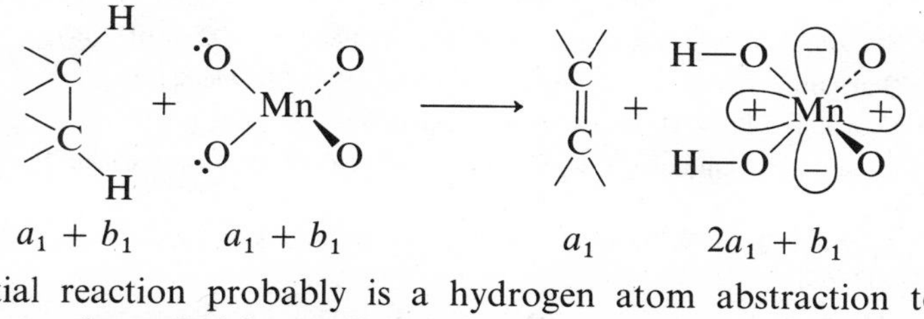

$$\text{H-C-C-H} + \text{MnO}_4 \longrightarrow \text{C}{=}\text{C} + (\text{HO})_2\text{MnO}_2 \tag{187}$$

$a_1 + b_1$ $\quad a_1 + b_1$ $\quad a_1$ $\quad 2a_1 + b_1$

The initial reaction probably is a hydrogen atom abstraction to form $HMnO_4^-$ and an alkyl free radical.

## REFERENCES

1. N. D. Epiotis, *J. Am. Chem. Soc.*, **95**, 5624 (1973).
2. J. D. Roberts and C. M. Sharts, *Org. React.*, **12**, 1 (1962).
3. S. Proskow, H. E. Simmons, and T. L. Cairns, *J. Am. Chem. Soc.*, **88**, 5254 (1966).
4. R. Huisgen and G. Steiner, *J. Am. Chem. Soc.*, **95**, 5054, 5055, 5056 (1973).

5. S. Inagaki, S. Yamabe, H. Fujimoto, and K. Fukui, *Bull. Chem. Soc. Jap.*, **45**, 3510 (1972).
6. R. Hoffmann, A. Imamura, and W. J. Hehre, *J. Am. Chem. Soc.*, **90**, 1499 (1968).
7. S. Inagaki and K. Fukui, *Bull. Chem. Soc. Jap.*, **46**, 2240 (1973).
8. N. C. Yang, W. Eisenhardt, and J. Libman, *J. Am. Chem. Soc.*, **94**, 4030 (1972).
9. D. A. Bittker and W. D. Walters, *J. Am. Chem. Soc.*, **77**, 1429 (1955); H. E. O'Neal and W. H. Richardson, ibid., **92**, 6553 (1970).
10. R. B. Woodward and R. Hoffmann, *Angew. Chem. Int. Ed. Engl.*, **8**, 781 (1969).
11. R. Sustmann, A. Ansmann, and F. Vahrenholt, *J. Am. Chem. Soc.*, **94**, 8099 (1972).
12. R. W. Rudolph, *J. Am. Chem. Soc.*, **89**, 4216 (1967); M. Zeldin, A. R. Gatti, and T. Wartik, ibid., 4217.
13. C. S. Foote, *Acc. Chem. Res.*, **1**, 104 (1968).
14. K. Fukui, *Bull. Chem. Soc. Jap.*, **39**, 498 (1966).
15. H. Basch, M. B. Robin, N. A. Kuebler, C. Baker, and D. W. Turner, *J. Chem. Phys.*, **51**, 52 (1969).
16. N. J. Turro, S. Edelson, J. R. Williams, T. R. Darling, and W. B. Hammond, *J. Am. Chem. Soc.*, **91**, 2283 (1969).
17. R. Hoffmann, *J. Am. Chem. Soc.*, **90**, 1475 (1968).
18. R. Huisgen, *Angew. Chem. Intern. Ed. Engl.*, **2**, 633 (1963).
19. K. N. Houk, J. Sims, C. R. Watts, and L. J. Luskus, *J. Am. Chem. Soc.*, **95**, 7301 (1973).
20. R. Criegee, *Rec. Chem. Prog.*, **18**, 111 (1957); for a modern revision, see R. P. Lattimer, R. L. Kuczkowski, and C. W. Gillies, *J. Am. Chem. Soc.*, **96**, 348 (1974).
21. R. A. Rouse, *J. Am. Chem. Soc.*, **95**, 3460 (1973).
22. H. D. Hartzler, *J. Am. Chem. Soc.*, **92**, 1412, 1413 (1970).
23. J. S. Wright, *J. Am. Chem. Soc.*, **96**, 4753 (1974).
24. E. Hayon and M. Simic, *J. Am. Chem. Soc.*, **92**, 7486 (1970).
25. K. Clusius and H. Schumacher, *Helv. Chim. Acta*, **41**, 972 (1958).
26. H. Kwart and J. Slutsky, *J. Chem. Soc. Chem. Commun.*, 1182 (1972).
27. C. D. Hurd and F. H. Blunck, *J. Am. Chem. Soc.*, **60**, 2421 (1938).
28. B. P. Stark and A. J. Duke, *Extrusion Reactions*, Pergamon, Oxford, 1967.
29. W. E. Parham and V. J. Traynelis, *J. Am. Chem. Soc.*, **76**, 4960 (1954).
30. W. L. Mock and P. A. H. Isaccs, *J. Am. Chem. Soc.*, **94**, 2747 (1972).
31. R. Hoffmann, H. Fujimoto, J. R. Swenson, and C. C. Wan, *J. Am. Chem. Soc.*, **95**, 7644 (1973).
32. F. G. Bordwell, J. M. Williams, Jr., E. B. Hoyt, Jr., and B. B. Jarvis, *J. Am. Chem. Soc.*, **90**, 429 (1968).
33. W. L. Mock, *J. Am. Chem. Soc.*, **92**, 3807, 6918 (1970).
34. C. J. Halstead and B. A. Thrush, *Proc. Roy. Soc. (Lond.)*, **295**, 363 (1966).
35. J. P. Linsky and C. G. Pierpont, *Inorg. Chem.*, **12**, 2959 (1973).
36. D. J. Hodgson, N. C. Payne, J. A. McGinnety, R. G. Pearson, and J. A. Ibers, *J. Am. Chem. Soc.*, **90**, 4486 (1968).
37. M. R. Snow, J. McDonald, F. Basolo, and J. A. Ibers, *J. Am. Chem. Soc.*, **94**, 2526 (1972).
38. E. J. Woodhouse and T. H. Norris, *Inorg. Chem.*, **10**, 614 (1971).
39. N. N. Lichtin, J. Laulicht, and S. Pinchas, *Inorg. Chem.*, **3**, 537 (1964).

40. An informative discussion of diradicals has been given by L. Salem, *Angew. Chem. Int. Ed. Engl.*, **11**, 92 (1972).
41. For a review, see R. G. Bergmann, in J. Kochi (ed.), *Free Radicals*, Wiley, New York, 1973, Vol. I.
42. T. S. Chambers and G. B. Kistiakowsky, *J. Am. Chem. Soc.*, **56**, 399 (1934); B. S. Rabinowitch, E. W. Schlag, and K. B. Wiberg, *J. Chem. Phys.*, **28**, 504 (1958).
43. John Horsley, Y. Jean, C. Moser, L. Salem, R. M. Stevens, and J. S. Wright, *J. Am. Chem. Soc.*, **94**, 279 (1972); P. J. Hay, W. J. Hunt, and W. A. Goddard, III, ibid., 639.
44. S. W. Benson, *J. Chem. Phys.*, **40**, 105 (1964).
45. J. S. Wright and L. Salem, *J. Am. Chem. Soc.*, **94**, 322 (1972).
46. (a) R. Hoffmann, S. Swaminathan, B. G. Odell, and R. Gleiter, *J. Am. Chem. Soc.*, **92**, 7091 (1970); (b) P. D. Bartlett et al., *J. Am. Chem. Soc.*, **94**, 2899 (1972).
47. This point has recently been emphasized; see M. J. S. Dewar and S. Kirschner, *J. Am. Chem. Soc.*, **96**, 5246 (1974).
48. W. D. Stohrer and R. Hoffmann, *J. Am. Chem. Soc.*, **94**, 779 (1972).
49. J. R. McCabe and C. A. Eckert, *Acc. Chem. Res.*, **7**, 251 (1974).
50. K. Hsu, R. J. Buenker, and S. D. Peyerimhoff, *J. Am. Chem. Soc.*, **93**, 2117 (1971).
51. P. Merlet, S. D. Peyerimhoff, R. J. Buenker, and S. Shih, *J. Am. Chem. Soc.*, **96**, 959 (1974).
52. G. Greig and J. C. J. Thynne, *Trans. Faraday Soc.*, **62**, 3338 (1966); ibid., **63**, 1369 (1967).
53. W. R. Dolbier, Jr., I. Nishiguchi, and J. M. Riemann, *J. Am. Chem. Soc.*, **94**, 3642 (1972).
54. H. Hogeveen and H. C. Volger, *Chem. Commun.*, 1133 (1967).
55. M. D. Newton, J. M. Schulman, and M. M. Marius, *J. Am. Chem. Soc.*, **96**, 17 (1974).
56. W. Adam, *Int. J. Chem. Kinet.*, **1**, 487 (1969).
57. (a) K. B. Wiberg, *Tetrahedron*, **24**, 1083 (1968); (b) G. L. Closs and P. E. Pfeffer, *J. Am. Chem. Soc.*, **90**, 2452 (1968); (c) M. J. S. Dewar and S. Kirschner, ibid., **97**, 2931 (1975).
58. M. J. S. Dewar, *Angew. Chem. Intern. Ed. Engl.*, **10**, 761 (1971).
59. H. J. Hansen and H. Schmid, *Chem. Brit.*, **5**, 111 (1969).
60. F. C. Whitmore, *J. Am. Chem. Soc.*, **54**, 3274 (1932).
61. W. von E. Doering and W. R. Roth, *Angew. Chem. Inter. Ed. Engl.*, **2**, 115 (1963).
62. F. A. Cotton, *Acc. Chem. Res.*, **1**, 257 (1968).
63. F. A. Cotton and T. J. Marks, *J. Am. Chem. Soc.*, **91**, 7523 (1969); *Inorg. Chem.*, **9**, 2802 (1970).
64. R. B. Larrabee, *J. Organometal. Chem.*, **74**, 813 (1974).
65. A. Davison and P. E. Rakita, *J. Organometal. Chem.*, **23**, 407 (1970).
66. J. L. Calderon, F. A. Cotton, and J. Takats, *J. Amer. Chem. Soc.*, **93**, 3587 (1971).
67. For example, see D. J. Cram, *Fundamentals of Carbanion Chemistry*, Academic, New York, 1965, Chapter 5.
68. J. E. Baldwin and J. E. Patrick, *J. Am. Chem. Soc.*, **93**, 3556 (1971); W. D. Ollis, J. O. Sutherland, and Y. Thebtaranonth, *J. Chem. Soc. Chem. Commun.*, 653, 654, 657 (1973).
69. C. K. Ingold, *Structure and Mechanism in Organic Chemistry*, 2nd ed., Cornell U. P., Ithaca, N. Y., 1969, Chapter 12.
70. J. A. Berson, *Acc. Chem. Res.*, **1**, 152 (1968).
71. H. Kwart and N. Johnson, *J. Am. Chem. Soc.*, **92**, 6064 (1970).

72. J. H. Brewster and M. W. Kline, *J. Am. Chem. Soc.*, **74**, 5179 (1952).
73. A. T. Blades and G. W. Murphy, *J. Am. Chem. Soc.*, **74**, 1039 (1952).
74. G. Schrauzer (ed.), *Transition Metals in Homogeneous Catalysis*, Dekker, New York, 1971.
75. D. H. Busch, *Record Chem. Prog.*, **25**, 107 (1964).
76. W. Slegeir, R. Case, J. S. McKennis, and R. Pettit, *J. Am. Chem. Soc.*, **96**, 287 (1974).
77. W. Merck and R. Pettit, *J. Am. Chem. Soc.*, **89**, 4788 (1967).
78. (a) F. D. Mango, *Tetrahedron Lett.*, 1509 (1973); (b) W. Th. A. M. van der Lugt, *Tetrahedron Lett.*, 2281 (1970).
79. F. D. Mango and J. Schachtschneider, *J. Am. Chem. Soc.*, **89**, 2484 (1967).
80. H. Hogeveen and H. C. Volger, *J. Am. Chem. Soc.*, **89**, 2486 (1967).
81. G. N. Schrauzer, *Adv. Catal.*, **18**, 373 (1968).
82. C. A. Tolman, *Chem. Soc. Rev.*, **1**, 337 (1972).
83. D. H. Gerlach, A. R. Kane, A. W. Parshall, J. P. Jesson, and E. L. Muetterties, *J. Am. Chem. Soc.*, **93**, 3543 (1971).
84. J. P. Collman, S. R. Winter, and D. R. Clark, *J. Am. Chem. Soc.*, **94**, 1788 (1972).
85. J. S. Bradley, D. G. Connor, D. Dolphin, J. A. Labinger, and J. A. Osborn, *J. Am. Chem. Soc.*, **94**, 4043 (1972).
86. P. B. Chock and J. Halpern, *J. Am. Chem. Soc.*, **88**, 3511 (1966); M. Kubota, *Inorg. Chim. Acta*, **7**, 1195 (1973).
87. R. G. Pearson and W. R. Muir, *J. Am. Chem. Soc.*, **92**, 5519 (1970).
88. J. P. Collman and M. R. MacLaury, *J. Am. Chem. Soc.*, **96**, 3019 (1974).
89. L. Vaska, *J. Am. Chem. Soc.*, **88**, 4100 (1966); L. H. Sommer, J. E. Lyons, and H. Fujimoto, ibid., **91**, 7051 (1969); H. M. Walborsky and L. E. Allen, *Tetrahedron Lett.*, **11**, 823 (1970); G. M. Whitesides, J. San Fillipo, E. R. Stedronsky, and C. P. Casey, *J. Am. Chem. Soc.*, **91**, 6542 (1969); C. Eaborn, D. J. Tune, and D. M. R. Walton, *J. Chem. Soc. Dalton*, 2255 (1973).
90. B. R. James, F. T. T. Ng, *J. Chem. Soc.*, **A**, 355 (1972).
91. J. A. Labinger, A. V. Kramer, and J. A. Osborn, *J. Am. Chem. Soc.*, **95**, 7908 (1973); M. F. Lappert and P. W. Ledner, *Chem. Commun.*, 948 (1973).
92. (a) J. K. Stille and R. W. Fries, *J. Am. Chem. Soc.*, **96**, 1514 (1974); (b) K. S. Y. Lau, R. W. Fries, and J. K. Stille, ibid., 4983; (c) see S. Otsuka, A. Nakamura, T. Yoshida, M. Naruto, and K. Ataka, ibid., **95**, 3180 (1973) for a free-radical mechanism with Pd(0).
93. D. Hopgood and R. A. Jenkins, *J. Am. Chem. Soc.*, **95**, 4461 (1973).
94. F. R. Jensen, V. Madan, and D. H. Buchanan, *J. Am. Chem. Soc.*, **92**, 1415 (1970).
95. G. M. Whitesides and D. J. Boschetto, *J. Am. Chem. Soc.*, **91**, 4313 (1969); R. W. Johnson and R. G. Pearson, *Chem. Commun.*, 986 (1970).
96. D. M. Blake and M. Kubota, *Inorg. Chem.*, **9**, 989 (1970); see also L. M. Haines, ibid., **10**, 1693 (1971), for rhodium(I).
97. J. P. Collman, *Acc. Chem. Res.*, **1**, 136 (1968); L. Vaska, L. S. Chen, and W. U. Miller, *J. Am. Chem. Soc.*, **93**, 6671 (1971).
98. D. R. Fahey, *J. Am. Chem. Soc.*, **92**, 402 (1970); B. L. Shaw and N. L. Tucker, *J. Chem. Soc.*, **A**, 185 (1971).
99. J. Rajaram, R. G. Pearson, and J. A. Ibers, *J. Am. Chem. Soc.*, **96**, 2103 (1974).
100. E. J. Corey and G. Posner, *J. Am. Chem. Soc.*, **89**, 3911 (1968).

101. C. R. Johnson and G. A. Dutra, *J. Am. Chem. Soc.*, **95**, 7777, 7783 (1973).

102. C. R. Gregory and R. G. Pearson, *J. Am. Chem. Soc.*, **98**, 4100 (1976).

103. J. Halpern and M. Pribanić, *Inorg. Chem.*, **9**, 2616 (1970).

104. P. B. Chock and J. Halpern, *J. Am. Chem. Soc.*, **91**, 582 (1969).

105. N. Kornblum, W. J. Jones, and D. E. Hardies, *J. Am. Chem. Soc.*, **88**, 1704, 1707 (1966).

106. H. Gilman and S. D. Rosenberg, *J. Am. Chem. Soc.*, **74**, 531 (1952); L. Waterworth and I. J. Worrall, *J. Chem. Soc. Chem. Commun.*, 569 (1971); E. C. Taylor and A. McKillop, *Acc. Chem. Res.*, **3**, 338 (1970).

107. M. D. Johnson, *Chem. Commun.*, 1037 (1970).

108. R. J. Buehler, R. B. Bernstein, and K. H. Kramer, *J. Am. Chem. Soc.*, **88**, 5331 (1966); P. R. Brooks and E. M. Jones, *J. Chem. Phys.*, **45**, 3449 (1966).

109. (a) P. S. Skell and J. E. Girard, *J. Am. Chem. Soc.*, **94**, 5518 (1972); (b) G. M. Whitesides, private communication.

110. J. F. Garst and R. H. Cox, *J. Am. Chem. Soc.*, **92**, 6389 (1970).

111. R. L. Letsinger, *J. Am. Chem. Soc.*, **72**, 4842 (1950).

112. E. Eliel and J. P. Freeman, *J. Am. Chem. Soc.*, **74**, 923 (1952).

113. B. Czochralska, *Chem. Phys. Lett.*, **1**, 239 (1967).

114. H. A. O. Hill, J. M. Pratt, M. P. O'Riordan, F. R. Williams, and R. J. P. Williams, *J. Chem. Soc.*, **A**, 1859 (1971).

115. (a) F. Basolo and R. G. Pearson, *Mechanisms of Inorganic Reactions*, Wiley, New York, 1967, Chapter 7; (b) J. P. Candlin, K. A. Taylor, and D. T. Thompson, *Reactions of Transition Metal Complexes*, Elsevier, Amsterdam, 1968.

116. K. Noack and F. Calderazzo, *J. Organometal. Chem.*, **10**, 701 (1967).

117. R. W. Glyde and R. J. Mawby, *Inorg. Chim. Acta*, **5**, 317 (1971); *Inorg. Chem.*, **10**, 854 (1971).

118. R. G. Pearson, *Fortschr. Chem. Forsch.*, **41**, 75 (1973).

119. B. L. Shaw, *Chem. Commun.*, 464 (1968).

120. (a) R. F. Heck, *J. Am. Chem. Soc.*, **90**, 5518, 5535 (1968); ibid., **91**, 6707 (1969); (b) P. M. Henry, *J. Org. Chem.*, **37**, 2443 (1972); (c) B. E. Mann, B. L. Shaw, and N. J. Tucker, *J. Chem. Soc.*, **A**, 2667 (1971); (d) R. Kemmitt, B. Y. Kimura, and G. W. Littlecott, *J. Chem. Soc. Dalton*, 636 (1973).

121. G. M. Whitesides and D. J. Boschetto, *J. Am. Chem. Soc.*, **91**, 4313 (1969); L. F. Hines and J. K. Stille, ibid., **94**, 485, 1798 (1972); J. K. Stille, F. Huang, and M. T. Regan, ibid., **96**, 1518 (1974); M. A. Haas, *Organometal. Chem. Rev.*, **4**, 307 (1969).

122. P. M. Henry, *Acc. Chem. Res.*, **6**, 16 (1973); A. Segnitz, P. M. Bailey, and P. M. Maitlis, *Chem. Commun.*, 698 (1973); J. K. Stille, D. E. James, and L. F. Hines, *J. Am. Chem. Soc.*, **95**, 5062 (1973).

123. T. H. Whitesides and R. W. Ashart, *J. Am. Chem. Soc.*, **93**, 5297 (1971).

124. A. J. Wojcicki, *Acc. Chem. Res.*, **4**, 344 (1971).

125. G. M. Whitesides and D. J. Boschetto, *J. Am. Chem. Soc.*, **93**, 1529 (1971).

126. A. J. Wojcicki and S. E. Jacobson, *J. Am. Chem. Soc.*, **95**, 6962 (1973).

127. See G. A. Olah and P. R. Clifford, *J. Am. Chem. Soc.*, **95**, 6067 (1973), for identification of the mercurinium ion.

128. W. Kitching, *Organometal. Chem. Rev.*, **3**, 61 (1968).

129. H. C. Brown, *Hydroboration*, Benjamin, New York, 1962.

130. P. R. Jones, *J. Org. Chem.*, **37**, 1886 (1972).
131. K. W. Egger and A. T. Cocks, *J. Am. Chem. Soc.*, **94**, 1810 (1972).
132. Y. Takahashi, S. Sokai, and Y. Ishii, *J. Organometal. Chem.*, **16**, 177 (1969); R. P. Hughes and J. Powell, *J. Am. Chem. Soc.*, **94**, 7723 (1972).
133. See reference 74, p. 73 (article by W. Keim).
134. P. S. Braterman and R. J. Cross, *J. Chem. Soc. Dalton*, 657 (1972).
135. J. X. McDermott, J. F. White, and G. M. Whitesides, *J. Am. Chem. Soc.*, **95**, 4451 (1973).
136. R. Cramer, *Acc. Chem. Res.*, **1**, 186 (1968); C. A. Tolman and L. H. Scharpen, *J. Chem. Soc. Dalton*, 584 (1973).
137. For a review of metal insertion into C—H bonds, see G. W. Parshall, *Acc. Chem. Res.*, **3**, 139 (1970).
138. J. F. Harrod and A. J. Chalk, *J. Am. Chem. Soc.*, **88**, 3491 (1966); C. P. Casey and C. R. Cyr, ibid., **95**, 2248 (1973).
139. W. Hübel and E. H. Braye, *J. Inorg. Nucl. Chem.*, **10**, 250 (1959); J. P. Collman, J. W. Kang, W. F. Little, and M. F. Sullivan, *Inorg. Chem.*, **7**, 1298 (1968); J. Ashley Smith, M. Green, and F. G. A. Stone, *J. Chem. Soc.*, **A**, 3019 (1969); R. Burt, M. Cooke, and M. Green, *J. Chem. Soc.*, **A**, 2975, 2981 (1970).
140. R. H. Grubbs and T. K. Brunck, *J. Am. Chem. Soc.*, **94**, 2538 (1972).
141. N. Rösch and R. Hoffmann, *Inorg. Chem.*, **13**, 2656 (1974).
142. M. A. Shaw, J. C. Tebby, R. S. Ward, and D. H. Williams, *J. Chem. Soc.*, **C**, 2795 (1968); N. E. Waite, J. C. Tebby, R. S. Ward, M. A. Shaw, and D. H. Williams, ibid., 1620 (1971).
143. P. M. Maitlis, *Pure Appl. Chem.*, **30**, 427 (1972).
144. E. W. Turnblom and T. J. Katz, *J. Am. Chem. Soc.*, 4292 (1973).
145. H. Breil, P. Heimbach, M. Kröner, H. Müller, and G. Wilke, *Makromol. Chem.*, **69**, 18 (1963); P. Heimbach, P. W. Jolly, and G. Wilke, *Advances in Organometallic Chemistry*, Academic, New York, 1970, Vol. 8, p. 29.
146. (a) A. Bond, M. Green, B. Lewis, and S. F. W. Lowrie, *Chem. Commun.*, 1230 (1971); (b) R. E. Davis et al., *J. Am. Chem. Soc.*, **96**, 7562 (1974).
147. (a) L. S. Meriwether, M. F. Leto, E. C. Colthup, and G. W. Kennerly, *J. Org. Chem.*, **27**, 3930 (1962); (b) H. Dietl, H. Reinheimer, J. Moffat, and P. M. Maitlis, *J. Am. Chem. Soc.*, **92**, 2276 (1970).
148. L. Malatesta, G. Santarella, L. M. Vallarino, and F. Zingales, *Angew. Chem.*, **72**, 34 (1960).
149. F. D. Mango and J. H. Schachtschneider, *J. Am. Chem. Soc.*, **91**, 1030 (1969).
150. R. Ugo, *Aspects of Homogeneous Catalysis*, Manfredi, Milan, 1970, Vol. 1; G. Dolcetti and N. W. Hoffman, *Inorg. Chim. Acta*, **9**, 269 (1974).
151. J. A. Osborn, F. H. Jardine, J. F. Young, and G. Wilkinson, *J. Chem. Soc.*, **A**, 1711 (1966), 1754 (1967); A. Nakamura and S. Otsuka, *J. Am. Chem. Soc.*, **95**, 7262 (1973).
152. J. Halpern, *Disc. Faraday Soc.*, **46**, 7 (1968).
153. E. W. Frankel, E. Selke, and C. A. Glass, *J. Am. Chem. Soc.*, **90**, 2446 (1968).
154. J. Halpern, *Acc. Chem. Res.*, **3**, 386 (1970).
155. L. A. Paquette, *Acc. Chem. Res.*, **4**, 280 (1971).
156. L. Cassar, P. E. Eaton, and J. Halpern, *J. Am. Chem. Soc.*, **92**, 6366 (1970).
157. L. Cassar, P. E. Eaton, and J. Halpern, *J. Am. Chem. Soc.*, **92**, 3515 (1970).

158. A. R. Fraser, P. H. Bird, S. A. Bezman, J. R. Shapley, R. White, and J. A. Osborn, *J. Am. Chem. Soc.*, **95**, 598 (1973).

159. J. Wristers, L. Brener, and R. Pettit, *J. Am. Chem. Soc.*, **92**, 7499 (1970); P. G. Gassman and T. J. Atkins, ibid., **93**, 4597 (1971).

160. C. H. DePuy and R. H. McGirk, *J. Am. Chem. Soc.*, **96**, 1121 (1974).

161. K. L. Kaiser, R. F. Childs, and P. M. Maitlis, *J. Am. Chem. Soc.*, **93**, 1270 (1971).

162. F. D. Mango and J. H. Schachtachneider, *J. Am. Chem. Soc.*, **93**, 1123 (1971).

163. N. Calderon, E. A. Ofstead, J. P. Ward, and K. W. Scott, *J. Am. Chem. Soc.*, **90**, 4133 (1968); W. B. Hughes, ibid., **92**, 532 (1970); G. S. Lewandos and R. Pettit, ibid., **93**, 7087 (1971); J. C. Mol and A. J. Moulijn, *Adv. Catal.*, **24**, 131 (1974).

164. G. L. Caldow and R. A. MacGregor, *J. Chem. Soc.*, **A**, 1654 (1971).

165. A. Montreux and M. Blanchard, *Bull. Soc. Chim. Fr.*, 1641 (1972); *J. Chem. Soc. Chem. Commun.*, 786 (1974).

166. (a) E. O. Fisher and A. Maasbol, *Angew. Chem.*, **76**, 645 (1964); (b) D. J. Cardin, M. J. Doyle, and M. F. Lappert, *Chem. Commun.*, 927 (1972).

167. J. L. Herrison and G. Chauvin, *Makromol. Chem.*, **141**, 161 (1970); C. P. Casey and T. J. Burkhardt, *J. Am. Chem. Soc.*, **96**, 7807 (1974); E. L. Muetterties, *Inorg. Chem.*, **14**, 951 (1975).

168. T. J. Katz and J. McGinnis, *J. Am. Chem. Soc.*, **97**, 1592 (1975); R. H. Grubbs, P. L. Burk, and D. D. Carr, ibid., 3265.

169. B. F. G. Johnson and S. Bhaduri, *Chem. Commun.*, 650 (1973); B. Haymore and J. A. Ibers, *J. Am. Chem. Soc.*, **96**, 3325 (1974); J. Reed, Jr. and R. Eisenberg, *Science*, **184**, 568 (1974).

170. The formation of coordinated $N_2O_2^{2-}$ from NO by a reducing metal complex is a known reaction. S. Cenini, R. Ugo, G. LaMonica, and S. D. Robinson, *Inorg. Chim. Acta*, **6**, 182 (1972).

171. D. E. Morris and F. Basolo, *J. Am. Chem. Soc.*, **90**, 2536 (1968).

172. Reference 115a, Chapter 6; H. Taube, *Electron Transfer Reactions*, Academic, New York, 1970.

173. E. F. Gurnee and J. L. Magee, *J. Chem. Phys.*, **26**, 1237 (1957).

174. R. P. Van Duyne and S. Fischer, *Chem. Phys.*, **5**, 183 (1974).

175. M. Mohammad, J. Hajdu, and E. M. Kosower, *J. Am. Chem. Soc.*, **93**, 1792 (1971).

176. C. A. Coulson, *Chem. Brit.*, **4**, 113 (1968).

177. M. Z. Hoffman and M. Simic, *J. Am. Chem. Soc.*, **94**, 1757 (1972).

178. H. Taube and E. S. Gould, *Acc. Chem. Res.*, **2**, 321 (1969).

179. H. Cohen and D. Meyerstein, *J. Am. Chem. Soc.*, **94**, 6944 (1972).

180. J. S. Littler, *Tetrahedron*, **27**, 81 (1971).

181. K. Wiberg and C. J. Deutsch, *J. Am. Chem. Soc.*, **95**, 3034 (1973).

# CHAPTER 6

# PHOTOCHEMICAL REACTIONS

The chemical behavior of molecules in excited electronic states is an area extremely broad in scope and much more difficult to understand than ground-state chemistry. The difficulty lies in the greater number of reactions that become possible because thermodynamic barriers are removed and in the transitory nature of the intermediates and the reactants themselves. A related problem is that we often do not know which excited states are present in a system, nor which state is reactive.

It will be easiest to start with a general review of what is known about the primary physical processes involved in photochemical excitation, using the well-known Jablonski diagram of Fig. 1. It is assumed that the ground state of the molecule, $S_0$, is a singlet state. Absorption of a quantum of radiation converts to excited singlet states, $S_1$, $S_2$, and so on, depending on the frequency of radiation. Because of the Franck–Condon principle, the transition produces a vibrationally excited state as well. In solution, or in a dense gas, several things can now happen in times of the order of $10^{-13}$–$10^{-10}$ s.[1] The vibrational excitation is quickly lost, by way of collisions, and in addition any molecule in an $S_2$, or higher state, will be deactivated to the $S_1$ state (Kasha's rule). These processes are nonradiative, the latter being called "internal conversion."

The molecule in the $S_1$ state usually in a somewhat longer time, $10^{-10}$ to $10^{-6}$ s, can do one of several things:

(a) revert to $S_0$ with emission of radiation, fluorescence,
(b) revert to $S_0$ in a nonradiative way, quenching,
(c) convert to an excited state of different multiplicity, $T_1$, $T_2$, and so on, intersystem crossing,
(d) undergo a chemical reaction.

The higher triplet states will undergo internal conversion to $T_1$, which will either revert to $S_0$ by quenching or by emission of radiation (phosphorescence), or react chemically. Lifetimes for the triplet state are much longer

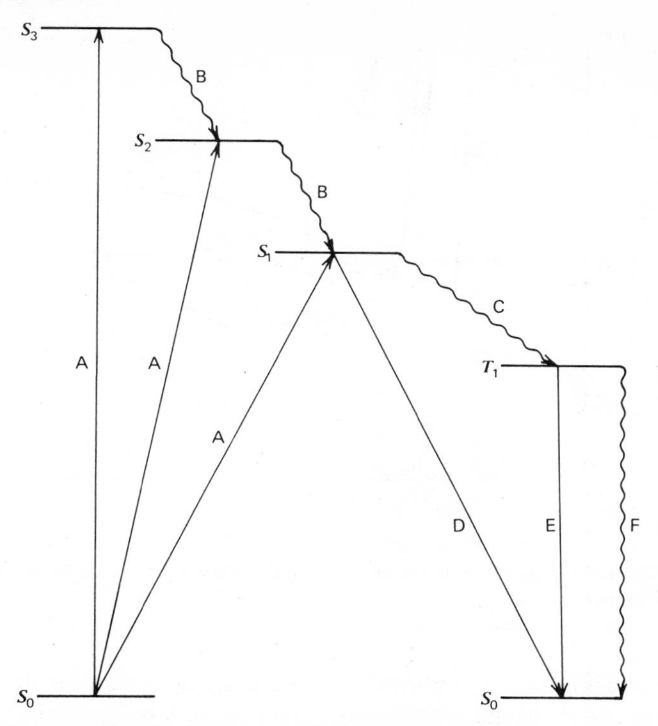


**Figure 1.** Jablonski diagram showing various photophysical processes: A = absorption of radiation; B = internal conversion, nonradiative; C = intersystem crossing, nonradiative; D = fluorescence,; E = phosphorescence; F = quenching, nonradiative, also from $S_1$ state to $S_0$.

than for the singlet state, because of the change in spin multiplicity required to reach $S_0$. Typical values for light atoms are $10^{-3}$–1 s. This creates a greater probability of chemical reaction. However heavy atoms in the molecule can shorten the lifetimes down to $10^{-7}$ s.

The above generalizations are not meant to be absolute, since exceptions do occur. For example, some photochemical reactions do take place in higher excited states than $S_1$ or $T_1$.[2] This suggests that there may be excursions from $S_1$ to $S_2$ or from $T_1$ to $T_2$. Indeed there is good reason to believe that the potential energy surfaces for excited states may cross and recross each other as nuclear configurations change. Figure 2 shows only a few of the states of the $S_2$ molecule. Even in this simplified case there are several crossings evident as the sulfur—sulfur distance changes. Polyatomic molecules, with many more electronic states and many degrees of vibrational freedom, would have many more crossings.

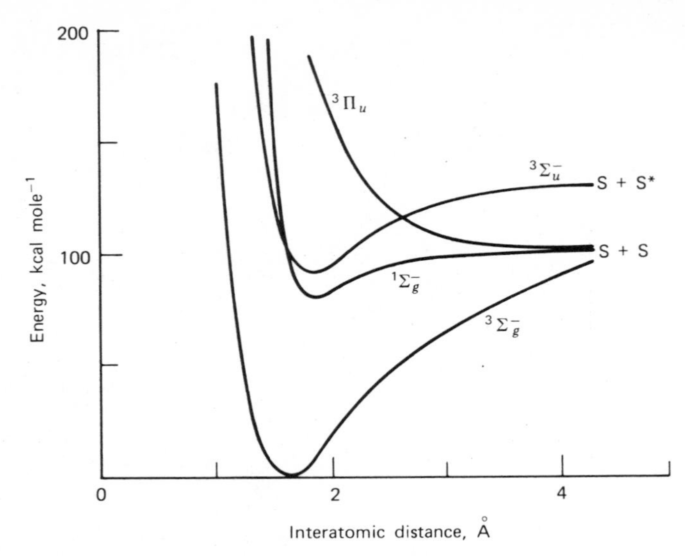


**Figure 2.** Potential energy curves for some of the states of the $S_2$ molecule, showing crossings of the S—S distance changes.

Photochemical excitation normally puts energy equivalent to 50–100 kcal mole$^{-1}$ into the molecule. This is more than enough to cause the observed reaction in most cases. Even so, there may be a modest temperature dependence for the rate constants of elementary reactions of the excited molecule; 5–10 kcal of vibrational energy can help distort the molecule to some reactive configuration. Another observation of importance is that the products of photochemical reactions of polyatomic molecules are formed in their ground states. They do *not* form in their excited states and then return to ground by mechanisms (a)–(c) above. Exceptions to this rule are rather rare, usually involving proton transfer reactions of weak acids and bases.[2,3] The reverse situation of ground-state reactants giving excited state products is actually more common. It is known as "chemiluminescence," if light is emitted.

The general picture of a photochemical reaction that emerges from the above collection of observations is that of an activated molecule that wanders about on several closely spaced potential energy surfaces, moving readily from one to another. Eventually it will move back to the $S_0$ surface. Since the same ground-state hypersurface suffices for both reactants and products, this can lead to chemical reaction or return to the original reactants.

Because of the conservation of energy, return to the ground-state surface necessarily is to highly excited vibrational states. In a gas at low pressure a vibrationally excited molecule will have a long enough life, before collisional deactivation, so that it can undergo its reaction. In solution there will be rapid deactivation, usually in $10^{-11}$ s or less. While photochemical changes usually occur more readily in a dilute gas, solution photochemistry with high quantum yields is still common. It appears that transfer to the $S_0$ surface must occur at a nuclear configuration close to that for the activated complex of the ground-state reaction. This gives the reaction a chance to compete with deactivation.

A process in which a jump from one potential hypersurface to another occurs is known as a "nonadiabatic (or diabatic) reaction." The general theory for such processes will be touched on later. For the moment it is enough to know that a requirement is that the two surfaces be very close; that is, that they have nearly the same energy at a given nuclear configuration. They can be surfaces corresponding to electronic states of different symmetry or multiplicity that are actually crossing each other, as in Fig. 2. Or they can be surfaces for states of the same symmetry and multiplicity that have attempted to cross.

The photochemical problem in this view becomes one of finding the regions of configuration space where the $S_1$ and $T_1$ surfaces (usually) come close to the $S_0$ surface. These will define photochemical products and also reaction paths or mechanisms, providing it can be shown that no barriers exist on the $S_1$ or $T_1$ surface en route to the critical regions. That is, we are interested in the geometric arrangements of nuclei that create energy maxima and minima on the excited-state surfaces, just as we were for the ground-state surface. It will be found that the minima on the excited-state surface very often correspond to maxima on the ground-state surfaces. These then will be "holes" or "funnels" through which the reactions will occur.[4]

## Excited-state Properties

The idea that an energy minimum may occur on an excited hypersurface at the same geometry where the ground state has a maximum, has an important message. An excited-state molecule is not just a ground-state molecule with a lot of energy. It is a quite different molecule, with its own physical and chemical properties. It has a different equilibrium shape, a different dipole moment (reflecting a different electron distribution), different acid and base strengths, and so on. In fact there is a "Dr. Jekyll–Mr. Hyde" relationship: the excited state behaves in ways that the ground state would never do.

Many of the differences are readily understandable, using simple MO theory to describe the excited-state molecule. For example, the lowest excited state of aldehydes and ketones arises from an $n, \pi^*$ transition. The electron goes from a lone-pair orbital on oxygen to the antibonding

$\pi^*$ orbital of the carbonyl group. This orbital is concentrated on carbon, just as the bonding $\pi$ is concentrated on oxygen. The resultant molecule may be written in a Lewis diagram sense as

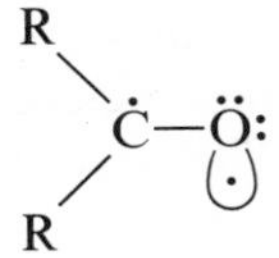

$$\begin{array}{l} R \\ \quad \diagdown \\ \quad\quad \dot{C}{-}\ddot{O}\!: \\ \quad \diagup \quad\ \ (\cdot) \\ R \end{array} \qquad\qquad \text{A}$$

The single lobe containing the odd electron on oxygen indicates that it is in a $\sigma$-type orbital, in the plane of the page. The odd electron on carbon is in a $\pi$-type orbital, perpendicular to the page.[5]

We see that molecule A is a diradical. It will undergo free-radical reactions, whether it is a singlet or a triplet, although we may expect that the triplet will be more likely to behave as a free radical. This follows simply from the argument that a triplet will be less likely to give alternate reactions, where the products are spin-paired. Furthermore the electron density in A is such that the oxygen atom is quite electron deficient. It will be an electrophilic free-radical center.[6] Its best reactions will be attack at electron-rich centers such as C—H bonds.

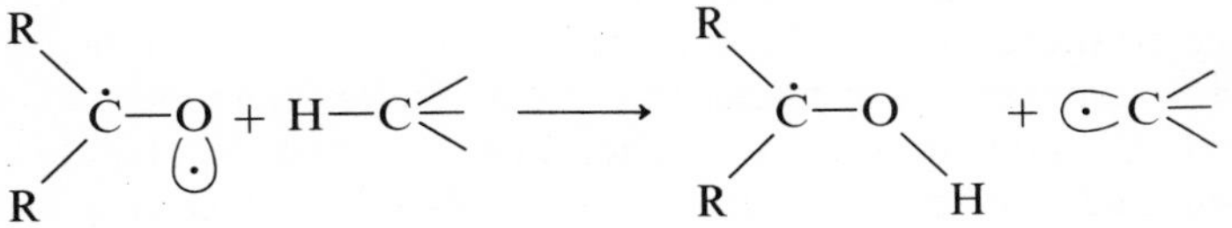

$$R_2\dot{C}{-}O(\cdot) + H{-}C\!\lessdot \longrightarrow R_2\dot{C}{-}O{-}H + (\cdot)C\!\lessdot \qquad (1)$$

The carbon atom is a nucleophilic free radical. It would be expected to attack electron-poor centers, such as olefins with electron-withdrawing substituents.

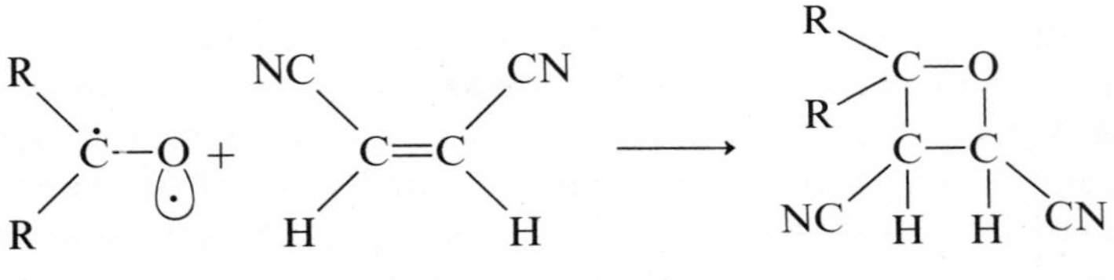

$$R_2\dot{C}{-}O(\cdot) + (NC)HC{=}CH(CN) \longrightarrow \begin{array}{c} R_2C{-}O \\ | \quad\ | \\ (NC)(H)C{-}C(H)(CN) \end{array} \qquad (2)$$

Experimentally the oxetane formed in (2) is the *cis* dicyano, so that initial attack and ring closure are concerted, or nearly so.[6]

The remarkable differences in acid and base strengths between ground-state and excited-state molecules was revealed by the pioneering work of Förster and Weller.[7] For example, the $pK_a$ values for $\beta$-naphthylamine and $\beta$-naphthol in $S_0$, $S_1$, and $T_1$ are as follows:

| | $pK_{S_0}$ | $pK_{S_1}$ | $pK_{T_1}$ |
|---|---|---|---|
| β-Naphthylamine | 4.1 | −2 | 3.3 |
| β-Naphthol | 9.5 | 2.8 | 8.1 |

In these two cases, and in others, it is found that the $S_1$ state is a very strong acid, compared to $S_0$, but that $T_1$ is not greatly different.

The explanation is that the $S_1$ state arises from an $n, \pi^*$ transition. Picking aniline as a simpler example, the excited state would be

H

—N—H B

Clearly B has a low electron density on the nitrogen atom, which would increase its acidity. The triplet state may not be an $n, \pi^*$ state, since its acidity is little affected. It may be a $\pi, \pi^*$ state.[8] Note that there is no inconsistency in $S_1$ and $T_1$ corresponding to different configurations. They are always defined energetically.

Other chemical effects follow from the changed electron distribution of excited state molecules. One of the earliest examples interpreted in this way was the thermal and photolytic hydrolysis of *m*- and *p*-nitrophenyl phosphates.[9]

$$p\text{-}NO_2C_6H_4OPO_3H_2 + H_2O \xrightarrow{\Delta} p\text{-}NO_2C_6H_4OH + H_3PO_4 \quad (3)$$

fast reaction

$$m\text{-}NO_2C_6H_4OPO_3H_2 + H_2O \xrightarrow{h\nu} m\text{-}NO_2C_6H_4OH + H_3PO_4 \quad (4)$$

fast reaction

It is well known that *para* and *ortho* nitro groups are activating for nucleophilic displacement reactions of ground-state molecules. Simple resonance diagrams are used to explain this and also why the *meta* nitro group is ineffective. The same kind of analysis will explain why the situation is reversed for the excited state, where it is the *meta* substituent that is an effective activator.

Inorganic examples are known as well. Metal carbonyls are particularly susceptible to photochemical substitution reactions.[10] A CO ligand is expelled, followed by pickup of another ligand, L. The net reaction is, for instance,

$$Cr(CO)_5(THF) + L \xrightarrow{h\nu} Cr(CO)_4L(THF) + CO \quad (5)$$

where THF is tetrahydrofuran. This oxygen donor is held very weakly and is easily displaced in a thermal process.

$$Cr(CO)_5(THF) + L \xrightarrow{\Delta} Cr(CO)_5L + THF \quad (6)$$

Consequently the ground-state and excited-state reactions are reversed in terms of which ligand is lost most readily. An explanation can again be given by considering the charge density in the excited state.[10]

The redox properties of excited-state molecules will also differ from those of the ground state. An electron that has been promoted to an antibonding orbital is much more likely to be lost in an electron-transfer process. Conversely, the vacancy created in the valence shell will more readily accept an electron from an external source.

## NONADIABATIC PROCESSES

Rather than pursue this line of exposition further,[11] we turn to the more specific problem of how symmetry enters in to explain excited state behavior. Symmetry enters, as one instance, in the problem of a reacting system jumping from one potential surface to another, a nonadiabatic process. The adiabatic assumption is that electrons move very rapidly compared to nuclear motion. An equivalent statement is that the electronic wavefunction can change instantaneously, for every change in nuclear positions. The consequence of this is that a system which originates on the lowest sheet of a collection of energy hypersurfaces should always stay on the lowest sheet, despite changes in the nuclear configuration.

Figure 3 shows two cases of crossing potential energy surfaces; Fig. 3*a* is for states differing in symmetry, spin multiplicity, or both, and Fig. 3*b* is for two states of the same symmetry and spin species. In the latter case, as already discussed (p. 36), the two surfaces do not cross, but instead are moved apart by an amount of energy, $2\varepsilon_{12}$, which is quite large. In general a system on the lower surface can be expected to stay there, unless activated externally. There is a small chance, however, that the system will jump.[12]

The case of Fig. 3*a* is more interesting. Here are two surfaces that one might normally expect to cross, since they do not interact with each other at the lowest level of approximation. However there is always some level at which they do interact, producing an energy splitting, $2\varepsilon_{12}$, which is very small compared to that of Fig. 3*b*. Now the question becomes, will the system stay on the lowest sheet, in other words, behave adiabatically, or will it jump the very small gap between the two sheets? An approximate answer to this question was given by Landau, Zener, and Stueckelberg.[13] The probability

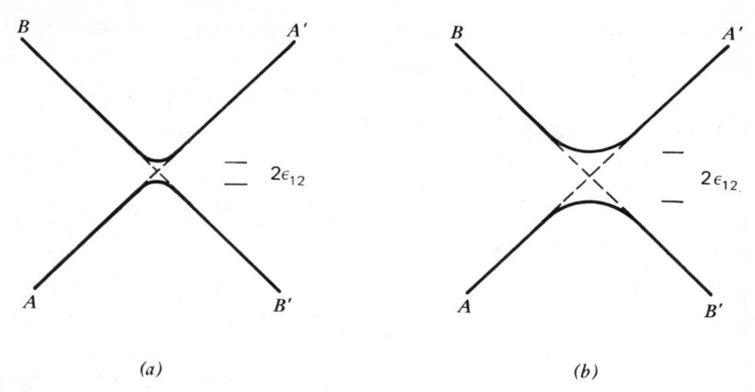


**Figure 3.** (*a*) Small splitting, $\varepsilon_{12}$ for two states differing in symmetry or multiplicity; (*b*) large splitting for two states of same symmetry and multiplicity.

of jumping is given by

$$P = \exp - \left( \frac{4\pi^2 \varepsilon_{12}^2}{hv|s_1 - s_2|} \right) \tag{7}$$

where $v$ is the velocity along the reaction coordinate, and $s_1$ and $s_2$ are the slopes with which the two surfaces would cross if there were no splitting.

The probability of crossing (nonadiabatic behavior) approaches unity as $\varepsilon_{12}$ becomes very small, and as the velocity becomes very large. The order of magnitude of $|s_1 - s_2|$ would be $2 \times 10^8$ eV cm$^{-1}$, and of $v$ about $2 \times 10^4$ cm s$^{-1}$. If $\varepsilon_{12}$ is $2 \times 10^{-3}$ eV, the calculated value of $P$ is $7 \times 10^{-3}$, which is a fair probability for jumping. The derivation of equation (7) depends on solving the time-dependent wave equation. That is, the wavefunction, $\psi$, must change rather dramatically, from $\psi_A$ to $\psi_B$, over a short interval of configuration space. Wavefunctions do not change instantaneously, but must obey the equation

$$\frac{\partial \psi_{(Q,t)}}{\partial t} = -\frac{i2\pi}{h} H\psi_{(Q,t)} \tag{8}$$

If the nuclei are moving fast, $v$ is large, and if the gap is small, the wavefunction does not have enough time to change from $\psi_A$ to $\psi_B$, and must perforce stay on the A surface.

Equation (7) is not very accurate, since it is based on an overly simple model; in particular, only one coordinate is considered. There has been a great deal of recent work on expanding the theory, to get rates of intersystem crossing and lifetimes of excited states.[14] The results are very complicated, because it is necessary to consider vibrational motions under conditions where two electronic states are very close. Under these conditions the Born–Oppenheimer approximation breaks down. Vibrational states and electronic states can no longer be separated, but are strongly coupled. Also the theories must evaluate $\varepsilon_{12}$, or its various equivalents.

There are two major mechanisms for creating an interaction energy between states of different symmetries and multiplicities. The first is vibronic coupling, often called Herzberg–Teller coupling. This is the mixing into the ground state of excited states of other symmetries as a result of nonsymmetric vibrations of the molecule. It is clearly a mechanism for interaction of states of different species. The theory of Herzberg–Teller coupling has already been given, since it is the same as the general PT for chemical change. However the vibrations are now the normal modes that are orthogonal to the reaction coordinate.

The second important mechanism is spin–orbit coupling, which changes both the spin multiplicity and the state symmetry in a synchronous manner. It depends on the inclusion in the Hamiltonian of a magnetic energy due to the spin of an electron and its orbital motion, each of which can produce a magnetic dipole. For a single electron in an atom the spin–orbit operator is

$$\mathbf{H}_{\text{so}} = \frac{1}{2m^2c^2}\left(\frac{1}{r}\frac{\partial U}{\partial r}\right)\mathbf{l}\cdot\mathbf{s} = \xi(r)\mathbf{l}\cdot\mathbf{s} \tag{9}$$

where $r$ is the distance of the electron from the nucleus and $\mathbf{l}$ and $\mathbf{s}$ are the orbital and spin angular momentum operators. For a many-electron, many-atom molecule, the expression becomes

$$\mathbf{H}_{\text{so}} = \sum_i \sum_K \xi(r_i)\mathbf{l}_{iK}\cdot\mathbf{s}_i \tag{10}$$

where $i$ refers to an electron and $K$ to a nucleus. The spin–orbit coupling term, $\xi$, is usually considered to be a constant, derivable from atomic data. This is certainly not very accurate, since the value would at least depend on the charge on the atom in each molecule.

The effect of $\mathbf{H}_{\text{so}}$ is that every state wavefunction, say for $S_0$, is always mixed somewhat with states of different multiplicity.

$$\psi_S = \psi_S^0 + \sum_k \frac{\langle\psi_S^0|\mathbf{H}_{\text{so}}|\psi_T^k\rangle}{(E_S^0 - E_T^k)}\psi_T^k \tag{11}$$

The effect of $\mathbf{s}$ is essentially that of changing the spin of an electron. The effect of $\mathbf{l}$ is to change the orbital angular momentum of an electron. This means that the electron must move from one orbital to another. In an atom the value of $m_l$ would change by one unit. For a $p$ orbital, the effect would be that of a counterclockwise rotation by 90°, say from $p_y$ to $p_z$. The spin operator, **s**, operates on the spin part of the wavefunction. There are symmetry-based selection rules for this, but the net effect is that the spin functions of the triplet states are transformed so that they no longer are orthogonal to those of the singlet state.[15]

The angular momentum operator **l** has three components that transform as $R_x$, $R_y$, and $R_z$, rotation about the three Cartesian axes of the molecule. In an atom its effect would be to rotate the orbital to a new value of $m_l$. For example, a $p_x$ orbital would become a $p_y$ orbital by the $R_z$ operation. In a molecule the **l** operator will couple together all excited states which obey the rule

$$\Gamma_S \times \Gamma_T \subset \Gamma_{R_{x,y,z}} \tag{12}$$

The symmetries of $R_x$, $R_y$, and $R_z$ for each point group are found in tables.

As an illustration of spin–orbit coupling and of intersystem crossing, we will take the interconversion of singlet

$$^1A_1\,CH_2 \longrightarrow {}^3B_1\,CH_2 \tag{13}$$

methylene, which is an excited state, into triplet ground-state methylene.[16] Both states have $C_{2v}$ structures, but the angle at equilibrium is 136° for the triplet, and 102° for the singlet. Consequently, the singlet $CH_2$ molecule at its stable geometry is equal in energy to a vibrationally excited form of the ground state. This leads to the curve crossing shown in Fig. 4, where it is estimated that the $v = 1$ state of $^1A_1\,CH_2$ has the same energy as $v = 7$ of $^3B_1$.

By symmetry, the $B_1$ and $A_1$ states can only be mixed by the $R_y$ rotation operator, which is $B_1$ in $C_{2v}$. It can be seen in Fig. 4 that rotation about the $y$ axis will take an electron from the $a_1$ orbital in the $^1A_1$ state and put it in the $b_1$ orbital of the $^3B_1$ state. At the same time the spin of one electron is changed by the **s** operator. Angular momentum is conserved by this joint operation.

The singlet and triplet states are strongly mixed near the crossing point and (11) is no longer valid. It applies only to weak mixing. Instead we have a two-state system. At the crossing point the energies given by the two values

$$E = \tfrac{1}{2}(E_S + E_T) \mp \varepsilon_{12} \tag{14}$$

where $\varepsilon_{12} = \langle \psi_S \chi_S | H_{so} | \psi_T \chi_T \rangle$ and is estimated to be only 1.4 cm$^{-1}$. This small value arises in part because the calculation must include vibrational wavefunctions, $\chi$, as well as electronic wavefunctions. The overlap of the two vibrational functions ($v = 1$ and $v = 7$) is only about 0.1 (the Franck–Condon factor is the square of this overlap).

If the nuclei moved infinitely slowly, the wavefunction would be an exactly 50–50 mixture of $\psi_S$ and $\psi_T$, at the crossing point. The probability

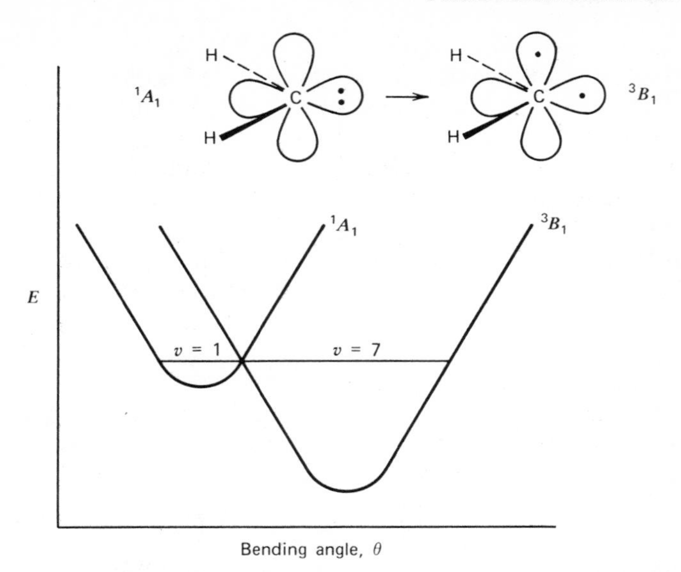


**Figure 4.** Potential energy diagram showing crossing of $^1A_1$ and $^3B_1$ states of methylene as bending angle, $\theta$, changes.

of being on either surface would be one half. Since singlet methylene lasts long enough to have its own chemistry, it must be that the system usually behaves nonadiabatically and stays on the singlet surface. Experimentally it is found that it is collision with other molecules which induce the singlet–triplet crossing for $CH_2$.[17]

As shown above, magnetic energies are quite small for light atoms. For heavy atoms they can become very large, because of the strong dependence on the nuclear charge, indicated in (9). Eventually they become as large as interelectronic repulsion energies, and we must use *j–j* coupling rather than Russell–Saunders coupling. This also means high rates of intersystem crossing. For example, iron(II) complexes, $FeL_6^{2+}$, can exist as low-spin singlet states, or as high-spin quintet states, with four unpaired electrons. The rate of intersystem crossing for some iron(II) and cobalt(II) complexes has been directly measured by relaxation methods.[18] The lifetimes are in the range of 10–30 ns (nanoseconds) for the various spin states ($1 \text{ ns} = 10^{-9}$ s).

It should also be mentioned that there are other, less efficient mechanisms for intersystem crossing. The next most common is the hyperfine interaction between electron spins and nuclear spins. This is the Fermi contact term, which produces fine structure both in n.m.r. and e.s.r. spectra.

## APPLICATION OF PERTURBATION THEORY

It is now appropriate to apply the methods developed earlier for ground-state reactions to those of the excited state. We will start with PT. Figure 5 shows a schematic MO diagram for a molecule in its ground state, $S_0$, and in its lowest excited states, $S_1$ or $T_1$. On general principles, we can see that the excited state will be much more reactive. The guiding principle is that low-lying excited states will facilitate easy nuclear motions, corresponding to various chemical reactions. $S_1$ or $T_1$ have numerous other states available to which they are close in energy.

For example, the promoted electron is in an orbital close in energy to a number of other orbitals. Moving the electron to any one of these creates new states that are excited states of $S_1$ and $T_1$. Also an electron can be promoted from any of the doubly filled orbitals to the half-filled orbital among the lower set. This generates more excited states. We can expect that the top electron is the most active in promoting chemical change or nuclear motion. For one thing, there are a very large number of orbitals open to it. This can be seen by drawing in the first IP of the molecule, on the energy scale. In principle, there are an infinite number of orbitals below the IP. The energy gap between the HOMO and the LUMO is 5–8 eV in most molecules, whereas the IP is 9–12 eV. Therefore the density of states is very large in the upper manifold.

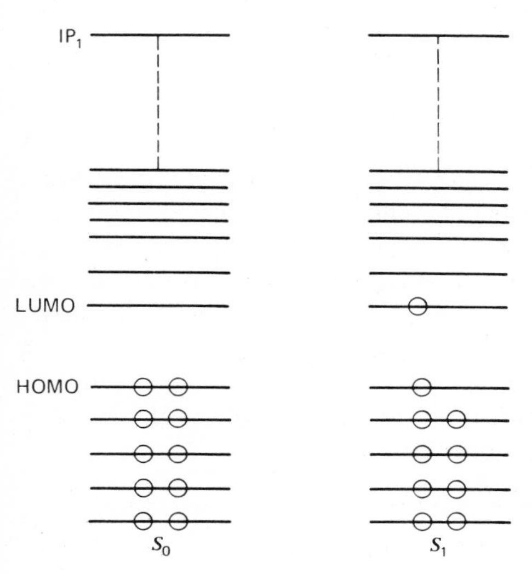


**Figure 5.** Molecular orbital configurations for a molecule in its ground state and in its first excited state; the dashed lines represents many virtual orbitals up to the first ionization potential.

Now what is the role of $S_0$ in the perturbational treatment of $S_1$ or $T_1$? It plays a special part, as can be seen by writing the equation for the energy of, say, $S_1$ due to a perturbation caused by nuclear motion.

$$E = E_1 + \left\langle \psi_1 \left| \frac{\partial U}{\partial Q} \right| \psi_1 \right\rangle Q + \left\langle \psi_1 \left| \frac{\partial^2 U}{\partial Q^2} \right| \psi_1 \right\rangle \frac{Q^2}{2} + \sum_k \frac{\left\langle \psi_1 \left| \frac{\partial U}{\partial Q} \right| \psi_k \right\rangle^2}{(E_1 - E_k)} Q^2 \tag{15}$$

We must still use a complete set of wavefunctions, and so $\psi_0$ becomes one of the $\psi_k$ in the last term of (13). However unlike all other states, $(E_1 - E_0)$ is a positive number. Therefore mixing in of $\psi_0$ to $\psi_1$ is accompanied by an *increase* in energy, rather than the usual decrease. This leads to the prediction that *a reaction that is allowed or favored in the ground state is forbidden or disfavored in the excited state.*[19] The assumption behind this prediction is that the real excited state is also that hypothetical excited state that strongly favors the ground-state reaction.

This seems paradoxical, since it seems reasonable that if a small amount of excitation favors a given reaction, a large amount should certainly cause it to go spontaneously. Yet the prediction agrees remarkably well with experimental results, and also can be rationalized in a number of ways. Let us examine some allowed ground-state reactions and see what happens if the excited state is formed, using the various HOMOs and LUMOs as a basis. Consider the conrotatory ring opening of cyclobutene.

```
HC=CH*                      HC—CH
|   |        ——→          //     \\          (16)
H2C—CH2               H2C          CH2
```

(*refers to an excited-state molecule)

The necessary excited states that drive the reaction (p. 72) are the $\pi, \sigma^*$ and $\sigma, \pi^*$. Suppose the molecule is fully excited by either of these transitions, or even better, by the $\pi, \pi^*$ process. Then the molecule has one of the necessary bonds broken to give the reaction, but the remaining processes become more difficult. Both the $\pi, \sigma^*$ and $\sigma, \pi^*$ transitions become of high energy because $\pi$ is only half occupied and $\pi^*$ is already half filled. What actually happens in (16) is that a disrotatory ring opening occurs.[20]

Suppose a nucleophilic substitution reaction at some stage has a complete transfer of one electron from the HOMO of the nucleophile to the LUMO of the electrophile, by photochemical or other means.

$$\begin{aligned} Br^- + CH_3Cl &\longrightarrow Br\cdot + CH_3Cl^{\overline{\cdot}} \\ CH_3Cl^{\overline{\cdot}} &\longrightarrow CH_3\cdot + Cl^- \end{aligned} \tag{17}$$

The bromine atom now formed is an electrophile, and not likely to give another electron to $CH_3Cl^{\overline{\cdot}}$. This radical anion will quickly dissociate to a methyl radical and chloride ion. The net reaction is quite different from nucleophilic substitution.†

The lesson is obvious; one can get too much of a good thing. A little mixing of an excited state wavefunction can help a reaction along. Complete conversion to the excited-state wavefunction is going too far. The system will be very reactive, but some other reaction will probably result. It should not be concluded that an excited state and a ground state can never give the same reaction, since they sometimes do. This is often the case for reactions that are quite difficult in the ground state, requiring a large activation energy. An example would be dissociation into free radicals.

To keep the discussion simple, consider first two diatomic molecules, $H_2$ and HI.

$$H_2 \xrightarrow{h\nu} H(1s) + H(2p) \quad (18)$$

$$\begin{aligned} HI &\xrightarrow{h\nu} H(1s) + I(5p) \\ HI &\xrightarrow{h\nu} H^+ + I^- \end{aligned} \quad (19)$$

Figure 6 gives the potential energy curves for the ground state and two lowest excited singlet states for each molecule. Exciting the hydrogen molecule from $(\sigma_g)^2$ to $(\sigma_g)(\sigma_u)$ produces a $\Sigma_u^+$ state, which has a rather shallow minimum and can dissociate. This is consistent with $\sigma_u$ being an antibonding orbital. Yet note that the dissociation is to an excited state of the products. One H atom must be in a $2p$ configuration. The same products are found for dissociation from the $\Pi_u$ state, the $(\sigma_g)(\pi_u)$ configuration.

For HI, on the other hand, dissociation from the lowest excited singlet, $^1\Pi$ is into the ground state atom products. In this case thermal and photolytic decomposition yield the same products. The excited $\Sigma^+$ state, corresponding to a configuration $(\sigma)(\sigma^*)$, dissociates into excited products. Because HI is a polar molecule, the correlation is to the ions. The $\sigma$, $\sigma^*$ transition is called the charge-transfer band for this reason. It is the transition that facilitates the thermal reaction, whereas the $\pi$, $\sigma^*$ transition that creates the $^1\Pi$ state does not.

For both $H_2$ and HI, the excited states which promote the thermal, or ground-state, reaction, are not effective in causing the same reaction to occur photochemically. Instead another reaction occurs. This emphasizes the point that the excited states that are the basis of PT are hypothetical states. Their relationship to real excited states is largely a formal one in that MO theory describes them by the same configuration.

† Unless $CH_3\cdot$ and $Br\cdot$ combine to give $CH_3Br$. This can readily happen in solution because of the cage effect of the solvent.

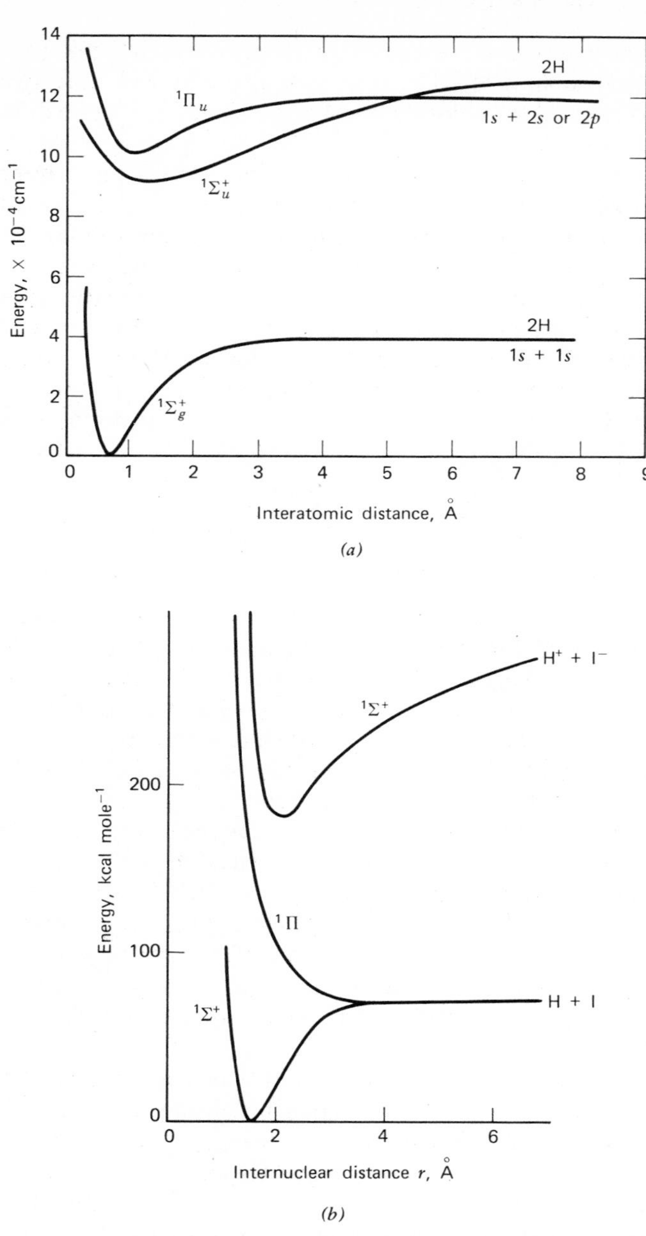


**Figure 6.** Potential energy curves for lowest singlet states of: (*a*) $H_2$ and (*b*) HI (modified from reference 46).

Another general conclusion can be made from (15), if we limit it to maxima or minima on the potential energy surface. Then the quadratic terms give the force constant for the normal mode, $Q$. We must assume that the ground state and the excited state belong to the same point group and have similar angles and distances. Under this rather severe assumption, we can deduce the rule: "If one force constant in the ground-state molecule has an unusually small value, because of a low-lying excited state, then that same force constant in the corresponding excited state will have an unusually large value." This rule is another consequence of $(E_1 - E_0)$ being positive. An example is the linear $C_3$ molecule. In the ground-state the frequency, $\nu_2$, for the bending mode is 63 cm$^{-1}$. This low value is due to a low-lying electronic state (p. 171). As expected this same excited state has $\nu_2 = 308$ cm$^{-1}$, a very much larger number.[21]

## CORRELATION DIAGRAMS

At this point PT, in the sense of (15), can do little more for us. The conclusions are that many low-lying excited states are available, and that something will surely happen, but the very number of possibilities makes prediction difficult. A negative prediction can be made, however, that a favorable ground-state mechanism will be adversely affected. A logical next step is to see if unfavorable ground-state mechanisms are benefited. The most unfavorable mechanisms are those forbidden by symmetry rules. We can see in selected cases that they will be benefited, but the question of generality remains. The most useful procedure at this point is to use correlation diagrams, since they can be generalized.

A good example is the ground state forbidden four-center $H_2$—$D_2$ exchange reaction

$$\begin{matrix} \text{H—H*} \\ + \\ \text{D—D} \end{matrix} \longrightarrow \begin{matrix} \text{H---H} \\ \vdots \quad \vdots \\ \text{D---D} \end{matrix} \longrightarrow \begin{matrix} \text{H} \quad \text{H*} \\ | \; + \; | \\ \text{D} \quad \text{D} \end{matrix} \tag{20}$$

$$a_1 + \tfrac{1}{2}(a_1 + b_2) \qquad\qquad\qquad a_1 + \tfrac{1}{2}(a_1 + b_2)$$

The point group can be taken as $C_{2v}$, but there is a degeneracy at the TS state that simplifies the argument. Figure 7*a* shows the familiar orbital-correlation scheme for (19). Suppose one reactant molecule is excited to the $(\sigma_g)(\sigma_u)$ configuration, or the $^1\Sigma_u^+$ state, and that the products are also formed with one HD molecule in an excited $\Sigma$ state, as shown. In the $C_{2v}$ point group of the reacting molecules, the configuration would be $(a_1)^2(2a_1)(b_2)$. In (20) this is indicated by writing 1/2 in front of $a_1 + b_2$, to show that these orbitals

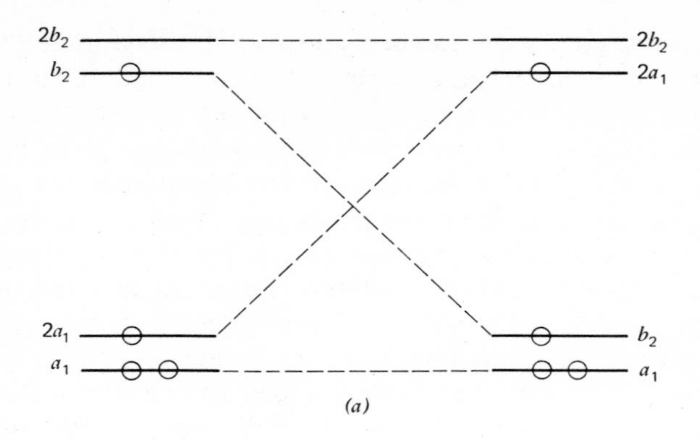


(a)

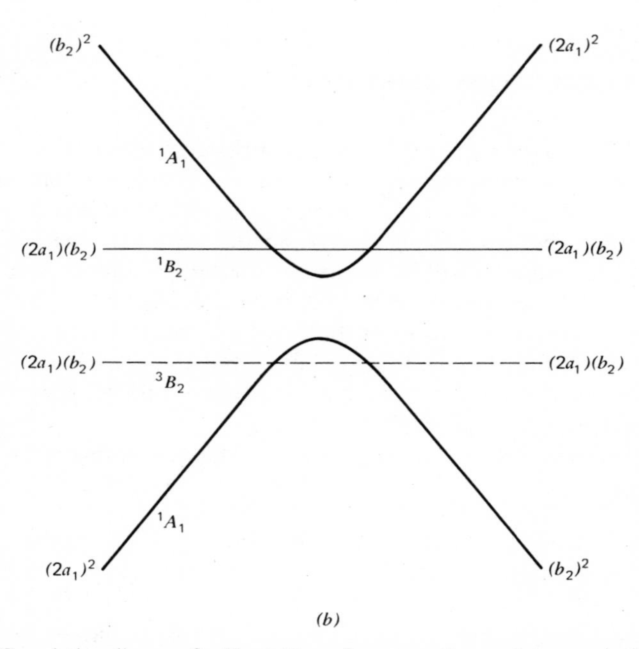


(b)

**Figure 7.** Correlation diagram for $H_2 + D_2$ exchange reaction, point group is $C_{2v}$: (*a*) orbital correlation; (*b*) state correlation—the configurations include $(a_1)^2$, in addition to those shown.

are each half occupied. The product also has the same $(a_1)^2(2a_1)(b_2)$ configuration. There would be no barrier to the exchange reaction, since as one orbital energy, $2a_1$, goes up, an equally half-occupied orbital, $b_2$, goes down in energy. Thus (2) becomes allowed as an excited-state reaction.

However this orbital picture is oversimplified since it ignores the configuration interaction which avoids the intended crossing at the state level. Figure 7*b* shows the state-correlation diagram. The critical new feature is configuration interaction. At the previous crossover point, the four nuclei forming a square, the molecular orbitals $2a_1$ and $b_2$ become equal in energy.

$$\begin{matrix} + & + \\ - & - \\ & 2a_1 \end{matrix} \qquad \begin{matrix} + & - \\ + & - \\ & b_2 \end{matrix}$$

Mixing of the two configurations $(a_1)^2(2a_1)^2$ and $(a_1)^2(b_2)^2$ becomes strong. Since both configurations belong to the totally symmetric representation, there will be an avoided crossing as shown in the Fig. 7*b*. There are two other states of importance, the singlet and triplet of the $(2a_1)(b_2)$ configuration. These are $^1B_2$ and $^3B_2$ and do not mix with the $^1A_1$ states.

At the one-electron level, all four of these states are equal in energy at the crossover point. Actually electron repulsion energies will split them apart, but not by large amounts. The triplet state will usually lie lowest, as shown in Fig. 7*b*. The initially formed excited state will correspond to $(2a_1)(b_2)$ in a collision complex of $H_2^*$ and $D_2$. However there can be an easy crossing into the $(b_2)^2$ state, by vibrational coupling, and at the crossing point, there can be a nonadiabatic jump from the upper $^1A_1$ to the lower $^1A_1$ surface. This will explain the formation of products in their ground states, as experimentally observed.

The diagrams of Fig. 7 and similar arguments can be repeated for most reactions forbidden by orbital symmetry. The chief requirement is that the excited state is formed by taking an electron from an orbital that goes up in energy, along the reaction coordinate, and putting it in an orbital that goes down in energy. These orbitals need not be the HOMO and LUMO, but often they will be. A second requirement is a geometric one, that the shape of the excited-state molecule be sufficiently similar to the ground-state molecule so that Fig. 7 has meaning. We will return to this point later.

The above explanation for photochemical reactions was first given by Van der Lugt and Oosterhoff,[22] who were considering the reverse of reaction (16), the photochemical cyclization of butadiene to cyclobutene. There is an energy problem in this reaction, if it is postulated that the reaction proceeds entirely on the excited-state potential surface.[23] The problem is shown in Fig. 8*a*. Firstly, cyclobutene is some 20 kcal higher in energy than butadiene.

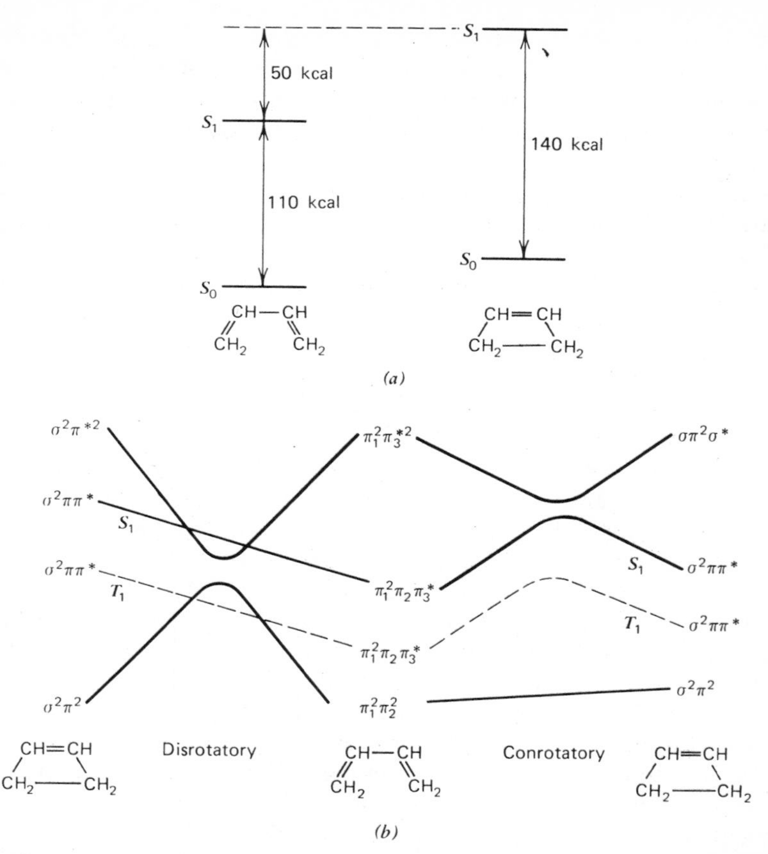


**Figure 8.** (*a*) Energy shortage in photochemical cyclization of butadiene to cyclobutene, if product is in $S_1$ state; (*b*) state correlations for disrotatory and conrotatory ring-opening reactions of cyclobutene. (After J. Michl, *in* G. Klopman, (ed.), *Chemical Reactivity and Reaction Paths*, Wiley-Interscience, New York, 1974, p. 322.)

Secondly, a $\pi,\pi^*$ excitation in conjugated butadiene requires some 30 kcal less energy than the $\pi,\pi^*$ excitation of unconjugated cyclobutene. Thirdly, it would require some 50 kcal of activation energy to go from $S_1$ of reactant to $S_1$ of product. This is prohibitive. Figure 8*b* shows the state-correlation diagram for both the conrotatory and disrotatory ring-closing reactions of butadiene. The similarity of the photochemical disrotatory process to Fig. 7*b* is obvious. Also the difficult nature of the conrotatory process in the excited state is shown.

So far nothing has been said of the $T_1$ triplet states shown in Figs. 7 and 8. They could undergo an intersystem crossing to the ground-state potential surface, since they are energetically suitable. However this is less probable than for the excited singlet states. Another possibility is that the triplet state will simply give some other reaction, characteristic of a free radical, or a diradical.[24] This does happen, as shown by the photochemical behavior of 1,3-pentadiene. The singlet undergoes the normal ring closing reaction. The triplet state, however, gives 1,3-dimethylcyclopropene.[25] Presumably an intermediate diradical is formed first.

$$H_2C{=}CH{-}CH^*{=}CHCH_3 \longrightarrow \cdot CH_2{-}HC\langle CH(CH_3){-}\dot{C}H\rangle \longrightarrow (CH_3)C{=}CH{-}CH(CH_3) \text{ (cyclopropene ring)} \quad (21)$$

The common pericyclic reactions, including cycloadditions, ring openings, and sigmatropic reactions seem to be affected in the same way upon photochemical excitation. Figure 9 shows the orbital-correlation diagram for a 1,3-sigmatropic shift, for instance,

$$CH_3{-}CH{=}CH_2 \longrightarrow CH_2{=}CH{-}CH_3 \quad (22)$$

The orbitals are correlated by following their nodal properties (see p. 93). It can be seen that the ground-state forbidden suprafacial reaction becomes allowed in the excited state. Furthermore the excited-state energy surface goes down as one approaches the TS for the ground-state reaction. Therefore a crossing, or near crossing, is ensured, and the products can be formed in their ground states by an intersystem crossing.

Conversely, the antarafacial process, which is allowed in the ground state, becomes forbidden in the excited state, as shown in Fig. 9*b*. The $\pi^*$ orbital of the reactant correlates with the very high energy $\sigma^*$ orbital of the product. The same correlation (Fig. 9*b*) can be used for a suprafacial shift with inversion of a *p* orbital. Considering Figs. 7–9, it is not surprising that the rules for pericyclic reactions become inverted for excited-state reactions, compared to ground state.[20] Table 1 shows the Dewar–Zimmerman rules summarized.[26]

An application of Table 1 is the observation that [1,5]-sigmatropic changes occur photochemically by an antarafacial process.[27] These are six-electron Hückel systems in the ground state, which react suprafacially, and six-electron

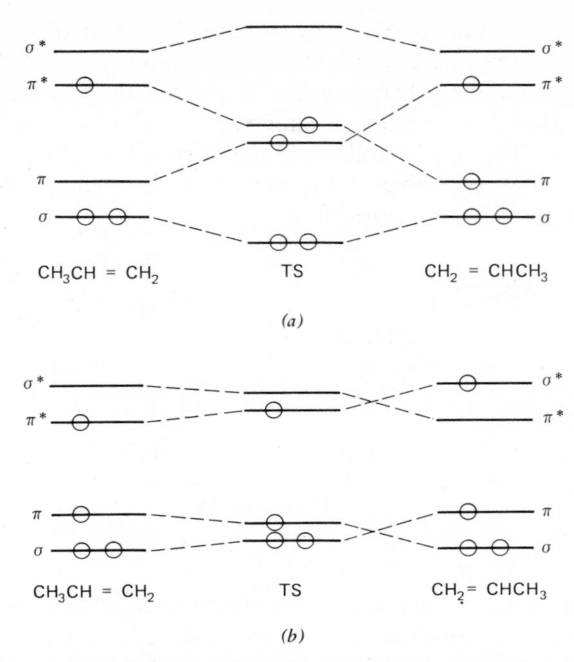


**Figure 9.** (*a*) Suprafacial [1,3] sigmatropic shift in propylene is allowed in $\pi$, $\pi^*$ excited state; (*b*) antarafacial [1,3] shift is forbidden in $\pi$, $\pi^*$ state of propylene.

**Table 1 Dewar–Zimmerman Rules for Pericyclic Reactions**[a]

| | Thermal | Photochemical |
|---|---|---|
| Hückel systems | $4n + 2$ | $4n$ |
| Möbius systems | $4n$ | $4n + 2$ |

[a] The rule states: "A pericyclic reaction is allowed if the transition state is aromatic."

[b] A Hückel system has an even number (including zero) of orbital phase-sign inversions. A Möbius has an odd number. The number of electrons needed for aromatic character is shown in each case.

Möbius systems in the excited state. Many more examples of the successful use of Table 1 may be cited. Woodward and Hoffmann were the first to point out the inversion in behavior for ground and excited states.[20] Their method was the use of correlation diagrams.

## Correlation to Ground-state Products

In addition to nonadiabatic processes leading to ground-state products, there also are photochemical processes in which correlation diagrams show that excited-state reactants lead adiabatically to ground-state products. Let us reconsider (1), hydrogen-atom abstraction by an $n,\pi^*$ excited ketone. A useful symmetry element is the plane containing the carbonyl group and the C—H bond.[5] The critical electrons of the $n$ orbital, the $\pi$ orbital, and the C—H$\sigma$ orbital must be considered. They are all $\sigma$ or $\pi$ with respect to the indicated plane. Note that this plane is not a true symmetry element for the entire reacting system, but it still is extremely useful.

Reactant, ground state

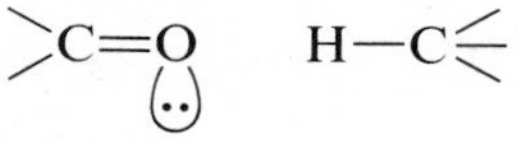

>C=O H—C< $4\sigma, 2\pi$

Reactant, $n, \pi^*$ excited state

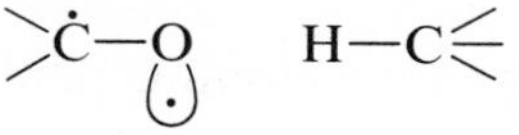

>Ċ—O H—C< $3\sigma, 3\pi$

Product, ground state

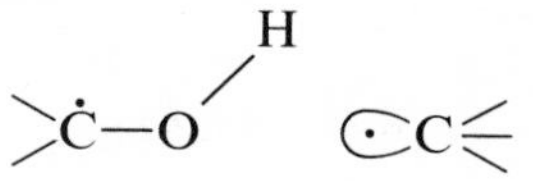

>Ċ—O—H (·)C< $3\sigma, 3\pi$

Product, excited state

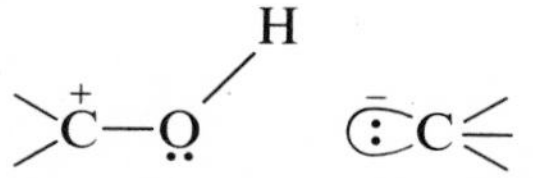

>C⁺—O—H (:)⁻C< $4\sigma, 2\pi$

A lone pair on oxygen not enclosed in a single lobe is in a $\pi$ type orbital. Figure 10 shows the energies of the ground and excited states for acetone. The product ground state energy would be near the latter.

It is interesting that the symmetry rule shows that the allowed reaction of the ground state leads to *proton* transfer from the hydrocarbon to the ketone. Normally this would be a very high energy process. Of course if the carbon—hydrogen bond were in a molecule activated by a nitro group, or similar, and if a polar solvent were present, the ionic products could become the ground-state products.

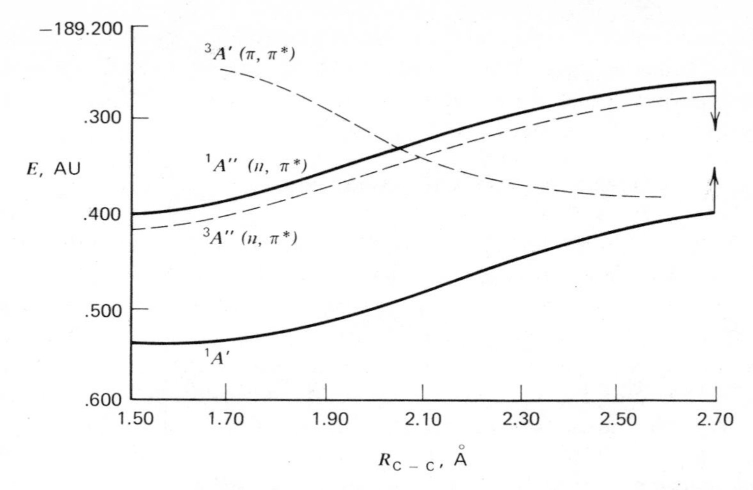


**Figure 10.** Potential energy surfaces for α cleveage of acetone; calculations are for $CH_3CO\cdot$ in a bent geometry. The arrows at $R_{CC}$ = 2.70 Å show the energy changes upon going to a linear geometry. [Reprinted with permission from L. Salem, *J. Am. Chem. Soc.*, **94**, 322 (1972). Copyright by the American Chemical Society.]

An analogous situation to that of the ketones exists for proton transfer and hydrogen-atom abstraction reactions of aza aromatics.[5] Pyridine serves as an illustration. The $n$ electrons, the $\pi$ electrons, and the $\sigma$ electrons of the R—H bond are considered.

Reactant, ground state

N H—R $4\sigma, 6\pi$

Reactant, $n,\pi^*$ excited state

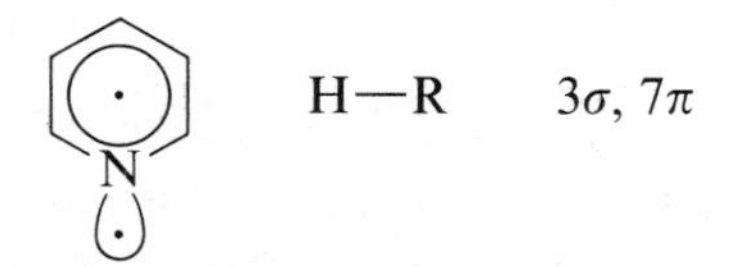


Product, excited state

N⁺ H :R⁻ $4\sigma, 6\pi$

Product, ground state

$$\text{(pyridinyl radical, N–H)} + \odot R \quad 3\sigma, 7\pi$$

The best-studied photochemical reactions of organic chemistry are probably the Norrish types I and II reactions of ketones.[28]

$$R-\overset{\overset{\displaystyle O}{\|}}{C}-R \xrightarrow{h\nu} R-\overset{\overset{\displaystyle O}{\|}}{C}\cdot + R\odot \quad \text{Type I} \qquad (23)$$

$$R-\overset{\overset{\displaystyle O}{\|}}{C}-CH_2-CH_2-CH_2-H \xrightarrow{h\nu} R-\overset{\overset{\displaystyle OH}{|}}{C}\cdot \;\; CH_2-CH_2-\odot CH_2 \quad \text{Type II} \quad C \qquad (24)$$

$$C \longrightarrow R-\overset{\overset{\displaystyle OH}{|}}{C}=CH_2 + CH_2=CH_2 \quad \text{or} \quad \text{(1-R-cyclobutanol: HO, R on C; ring } CH_2, CH_2, CH_2)$$

The type II reaction is no different from (1). The intermediate diradical is formed in its ground state. A conformational change is necessary to form any of the final products, but this would require little energy. Both singlet and triplet states can give the type II process.

The type I reaction, or $\alpha$ cleavage, is more common, both for singlet and triplet $n,\pi^*$ states. The count of $\sigma$ and $\pi$ electrons reveals that a ground-state product is not directly formed.

$$R_2\dot{C}-O(\odot) \longrightarrow R\odot + R-\dot{C}=O \quad \text{not} \quad R-(\odot)C=O \qquad (25)$$

$$3\sigma, 3\pi \qquad \sigma \qquad 2\sigma, 3\pi \qquad 3\sigma, 2\pi$$

The ground-state acyl radical has a bent structure with its unpaired electron in a $\sigma$ orbital on carbon. The product which is given by the correlation is linear, with an unpaired electron in a $\pi$ orbital on carbon. The double bond shown in the linear radical of (24) lies in the plane of the page.

Figure 10 shows a calculated set of energy surfaces for $\alpha$ cleavage of acetone.[5] The excited-state product lies about 1 eV above the ground-state radical, and at no point do their surfaces cross. However the linear structure for the acyl radical is identical with a vibrationally excited form of the bent radical, differing only in rotation about the internuclear axis. Thus it will become the ground-state radical product by a simple loss of vibrational energy. Examples of this kind will be fairly common, and provide another route for formation of ground-state products from an excited-state surface.

It will not often be true that the two products become identical, as in the above example. However if the electronic energies become nearly equal, then surface crossing becomes easy in any case. The real requirement, then, is a substantial change in geometry between the ground-state and excited-state products. This will provide a way for the surfaces to approach each other and also the mean whereby the stable geometry of one state becomes a vibrationally excited form of the other. It is accordingly time to look at the changes in molecular shape induced by electronic excitation. Other important reasons exist: (1) the shape of a molecule influences the kind of elementary reactions it will undergo, and (2) large changes in geometry will invalidate any correlation diagrams which are drawn, assuming the point group of the molecule to remain constant.

## THE STRUCTURES OF MOLECULES IN EXCITED ELECTRONIC STATES

The complete transfer of an electron from one MO to another, to form an excited state, creates a substantial change in the electron-density pattern of the molecule. Naturally the nuclei will rearrange to accommodate to this new pattern. The direction of nuclear change cannot be simply found from the product of $\varphi_i \times \varphi_f$, since ground-state PT does not apply. Instead we must consider the probable effect of the structure of first removing an electron from $\varphi_i$, and then replacing it in $\varphi_f$. This becomes equivalent to predicting the structure of molecules with one less electron and with one more electron, than the original. In Chapter 3 a number of structural theories were presented in which the number of electrons for a given number of nuclei was just the information needed. Unfortunately these theories are easy to use only for rather simple molecules.

We could attempt to use excited-state PT, as expressed in (15).[29] This is impractical for any but the simplest molecules, because it requires detailed knowledge of many excited states, and rarely is this information available. However one example may be tried, that of excited formaldehyde. The sequence of singlet states is known to be[30]

| | | | |
|---|---|---|---|
| $S_0$ | $^1A_1$ | 0.0 eV | — |
| $S_1$ | $^1A_2$ | 4.3 | $n,\pi^*$ |
| $S_2$ | $^1B_1$ | 7.1 | $\sigma,\pi^*$ |
| $S_3$ | $^1A_1$ | 8.0 | $\pi,\pi^*$ |
| $S_4$ | $^1B_2$ | 10 | $n,\sigma^*$ |

The structure of the $^1A_2$ state should be predicted from the direct product $(A_2 \times B_1) = B_2$. However this reaction coordinate is just an in-plane distortion. Figure 11 shows some nuclear motions for $CH_2O$, with symmetry labels. The experimental structure for the $^1A_2$ state is nonplanar.[30] The correct reaction coordinate has $B_1$ symmetry, an out-of-plane bending. It is necessary to invoke the $S_4$ state of $B_2$ symmetry in order to predict this correctly. This transition, $S_1 \to S_4$, which dominates the geometry change, is shown in Fig. 11*b* in terms of MOs. It corresponds to a further excitation to a $\sigma^*$ orbital of the excited electron.

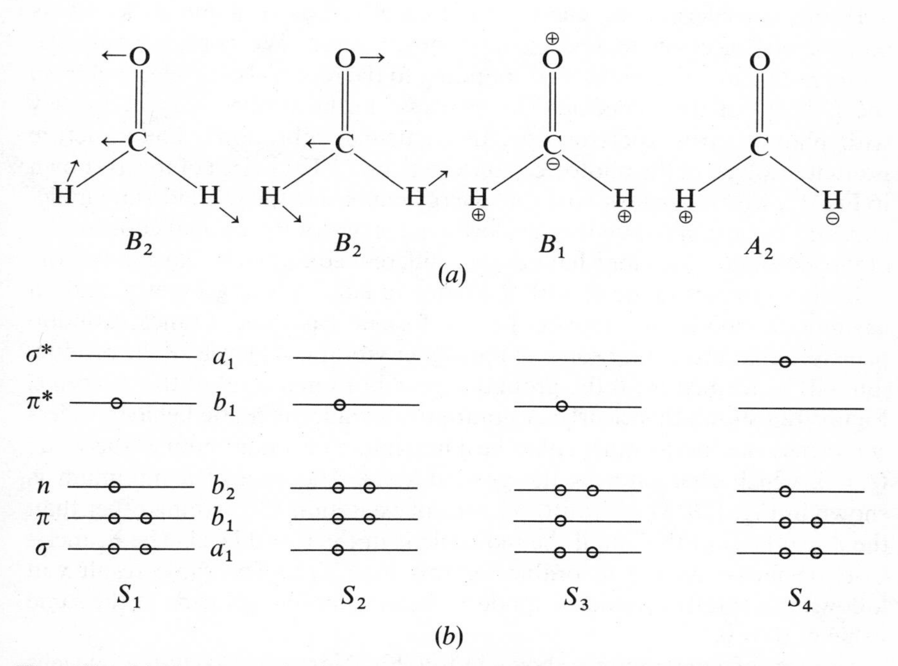


**Figure 11.** (*a*) Nonsymmetric vibrational modes of formaldehyde in $C_{2v}$ point group, the $A_2$ mode is a rotation of the molecule; (*b*) the first four excited singlet states of $CH_2O$ in terms of occupied MOs.

This result suggests that adding an electron to an empty orbital has a greater structural effect than removing an electron from a doubly occupied one. This seems to be the case. For example, the ion $CH_2O^+$ would be a 15-electron molecule. According to the various methods of determining structure (p. 170), this would still be a planar molecule. However $CH_2O^-$, a 17-electron molecule, would be predicted to have a nonplanar structure, like $CH_2F$. Accordingly, Walsh[31] was led to state the postulate: "The first excited state of a molecule with $n$ electrons should belong to the same point group as the ground state of a similar molecule having $\underline{n + 1}$ or $\underline{n + 2}$ electrons." The extra electrons must be in the orbital which becomes occupied in the excited state.

Exactly the same rule can be derived from PT, or the SOJT effect.[32] The assumption is that the promoted electron, being in a very unstable orbital, is the one which is active in causing rearrangement. The vacancy in the stable orbital is presumed to be less active, since even more stable electrons must be disturbed to utilize it. Regardless of the validity of any theoretical justification, it can still be treated as an empirical rule, and one that is relatively easy to apply.

Before considering the effect of adding electrons to a molecule, let us remove one electron to form a monopositive ion. We consider only the ground-state ion, normally corresponding to the removal of an electron from the HOMO of the molecule. This process can be studied very accurately with photoelectron spectroscopy. In particular, vibrational fine structure is often resolved in the photoelectron spectrum.[33] The origin of this is shown in Fig. 12, which displays potential energy curves for the ground-state molecule and the ground-state ion. The ordinate is one of the normal coordinates of the molecule which may have quite a different equilibrium value in the ion.

Both a symmetric mode with a change in equilibrium geometry, and an asymmetric mode are shown. In the former case, the Franck–Condon principle guarantees that some of the higher vibrational levels of the product ion will be formed from the ground-state vibrational level of the molecule. Figure 12*a* shows this clearly. A nontotally symmetric mode behaves differently since the energy must either be a maximum or a minimum at the value $Q = 0$, which characterizes the ground state. The case of a minimum is shown in Fig. 12*b*. There is little chance of excitation to anything other than the $v = 0$ level of the ion, if the molecule is in the $v = 0$ level. The Franck–Condon factor is very favorable for this transition. The same result will follow for a totally symmetric mode if the two minima coincide at the same value of $Q = 0$.

The case of a maximum is shown in Fig. 12*c*. Normally two potential wells will exist in the upper state, and there will be a coupling of the vibrational levels in each well. As a result there will be a large number of allowed vibra-

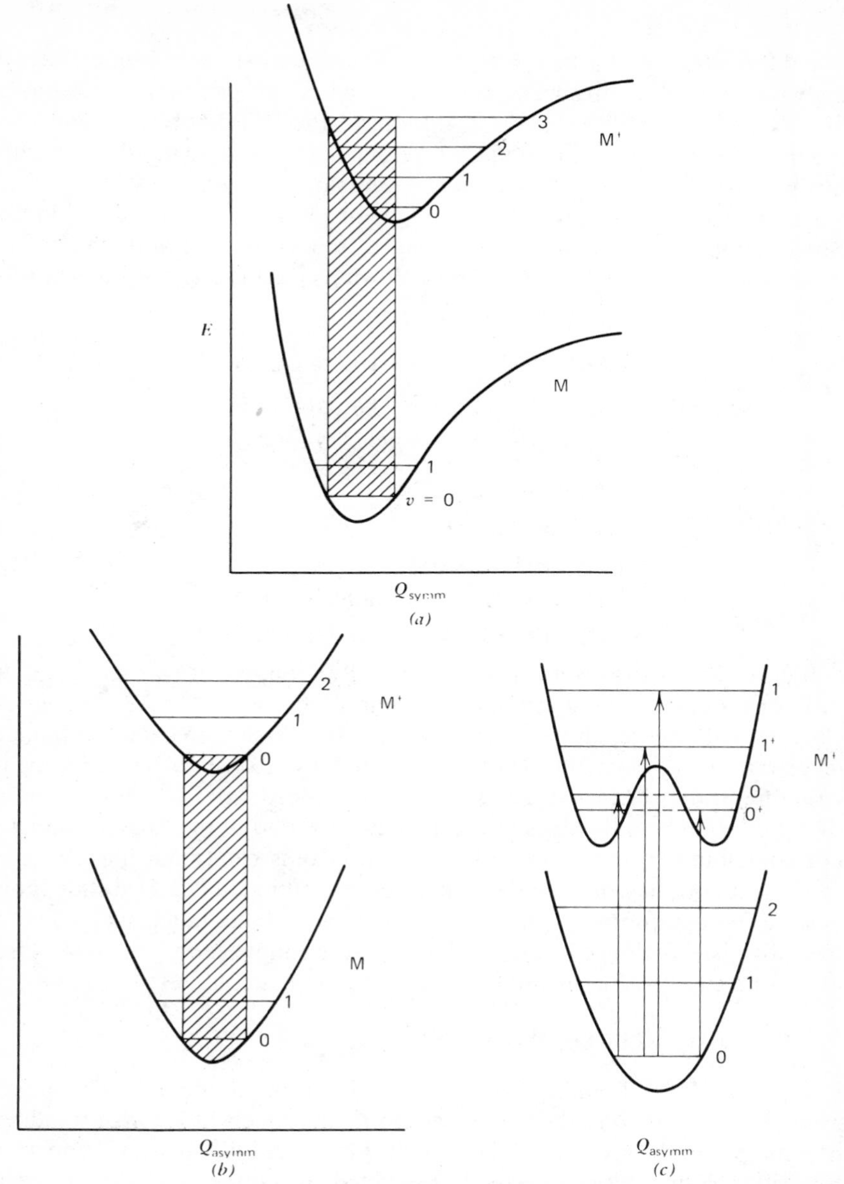


**Figure 12.** Potential energy curves for a molecule M and its ion $M^+$, as a function of $Q$ (the shaded area is the region of maximum probability for the vibrational wavefunction of the molecule): (*a*) $Q$ is symmetric and changes its equilibrium value in the ion; (*b*) $Q$ is asymmetric but corresponds to an energy minimum in the ion; (*c*) $Q$ is asymmetric and corresponds to an energy maximum in the ion—numerous transitions from $v = 0$ in the molecule are found (shown by the arrows).

tional transitions from the $v = 0$ level of the ground-state molecule. Accordingly the photoelectron spectrum will show a long progression of vibrational fine structure in the mode corresponding to the coordinate $Q$.[34] Since the observed frequency for this mode will belong to the ion, and not the molecule, it is not easy to identify $Q$ from the experimental frequency.

As a result, the structures of only a few ions have been determined in this way. Using structural theory as a guide, and consistent with the photoelectron spectroscopic results,[33] the following conclusions can be drawn:

1. The ion remains linear:

$$CO_2^+, CS_2^+, N_2O^+, HCN^+, C_2H_2^+$$

2. The ion remains bent (angle may change):

$$H_2O^+, H_2S^+, SO_2^+, O_3^+$$

3. The ion changes shape:

$NO_2^+$ becomes linear
$NH_3^+$, $PH_3^+$ becomes planar
$C_2H_4^+$ twists out-of-plane, like $B_2H_4$
$C_2H_6^+$ distorts toward $B_2H_6$ structure

Figure 12 was drawn for a molecule and its ion, but it can serve equally well for a molecule in its ground-state and an excited-state. The same conclusions will emerge. Those normal coordinates which correspond to changes in geometry between the ground state and excited state will create fine structure in the electronic spectrum. Normal coordinates which have about the same equilibrium values in the ground state and in the excited state will not contribute. This vibrational information plus rotational fine structure when available, has enabled the structures of a number of excited state molecules to be determined. A few other spectroscopic methods, such as e.p.r., have also been helpful. It is clear that only the simplest, or most symmetric, molecules can be experimentally studied by these methods.

## Application of Walsh's Rule

It will be helpful if simple rules can be found for predicting structures of excited states, such as the one given above in which electrons are added, in principle, to the ground state molecule. Let us see how such a rule operates by testing some systems correctly predicted by Walsh in 1953.[31] Excited-state molecules will be marked by an asterisk.

$$N_2O^* \text{ or } CO_2^* = NO_2 \text{ or } SO_2 \qquad NO_2^{-*} = SO_3^{2-}$$
$$C_2H_4^* = N_2H_4 \qquad CH_2O^* = NH_2F$$
$$C_2H_2^* = N_2H_2 \qquad HCN^* = HCO$$

Linear $N_2O$ or $CO_2$ become bent in the excited state. The models taken for the structure need not be $N_2O^-$ or $CO_2^-$, but $NO_2$ or $SO_2$. The reason is that it is the number of electrons and number of nuclei that determine structure, and not the exact nature of the nuclei. Thus $C_2H_4$* is equated to $N_2H_4$, and predicted to have both twisting about the C—C bond and bending of the $CH_2$ groups (p. 217). HCN* becomes bent, like the formyl radical.

The structure of $NH_3$* was predicted to be planar by Walsh. The model used should be $ClH_3$, which is unstable. Nevertheless we expect that $ClH_3$ would not only be planar, but also T-shaped. This last prediction has not been verified for $NH_3$*. It should be noted that the rule being used is valid only for the lowest excited state. Also it makes no distinction between singlet and triplet states, the same point group being predicted for each.

The following excited state molecules and radicals have their structures correctly predicted by the rule:[35]

$SO_2$, $NO_2$, $CO_2^+$, $CS_2^+$, $C_3$, NCO, NCS, $N_3$, $ClO_2$, NCN, CNC, CCN, SiCC, HCF, HCCl, HSiCl, HSiBr, HPO, HNO, $BO_2$, $NF_2$, $CF_2$, $NH_2$, $PH_2$, $SiF_2$, $S_2O$, $HCO_2$, $C_2N_2$, $NO_3$, $F_2BO$, $Cl_2CO$, $Cl_2CS$, $CNC_2H$, $HC_2C_2H$, $CH_2CO$

In the case of excited free radicals, only one electron is added to make the prediction; thus $ClO_2^* = ClO_2^-$(bent) and $NO_3^* = NO_3^-$(planar). This is based on the assumption that the lowest excited state corresponds to excitation of an electron to the half-filled MO. If this is not true, the prediction fails. For example, $CH_3$* is planar, like the ground state, because the observed excited state is Rydberg-like. That is, the unpaired electron is promoted to a very diffuse higher orbital. There is an unobserved lowest excited state of $CH_3$* that is calculated to be pyramidal.[36]

Many of the observed transitions in the UV spectra of simple molecules are to Rydberg states. The excited electron is in an orbital that is like an AO, with the entire molecule acting as a pseudonucleus, or a single heavy atom can take the role of the nucleus. For these states, MO theory has little significance and the valence shell plays no special role.

The rule fails for the radicals $BH_2$*, $AlH_2$*, and HCO*. In all three cases the ground state is bent and the excited state is linear. The prediction is that the excited state should resemble triplet $CH_2$ or CHF, since in these cases the odd electron is promoted to the next higher orbital. Both $CH_2$ and CHF are bent. However $AlH_2$, $BH_2$ and HCO are subject to a strong Renner–Teller effect. This is a coupling between the orbital angular momentum of the electron and the rotational angular momentum of the molecule. It causes the ground state to be strongly bent and the first excited state to be linear.[37] The linear forms of $CH_2$ and CHF are not subject to the Renner–Teller effect.

Other molecules whose excited-state structures are known and correctly predicted[32] include: (1) $NCO^{-*}$, $NCS^{-*}$, $N_3^{-*}$, and $NO_2^{-*}$, all bent, (2) organic azides, where the $N_3$ group is bent, (3) $PH_3$* and $CH_3NH_2$*, planar and (4) $(CH_3)_2CO$*, nonplanar like formaldehyde.

The excited state of allene is not known experimentally, except that it differs greatly from the ground state. Detailed MO calculations predict that it would be bent at the central carbon atom, and nearly coplanar, except possibly for the hydrogen atoms being slightly buckled.[38] This is the structure predicted by either of the model molecules, diaminocarbene $NH_2CNH_2$, or the hydrazone of formaldehyde $NH_2NCH_2$. For excited butadiene two possible models, vinylhydrazine $CH_2CHNHNH_2$, and 1,2-diaminoethylene $NH_2CHCHNH_2$, give different structural predictions. The actual structure is not known, but detailed MO calculations favor the vinylhydrazine structure.[39] Simple MO theory would favor the diaminoethylene structure, with its 2,3-double bond.

The excited state of ketene, $CH_2{=}C{=}O$*, should have the planar structure of nitrosamine, D,[40]

```
H         O
 \       /
  N=N          D
 /
H
```

This is the calculated[41] structure for the lowest singlet state, which is $^1A''$. The lowest triplet, however, arises from a different transition and has a different structure, E.

```
H         O
 ·.      /
  C=C          E
 ◢
H
```

This is the $^3A'$ state, lower than $^3A''$. The corresponding $^1A'$ state is very high in energy.

Ketene is important photochemically as a source of methylene.

$$CH_2{=}C{=}O \xrightarrow{h\nu} CH_2 + CO \tag{26}$$

The ground-state reaction is strongly forbidden as a LM process, and occurs by a $C_s$ reaction path instead. As expected, 2700 A radiation gives chiefly singlet methylene, and 3800 A radiation gives chiefly triplet methylene.

The concept of chemical reduction is useful for more extended $\pi$ systems. Usually a predominant resonance structure is indicated, which should then resemble the excited state most closely. For example the excited state of acrolein should have the structure of the dianion of propionaldehyde.

$$CH_2{=}CH{-}CH{=}O^* = {}^-CH_2{-}CH{=}CH{-}CHO^-$$

The dianion would have the vinylic form, as if 1,4 reduction had occurred. Figure 13 shows the predicted excited-state structures for some aromatic compounds. The two CO groups of an excited α-diketone will be coplanar. In the case of excited benzil, the two CO groups are coplanar, but the phenyl groups need no longer be coplanar with them.

The excited state of benzene is predicted by the rule to have the same structure as $C_6H_6^-$. The ${}^3B_{1u}$ state of benzene is known to have an elongated structure of $D_{2h}$ symmetry with 1,4-hydrogen atoms bent out of the plane.[42] Three equivalent, fluctuating forms would exist. Less is known about $C_6H_6^-$, but it is believed to have $D_{2h}$ symmetry and the broadness of the e.p.r. spectral lines shows a fluctuating structure.[43] It is predicted to have a $D_{2h}$

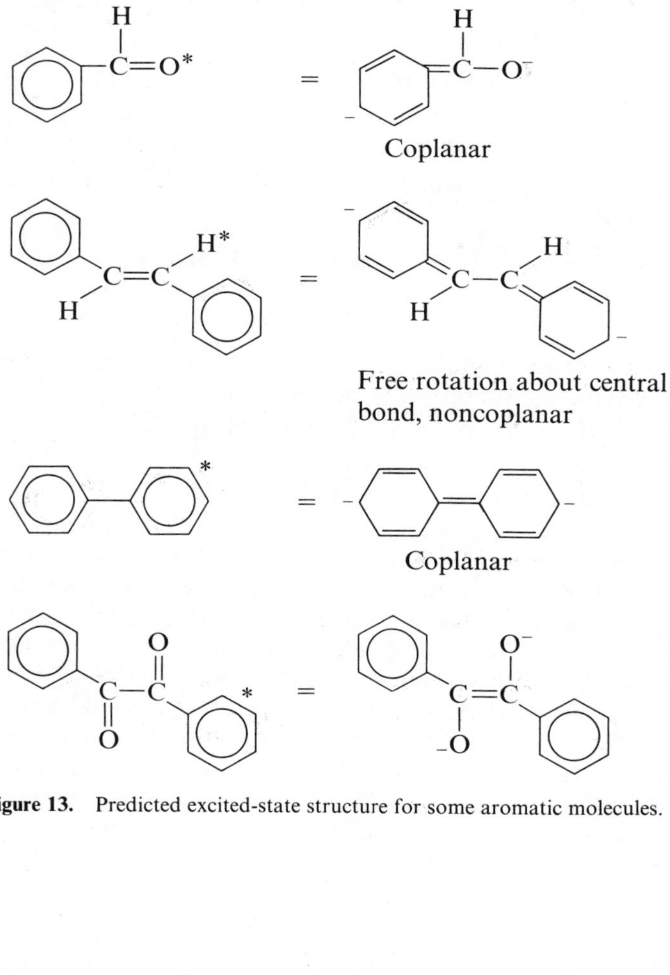

Coplanar

Free rotation about central bond, noncoplanar

Coplanar

**Figure 13.** Predicted excited-state structure for some aromatic molecules.

structure by virtue of a FOJT effect. One noteworthy feature is the smallness of the distortion from $D_{6h}$ symmetry in the case of excited benzene. The lowest singlet, ${}^1B_{2u}$, appears to be undistorted from $D_{6h}$ symmetry.

From a chemical point of view, adding electrons to a planar aromatic ring should usually cause some atoms to bend out of the plane. Essentially the ring is being reduced so that benzene becomes the dianion of cyclohexadiene, for example. Evidence is accumulating for out-of-plane deformation for the excited states of molecules such as 1,4-pyrazine and *p*-dibromobenzene and pyridine.[44] Polycyclic aromatic molecules are harder to distort from planarity.[45]

Addition of an electron to some organic molecules causes immediate dissociation. Examples are the organic halides (p. 413). Alkyl halides in their first excited state also dissociate with a quantum yield of unity.[46]

$$RX \xrightarrow{h\nu} R\cdot + X\cdot \tag{27}$$

Strained small rings, such as ethylene oxide and ethylene sulfide, would open up upon reduction.

$$H_2C\text{—}CH_2 \text{ (bridged by O)} \xrightarrow{2e} {}^-CH_2\text{—}CH_2\text{—}O^- \tag{28}$$

The excited states of ethylene sulfide have been calculated to have a ring-opened diradical structure (p. 337).

$$H_2C\text{—}CH_2 \text{ (bridged by S)} \xrightarrow{h\nu} S\cdot\text{—}CH_2\text{—}CH_2\cdot \tag{29}$$

A similar ring opening is indicated by the photochemistry of ethylene oxide.[46] The same products are observed as for the thermolysis, in which the diradical, $\cdot O\text{—}CH_2\text{—}CH_2\cdot$, is the presumed intermediate (p. 377).

### Transition-metal Complexes

The structures of excited states of simple complexes of the transition metals are easily predicted.[32] Consider octahedral complexes, such as $Cr(NH_3)_6{}^{3+}$. Both the *d–d* bands and the charge-transfer bands put an electron into the $e_g^*$ orbitals. The excited state should then have the structure of a similar complex with one electron more in these orbitals. $Cr(NH_3)_6{}^{3+}$ in its ${}^4T_{1g}$ or ${}^4T_{2g}$ states would then resemble $Cr(NH_3)_6{}^{2+}$. It would have a strong tetragonal distortion with two *trans* $NH_3$ molecules held more weakly than the other four. There is much evidence from photochemical studies that this is so, and some spectroscopic evidence. For the complex $MnF_6{}^{2-}$, also a

$d^3$ system, there is spectral evidence that the excited state has a tetragonal distortion, as predicted.

There is also spectroscopic evidence that a tetragonal distortion occurs in the excited states of $Fe(H_2O)_6{}^{2+}$, $CoF_6{}^{3-}$, and $Fe(H_2O)_6{}^{3+}$. The general rule is that a complex with 0 or 2 electrons in the $e_g^*$ orbital would become distorted in this way. However if 1 or 3 electrons are initially present [as in $Cu(H_2O)_6{}^{2+}$], the distortion already exists in the ground state and would be *removed* in the excited state.

A square planar $d^8$ complex such as $PtCl_4{}^{2-}$ or $Ni(CN)_4{}^{2-}$ will upon excitation take up the structure of a $d^9$, or possibly $d^{10}$, complex. It will then go to a $D_{2d}$ (squashed tetrahedron) or $T_d$ structure. There is much evidence, both of photochemical, and spectroscopic nature to support this prediction.[47] Linear $MX_2$ molecules will remain linear if the excited state is due to a $d$—$d$ band,[48] since filling the $d$ orbitals with electrons does not change the linearity. However a metal–ligand transition should cause bending. Thus the excited state of $HgCl_2{}^*$ should be bent, since the reference molecule is $PbCl_2$, which is bent. However $HgCl_2{}^*$ is reported to be linear.[35]

### Simple Excited-State Molecules

The structures of simple molecules $XY_n$ or $X_2Y_n$ can be accurately predicted from the number of electrons. This should make it possible to predict not only the structures of their excited states, but also their probably photochemistry. For example, $H_2O^*$ would have the structure of $HeH_2$. It should be linear and prone to dissociate. The lowest singlet of $H_2O^*$ is dissociative, and the lowest triplet, $^3\Pi_u$, is linear, and unstable.[49] The UV absorption spectrum of $H_2S$ shows an 1118 $cm^{-1}$ progression band (822 $cm^{-1}$ for $D_2S$).[50] This can only correspond to $\nu_2$, the bending mode. Therefore the bond angle in $H_2S^*$ is quite different from that of $H_2S$, but we cannot say it is linear.

The prediction that $NH_3{}^*$ is not only planar but T-shaped, should probably be modified. Since there is only a single electron in the orbital that promotes the $E'$ distortion,

$$(a_1')^2(e')^4(a_2'')^1(2a_1')^1(2e')^0$$

the tendency is reduced. There are two possible reactions of $NH_3{}^*$ to consider. The following correlations can be made considering a reaction coordinate that keeps the plane of symmetry.

$$^{1,3}A''\,NH_3{}^* \longrightarrow {}^2A''\,NH_2\cdot + {}^2A'\,H\cdot \tag{30}$$

$$\begin{aligned} ^3A''\,NH_3{}^* &\longrightarrow {}^3A''\,NH + {}^1A\,H_2 \\ ^1A''\,NH_3{}^* &\longrightarrow {}^1A''NH + {}^1A'\,H_2 \end{aligned} \tag{31}$$

That is, both singlet and triplet $NH_3{}^*$ can dissociate to ground-state $NH_2\cdot$ and H·, but only triplet $NH_3{}^*$ can give ground-state nitrene and molecular hydrogen. For the lowest excited state, (30) dominates experimentally.[46] A similar situation exists for $H_2O^*$ in that the lowest excited state gives only H and OH, just as $H_2S$ gives H and SH.

A different situation exists for $CH_4{}^*$, where even the lowest excited state gives predominantly carbene.[51]

$$CH_4{}^* \longrightarrow CH_2 + H_2 \tag{32}$$

This reaction is strongly forbidden in the ground state. Figure 14*a* shows the correlation diagram for (32) assuming a LM departure of $H_2$, maintaining a $C_{2v}$ point group. In principle the photochemical reaction becomes allowed in a $(t_2) \rightarrow (2t_2)$ excited state. The two orbitals are those which become $b_1$ and $3a_1$ in the products. This is a $T_2$ state in the tetrahedral point group. The mechanism would be one of crossover from the excited state surface to the ground-state surface at some intermediate point where these two orbitals cross.

There is a complication in that the structure of $CH_4{}^*$ is predicted to be that of $SH_4$, presumably the same as that of $SF_4$. A detailed calculation shows that the lowest triplet of methane is actually planar $D_{4h}$ in structure.[52] This is quite reasonable in view of Fig. 14*b*, which shows the correlation diagram for loss of $H_2$ from planar $CH_4{}^*$ by either a *cis* or *trans* departure of the two hydrogen atoms. The configuration would be

$$(a_{1g})^2(e_u)^4(a_{2u})^1(b_{1g})^1(2a_{1g})^0 \quad {}^3B_{2u}$$

with a large gap between the $a_{2u} - b_{1g}$ orbitals and the empty $2a_{1g}$. This minimizes the distortions leading to the $SF_4$ structure (see p. 187). The $(a_{2u}) \rightarrow (b_{1g})$ excitation, leading to a $D_{2d}$ structure, is forbidden because the single electrons already in each orbital have the same spin.

It can be seen in Fig. 14*b* that orbital crossings still exist for the ${}^3B_{2u}$ state of $CH_4{}^*$, both for *cis* and *trans* loss of $H_2$. This is consistent with the fact that triplet methylene reacts with $H_2$ only by abstraction.[52]

$${}^3B_1\,CH_2 + H_2 \longrightarrow CH_3\cdot + H\cdot \tag{33}$$

Presumably the methylene observed in the photochemical decomposition of $CH_4$ is the ${}^1A_1$ state coming from the excited singlet state of $CH_4{}^*$. The most probable course of events is that the ${}^1T_2$ state, which is Jahn–Teller unstable, decomposes along the Jahn–Teller allowed $T_2$ reaction coordinate, rather than the equally allowed $E$ coordinate (see p. 78). The $T_2$ coordinate can give rise to $CH_3 + H$, an observed set of products, or to $CH_2 + H_2$, by generation of the $C_{2v}$ point group.[53] Crossover to the ground state potential surface would occur en route to product.

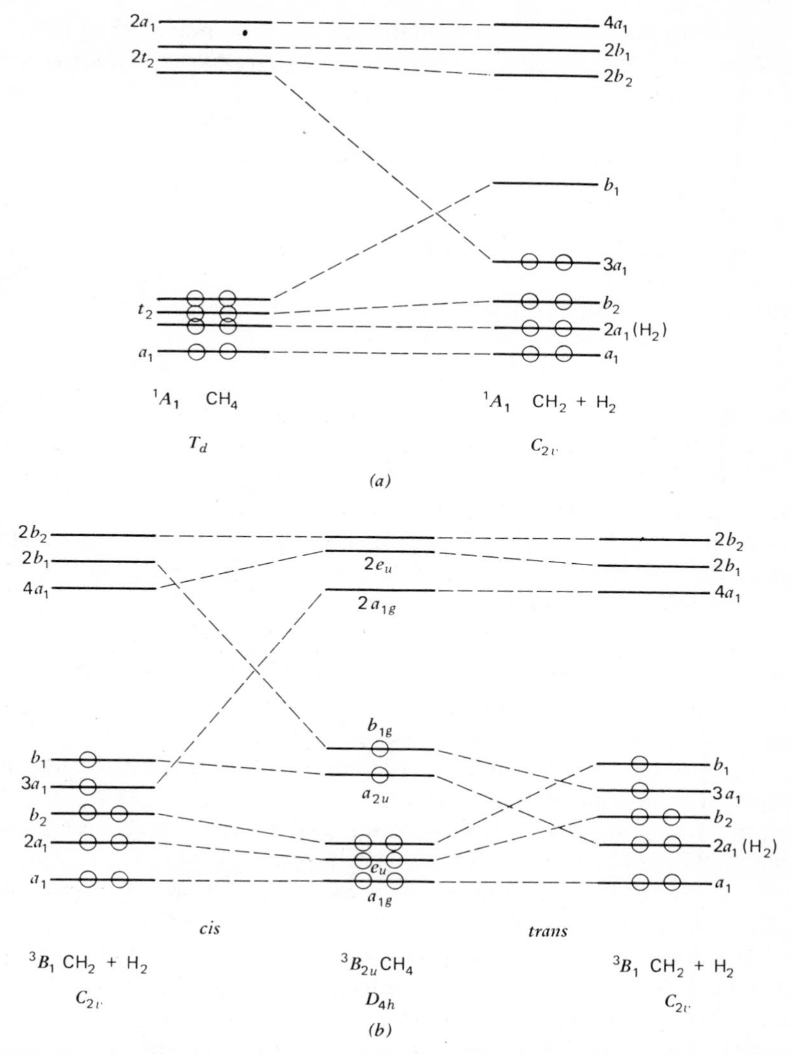


**Figure 14.** Orbital-correlation diagrams for loss of $H_2$ from $CH_4$ by a $C_{2v}$ LM path: (*a*) for the ground-state reaction giving singlet methylene; forbidden, but allowed in first singlet excited state; (*b*) for the reaction of the planar $^3B_{2u}$ state of methane: forbidden for loss of either two *trans* H atoms or two *cis* atoms (after reference 52).

Following the simple electron-addition rule, it is predicted that $SF_4^*$ will be planar like $XeF_4$, and $PF_5^*$ will have a square pyramidal structure like $IF_5$. An octahedral molecule such as $SF_6^*$ will have a fluctuating structure ($C_{3v}$ symmetry) like $XeF_6$. Unfortunately all of these excited states are unstable and dissociate immediately, so that structures are not known. However dissociation is not unreasonable in light of the properties of $XeF_4$ and $XeF_6$ as fluorine-atom donors.

Application of the rule to $C_2H_6^*$ leads to a surprise. The structure predicted is that of $N_2H_6$, which surely should dissociate into two ammonia molecules. Accordingly we predict

$$C_2H_6 + h\nu \longrightarrow 2CH_3. \tag{34}$$

This prediction is strengthened by examining the frontier MOs (p. 220). The first excited state is of $E_u$ species [54] and should come from an $e_g$ to $a_{2u}$ transition. The former orbital is weakly antibonding and the latter strongly antibonding in the carbon—carbon bond.

Indeed the $E_u$ state of ethane is a dissociative one. However the products are not methyl radicals (except in minor amount), but instead methyl carbene and hydrogen, primarily,

$$C_2H_6^* \longrightarrow CH_3CH + H_2 \tag{35}$$

In other words, ethane behaves like a substituted methane molecule, rather than as the prototype of its own series of molecules. Other saturated hydrocarbons also give carbenes, but a variety of reactions become possible photochemically, and all occur to some extent.[46]

It must be pointed out that we do not really know which MO transition generates the observed $E_u$ state. In addition to $(e_g) \rightarrow (a_{2u})$, $(a_{1g}) \rightarrow (e_u)$ and $(e_g) \rightarrow (e_u)$ are also possibilities. The latter transition corresponds to a strengthening of the C—C bond and a weakening of the C—H bond. This makes sense in explaining the photochemistry of (35). The excited state of ethane is unusual in showing a vibrational progression of 1150 $cm^{-1}$, corresponding to the $E_g$ vibrational mode of $C_2H_6$. This is the mode that converts into the diborane structure (p. 219). However such a structure does not explain the photochemistry. Perhaps intersystem crossing into a second state precedes photochemical decomposition.

The positive ion, $C_2H_6^+$, is formed in an $E_g$ state. In its formation by photoelectron spectroscopy it shows an 1170 $cm^{-1}$ vibrational progression. Both $E_u$ and $E_g$ states are unstable by virtue of a Jahn–Teller effect, the active mode being $E_g$. The structure of $C_2H_6^+$ has been probed theoretically.[55] It apparently has a $C_{2h}$ structure, somewhere between that of $C_2H_6$ and $B_2H_6$, but closer to the former.

Cyclopropane does not undergo ring opening upon illumination as ethylene oxide does, but instead eliminates $CH_2$ preferentially.[46]

$$\underset{H_2C\text{——}CH_2}{\overset{CH_2}{\diagup\ \diagdown}} \xrightarrow{h\nu} CH_2{=}CH_2 + {}^1A_1CH_2 \qquad (36)$$

Hydrocarbons normally do not eject a carbene from a nonterminal position, but polysilanes undergo the analogous reaction[56]

$$R_3Si{-}SiR_2{-}SiR_3 \xrightarrow{h\nu} R_3Si{-}SiR_3 + R_2Si \qquad (37)$$

Just as for (32), this kind of reaction becomes allowed photochemically, if there is no large change in geometry to complicate the process.

In spite of a few failures, it does seem that the Walsh rule for excited-state structures works quite well, especially when other methods do not seem to be available. Also photochemistry is predicted to some extent. One may worry about the half-filled orbital, previously the HOMO, which is ignored in this method. Indeed sometimes it should not be. In some polycyclic aromatic hydrocarbons, it appears that the excited-state structures depend primarily on this orbital.[57] The examples are all cases where the HOMO is one orbital of a degenerate set.

### Effects of Structural Change

Accepting that an excited-state molecule may have a structure quite different from that of the ground-state molecule, we must now ask how this affects the photochemical reactivity. More specifically, how does the photochemistry differ from what we might predict assuming no structural change? Before answering this question, a preliminary one to consider is: "Do structural changes have time to occur before photochemical reactions of a more extensive nature do?" The answer is affirmative, in the majority of cases.

Many structural changes take place in times of vibration, since they are essentially due to vibrational modes where the force constant has become negative. Times are of the order of $10^{-12}$ s, whereas most photochemical reactions require $10^{-9}$–$10^{-6}$ s. Other structural changes may have a small energy barrier, and be slower. For example, conformational changes in excited alkyl phenyl ketones occur in times of the same order as do $\alpha$-cleavage and $\gamma$-hydrogen abstraction.[58] The processes are competitive. Some unimolecular photochemical reactions, such as dissociation, may also occur in $10^{-12}$ s, since they also are in the nature of vibrations. This is likely only for very simple molecules. Even a bimolecular reaction of one excited molecule with a ground-state molecule could occur in $10^{-12}$ s, if the second molecule were the solvent.

A molecule with a different molecular shape may react differently. Nitro compounds in pyrolysis split the carbon—nitrogen bond, as a rule.

$$C_2H_5NO_2 \xrightarrow{\Delta} C_2H_5\cdot + NO_2\cdot \tag{38}$$

In photolysis, hydrogen atom abstraction reactions occur as well.

$$C_2H_5NO_2 \xrightarrow{h\nu} C_2H_4 + HNO_2 \tag{39}$$

In the ground state, the nitro group is coplanar with the carbon atom to which it is bound. In the excited state, the structure is puckered, similar to $RNF_2$. This allows an oxygen atom to get near enough to a $\beta$ hydrogen atom to abstract it. Similarly, the nonplanarity of excited ketone molecules facilitates intramolecular hydrogen abstraction.

Changes in shape may produce unexpected barriers to ground-state reactions. For example, bromoacetylene undergoes the same reaction both thermally and photolytically.

$$H{-}C{\equiv}C{-}Br \xrightarrow{\Delta} H{-}C{\equiv}C\cdot + Br\cdot$$

$$\begin{matrix} H & & \\ & \diagdown & \\ & & C{\equiv}C^* \\ & & & \diagdown \\ & & & & Br \end{matrix} \longrightarrow H{-}C{\equiv}C\cdot + Br\cdot \tag{40}$$

However the rate of the reaction for the bent excited state molecule is surprisingly small.[59] The explanation is a dynamic one—it is difficult to simultaneously conserve energy, linear momentum, and angular momentum in the nonlinear case. In effect there is a rotational barrier to (40).

A serious problem connected with geometric changes of excited states comes in the construction of correlation diagrams. It is customary to use the ground-state geometry to make correlation diagrams, even for excited-state reactions. This procedure may have no validity, if a large structural change occurs prior to the photochemical reaction. A moderate change in geometry, which leaves the point group unaltered, or which simply lowers the symmetry, will not invalidate the correlation diagrams. Such a change corresponds to a deformation of the MOs, without changing their essential topology. Symmetry restrictions will be relaxed somewhat, but not wiped out.

Large structural changes, particularly to a point group with new elements of symmetry, will cause the largest errors. We have seen an example of this in Fig. 14, where it is shown that a correlation diagram for $CH_4{}^*$ going to $(CH_2 + H_2)$ shows crossings, if the square planar structure for excited methane is used. In the original tetrahedral structure, the reaction is predicted to be allowed in the excited state. In the correct structure, at least for the triplet state, the reaction is still forbidden, just as it is for the ground state.

Another example is found in 1,2-elimination reactions of vinyl halides.

$$\begin{matrix} H & & H \\ & C{=}C & \\ H & & Cl \end{matrix} \xrightarrow{h\nu} \begin{matrix} H{-}C{\equiv}C{-}H \\ H{-}Cl \end{matrix} \tag{41}$$

The ground-state reaction is forbidden as a typical four-center process. The excited state that reacts in (41) is the usual $\pi,\pi^*$ state. If the correlation diagram is made using the planar geometry of the ground state, it will be seen that the excitation of a $\pi$ electron does nothing to make the reaction an allowed one. The forbidden character of (40) lies in the $\sigma$ electrons. If the correct twisted structure of $C_2H_3Cl^*$ is used, the correlation diagram will show that (41) is allowed as an excited state reaction.

## SOME PHOTOCHEMICAL REACTIONS

The behavior of olefins upon illumination has been extensively studied. There are fragmentation reactions in which H atoms and $H_2$ molecules are produced, just as for saturated hydrocarbons. Of more interest, however, are *cis-trans* isomerization of olefins and cycloaddition of two olefin molecules. The former reaction is a direct consequence of the twisted geometry characteristic of an excited olefin molecule. Figure 15 shows a potential energy diagram for some states of ethylene as a function of the twist angle, $\theta$, about the carbon—carbon bond.[60] The height of the ground-state barrier at $\theta = 90°$, twisted form, comes from the activation energy for thermal *cis–trans* isomerization.

According to Fig. 5, excitation to either $S_1$ or $T_1$ should lead to efficient isomerization. The quantum yield should be $\varphi = 0.50$, assuming that deactivation cannot occur before the stabilizing geometry change. Irradiation of 1,2 dideuterioethylenes, either *cis* or *trans*, does lead to isomerization from both $S_1$ and $T_1$. The quantum yields from the triplet state are close to, but not exactly, 0.50, These results are completely consistent with Fig. 15. However the behavior of substituted olefins is not so simple. The isomerization of *cis* and *trans* stilbene has been very widely studied, and can be used as an example of the problems encountered.

In hydrocarbon solvents, either isomer upon irradiation gives a photostationary state which favors the *cis* isomer.

$$\underset{trans}{\begin{matrix} H_5C_6 & & H \\ & C{=}C & \\ H & & C_6H_5 \end{matrix}} \overset{h\nu}{\rightleftharpoons} \underset{cis}{\begin{matrix} H_5C_6 & & C_6H_5 \\ & C{=}C & \\ H & & H \end{matrix}} \tag{42}$$

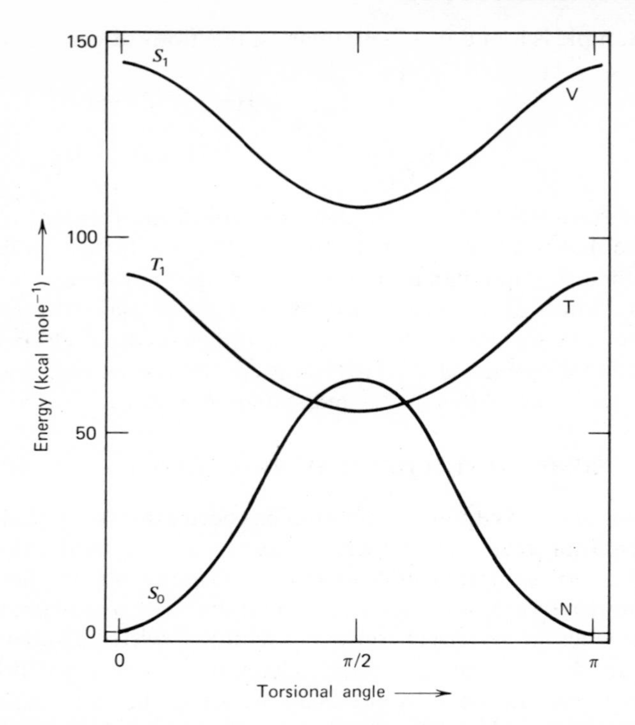


**Figure 15.** Energy of three lowest states of ethylene as a function of the twist angle, $\theta$ (the labels N, T, and V are commonly used for the three states; in $D_{2h}$, the labels are $^1A_g$, $^3B_{1u}$ and $^1B_{1u}$, respectively) (based on reference 60).

In the absence of light, the *trans* isomer is more stable. The singlet excited-state from the *trans* isomer fluorescences, and that from the *cis* does not. Fluorescence competes with isomerization and at low temperatures only fluorescence is found. At high temperatures, only isomerization occurs, with $\varphi = 0.50$. The evidence is that isomerization occurs directly from either *cis* or *trans* excited singlet and does not require intersystem crossing.[61]

Thus it appears that different, $S_{1c}$ and $S_{1t}$, states exist for each isomer, rather than a single, twisted state, as suggested by Fig. 15. This figure is drawn assuming no bending of the methylene groups after twisting. The structure of $C_6H_5CH{=}CHC_6H_5$* is expected to be that of hydrazobenzene, which is definitely not all coplanar. Approximate MO calculations predict that the excited state is twisted about the double bond,[63] although of course the minimum does not occur at a 90° twist. Two explanations, at least, exist for the failure of $S_{1c}$ to be identical with $S_{1t}$.

One is that, because of $\pi$ conjugation to the phenyl rings, there is a barrier to twisting, even though the twisted form is more stable. The barrier must be larger for the *trans* isomer than the *cis*.

The other possibility is to consider the bending of the groups attached to carbon, as well as the twist. This gives rise to two stereoisomeric forms. These are shown as Newman projections of one possible pair of conformers.

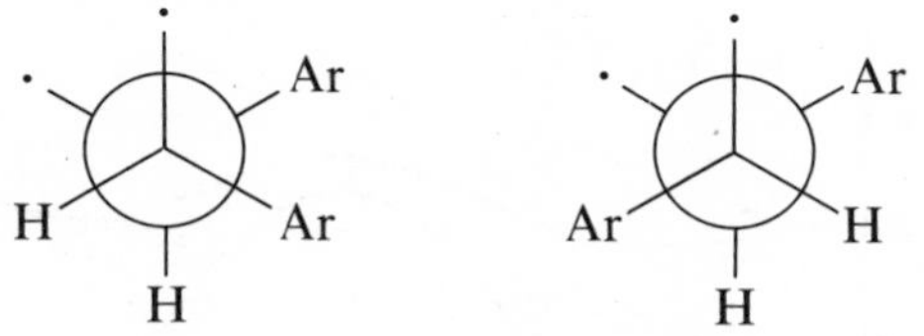

Ar Ar H H Ar Ar H H

$S_{1c}$ $S_{1t}$

Interconversion would require not only rotation about the carbon—carbon bond, but also inversion at one carbon atom. While these are fairly rapid processes for free radicals, such as we essentially have here, they could well have lifetimes of $10^{-8}$–$10^{-9}$ s, needed to explain the results. The $S_{1c}$ structure would have to be the least stable to be consistent. While different triplet state conformers would also exist, the long lifetime of the triplet state would guarantee that isomerization would occur. A plausible energy diagram is shown in Fig. 16.

An objection to the above explanation is that one would expect each aryl group to be coplanar with the carbon atom to which it is bound because of delocalization of the odd electron into the ring. This would be similar to the benzyl radical, for example. However it is not neceesary to have complete coplanarity in order to have substantial $\pi$ overlap. As an illustration, while excited aniline is coplanar, ground-state aniline is not.[64] The $NH_2$ group is depressed 46° below the phenyl plane. Since the overlap of two $p$ orbitals on adjacent atoms depends on $\cos\theta = 0.69$,

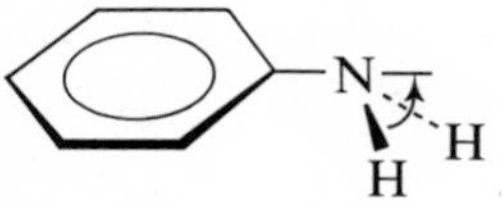

N H H $\theta = 46°$

there is still considerable $\pi$ bonding.

## Cycloaddition Reactions

The explanation for *cis–trans* isomerization based on out-of-plane bending is not original, since it was first postulated to explain the stereospecific cycloaddition reactions of *cis* and *trans*-2-butene.[65] If a liquid 2-butene is

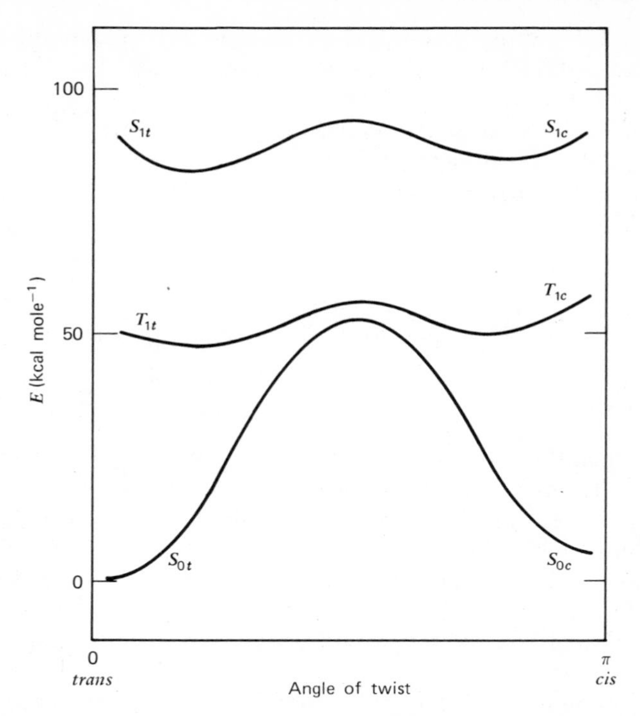


**Figure 16.** Potential energy of three lowest states of stilbene as a function of the twist angle, $\theta$; the assumption is that the excited state have pyramidal structures at each carbon atom (energy is a maximum at $\theta = 90°$).

irradiated, there is a large amount of *cis–trans* isomerization, some fragmentation, and some dimerization to 1,2,3,4-tetramethylcyclobutane.[66] Four isomers are possible, F, G, H, and I.

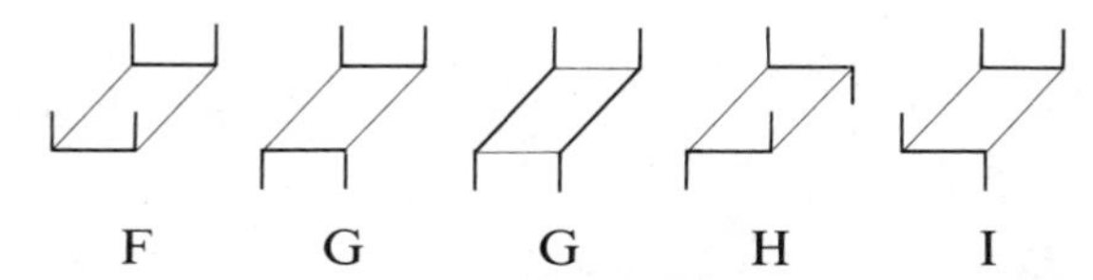


Only F and G are formed by dimerization of pure *cis*-2-butene. Only G and H are formed from pure *trans*-2-butene, and I can only be formed from a mixture of *cis* and *trans*-2-butene. In other words, the cyclodimerization is completely stereospecific, with retention of configuration of the original olefin. Condensation is suprafacial in both olefin molecules.

Such stereospecificity is not expected if a twisted structure is formed in the excited state that is the same for both isomers. Different structures, $S_{1c}$ and

$S_{1t}$, must be formed first, and these must react with a second molecule of olefin (ground state) before the barriers to interconversion are overcome. This is not unlikely when it is noticed that the neat liquid olefin is used, and that only three or four molecules of cyclodimer are formed for every 100 molecules that simply isomerize.

The dimerizations that are found result from the $S_1$ state, since direct formation of $T_1$, or intersystem crossing, is not efficient for simple olefins. However the triplet state can be specifically formed by the use of sensitizers. These are excited-state atoms or molecules that transfer their excess energy to a ground-state molecule of another kind, for instance,

$$^3P_1\,\mathrm{Cd}^* + {}^1A_{1g}\,\mathrm{C_2H_4} \longrightarrow {}^1S_0\,\mathrm{Cd} + {}^3B_{1u}\,\mathrm{C_2H_4}^* \qquad (43)$$

A triplet sensitizer will usually give a triplet product. Also the excitation energy of the sensitizer can be selected so that it is large enough to form $T_1$, but not large enough to form $S_1$. Triplet-state ketones are often used as sensitizers, a technique developed by Hammond.

If the $T_1$ state of a simple acyclic olefin is formed in solution, the primary reaction is *cis–trans* isomerization. Cyclodimerization is not found. There is time for structural equilibration because of the long lifetime of the triplet state, but ring closure is inhibited by the presence of unpaired spins.

The dimerization of an olefin is a typical four-center reaction forbidden in the ground state and made allowed in the excited state. Figure 7 can serve as a correlation diagram for suprafacial–suprafacial addition. Such a diagram would have little meaning in cases where the excited olefin had become equilibrated before reacting. We must assume that similar situations exist for other cases where ground-state reactions of olefins that are forbidden, are observed photochemically.

This would be the case, for example, in the electrocyclic ring closure of butadienes, which must compete with isomerizations. In order for the terms "disrotatory" and "conrotatory" to have meaning, ring closure must occur while the original structure is essentially maintained, though badly distorted. It is not surprising that photochemical reactions are likely to give a mixture of products, rather than the single product predicted by symmetry rules.[67]

The (4 + 2) cycloaddition, which is the Diels–Alder reaction, is predicted to occur suprafacial on one component and antarafacial on the other in the photochemical case. Actually both kinds of product are formed, suprafacial–suprafacial and suprafacial–antarafacial, depending on the system.[68] A diradical mechanism is indicated, which is initiated by the excited-state diene adding to one end of the olefin. The diradical thus formed equilibrates by rotation about single bonds before ring closing.[69]

Of course, in many cases of cycloaddition, and so on, the olefinic bonds are part of a ring system in which twisting of the bond is not possible. An example would be the isomerization of norbornadiene to quadricyclane.[70]

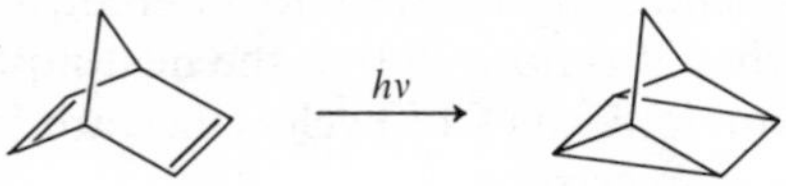

$$\xrightarrow{h\nu} \qquad (44)$$

In this case predictions based on symmetry rules are valid. In other cases where structural factors intervene, these factors may become dominant. Thus the same products may be formed in both thermal and photochemical reactions, contrary to the rules.[71] Or reactions which are forbidden photochemically may, in fact, occur.[72]

In short, rules based on either symmetry properties or aromatic–antiaromatic character in the TS, are not likely to be as reliable for photochemical reactions as they are for ground-state reactions. A change in geometry in the excited state may be the complicating factor. In other cases it is simply that photoexcited molecules are very energetic. They have too many options open to them, including that of becoming vibrationally excited ground-state molecules. The general prediction that can be made is that a reaction forbidden by symmetry in the ground state is very likely to occur in an excited state. However other reactions may still be more efficient.

Photolysis of ethylene as a vapor gives acetylene and hydrogen as major products.[73] One of the elementary reactions appears to be direct loss of $H_2$

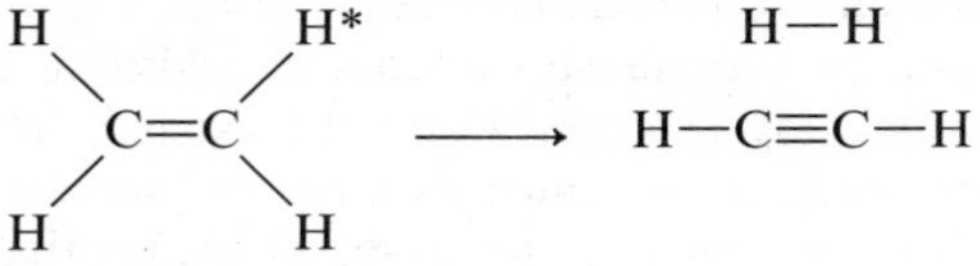

$$H_2C{=}CH_2^* \longrightarrow H{-}C{\equiv}C{-}H + H{-}H \qquad (45)$$

$$a_1 + b_1 \qquad\qquad 2a_1$$

In the ground-state reaction (45) is forbidden, as the bond symmetry analysis shows. In the $\pi,\pi^*$ excited state, an electron from a $b_1$ orbital (in $C_{2v}$) is excited to an $a_2$ orbital. This obviously does not help to overcome the ground state symmetry barrier, which resides in a $b_2$–$a_1$ crossing. However twisting to the $D_{2d}$ (or $D_2$ or $C_2$) state causes the reaction to be allowed. A twofold axis is preserved in the LM path.

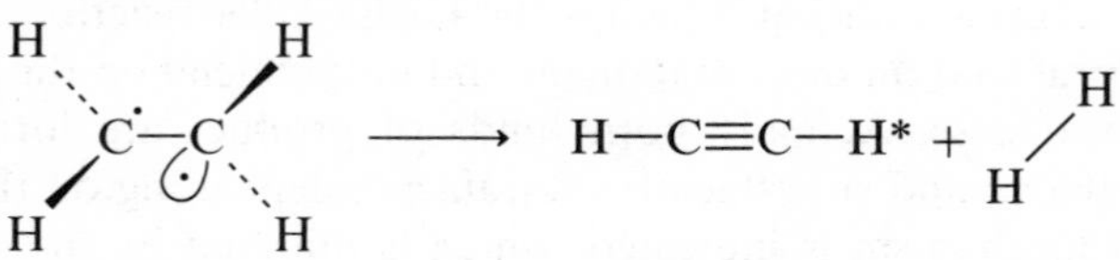

$$H_2\dot{C}{-}\dot{C}H_2 \longrightarrow H{-}C{\equiv}C{-}H^* + H{-}H \qquad (46)$$

$$3a + 3b \qquad\qquad 3b + a \qquad\qquad 2a$$

The number of electrons in orbitals of $a$ and $b$ species is shown. The acetylene is in its $\pi,\pi^*$ state. Reaction (46) can also occur by crossing over to the vibrationally excited ground-state surface.

### Exciplex Formation

Earlier it was mentioned that excited ketone molecules will add to electron-poor olefins by way of the odd electron on carbon. Conversely, electron-rich olefins will add by way of the odd electron on oxygen. In some cases, at least, the initial interaction is that of forming a $\pi$ complex.[74] This is an example of exciplex formation, a molecular complex that exists only in the excited state. If the complex is between an excited molecule and a ground-state molecule of the same kind, it is called an "excimer." The $\pi$ complex, if formed, rearranges in any case to a more stable diradical. The correlation between the original reactants and the diradical is as follows:

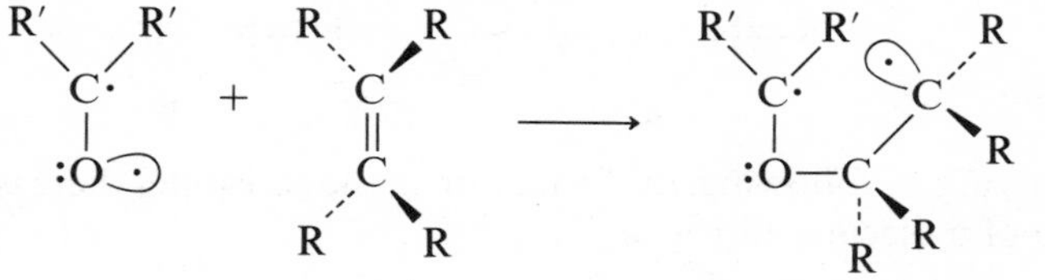

$$R'_2\dot{C}-\ddot{O}\cdot \;+\; R_2C{=}CR_2 \longrightarrow R'_2\dot{C}-\ddot{O}-CR_2-\dot{C}R_2 \quad (\text{J}) \qquad (47)$$

$$\sigma + 3\pi \qquad\qquad 2\sigma \qquad\qquad 3\sigma + 3\pi$$

The excited state reactants correlate directly with the ground-state diradical.[5] The ground-state reactants correlate with a zwitterionic excited state of the product.

The diradical J can now ring close to form an oxetane, or it can dissociate either into an excited state olefin and a ground-state ketone, or back into the reactants.

$$\text{J} \longrightarrow \begin{array}{c} R_2'C-CR_2 \\ |\quad\;\; | \\ O-CR_2 \end{array} \qquad (48)$$

$$\text{J} \longrightarrow R_2'CO + R_2C{=}CR_2{}^* \qquad (49)$$

If the excited ketone was originally in the $S_1$ state, the oxetane may be formed with some retention of configuration at the olefin moiety.[61,68] The $T_1$ state forms oxetane nonstereospecifically.

More important, the $T_1$ state of the ketone provides a way to form the excited triplet state of the olefin via reaction (49).[75] This in turn provides a mechanism for *cis–trans* isomerization of the olefin, which is commonly found. Of course triplet excitation of the olefin may also result from either simple collision or exciplex formation. Also isomerization of the olefin may result from bond rotations in J, followed by reversion back to excited-state ketone and ground-state olefin (the Schenck mechanism).

It appears that prior exciplex formation is quite common in the cycloaddition reactions of olefins, ketones, dienes, and enones. The concept can explain the stereoselectivity found in unsymmetrical cases. Many quenchers seem to act by deactivating the exciplex, causing it to dissociate. Also temperature effects on photochemical reactions can often be explained by invoking the reversible, exothermic formation of an exciplex.

### Azo Compounds

Aliphatic azo compounds also have an interesting photochemistry.[76] Normally only the *trans* isomers can be synthesized, but irradiation produces the *cis* isomer, with quantum yields normally less than 0.50.

$$\text{R—N=N—R } (trans) \xrightarrow{h\nu} \text{R—N=N—R } (cis) \qquad (50)$$

Accompanying, or following, the formation of the *cis* isomer there is extensive formation of molecular nitrogen

$$\text{R—N=N—R}^* \longrightarrow \text{N}\equiv\text{N} + \text{R—R} \quad \text{or} \quad 2\text{R}\cdot \qquad (51)$$

The *trans* isomer decomposes thermally according to (51), with no detectable isomerization intervening. The *cis* isomers, when isolable, are much more unstable to thermolysis.

In (51) there is much evidence that free radicals are formed, since the yield of coupled product, R—R, is considerably less than that of $N_2$. Nevertheless, it is possible that some R—R is formed directly. The structure of $R_2N_2^*$ is predicted to be that of a dialkylperoxide, with a skew $C_2$ shape, K.

$$\underset{\text{K}}{\text{R—}\dot{\text{N}}\text{—}\dot{\text{N}}\text{—R}} \longrightarrow \begin{array}{l}\text{isomerization or}\\ N_2 + 2\text{R}\cdot \quad \text{or}\\ N_2 + \text{R—R}\end{array} \qquad (52)$$

Intermediate K has several properties which explain the photochemistry. It is a logical intermediate for *cis–trans* isomerization, since it is halfway between each structurally. Also the concerted elimination of R—R, or of 2R·, becomes an allowed process, just as in (46).

The final advantage lies in the obliteration of the distinction between the $n,\pi^*$ and $\pi,\pi^*$ excited states of the parent azo compound. The MO configuration of the ground state would be, in $C_{2h}$,

$$\underset{\sigma}{(a_g)^2}\underset{\sigma}{(b_u)^2}\underset{\sigma}{(2a_g)^2}\underset{n}{(2b_u)^2}\underset{\pi}{(a_u)^2}\underset{n}{(3a_g)^2}\underset{\pi^*}{(b_g)^0}$$

In the new $C_2$ point group, both $A_u$ and $A_g$ become $A$, and the $\pi$ and $n$ MOs are mixed. The reactive state is thus a $^1B$ state. The $^3B$ state is also reactive, but normally quantum yields are less, both for isomerization and decomposition.[76] The interest in which state is photochemically active comes about because of a paradox. The $n,\pi^*$ is the only singlet state that is energetically available, yet it is the $\pi,\pi^*$ state that allows easy twisting about the double bond.

### The Photochemistry of Benzene

In recent years benzene has been found to have a rich photochemistry. Most products are formed only in small amounts and were difficult to detect. Substituted benzenes sometimes give much better yields, but we will discuss only benzene. The ground state of $C_6H_6$, $^1A_{1g}$, is $(a_{2u})^2(e_{1g})^4(e_{2u})^0(b_{2g})^0$, considering only the $\pi$ orbitals. The lowest excited states arise from the $(e_{1g}) \rightarrow (e_{2u})$ transition. Since $(E_{1g} \times E_{2u}) = (B_{1u} + B_{2u} + E_{1u})$, the lowest excited states are $^{1,3}B_{1u}$, $^{1,3}B_{2u}$, and $^{1,3}E_{1u}$.[77a] The $^1B_{2u}$ state is the lowest singlet and is centered at 2600 A (4.9 eV). The next singlet at 2000 A is usually considered to be the $^1B_{1u}$, but it may be the $^1E_{2g}$ from the $(e_{1g}) \rightarrow (b_{2g})$ transition. The $^1E_{1u}$ state lies at 1800 A.

These excited states are described by symmetry adapted combinations of configurations.[77b] The wavefunctions are shown in Table 2. The molecular orbitals $e_{1g}$, $e'_{1g}$, $e_{2u}$, and $e'_{2u}$, are also shown in Fig 28 of Chapter 4 (p. 343). The states differ in energy because of different interelectronic repulsions. The corresponding triplet states lie lower than their singlet counterparts, but it is the singlet states that seem to give the interesting photochemistry.

Knowing the nature of the excited states, the question is whether the photochemistry of benzene can be rationalized in terms of their symmetries. The naive approach is to construct a least motion reaction path for the formation of an observed product. If the reaction is forbidden, the orbitals or bonds responsible are identified. One electron is then promoted from a mismatched orbital in such a way as to reduce the orbital mismatch to a one electron one. Theoretically this allows the reaction to go unhindered from the excited state of reactant to the excited state of product. Also, as we have seen, it makes it very likely that a curve crossing or touching makes it possible to form the ground-state reactant in an energy rich condition. This can then react on the ground-state surface.

The following isomerization reactions have been observed for benzene, or its derivatives:[78]

$$\text{Benzene} \xrightarrow{h\nu} \text{Fulvene} \qquad (53)$$

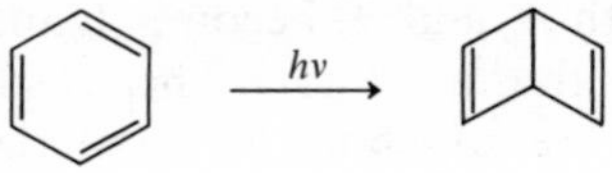

$\xrightarrow{h\nu}$ Dewar benzene (54)

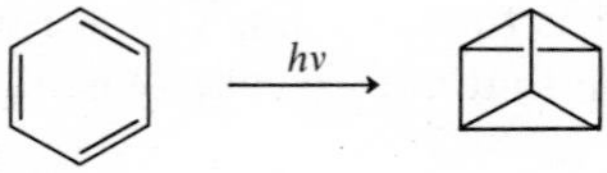

$\xrightarrow{h\nu}$ Prismane (55)

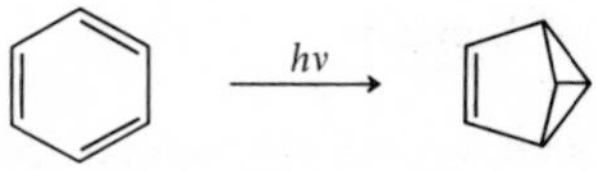

$\xrightarrow{h\nu}$ Benzvalene (56)

Concerted mechanisms can be envisaged for (53)–(56). The formation of Dewar benzene and of prismane could occur by way of a $C_{2v}$ reaction path. Figure 17 shows the geometric changes and the symmetries of the bonds made and broken. Both reactions are forbidden in the ground state, but (54) could occur by way of an excited state from a $(b_1) \rightarrow (a_1)$ transition, that is, a $B_1$ excited state.[79] The $B_{1u}$ state in $D_{6h}$ correlates with the $B_1$ state in $C_{2v}$, and $B_{2u}$ correlates with $B_2$. This can be checked by noting that the $e_{1g}$

**Table 2 Excited-state Wavefunctions for Benzene and Symmetries of Constituent Molecular Orbitals**

$$B_{1u} = \frac{1}{\sqrt{2}}(\psi_3 + \psi_4) \qquad B_{2u} = \frac{1}{\sqrt{2}}(\psi_1 - \psi_2)$$

$$E_{1u} = \frac{1}{\sqrt{2}}(\psi_1 + \psi_2) \quad \text{and} \quad \frac{1}{\sqrt{2}}(\psi_3 - \psi_4)$$

$$\psi_1 = (e_{1g})(e'_{2u}) \qquad \psi_3 = (e_{1g})(e_{2u})$$

$$\psi_2 = (e'_{1g})(e_{2u}) \qquad \psi_4 = (e'_{1g})(e'_{2u})$$

| | | |
|---|---|---|
| + <br> +    + <br> −    − <br> − | $e_{1g}$ | + <br> −    − <br> −    − <br> + | $e_{2u}$ |
| 0 <br> +    − <br> +    − <br> 0 | $e'_{1g}$ | 0 <br> +    − <br> −    + <br> 0 | $e'_{2u}$ |

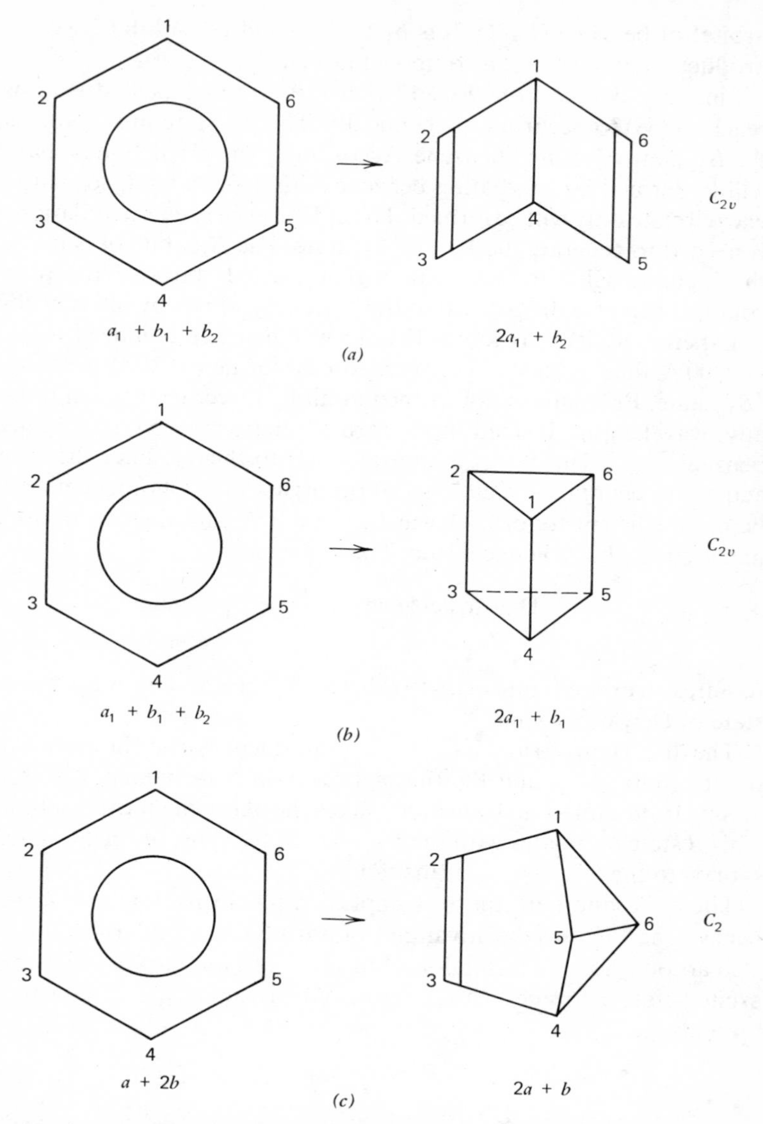


**Figure 17.** Bond- and orbital-symmetry correlations for some isomerizations of benzene—only the $\pi$ orbitals of benzene need be considered: (*a*) formation of Dewar benzene, $C_{2v}$ reaction path; (*b*) formation of prismane, $C_{2v}$ reaction path; (*c*) formation of benzvalene, $C_2$ reaction path—the $C_2$ axis bisects the 5—6 bond (see reference 74).

orbital of benzene (Table 2) is $b_1$ in $C_{2v}$, and $e_{2g}$ orbital becomes $a_1$. The product $(e_{1g})(e_{2u})$ is $\psi_3$, a component of $B_{1u}$.

Similarly we see from Fig. 17 that a $(b_2) \rightarrow (a_1)$ excited state will allow reaction (55) to occur photochemically. This $B_2$ state in $C_{2v}$ correlates with the $B_{2u}$ state of excited benzene. Accordingly the prediction is that prismane will be formed by irradiating benzene with 2600 A light, so that the lowest excited state only will be formed. To get Dewar benzene, irradiation at 2000 A is needed to generate the $B_{1u}$, or $S_2$, state. The $E_{1u}$ state of benzene, if it lies the highest, will correlate with higher excited states of the products. Of course it can also drop down to the $B_{1u}$ or $B_{2u}$ states by internal conversion.

Experimentally it is found that Dewar benzene is indeed formed better at 2000 A, than at 2600.[79] However some is formed at 2600, possibly from the $^3B_{1u}$ state. Prismane is not formed at all by direct irradiation of benzene at any wavelength. Instead it is formed indirectly by irradiating Dewar benzene.[78c, 80] This is not a contradiction to theory, since there may be an activation energy for reaction (55) occurring even photochemically. It may be more efficient to form Dewar benzene first, and then go on to prismane in a second photochemical step. The conversion

$$\begin{array}{ccc} \text{Dewar benzene} & \xrightarrow{h\nu} & \text{prismane} \\ 2a_1 + b_2 & & 2a_1 + b_1 \end{array} \tag{57}$$

requires an excited state of $(b_2) \rightarrow (b_1)$ or $A_2$ species. This is the lowest excited state of Dewar benzene.[79]

The direct conversion of $C_6H_6$* to benzvalene is also shown in Fig. 19. The point group is $C_2$, and the thermal reaction is forbidden. An excited state arising from a $(b) \rightarrow (a)$ transition makes the photochemical reaction allowed. This $B$ state correlates with the $B_{1u}$ state of benzene, or the $S_2$ state. It corresponds to the $(e'_{1g}) \rightarrow (e'_{2u})$ transition.

There is another, more complex mechanism for the formation of benzvalene.[81] It has the advantage of postulating an intermediate which can also account for the formation of fulvene, and for certain other reactions of excited-state benzene. This intermediate is an allylic biradical, L, called "prefulvene."

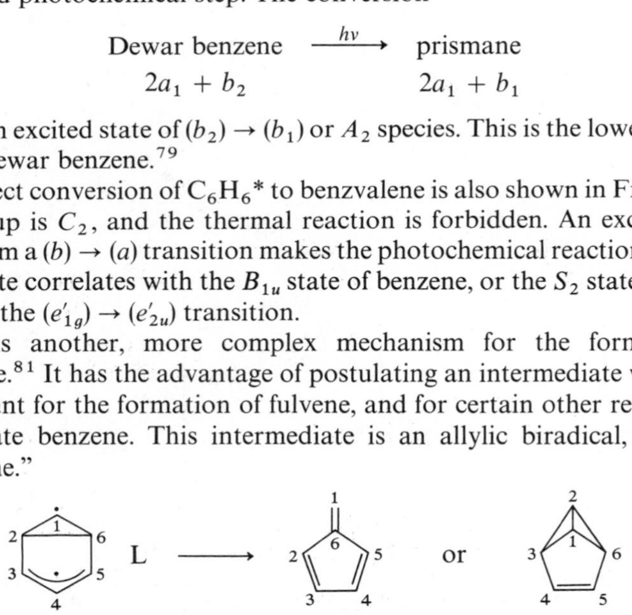

L ⟶ (fulvene) or (benzvalene) (58)

The formation of fulvene from L requires a hydrogen atom to shift from carbon atom 6 to carbon atom 1, and the breaking of the 1—2 bond. Also benzvalene can easily be formed from L by carbon atom 1 bonding to carbon atom 3. Each is a free-radical center.

An examination of the bond symmetries in benzene and in L in the $C_s$ point group, shows the ground-state reaction to be forbidden.

$$4a' + 2a'' \longrightarrow 5a' + a'' \tag{59}$$

The number of electrons in orbitals that are symmetric and antisymmetric is shown. An $(a'') \rightarrow (a')$ excited state would allow the photochemical reaction to occur. The resulting $A''$ state correlates with the $B_{2u}$ state of benzene, or $S_1$.[81] It comes from the $(e'_{1g}) \rightarrow (e_{2u})$ transition in benzene. Actually 2600 A radiation does not produce much fulvene or benzvalene, but this is explained by assuming that vibrationally excited $S_1$ benzene is needed.[78a] The vibrational excitation comes from deactivation of the $S_2$ or $S_3$ states.

### Reactions of Excited-state Benzene and Olefins

Irradiation of a mixture of benzene and olefin with 2540 A light leads to the formation of cycloaddition products.[78a, 82] Similar results occur with benzene and acetylenes. The products with olefins are the result of 1,2 and 1,3 addition mainly, with minor amounts of 1,4 product. The addition reactions are stereospecific for the olefin and result from suprafacial, or *cis*, addition. Simple olefins give chiefly 1,3 addition products, and olefins with either electron-attracting or electron-donating groups give chiefly 1,2-addition.[83] The singlet excited states of benzene are involved in cycloaddition and triplet states seem to give largely *cis–trans* isomerization of the olefin.[84]

The 1,2-addition of an olefin is forbidden as a ground-state reaction. The point group is $C_s$ for a LM reaction path.

$$2a' + a'' \quad + \quad a' \longrightarrow 2a' + 2a'' \tag{60}$$

(··· is plane of symmetry)

To make the reaction allowed photochemically, an $(a') \rightarrow (a'')$ excited state is needed. The occupied MO of benzene that is symmetric to the indicated plane of symmetry is $e'_{1g}$. The empty MO which is antisymmetric is $e'_{2u}$. Therefore the excited state corresponds to the wavefunction $\psi_4$, which relates to the $B_{1u}$ state. Therefore this reaction should not occur at 2540 A, where only the lower energy $B_{2u}$ state is excited. Reaction (55) would be allowed for excited ethylene, which is an $A''$ state. The $^1B_{1u}$ state of ethylene is centered at 1850 A, but in the liquid-state absorption occurs down to 2600 A.

The more abundant 1,3-addition should occur at 2540 A, however, if it postulated that prefulvene, L, is the form of $C_6H_6$* which actually reacts.[85]

(61)

Unlike isomerization to benzvalene or to fulvene, it appears that vibrationally excited L is not needed for this bimolecular reaction. The infrequent 1,4-addition of an olefin to benzene is allowed in the ground state. It could result from vibrationally excited $S_0$ benzene produced by radiationless decay of $S_1$.

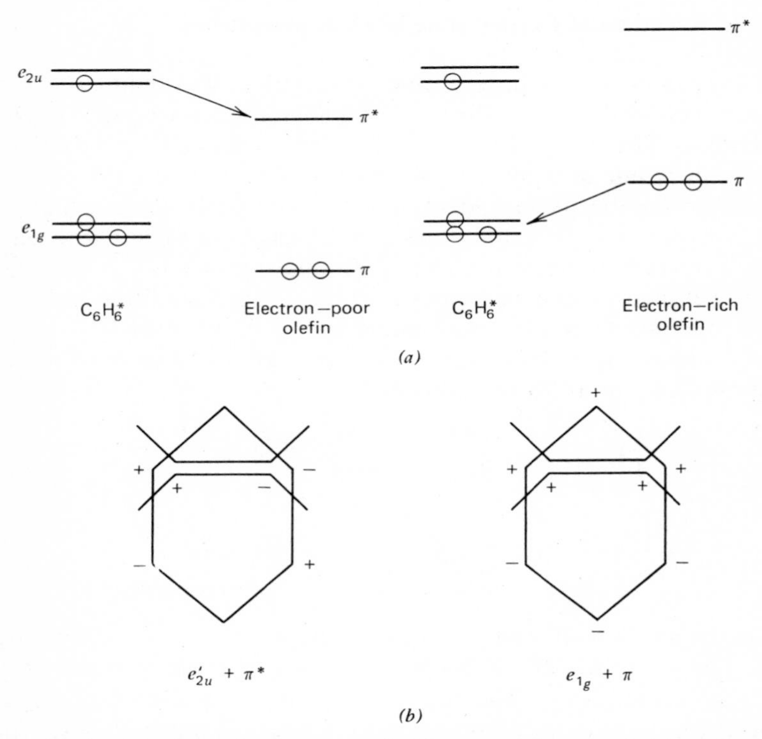


**Figure 18.** Exciplex formation between excited benzene and a ground-state olefin molecule: (*a*) major orbital interactions (arrows) when olefin is a good electron acceptor (left side) and when olefin is a good electron donor (right side); (*b*) favorable overlaps for a 1,3 orientation of olefin with a $^{1,3}B_{2u}$ benzene molecule; phases of relevant $\pi$ orbitals shown.

The cycloaddition reactions of benzene and olefins are complicated by the probable formation of exciplexes, which can then be the actual reactants for cyclization.[86] Formation of the $B_{2u}$ state of benzene creates both a good accepting orbital with a vacancy, and a good donating orbital with an electron in it. Figure 18 shows the major orbital interactions with olefins that are substituted so as to be good electron donors or acceptors. Orbital symmetry can be used to predict the most favorable orientations for exciplex formation, but substituents obviously play a major role, both electronically and sterically.

For benzene in the $B_{2u}$ excited state and for ground-state ethylene, the most stable exciplex would have a 1,3 orientation, as shown in Fig. 18. The $e_{1g}$ orbital of $C_6H_6{}^*$ acts as an acceptor for the $\pi^*$ orbital of $C_2H_4$. There is no 1,2 orientation of the olefin which has both good overlap for both kinds of interaction. However if the olefin were substituted to be a good donor or acceptor primarily, then a 1,2 complex could be formed.

The cycloaddition of *cis*-butadiene to benzene such that the bonded atoms are 1,4 in butadiene and 1,2 in benzene, is an allowed thermal reaction.

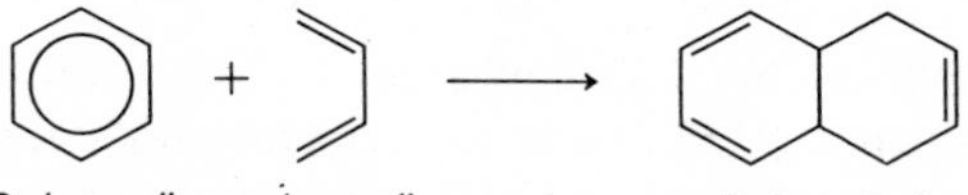

$2a' + a''$ + $a' + a''$ ⟶ $3a' + 2a''$ (62)

It is essentially a Diels–Alder reaction, with a six-electron TS. However 1,4-addition at both reactants is forbidden in the ground state, both for *cis* and *trans* butadiene

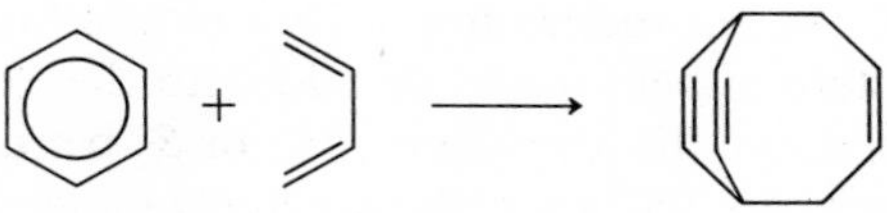

$2a' + a''$ + $a' + a''$ ⟶ $4a' + a''$ (63)

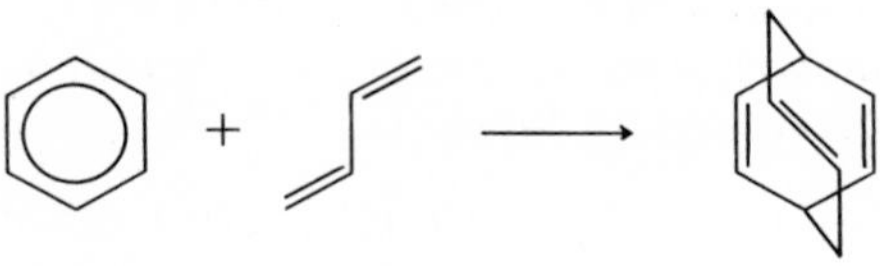

$a + 2b$ + $a + b$ ⟶ $3a + 2b$ (64)

It is assumed that a plane of symmetry is maintained in (63) and a $C_2$ axis in (64). The twofold axis passes through the center of both molecules, with the diene directly above the benzene ring. In (63), an excited state due to $(a'') \rightarrow (a')$ will have an allowed reaction course. This is the $(e_{1g}) \rightarrow (e_{2u})$ process that creates the $B_{1u}$ state, as shown in Table 2. The lowest-energy

$B_{2u}$ state still has (63) as a forbidden reaction. However *trans* butadiene can condense with $C_6H_6^*$ as in (64) both from the $B_{1u}$ state and the $B_{2u}$ state.[85] The required excited state must come from a $(b) \rightarrow (a)$ transition. However both $e_{1g}$ and $e'_{1g}$ are of $b$ species in $C_2$, and both $e_{2u}$ and $e'_{2u}$ are of $a$ species. Experimentally irradiation of dienes and benzene produces a complex mixture of products.[87] However the 1,4 *trans* adduct shown in (64) is the precursor to most of these.

### Chemiluminescence

An interesting sidelight on the photochemistry of benzene comes from a study of the thermal decomposition of Dewar benzene to form ordinary benzene. A phosphorescence due to the $^3B_{2u}$ state of benzene is observed. It is very weak, only one molecule in $10^3$–$10^4$ of reactant giving this product.

$$\text{Dewar benzene} \longrightarrow C_6H_6^* + C_6H_6 \qquad (65)$$

$$T_1 = 1 \quad S_0 = 10^3 - 10^4$$

This is an example of chemiluminescence. It is a result consistent with a mechanism in which $C_6H_6^*$ reacts to form ground-state Dewar benzene that is vibrationally excited, or hot. Obviously there is an intersystem crossing involved in (65).

One might be tempted to argue from microscopic reversibility that (65) proves that the $T_1$ state of benzene is the precursor to Dewar benzene. However this would be incorrect, except for one molecule in 1000. The activation energy for the isomerization of Dewar benzene is only enough to produce the $^3B_{2u}$ state of benzene when added to the exothermicity of the overall process. Therefore the formation of $S_1$ or $S_2$ benzene is energetically impossible, and we cannot tell where the other 999 molecules originate from. What we can say by applying microscopic reversibility to (65) is that very hot ground-state molecules of benzene can form Dewar benzene.

An interesting and well studied case of chemiluminescence is the thermal decomposition of tetramethyl-1,2-dioxetane.[89]

$$\underset{\text{O—O}}{(CH_3)_2C—C(CH_3)_2} \xrightarrow{\Delta} {}^1A_1(CH_3)_2CO + {}^3A''(CH_3)_2CO^* \qquad (66)$$

The results here show that every mole of dioxetane that decomposes gives one mole of ground-state acetone, and one mole of excited state $^3A''$ acetone. The amount of excited singlet is less than 1% of the triplet. The $^3A''$ state arises from an $n,\pi^*$ transition, which is lower in energy than the $\pi,\pi^*$. The various energies which are pertinent are shown in Fig. 19.

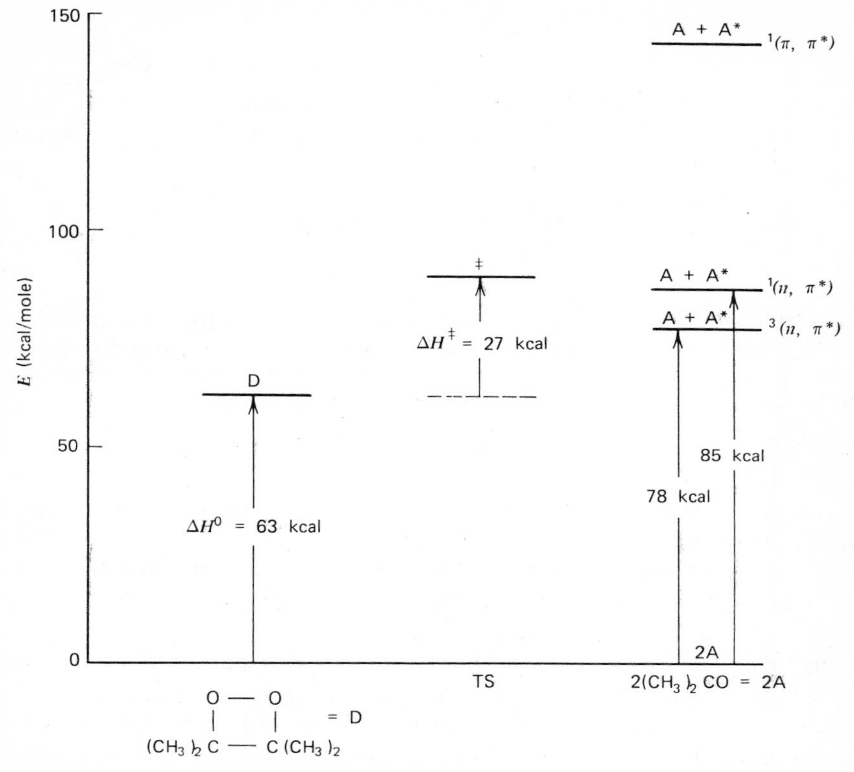


**Figure 19.** Energetics of the thermal decomposition of tetramethyl-1,2-dioxetane into acetone; the zero of energy is two molecules of ground state acetone; dioxetane is 63 kcal mole$^{-1}$ higher in energy in its ground state. [Reprinted with permission from N. J. Turro, P. Lechten, N. E. Schore, H. C. Steinmetzer, and A. Yetka, **7**, *J. Am. Chem. Soc.*, 97 (1974). Copyright by American Chemical Society.]

Activated dioxetane has enough energy to produce both the triplet and singlet $n,\pi^*$ states, but not enough to produce the $\pi,\pi^*$ states. This is of interest because (66) is forbidden as a ground-state reaction, if a least motion, $C_{2v}$ reaction coordinate is assumed.

$$(CH_3)C{-}C(CH_3)_2 \ (\text{ring } O{-}O) \xrightarrow{\Delta} (H_3C)(CH_3)C{=}O + (H_3C)(CH_3)C{=}O \qquad (67)$$

$$2a_1 \qquad\qquad a_1 + b_2$$

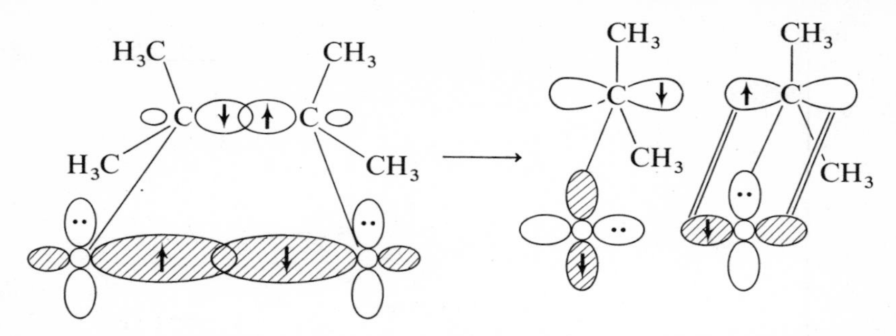


**Figure 20.** The concerted, nonadiabatic spin-forbidden cleavage of tetramethyl-1,2-dioxetane into one molecule of ground-state acetone and one molecule of acetone in its $^3(n, \pi^*)$ state. [Reprinted with permission from N. J. Turro, P. Lechten, N. E. Schore, H. C. Steinmetzer, and A. Yetka, *J. Am. Chem. Soc.*, **7**, 97 (1974). Copyright by American Chemical Society.]

Looking at the reverse reaction, we predict that a $(b_2) \rightarrow (a_1)$ excited state would be needed to photochemically condense acetone to dioxetane. These symmetries are in terms of the $C_{2v}$ point group of the combined molecules. In terms of one molecule, it means that a $\pi,\pi^*$ state, or the $^1A'$ state is needed. Spin conservation demands the singlet.

By microscopic reversibility we see that the wrong state, both by spin and by symmetry, seems to be the precursor of dioxetane formed photochemically from acetone. Indeed this must be one route, but (66) gives no information about the predicted reaction of $S_2$ acetone, since there is not enough energy to form it in the thermolysis of dioxetane. The unexpected efficiency of the $^3A''$ reaction arises from two factors: first there is an accidental degeneracy of energy at the transition state for dioxetane pyrolysis. This is implied in Fig. 19. Intersystem crossing can occur from one surface to another in this region.[90] Secondly, and most important, the changes in symmetry and in spin needed for intersystem crossing are just right to give strong spin–orbit coupling.[91] This is shown in Fig. 20, which focuses on the four electrons of the two bonds that are to be broken. A rotation of an electron on one oxygen atom produces both the triplet spin and the $n,\pi^*$ state. The change in orbital angular momentum produces the necessary change in spin angular momentum.

## Miscellaneous Reactions

Below are a number of reactions which occur photochemically, and which are forbidden as LM concerted processes in the ground state.

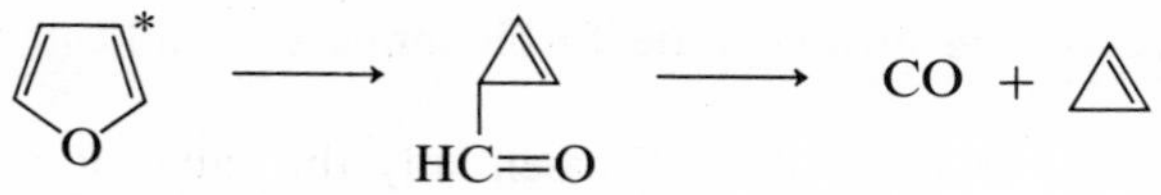

$$\text{furan}^* \longrightarrow \text{cyclopropene-CHO (HC=O)} \longrightarrow CO + \triangle \qquad (68)$$

6π electrons    4π electrons

$$CH_3OH^* \longrightarrow CH_2O + H_2 \qquad (69)$$

4-electron TS

$$SO_2^* + RH \longrightarrow RSO_2H \qquad (70)$$

4-electron TS

$$C_2H_5OH^* \longrightarrow C_2H_4 + H_2O \qquad (71)$$

4-electron TS

$$^3B_2SO_2^* + SO_2 \longrightarrow {}^3\Sigma\,SO + SO_3 \qquad (72)$$

ground-state spin and symmetry forbidden

$$\text{o-benzoquinone}^* + SO_2 \longrightarrow \text{C}_6\text{H}_4\text{O}_2\text{SO}_2 \qquad (73)$$

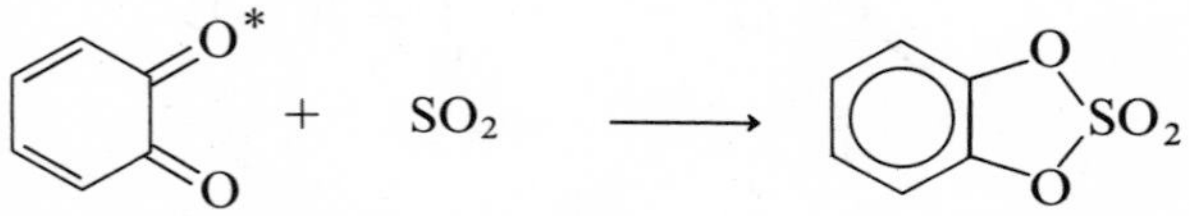

8π electrons    4π electrons    14π electrons

$$\text{o-benzoquinone}^* + C_2H_4 \longrightarrow \text{C}_6\text{H}_4\text{O}_2(CH_2)_2 \qquad (74)$$

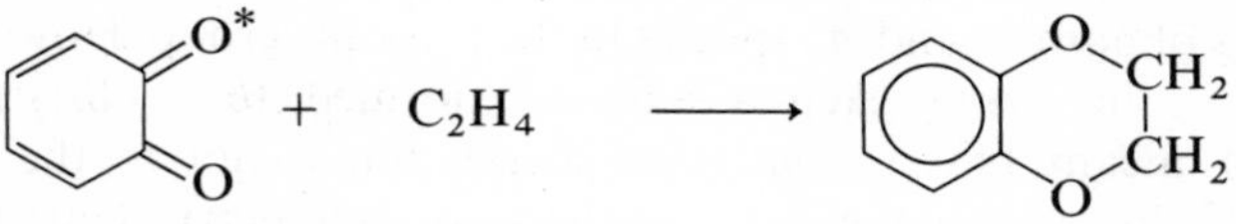

8π electrons    0π electrons    10π electrons

(The double bond of ethylene lies in the plane of the paper and is not counted as π.)

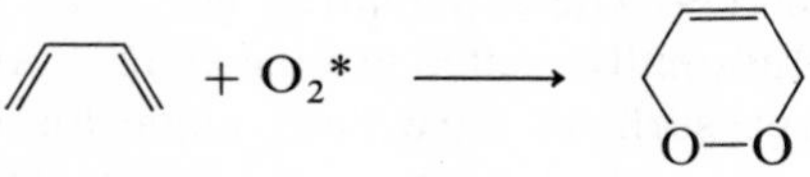

$$\text{butadiene} + O_2^* \longrightarrow \text{3,6-dihydro-1,2-dioxin (O–O)} \qquad (75)$$

ground-state spin and symmetry forbidden

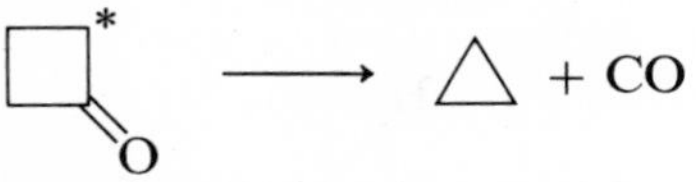

$$\text{cyclobutanone}^* \longrightarrow \triangle + CO \qquad (76)$$

$a_1 + b_2$    $a_1$    $a_1$

$$COBr_2^* \longrightarrow Br_2 + CO \qquad (77)$$

$a_1 + b_2$    $a_1$    $a_1$

$$H_2CO^* \longrightarrow H_2 + CO \qquad (78)$$

$a_1 + b_2$    $a_1$    $a_1$

The reasons for the ground-state forbiddenness is indicated below each equation.

The last three reactions are all essentially the same. Reaction (78) has been much studied, both experimentally and theoretically. It is instructive to examine it in detail, since it shows that it is not easy to predict just which excited state is needed to make a ground-state forbidden reaction occur photochemically. The naive approach predicts that a $(b_2) \rightarrow (a_1)$ excited state of formaldehyde is needed. Experimentally it is found that the molecular elimination of hydrogen comes from the first excited state, which arises from the familiar $n,\pi^*$ transition. In the $C_{2v}$ point group of the unexcited formaldehyde molecule, this is a $(b_2) \rightarrow (b_1)$ state of ${}^1A_2$ symmetry.

The change in geometry of the excited state must be taken into account. The excited state becomes nonplanar and the new symmetry is ${}^1A''$. Now the reaction is imagined to be completed.

$$\underset{{}^1A''}{H_2C{=}O} \longrightarrow \underset{{}^1A'}{H{-}H} + \underset{{}^1A''}{:C{=}O} \tag{79}$$

The state symmetries match, since $(A' \times A'') = A''$. The reaction appears allowed, since the lowest excited state of carbon monoxide, the ${}^1\Pi$ state, gives states of both $A'$ and $A''$ species in the $C_s$ point group. However the fact that in $C_{2v}$, the wrong excited state was formed, $(b_2) \rightarrow (b_1)$ rather than $(b_2) \rightarrow (a_1)$, means that the *intended* correlation is not to the lowest ${}^1A''$ state of CO, but to a higher state of the same symmetry. Thus there is still a large energy barrier to (79), even though the noncrossing rule forces a correlation to the lowest ${}^1A''$ state.[93]

Nevertheless, molecular hydrogen is eliminated with quantum yields of the order of 0.5 at 3000 A. The explanation lies in an internal conversion from $S_1$ to $S_0$, the vibrationally excited ground state.[94] Ab initio calculations of the potential energy surfaces have been made for the $S_0$, $S_1$, and $T_1$ states.[93,95] These show that the lowest-energy path for dissociation is on the $S_0$ surface, with a $C_s$ reaction path in which all four atoms of $CH_2O$ remain planar.

$$\underset{2a'}{H_2C{=}O} \longrightarrow H{-}C{=}O\,(\cdots H) \longrightarrow \underset{a' + a'}{H{-}H \quad C{=}O} \tag{80}$$

In this low-symmetry point group, the reaction is allowed, although a barrier of about 100 kcal mole$^{-1}$ exists. The overall reaction (80) is endothermic by 7 kcal/mole.

The high barrier is typical of what happens when one attempts to circumvent a barrier dictated by symmetry by going to a low symmetry, distorted reaction path. The symmetry barrier is reduced, but a price is paid in distortion energy. We see, however, that in photochemistry we often have enough energy in the excited state (in this case, 4–5 eV), so that a forbidden ground-state process can occur via crossing to the ground-state surface. After the hot $S_0$ molecule has been formed, there will be a competition between quenching, fluorescence, and the forbidden chemical reaction. The latter has the best chance of success in the dilute vapor phase.

The major competing reaction to molecular loss of $H_2$ from $CH_2O^*$ is loss of atomic H.

$$H_2\dot{C}-\dot{O} \longrightarrow H\cdot + H\dot{C}=O \qquad (81)$$

This is essentially the same as (23). Reaction (81) is allowed in the ground-state but has an activation energy of 87 kcal, the carbon—hydrogen bond energy. In addition the formyl radical must be formed with excess vibrational energy, as discussed earlier. Calculations on the potential energy surface show that (81) occurs by the sequence $S_1 \rightarrow T_1 \rightarrow$ ground-state of the products.[96] Shorter wavelengths increase $\varphi$ for reaction (81) by supplying the excess vibrational energy needed.

Even though enough energy may be added to a molecule to overcome a barrier due to symmetry, it will not always be the forbidden reaction that occurs. No chemical change may result, because of efficient quenching or other processes, or some other reaction may occur. For example, photolysis of sulfuryl chloride produces chlorine atoms, not chlorine molecules.

$$SO_2Cl_2^* \longrightarrow \cdot SO_2Cl + Cl\cdot \qquad (82)$$

not

$$SO_2Cl_2^* \longrightarrow SO_2 + Cl_2 \qquad (83)$$

The latter is a forbidden process in the ground-state. It is interesting that adding an electron to $SO_2Cl_2$ does result in the elimination of molecular chlorine.[97]

$$SO_2Cl_2^- \longrightarrow SO_2^- + Cl_2 \quad \text{and} \quad SO_2 + Cl_2^- \qquad (84)$$

In this case the rule that an excited-state molecule behaves like the ground-state molecule, with an added electron, is not followed.

## PHOTOCHEMISTRY OF METAL COMPLEXES

Inorganic complexes of the transition metals show several characteristic kinds of photochemical behavior.[98] Higher-energy photons can cause anions to eject an electron into the solution.

$$Fe(CN)_6{}^{4-} \xrightarrow{2200\text{ A}} Fe(CN)_6{}^{3-} + e^-(aq) \tag{85}$$

The formation of a hydrated electron can also occur from cations, if the central metal atom is easily oxidized. In other cases photons in the uv range cause redox reactions to occur within the complex.

$$Co(NH_3)_5Br^{2+} \xrightarrow{3700\text{ A}} Co(NH_3)_5{}^{2+} + Br\cdot \tag{86}$$

In this example the metal ion is reduced from +3 to +2 oxidation state, and the bromide ion is oxidized to the atom. In rarer cases the metal atom may be oxidized and the ligand reduced.

Irradiation with visible light or near UV most commonly causes ligand substitution to occur, sometimes with quite good quantum yields.

$$\underset{\varphi = 0.35}{Cr(NH_3)_5Br^{2+} + H_2O} \xrightarrow{4400\text{ A}} \textit{trans}\text{-}Cr(NH_3)_4(H_2O)Br^{2+} + NH_3 \tag{87}$$

With very inert metal complexes, this is often a convenient route to synthesis.[99]

$$W(CO)_6 + (C_6H_5)_3P \xrightarrow{4000\text{ A}} W(CO)_5P(C_6H_5)_3 + CO \tag{88}$$

Quantum yields in reactions like (88) are independent of the concentration of the entering group, which is taken to mean that a dissociation of the activated molecule is rate-determining. This is consistent also with the rearrangements that are often found.[100]

$$\textit{trans}\text{-}Cr(en)_2Cl_2 + H_2O \xrightarrow{5000\text{ A}} \textit{cis}\text{-}Cr(en)_2(H_2O)Cl^{2+} + Cl^- \tag{89}$$

(en = ethylenediamine)

Considering the various excited states that occur in transition metal complexes, it follows that charge-transfer states are likely to lead to redox reactions, and that *d–d* transitions are likely to lead to ligand substitutions. These are not unreasonable conclusions considering the nature of the two kinds of excited states. The redox reactions that can occur depend on the relative ease of oxidation or reduction of the various parts of the photoactivated molecule. The ligand-substitution reactions can be explained in a generally satisfactory way by using ligand field theory.[101]

The basic assumption for six-coordinated complexes is that the excited state reacts by dissociation of a ligand to give a five-coordinated intermediate. This reactive species rapidly picks up a solvent molecule, or other ligand from solution. The dissociation step, at least, occurs on the excited state potential surface. Energetically it is often possible to form the product in the excited state. There is little evidence from fluorescence or phosphorescence studies that this occurs. It is likely that return to the ground-state occurs in the five-coordinate intermediate, since the two potential surfaces are closest at this point.

The experimental observations for octahedral complexes are summed up by Adamson's rules:[102]

1. Consider the ligands to lie along three Cartesian axes. The axis having the weakest average crystal field will be the one that is labilized.
2. The ligand of greatest crystal field strength on the labilized axis will be the one that is preferentially dissociated. Crystal-field strength is empirically determined for each ligand by 10 Dq, or the magnitude of the $t_{2g} - e_g^*$ orbital separation in symmetrical $ML_6$ complexes.

To understand the basis for the rules, consider the excited quartet states of chromium(III) octahedral complexes. These are shown in Fig. 21, first for a regular octahedron of $O_h$ symmetry, such as $Cr(NH_3)_6{}^{3+}$, and then for a

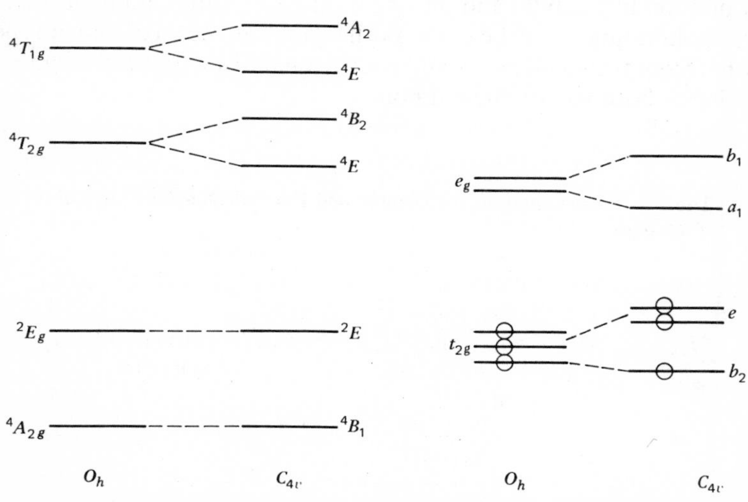


**Figure 21.** On the left are shown the state splittings for a $d^3$ $ML_6$ complex going from $O_h$ to $C_{4v}$ symmetry and on the right, the splittings of the $d$ orbitals in a complex such as $Cr(NH_3)_5Cl^{2+}$ compared to $Cr(NH_3)_6{}^{3+}$; the $d_{xz}$ and $d_{yz}$ orbitals, which are $e$, are raised in energy by $\pi$-bonding effects.

complex of lower symmetry, $C_{4v}$, such as $Cr(NH_3)_5Cl^{2+}$. The chloride ion has a lower crystal-field strength than $NH_3$. Accordingly, Adamson's rules predict that the photolysis of $Cr(NH_3)_5Cl^{2+}$ in water will lead to the loss of the $NH_3$ molecule *trans* to the chloride ion (cf. reaction 87). This is what is observed, the quantum yields for $NH_3$ loss being 0.36 and for $Cl^-$ loss being about 0.04, with little dependence on wavelength. That is, the states arising from either the ${}^4T_{2g}$ or ${}^4T_{1g}$ progenitors can be excited.

This is consistent with the reactive state being the lowest state in all cases. If a higher state is formed initially, internal conversion quickly carries it down to the ${}^4E$ state coming from the ${}^4T_{2g}$. It is possible that the ${}^2E$ state is also populated by intersystem crossing, and some photochemical reactivity could arise from this state.[103] However it is likely that the quartet states are chiefly involved, in this case. Table 3 shows the ground-and excited-state wavefunctions for the quartet states of an octahedral $d^3$ complex. Only the $d$ orbitals are considered and a crystal-field viewpoint is used.[104] Nevertheless, a more sophisticated molecular orbital approach would not change the symmetry features.[101]

The lowest ${}^4E$ state corresponds to taking an electron from the degenerate $d_{xz}$–$d_{yz}$ pair of orbitals and placing it in an orbital which in $C_{4v}$ is largely a $d_{z^2}$ orbital. This requires less energy than forming the ${}^4B_2$ state by the $(d_{xy}) \rightarrow (d_{x^2-y^2})$ transition, because the crystal field is less along the $z$ axis, which contains the chloride ion. The effect of taking an electron from a chiefly nonbonding orbital and putting it into a $\sigma^*$ antibonding orbital, is to weaken the bonding along the $z$ axis. Both ligands on the $z$ axis will be repelled somewhat from the metal atom, and a tetragonal distortion will result with a weakening of both metal—ligand bonds.

**Table 3 Wavefunctions for Ground and Excited Quartet States of a $d^3$ Complex**[a]

| | | |
|---|---|---|
| ${}^4A_{2g}$ | $\theta = (xz)(yz)(xy)$ | ${}^4B_1$ |
| | $\psi_1 = (xz)(yz)(x^2 - y^2)$ | ${}^4B_2$ |
| ${}^4T_{2g}$ | $\psi_2 = -1/2(yz)(xy)(x^2 - y^2) + \sqrt{3}/2(yz)(xy)(z^2)$ | ${}^4E$ |
| | $\psi_3 = -1/2(xy)(xz)(x^2 - y^2) - \sqrt{3}/2(xy)(xz)(z^2)$ | |
| | $\varphi_1 = (xz)(yz)(z^2)$ | ${}^4A_2$ |
| ${}^4T_{1g}$ | $\varphi_2 = -1/2(yz)(xy)(z^2) - \sqrt{3}/2(yz)(xy)(x^2 - y^2)$ | ${}^4E$ |
| | $\varphi_3 = -1/2(xy)(xz)(z^2) + \sqrt{3}/2(xy)(xz)(x^2 - y^2)$ | |

[a] The symmetry species for $O_h$ are shown on the right and those for $C_{4v}$, on the left. Only the $d$ orbitals are considered. In $C_{4v}$ the predominant orbital, $z^2$ or $x^2 - y^2$, contributes to each $E$ state even more than indicated.

This explains Adamson's first rule for $d^3$ complexes, and similar reasoning can be given for other numbers of $d$ electrons. The second rule is not so obvious and cannot be obtained from crystal-field theory. Why is the *trans* $NH_3$ released rather than the $Cl^-$, when in the ground state the reverse occurs? The explanation must lie in the covalency of the bonds to the two ligands. The more covalent the bond, the more the bond energy will suffer by putting an electron into a $\sigma^*$ orbital. Thus if $NH_3$ is more covalently bonded than $Cl^-$, it will be labilized the most in the excited-state.

The rule then should be that the most covalent ligand on the weak field axis is the one that is released. The crystal-field parameter can be used as an approximation to covalency, because in general 10 Dq, or the crystal-field strength, is determined largely by covalent bonding effects. Obviously, if $\pi$ bonding is an important factor in determining 10 Dq, its role in the excited state must be specifically considered. For example, removing an electron from a $t_{2g}$ orbital in $Cr(CO)_6$ and putting it into an $e_g^*$ orbital affects the bonding in two ways. It weakens the $\pi$-backbonding as well as the $\sigma$ bonding.

Other ligands, such as $F^-$ and $OH^-$, are $\pi$-donors, and not $\pi$-acceptors. Depopulating a $t_{2g}$ orbital in such cases may cause the ligand to be bound more strongly. The complex *trans*-$Cr(en)_2F_2^+$ is photolyzed to give *trans*-$Cr(en)(enH)(H_2O)F_2^{2+}$, a chromium–nitrogen bond being broken.[105] This is an exception to Adamson's rules, since $NH_3$ has a larger crystal field than $F^-$. However even if the excited state is the $^4E$ due to the usual $(d_{xz} - d_{yz}) \rightarrow (d_{z^2})$ excitation, it may well be that the fluoride-ion bonding is not weakened greatly, because of compensating $\sigma^*$ and $\pi$ effects. The excited orbital still has a $d_{x^2-y^2}$ component (Table 3) and the amine bonding is subject to $\sigma$ weakening only, unmitigated by $\pi$ effects.

The photolysis of pentacarbonyl amine complexes of Cr, Mo, and W produces the following results:[10, 106]

$$M(CO)_5(\text{amine}) \xrightarrow{\text{short } \lambda} M(CO)_4(\text{amine}) + CO \quad (90)$$

$$M(CO)_5(\text{amine}) \xrightarrow{\text{long } \lambda} M(CO)_5 + \text{amine} \quad (91)$$

The CO lost in (90) is from an equatorial position, that is, *cis* to the amine ligand. The interpretation is that higher energy quanta produce the $^{1,3}A_2$ state from the $(b_2) \rightarrow (b_1)$ excitation. This labilizes the ligands in the $xy$ plane, or the equatorial COs. Lower energy quanta produce the $^{1,3}E$ state which activates the groups on the $z$ axis.

While Adamson's first rule is followed, the second rule is not found in (91). Carbon monoxide has a greater crystal field strength than amines do, and covalent bonding is greater for CO. However the ground-state reactivity must also be considered, besides the bond-weakening effects of the photo-excitation. Thermally, the lability of the ligands in $M(CO)_5(\text{amine})$ follows the pattern amine > CO (equatorial) ≫ CO (axial). The bond strength of the metal to the *trans* carbonyl group is unusually strong. This is an example of a well-known phenomenon in which a hard base and a soft base *trans* to each other in a complex have a stabilizing effect.[107] The rationale is that one strong covalent bond can be formed, but not two, along a single axis.

The photochemistry of square planar complexes of platinum(II) has been studied in some detail.[98] A common reaction is *cis–trans* isomerization, for example,

$$cis\text{-Pt(pyridine)}_2\text{Cl}_2 \xrightarrow{h\nu} trans\text{-Pt(pyridine)}_2\text{Cl}_2 \qquad (92)$$

which appears to occur intramolecularly. Ligand substitution, usually hydrolysis, also occurs. In most cases the photochemical reaction is the same as the thermal one.

$$\text{Pt(NH}_3)_3\text{Cl}^+ + \text{H}_2\text{O} \xrightarrow{h\nu} \text{Pt(NH}_3)_3(\text{H}_2\text{O})^{2+} + \text{Cl}^- \qquad (93)$$

However at least one case of an antithermal reaction is known.[108]

$$\text{Pt(C}_2\text{H}_4)\text{Cl}_3^- + \text{H}_2\text{O} \xrightarrow{\Delta} \text{Pt(C}_2\text{H}_4)(\text{H}_2\text{O})\text{Cl}_2 + \text{Cl}^- \qquad (94)$$

$$\text{Pt(C}_2\text{H}_4)\text{Cl}_3^- + \text{H}_2\text{O} \xrightarrow{h\nu} \text{Pt(H}_2\text{O})\text{Cl}_3^- + \text{C}_2\text{H}_4 \qquad (95)$$

In (94) it is the chloride ion *cis* to ethylene that is aquated. The *trans* chloride ion is extremely labile and is kept on only by a high concentration of external chloride ion.

The ligand field bands of a $d^8$ planar complex all involve promoting an electron to the $b_{1g}$ or $d_{x^2-y^2}$ orbital. All four ligands should be activated, since they all lie in the $xy$ plane. The ligand most covalently bound should be activated most. However the geometry change in the photoexcited state is important. A $d^8$ complex should take up the structure of a $d^9$ complex. For an $ML_4$ molecule this should give a $D_{2d}$ point group, or a flattened tetrahedron. An $ML_3X$ molecule would give the $C_s$ point group. In any case an electron is removed from a $\pi$-bonding orbital and put in a $\sigma^*$ antibonding orbital in the lowest excited-state. Ethylene bonding would be most seriously affected, since $\pi$ bonding is greatest for this ligand.

The assumption that the excited-state has nearly a tetrahedral structure offers a ready explanation for *cis–trans* isomerization. Figure 22 shows that a pseudotetrahedral structure for an $ML_2X_2$ complex allows either a *cis* or *trans* isomer to be formed on returning to the planar form. The process

$$^1A_{1g}\ \text{ML}_4(D_{4h}) \rightleftharpoons {}^3E\ \text{ML}_4(T_d) \qquad (96)$$

is forbidden as a ground-state reaction, by both spin and orbital symmetry. However excitation to the $^3E_g$ state of the planar form causes the reaction to

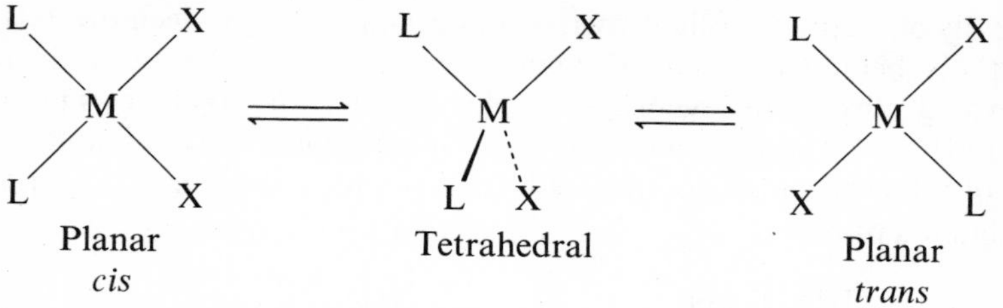

$$\mathrm{L_2MX_2}\ (\text{Planar, } cis) \rightleftharpoons \mathrm{L_2MX_2}\ (\text{Tetrahedral}) \rightleftharpoons \mathrm{L_2MX_2}\ (\text{Planar, } trans)$$

Planar *cis* — Tetrahedral — Planar *trans*

**Figure 22.** Isomerization of a square planar complex by way of a tetrahedral intermediate [cf. D. R. Eaton, *J. Am. Chem. Soc.*, **90**, 4272 (1968)].

become allowed.[109] Because of the large spin–orbit coupling, the spin-state restriction is of little consequence.

Charge-transfer excited states of metal complexes most commonly take electron density from a ligand and put it on the central metal, largely in a *d* orbital. This leads to an internal redox reaction as typified by (86). The release of the ligand as a free atom or radical is a natural follow-up. The reverse process, in which an electron from the metal goes to the ligand, should not lead to dissociation, since the increased negative charge of the ligand and the increased positive charge of the metal give increased ionic bonding.

There can still be chemical consequences. For example, olefins coordinated to certain electron-rich metal atoms undergo *cis–trans* isomerization.[10,110]

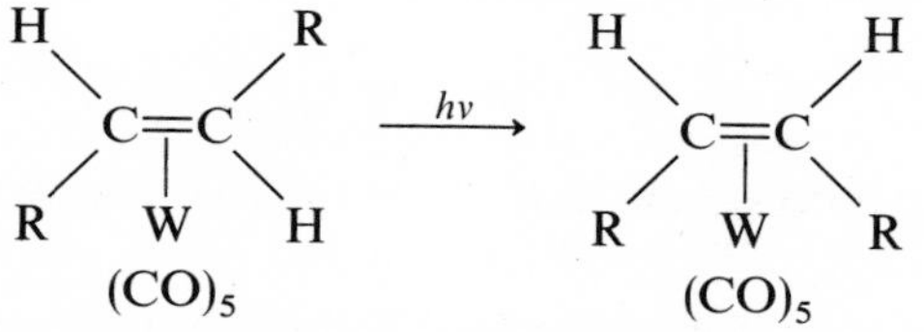

$$\text{HRC=CHR}\,[\mathrm{W(CO)_5}]\ (trans) \xrightarrow{h\nu} \text{HRC=CHR}\,[\mathrm{W(CO)_5}]\ (cis) \qquad (97)$$

An explanation is that an electron is taken from a *d* orbital on the metal and placed in the antibonding $\pi^*$ orbital of the metal–ligand bond. The effect would be the same as for the ligand migration reaction (p. 413), except that no ligand follows the electron. A coordinated radical ion would be formed, which is only $\sigma$-bonded,

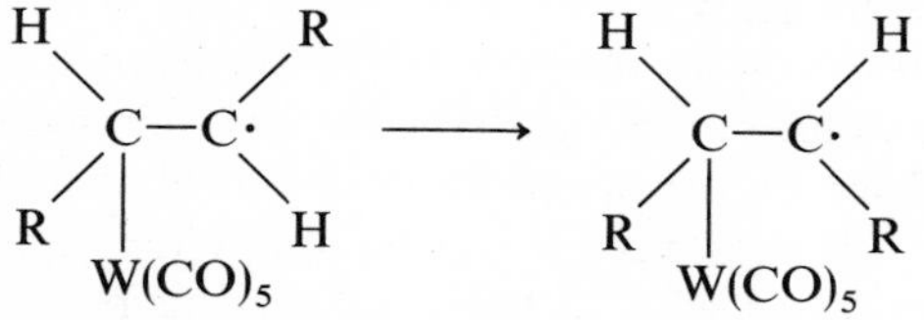

$$\mathrm{HRC[W(CO)_5]{-}\dot{C}HR} \longrightarrow \mathrm{HRC[W(CO)_5]{-}\dot{C}HR}\ (\text{rotated}) \qquad (98)$$

This species would have free rotation about the carbon—carbon bond, and isomerization results upon return to the ground state.

Finally absorption of light by a complex can occur between orbitals that essentially belong to a ligand, and where the metal atom exerts only a perturbing effect. This seems to be the case for the azido complexes of iridium(III).[111] The products formed on irradiation are explainable on the basis of the cleavage of coordinated azide ion to molecular nitrogen and a coordinated nitrene

$$Ir(NH_3)_5N_3^{2+} \xrightarrow{h\nu} Ir(NH_3)_5N^{2+} + N_2 \tag{99}$$

This is a reaction characteristic of the azide ion, or organic azides.

$$\begin{aligned} N_3^- &\xrightarrow{h\nu} N^- + N_2 \\ N^- + H_2O &\longrightarrow HN + OH^- \end{aligned} \tag{100}$$

The excited state involved in (100) is probably that from the $(\pi_g) \rightarrow (2\pi_u)$ transition. This state would correlate with an excited state of the $N_2$ molecule. Since there is not enough energy to form the latter state, the reaction must occur from the vibrationally excited ground state formed by internal conversion. Reaction (100) is allowed in the ground state, but is quite endothermic. For comparison, the reaction

$$HN_{3(g)} \longrightarrow N_{2(g)} + NH_{(g)} \tag{101}$$

is endothermic by 14 kcal.[113] Reaction (100) in aqueous solution would be even more energy demanding, because nitrene is a much weaker acid than hydrazoic acid.

## PARTITIONING OF THE ACTIVATION ENERGY

In this chapter we have been concerned with the way in which excess electronic energy is used to cause chemical change. The question may also be raised, as to what extent translational, or vibrational, or even rotational energy can be used to promote specific chemical reactions. Given that a certain activation energy exists, can any form of molecular energy be used with equal efficiency to supply it? For reactions in solution, this question is rarely asked, since the solvent serves as a heat bath. The activated complex can draw on the heat bath for the energy it needs in any form that is effective.

In the dilute gas this is no longer true. A molecule with sufficient energy will receive it in different forms, and not all of it will be useful in overcoming the energy barrier. In such cases it can be of practical importance to know what kind of energy is useful to promote a given reaction. For example, if vibrational energy corresponding to a certain vibrational level is most effective, selective radiation with a laser may give a selective chemical reaction. Perhaps measurements of the kinds of energy released in a reaction can tell us something about the mechanism. In any case, our knowledge of the mechanisms of reactions is incomplete if we do not understand the details of activation for any process.

The hydrogen–iodine reaction has been studied by preparing DI and HI molecules with very high translational kinetic energies (this is done by nozzle expansion of compressed helium into a vacuum).[114] Even though energies in excess of 200 kcal mole$^{-1}$ were reached, there was no evidence for reaction.

$$HI + DI \not\longrightarrow HD + I_2 \tag{102}$$

This negative result is in agreement with the conclusion that the four-center bimolecular mechanism is forbidden. Obviously the activation energy of 44 kcal must be supplied in some other way, or at least some large portion of it must be. Vibrational energy is the only plausible alternative.

A mechanism has been given for the $H_2$—$I_2$ reaction, in which free I atoms are involved (p. 56). There is another mechanism, which is closely related. Suppose $I_2$ molecules with a high degree of vibrational excitation, about 20 kcal, were formed, thermally or otherwise. Such molecules already have an internuclear separation of some 5 Å, at the outer limits of their vibration. It would be possible for a hydrogen molecule to insert into this extended iodine.[115]

$$\begin{matrix} \overset{5.0\ \text{Å}}{\text{I}\text{- - - - - - - - - -}\text{I}} \\ \underset{0.74\ \text{Å}}{\text{H—H}} \\ 2a' \end{matrix} \longrightarrow \begin{matrix} \text{I- - -H- - -H- - -I} \\ \\ a' + a'' \end{matrix} \longrightarrow 2\text{HI} \tag{103}$$

This reaction is still forbidden, as shown by the bond symmetry analysis. However one of the original bonds is now so weak that the restriction is greatly relaxed. The energetics of (103) would resemble those of the reaction of two free I atoms with an $H_2$ molecule, and kinetic studies cannot differentiate between the two mechanisms.

There are two complementary ways of studying the partitioning of activation energy. One is to test experimentally what kinds of energy must be supplied to cause reaction. The second is to examine the kinds of energy contained in molecules which are products of exothermic reactions. By microscopic reversibility, these are the same energies which will cause the reverse reaction to occur. Fortunately there are several techniques by which product energies can be probed: (1) molecular beam reactive scattering, (2) infrared chemiluminescence, (3) fluorescence, and (4) chemical laser methods.[116] Results can be very detailed, with populations of individual energy levels being determined.

**Table 4 Energy Released in Exothermic Reactions**

| Reaction | $E$ (in kcal mole$^{-1}$) | $\langle f_v \rangle^b$ | $\langle f_T \rangle^c$ |
|---|---|---|---|
| $Cl + HI \rightarrow I + HCl$ | 34.0 | 0.71 | 0.16 |
| $F + H_2 \rightarrow H + HF$ | 34.7 | 0.67 | 0.26 |
| $K + I_2 \rightarrow KI + I$ | 44.3 | | 0.03 |
| $Ba + HF \rightarrow BaF + H$ | 12.7 | 0.12 | 0.75 |
| $D + Cl_2 \rightarrow DCl + Cl$ | 55.0 | | 0.44 |
| $K + CH_3I \rightarrow KI + CH_3$ | 26.9 | | 0.55 |

[a] Data from reference 118.
[b] The fraction of the total energy $E$ in the product vibration.
[c] The fraction of the total energy $E$ in the product translation.

Table 4 shows some results of studies on the energy content of product molecules of exothermic reactions. It can be seen that the results are quite specific. For some reactions, the activation energies for the reverse process must be largely vibrational and for others, largely translational. Rotational energy is not shown in Table 4 but can be measured in some cases. For example, the reaction

$$Cl + HI \longrightarrow HCl + I \tag{104}$$

has 0.71 of the total energy release in vibration of the HCl molecule, 0.16 in translational energy of HCl and I, and the remaining 0.13 fraction as rotational energy of HCl.[117]

Certain features of Table 4, and similar results, are reasonably well understood. For example, the reaction

$$K + I_2 \longrightarrow KI + I \tag{105}$$

has almost all of its exothermicity appearing as vibration because it is a reaction where the TS comes early, when the K—I bond distance is great (see p. 147). Also, in reaction (104) the heavy I atom acts as a spectator during the transfer of H to Cl.[118] That is, its momentum scarcely changes. Since the momentum of HCl must equal in magnitude, but be of opposite direction to that of the iodine atom, neither particle can pick up much translational energy as a result of the reaction.

Consider an activated complex, ABC, with no excess energy in the transition state. Now allow it to decompose along the reaction coordinate.

$$AB + C \longrightarrow A\text{---}B\text{---}C \longrightarrow A + BC \tag{106}$$

The electronic energy drops by an amount equal to the energy barrier for the reverse process, with due allowance for zero-point energies. In a condensed medium this change in energy is converted into heat energy of the medium. In a dilute gas the energy is converted into increased vibration, rotation, and translation of the products, A and BC. The drop in electronic energy has two main components: (1) increased bonding in BC and (2) repulsion between A and BC. The increased bonding in BC will show up as vibrational energy and the repulsion between A and BC will appear as translational energy.[119]

If ABC is linear, there will be no rotational energy. However a nonlinear arrangement will cause some of the repulsion energy to appear as rotation of BC. Conversely, in the reaction of A with BC, rotation of BC can supply some of the activation energy, but only for collisions that are not end-on.

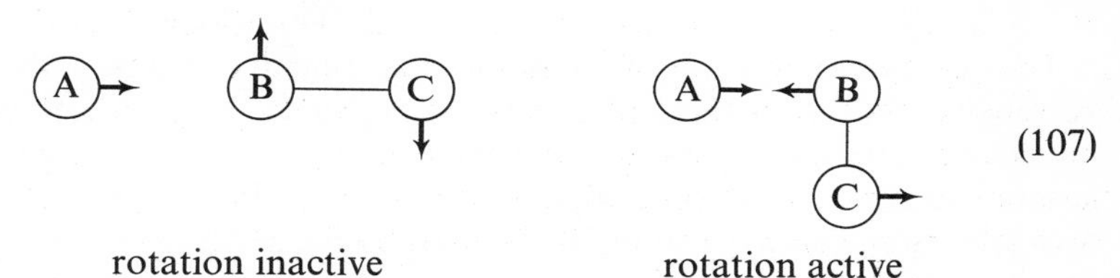


Rotational energy is of the same nature as translational energy. It does not cause any appreciable change in the internuclear coordinates, and is almost purely kinetic in nature. Because of its restrictive form, it is normally less useful than translational energy in overcoming energy barriers.

The above analysis is helpful in determining the energy characteristics of reactions forbidden by orbital symmetry, and related effects. A bimolecular reaction that is forbidden will have a strong repulsion between the approaching molecules. Therefore the activation energy must contain a large contribution of translational energy. In a unimolecular reaction which is forbidden, there will be a large kinetic energy release if two or more particles are formed in the reaction.[120] Detection of this energy can serve as a diagnostic criterion.

There can also be considerable vibrational energy release as well. This comes from those atoms which remain bonded together (or become bonded) during the reaction. The requirement is that the geometries be quite different in both the TS and ground state of the products. For example, in the concerted reaction

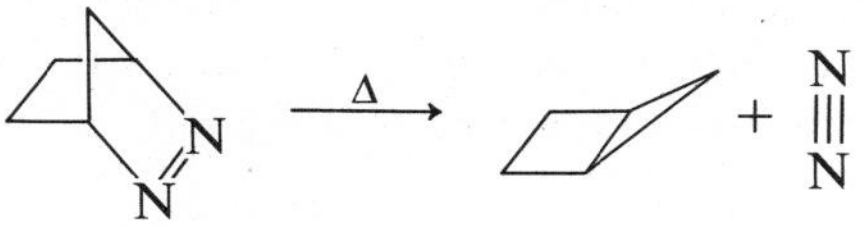

(108)

the nitrogen molecule would be formed in a vibrationally excited state, since a double bond is converted to a triple bond. The hydrocarbon molecule would also have vibrational excitation, but this would rapidly randomize over the entire molecule. The nitrogen excitation has nowhere to go, especially since the stretching mode is normal to the reaction coordinate. An experimental detection of highly excited nitrogen would support a concerted mechanism.[121] A stepwise mechanism, with the intermediate biradical M being formed,

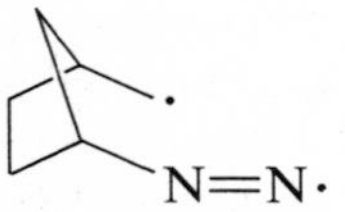

N=N· M

would dissipate more of the activation energy over the entire molecule. The $N_2$ molecule would be formed with less excess vibrational energy.

The energy requirement of a unimolecular reaction can be gathered from various sources. In a thermal process, it is usually by collision with other molecules. There can be interconversion of vibrational, rotational, translational, and even electronic energy during a collision.[122] However only its own vibrational energy can finally be used by the activated molecule to pass over the barrier. Translational and rotational energies have no effect, except for a small centrifugal influence of the latter. Internal rotations count as vibrations, since they include potential energy changes.

The rate of decomposition of an activated molecule is usually considered within the framework of the Rice–Ramsperger–Kassel–Marcus theory (RRKM).[123] All the vibrational modes of a molecule are assumed to be loosely coupled together, as a result of anharmonicity. This allows vibrational energy to flow from one mode to another. Eventually enough accumulates in the reaction coordinate mode to cause passage over the barrier. The greater the total vibrational energy in excess of the critical energy, $E_0$, or the barrier height, the greater is the rate of decomposition of the molecule. Normally the lifetime of an activated molecule before reaction is $10^{-5}$–$10^{-13}$ s. Collisional deactivation can then play a competing role, depending on the pressure.

## Reactions of Metastable Ions

A special way in which molecules can be activated is by electron impact. The mass spectrometer provides a sensitive method for studying the fate of such activated molecules. Primary ions are normally formed from collision with energetic electrons by simple loss of a second electron.

$$M + e \longrightarrow M^{\ddagger} + 2e \qquad (109)$$

The parent ion, $M^{\ddagger}$, formed in this reaction usually undergoes a series of secondary reactions in which smaller fragments are formed.

$$M^{+} \longrightarrow F_1^{+} + N \text{ (etc.)} \qquad (110)$$

N is often a stable small molecule. The charge and the odd electron would then stay with the larger fragment. At a later stage there may be bimolecular reactions between some of the ions formed and neutral molecules, either those formed from M, or added independently.

The parent ion can be formed with very large amounts of excess energy, depending on the voltage drop of the ionizing electrons. It can be created in an excited electronic state, corresponding to the second or higher ionization potential It usually has excess vibrational energy because of the Franck–Condon effect, and it can also take up translational and rotational energy from the electron impact. The most energetic ions will fragment first, in the ionization chamber. Lightly activated ions may reach the field-free region of the mass spectrometer before decomposing. In this case they will form so-called "metastable" ions. Such ions show up in the mass spectrum usually as broad peaks of low intensity, centered on nonintegral mass values.[124]

The peak shapes of metastable ions provide valuable information about energy distribution in exothermic reactions.[125] They are broad because of translational energy release. The newly created velocity has a random direction, some of which adds to the velocity created by the accelerating voltage of the mass analyzer, and some of which subtracts. Under suitable conditions, the amount of translational energy released can be accurately measured. Figure 23 shows some typical metastable peak shapes. The wider peaks are the ones with large amounts of translational energy release.

The reactions of molecular ions are similar in many ways to those of excited electronic states of the same molecules. Many of the reactions observed are close counterparts to photochemical reactions. For example, there are analogs in ion-fragmentation reactions to the Norrish types I and II processes for ketones.

$$C_6H_5{-}\underset{\|}{C}(=O^{\cdot +}){-}CH_3 \longrightarrow C_6H_5{-}\overset{+}{C}{=}O + CH_3\cdot \qquad (111)$$

$$R{-}C(=\overset{\cdot +}{O}\cdots H{-}C{<}){-}C{<}{-}C{<} \longrightarrow R{-}\underset{|}{C}(^{+}OH){-}\dot{C}{<} + {>}C{=}C{<} \qquad (112)$$

In mass spectroscopy, the Norrish type II reaction is called the "McLafferty rearrangement." One important difference for ion molecules is that there are

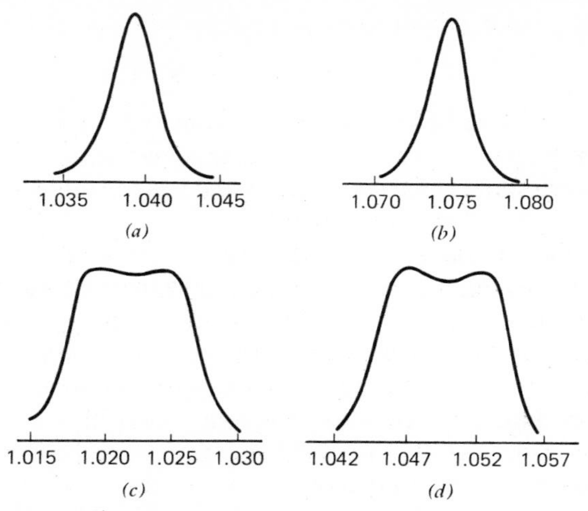


**Figure 23.** Metastable peaks for loss of molecular hydrogen from: (*a*) $C_6H_7^+$; (*b*) $C_2H_5^+$; (*c*) $C_7H_9^+$; (*d*) $C_3H_7^+$; the abscissa is in units of $(V/V^0)$, proportional to apparent mass. [Reprinted with permission from D. H. Williams and G. Hvistendahl, *J. Am. Chem. Soc.*, **96**, 6755 (1974). Copyright by the American Chemical Society.]

no mechanisms for return to the ground state of the molecule, except eventually by electron capture. Hence most fragmentation reactions of molecular ions are from vibrationally excited, but lowest electronic, states of the ion.

The interesting feature of metastable ion peaks, from the viewpoint of orbital symmetry, is the ability to measure the kinetic energy release. The reactions of molecular ions are subject to the same kind of symmetry restrictions as those of normal molecules, that is, they fall into allowed and forbidden categories.[126] A reaction that is forbidden, as we have pointed out, will have a large release of kinetic energy if two or more particles are formed from one. A reaction that is allowed will have a small kinetic energy release, even though there may be a large activation energy because of endothermicity.

Consider the two reactions of the parent ion of propane.[127]

$$C_3H_8^{\ddagger} \longrightarrow C_3H_6^{\ddagger} + H_2 \qquad T = 0.40 \text{ eV} \tag{113}$$

$$C_3H_8^{\ddagger} \longrightarrow C_3H_7^{+} + H\cdot \qquad T = 0.002 \text{ eV} \tag{114}$$

Reaction (113) has a large kinetic energy release, when it occurs in the field-free region, because the reverse process is a forbidden one, similar to the concerted addition of hydrogen to an olefin. The allowed reaction (114) has only a small kinetic energy release. As another example of a forbidden reaction, the loss of $H_2$ from the ion $CH_2OH^+$ has been studied.[128]

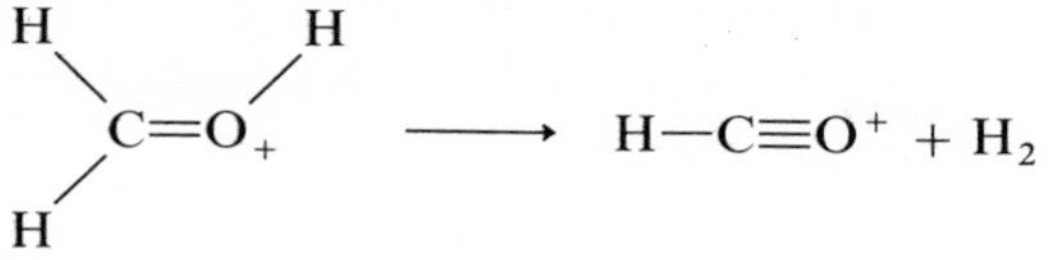

$$\mathrm{H_2C{=}O^+{-}H} \longrightarrow \mathrm{H{-}C{\equiv}O^+} + \mathrm{H_2} \qquad (115)$$

In this case some 33 kcal of kinetic energy are released. This may be compared with the allowed reaction (111), where only 0.2 kcal of kinetic energy is measured. Note that $CH_2OH^+$ is a simple ion and not a radical ion. The presence of an odd electron does not determine the results.

Kinetic energy release can also occur in unimolecular reactions which are symmetry allowed. The requirement is only that an activation energy exists for the reverse process which arises from a repulsion of the reactants. For example, the fragmentation of doubly charged ions usually has a large kinetic energy release. A double ion from *p*-xylene decomposes into two smaller ions.[129]

$$p\text{-}C_8H_8{}^{2+} \longrightarrow CH_3{}^+ + C_7H_5{}^+ \qquad T = 2.6\ \text{eV} \qquad (116)$$

In the hypothetical reverse reaction, there would be an electrostatic repulsion, given classically by $U = 14.39/R$ eV $= T$, where $R$ is the distance between the two positive charges in the $C_8H_8{}^{2+}$ reactant. From the measured kinetic energy release, $R$ is calculated as 5.6 Å, a not unreasonable value.

Useful deductions about mechanism can be made from the shape of metastable ion peaks. The reaction of the parent ion of ethoxybenzene gives ethylene and the radical ion of phenol. This reaction could occur directly,

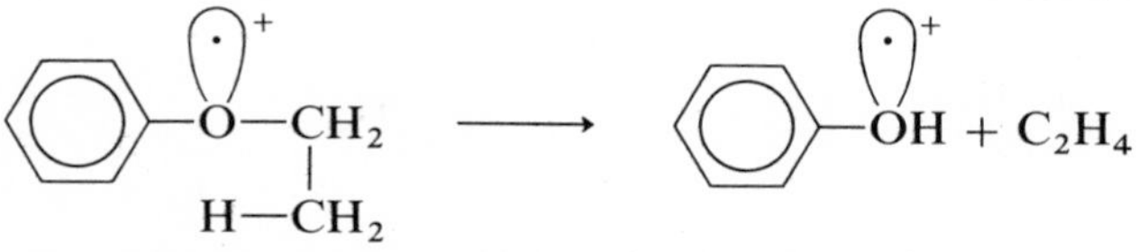

$$\mathrm{C_6H_5{-}\overset{\cdot +}{O}{-}CH_2{-}CH_2{-}H} \longrightarrow \mathrm{C_6H_5{-}\overset{\cdot +}{O}H} + C_2H_4 \qquad (117)$$

in which case it is forbidden (four-center reaction). Or it could occur as an allowed, but more involved, sequence of steps.

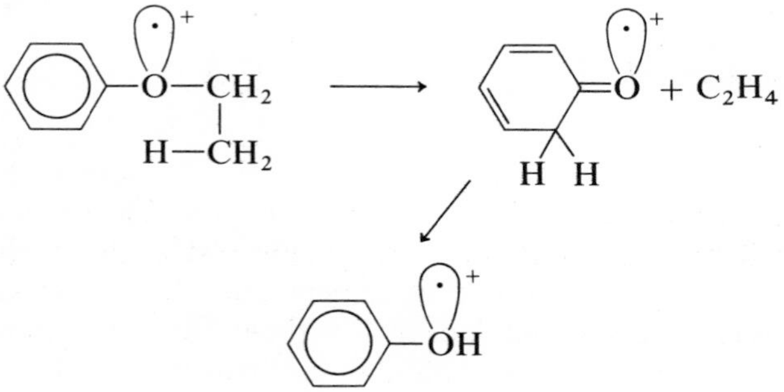

$$\mathrm{C_6H_5{-}\overset{\cdot +}{O}{-}CH_2{-}CH_2{-}H} \longrightarrow \mathrm{C_6H_6{=}\overset{\cdot +}{O}}\ (\text{H H}) + C_2H_4 \longrightarrow \mathrm{C_6H_5{-}\overset{\cdot +}{O}H} \qquad (118)$$

The very small kinetic energy release found experimentally is good evidence that (118) is the mechanism, and not (117).[129b]

Figure 23 shows two reactions with sharp peaks, or no kinetic energy release. These are for the reactions

$$\begin{aligned} C_2H_5^+ &\longrightarrow C_2H_3^+ + H_2 \\ C_6H_7^+ &\longrightarrow C_6H_5^+ + H_2 \end{aligned} \tag{119}$$

which can occur by allowed 1,1-eliminations.[130]

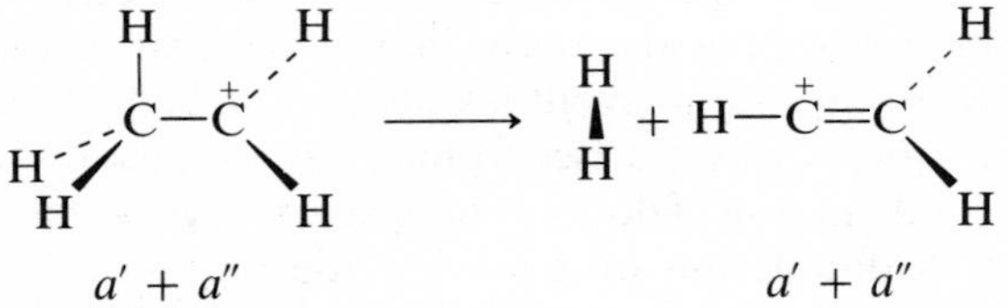

$$H_3C{-}\overset{+}{C}H_2 \longrightarrow H{-}H + H{-}\overset{+}{C}{=}CH_2 \tag{120}$$

$a' + a''$ $a' + a''$

The two C—H bonds become the H—H bond and the $\pi$ bond of the vinylium ion.

The two broad peaks of Fig. 23 are for the reactions

$$\begin{aligned} C_3H_7^+ &\longrightarrow C_3H_5^+ + H_2 \\ C_7H_9^+ &\longrightarrow C_7H_7^+ + H_2 \end{aligned} \tag{121}$$

In these two reactions the mechanism is probably one of forbidden 1,2 or 1,3 eliminations.

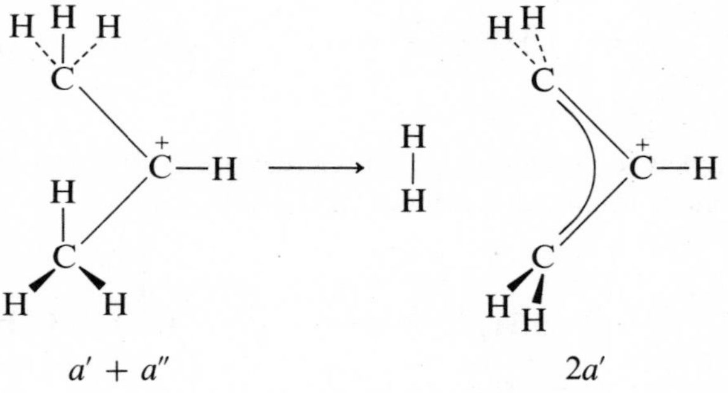

$$H_3C{-}\overset{+}{C}H{-}CH_3 \longrightarrow H{-}H + H_2C{\cdots}\overset{+}{C}H{\cdots}CH_2 \tag{122}$$

$a' + a''$ $2a'$

The stabilization energy of the allyl cation favors this reaction over 1,2-elimination. Similarly the dihydrotropylium ion, $C_7H_9^+$, forms the stabilized tropylium ion, $C_7H_7^+$. It should be remembered that rapid 1,2 hydrogen atom shifts occur in carbenium ions such as these; therefore, isotope labeling cannot be used as a diagnostic method of mechanism.

## REFERENCES

1. The behavior of excess vibrational energy in isolated excited molecules (dilute gas) is discussed by S. A. Rice, *Adv. Chem. Phys.*, **21**, 153 (1971).
2. J. Michl, *J. Am. Chem. Soc.*, **93**, 523 (1971); E. F. Ullman, *Acc. Chem. Res.*, **1**, 353 (1968).
3. C. A. Taylor, M. A. El-Bayoumi, and M. Kasha, *Proc. Natl. Acad. Sci. U.S.*, **63**, 253 (1969).
4. R. C. Dougherty, *J. Am. Chem. Soc.*, **93**, 7187 (1971); J. Michl, *Mol. Photochem.*, **4**, 243 (1972); see also H. E. Zimmerman, K. S. Kamm, and D. P. Werthemann, *J. Am. Chem. Soc.*, **96**, 7821 (1974) for earlier references.
5. This useful convention follows L. Salem, *J. Am. Chem. Soc.*, **96**, 3486 (1974); see W. G. Dauben, L. Salem, and N. J. Turro, *Acc. Chem. Res.*, **8**, 41 (1975), for a classification of photochemical reactions.
6. J. C. Dalton and N. J. Turro, *Ann. Rev. Phys. Chem.*, **21**, 499 (1970).
7. For a review, see A. Weller, *Disc. Faraday Soc.*, **27**, 28 (1959).
8. G. Jackson and G. Porter, *Proc. Roy. Soc. (Lond.)*, **A260**, 13 (1961).
9. E. Havinga, R. O. de Jongh, and W. Dorst, *Rec. Trav. Chim.*, **75**, 378 (1956); H. Somasekhara, *J. Am. Chem. Soc.*, **85**, 922 (1963).
10. M. Wrighton, *Chem. Rev.*, **74**, 401 (1974).
11. R. Daudel, *Adv. Quant. Chem.*, **5**, 1 (1970). This has a discussion of charge density effects in the reactions of excited states.
12. C. W. Bauschlicher, Jr., S. V. O'Neil, R. K. Preston, and H. F. Schaefer, III, *J. Chem. Phys.*, **59**, 1286 (1973).
13. L. Landau, *Phys. Z. Sowjetunion*, **1**, 88 (1932); **2**, 46 (1932); C. Zener, *Proc. Roy. Soc. (Lond.)*, **A137**, 696 (1932); **A140**, 666 (1933); E. G. C. Stueckelberg, *Helv. Phys. Acta*, **5**, 369 (1932).
14. For reviews, see J. Jortner, S. A. Rice, and R. M. Hochstrasser, *Adv. Photochem.*, **7**, 149 (1969); E. W. Schlag, S. Schneider, and S. Fischer, *Ann. Rev. Phys. Chem.*, **22**, 465 (1971).
15. For a clear discussion of spin–orbit coupling see S. P. McGlynn, L. G. Vanquickenborne, M. Kinoshita, and D. G. Carroll, *Introduction to Applied Quantum Chemistry*, Holt, New York, 1972, Chapter 11. The symmetries of spin functions can usually be ignored, except in the case of spin–orbit coupling. For a discussion see E. P. Wigner, *Group Theory*, Academic, New York, 1959.
16. T. Y. Chang and H. Basch, *Chem. Phys. Lett.*, **5**, 147 (1970).
17. T. W. Eder, R. W. Carr, Jr., and J. W. Koenst, *Chem. Phys. Lett.*, **3**, 520 (1969).
18. J. K. Beattie, N. Sutin, D. H. Turner, and G. W. Flynn, *J. Am. Chem. Soc.*, **95**, 2052 (1973).
19. L. Salem and J. S. Wright, *J. Am. Chem. Soc.*, **91**, 5947 (1969).
20. R. B. Woodward and R. Hoffmann, *The Conservation of Orbital Symmetry*, Academic, New York, 1970.
21. G. Herzberg, *Spectra and Structures of Simple Free Radicals*, Cornell U. Pr., Ithaca, 1971, p. 114.
22. W. T. A. M. van der Lugt and L. J. Oosterhoff, *Chem. Commun.*, 1235 (1968); *J. Am. Chem. Soc.*, **91**, 6043 (1969).

23. W. G. Dauben, *Reactivity of Photoexcited Organic Molecules*, Interscience, New York, 1967, p. 171.
24. K. Fukui, *Acc. Chem. Res.*, **4**, 57 (1971).
25. S. Boué and R. Srinivasan, *J. Am. Chem. Soc.*, **92**, 3226 (1970).
26. M. J. S. Dewar, *Angew. Chem. Int. Ed. Engl.*, **10**, 761 (1971).
27. W. G. Dauben, C. D. Poulter, and C. Suter, *J. Am. Chem. Soc.*, **92**, 7408 (1970).
28. For reviews, see reference 6 and P. J. Wagner, *Acc. Chem. Res.*, **4**, 168 (1971).
29. A. Devaquet, *J. Am. Chem. Soc.*, **94**, 5626 (1972).
30. J. C. D. Brand, *J. Chem. Soc.*, 858 (1956); G. W. Robinson and V. E. DiGiorgio, *Can. J. Chem.*, **36**, 31 (1958).
31. A. D. Walsh, *J. Chem. Soc.*, 2325 (1953).
32. R. G. Pearson, *Chem. Phys. Lett.*, **10**, 31 (1971).
33. D. W. Turner, C. Baker, A. D. Baker, and C. R. Brundle, *Molecular Photoelectron Spectroscopy*, Wiley-Interscience, New York, 1970.
34. G. W. King, *Spectroscopy and Molecular Structure*, Holt, New York, 1964, p. 429.
35. For experimental data see G. Herzberg, *Electronic Spectra of Polyatomic Molecules*, Van Nostrand, Princeton, N.J., 1967, Chapter 5.
36. R. McDiarmid, *Theor. Chim. Acta*, **20**, 282 (1971).
37. D. A. Ramsey, *Disc. Faraday Soc.*, **35**, 90 (1963).
38. R. Hoffmann, *Tetrahedron*, **22**, 521 (1966).
39. J. Langlet and J. P. Malrieux, *Theor. Chim. Acta*, **33**, 307 (1974).
40. P. Rademacher, P. Stalevik, and W. Lüttke, *Angew. Chem.*, **80**, 842 (1968).
41. J. E. Del Bene, *J. Am. Chem. Soc.*, **94**, 3713 (1972); H. Basch, *Theor. Chim. Acta*, **28**, 151 (1973).
42. A. M. Ponte Goncalves and C. A. Hutchison, Jr., *J. Chem. Phys.*, **49**, 4235 (1968).
43. H. M. McConnell and A. D. McLachlan, *J. Chem. Phys.*, **34**, 1 (1961).
44. For example, see J. P. Jesson, H. W. Kroto, and D. A. Ramsay, *J. Chem. Phys.*, **56**, 6257 (1972).
45. J. B. Berlman, *J. Phys. Chem.*, **74**, 3085 (1970).
46. The *magnum opus* for information on photochemical reactions is J. G. Calvert and J. N. Pitts, Jr., Wiley, New York, 1966, p. 524.
47. V. Balzani and V. Carassiti, *Photochemistry of Coordination Compounds*, Academic, New York, 1970, pp. 68 and 253.
48. H. Basch, G. Hollister, and J. W. Moskowitz, *Chem. Phys. Lett.*, **4**, 79 (1969).
49. R. A. Gangi and R. F. W. Bader, *J. Chem. Phys.*, **55**, 5369 (1971); *Chem. Phys. Lett.*, **6**, 312 (1970).
50. S. D. Thompson, D. G. Carroll, F. Watson, M. O'Donnell, and S. P. McGlynn, *J. Chem. Phys.*, **45**, 1367 (1966).
51. R. E. Rebbert, S. G. Lias, and P. Ausloos, *Chem. Phys. Lett.*, **12**, 323 (1971).
52. J. N. Murrell, J. B. Pedley, and S. Durmaz, *J. Chem. Soc., Faraday Trans.*, **II**, 1370 (1973).
53. S. Karplus and R. Bersohn, *J. Chem. Phys.*, **51**, 2040 (1969).
54. J. W. Raymonda and W. T. Simpson, *J. Chem. Phys.*, **47**, 430 (1967); E. F. Pearson and K. K. Innes, *J. Mol. Spectrosc.*, **30**, 232 (1969); R. Hoffmann, *Pure Appl. Chem.*, **24**, 567 (1970).

55. J. P. Lorquet, *Disc. Faraday Soc.*, **35**, 83 (1963); S. D. Peyerimhoff and R. J. Buenker, *J. Chem. Phys.*, **49**, 312 (1968).
56. B. G. Ramsey, *J. Organomet. Chem.*, **67**, C67 (1974).
57. From work based on SCF calculations, T. Nakajima, A. Toyota, and S. Fujii, private communication.
58. F. D. Lewis, R. W. Johnson, and D. E. Johnson, *J. Am. Chem. Soc.*, **96**, 6090 (1974).
59. K. Evans and S. A. Rice, *Chem. Phys. Lett.*, **14**, 8 (1972).
60. A. J. Merer and R. S. Mulliken, *Chem. Rev.*, **63**, 639 (1969).
61. J. Saltiel et al., in O. L. Chapman (ed.), *Organic Photochemistry*, Dekker, New York, 1973, Vol. 3, p. 1. This has an extensive review of photochemical *cis–trans* isomerization of olefins.
62. S. Malkin and E. Fischer, *J. Phys. Chem.*, **68**, 1153 (1964).
63. P. Borrell and H. H. Greenwood, *Proc. Roy. Soc. (Lond.)*, **A298**, 453 (1967).
64. J. C. D. Brand, D. R. Williams, and T. J. Cook, *J. Mol. Spectrosc.*, **20**, 359 (1966).
65. D. R. Arnold and V. Y. Abraitys, *Mol. Photochem.*, **2**, 27 (1970).
66. H. Yamasaki and R. J. Cvetanović, *J. Am. Chem. Soc.*, **91**, 520 (1969).
67. R. Srinivasan, *J. Am. Chem. Soc.*, **91**, 7557 (1969).
68. N. D. Epiotis, *J. Am. Chem. Soc.*, **94**, 1941 (1972).
69. B. D. Kramer and P. D. Bartlett, *J. Am. Chem. Soc.*, **94**, 3934 (1972).
70. W. G. Dauben and R. L. Cargill, *Tetrahedron*, **15**, 197 (1961).
71. J. A. Berson and S. S. Olin, *J. Am. Chem. Soc.*, **92**, 1086 (1970).
72. K. N. Houk and D. J. Northington, *J. Am. Chem. Soc.*, **93**, 6694 (1971).
73. See reference 46, p. 504.
74. R. A. Caldwell, G. W. Sovocool, and R. P. Gajewski, *J. Am. Chem. Soc.*, **95**, 2549 (1973); F. D. Lewis, C. S. Hoyle, and D. E. Johnson, ibid., **97**, 3267 (1975).
75. N. C. Yang, J. J. Cohen, and A. Shani, *J. Am. Chem. Soc.*, **90**, 3264 (1968).
76. For a review of the photochemistry of aliphatic azo compounds in solution, see P. S. Engel and C. Steel, *Acc. Chem. Res.*, **6**, 275 (1973).
77. (a) See reference 35, p. 555; (b) C. A. Sandorfy, *Electronic Spectra and Quantum Chemistry*, Prentice-Hall, Englewood Cliffs, N.J., 1964, p. 204.
78. (a) For fulvene, see D. Bryce-Smith, *Pure Appl. Chem.*, **16**, 47 (1968); (b) for Dewar benzene, see E. E. Van Tamelen and S. P. Pappas, *J. Am. Chem. Soc.*, **84**, 3789 (1962); (c) for prismane and benzvalene, see K. E. Wilzbach and L. Kaplan, *J. Am. Chem. Soc.*, **87**, 4004 (1965).
79. I. Haller, *J. Chem. Phys.*, **47**, 1117 (1967). The $B_1$ and $B_2$ states are interchanged due to the choice of axes.
80. D. M. Lemal and J. P. Lokensgard, *J. Am. Chem. Soc.*, **88**, 5934 (1966).
81. D. Bryce-Smith and H. C. Longuet-Higgins, *Chem. Commun.*, 593 (1966).
82. (a) K. E. Wilzbach and L. Kaplan, *J. Am. Chem. Soc.*, **88**, 2066 (1966); (b) ibid., **93**, 2073 (1971).
83. D. Bryce-Smith, A. Gilbert, B. Orger, and H. Tyrrell, *J. Chem. Soc. Chem. Commun.*, 334 (1974).
84. A. Morikawa, S. Brownstein, and R. J. Cvetanovic, *J. Am. Chem. Soc.*, **92**, 1471 (1970).
85. D. Bryce-Smith, *Chem. Commun.*, 806 (1969). This author has worked out the symmetry rules for all cycloadditions of olefins and dienes to benzene.

86. W. Ferree, Jr., J. B. Grutzner, and H. Morrison, *J. Am. Chem. Soc.*, **93**, 5502 (1971); J. Cornelisse, V. Y. Merritt, and R. Srinivasan, ibid., **95**, 6197 (1973).

87. G. Koltzenburg and K. Kraft, *Angew. Chem. Int. Ed. Engl.*, **4**, 981 (1965); K. Kraft and G. Koltzenburg, *Tetrahedron Lett.*, **44**, 4357, 4723 (1967).

88. P. Lechtken, R. Breslow, A. H. Schmidt, and N. J. Turro, *J. Am. Chem. Soc.*, **95**, 3025 (1973).

89. N. J. Turro, P. Lechtken, N. E. Schore, H. C. Steinmetzer, and A. Yetka, *Acc. Chem. Res.*, **7**, 97 (1974).

90. D. R. Kearns, *Chem. Rev.*, **71**, 395 (1971); D. R. Roberts, *J. Chem. Soc., Chem. Commun.*, 683 (1974).

91. N. J. Turro and P. Lechtken, *J. Am. Chem. Soc.*, **95**, 264 (1973).

92. R. D. McQuigg and J. G. Calvert, *J. Am. Chem. Soc.*, **91**, 1590 (1969).

93. W. H. Fink, *J. Am. Chem. Soc.*, **94**, 1079 (1972).

94. E. S. Yeung and C. B. Moore, *J. Chem. Phys.*, **58**, 3988 (1973).

95. R. L. Jaffe, D. M. Hayes, and K. Morukama, *J. Chem. Phys.*, **60**, 5108 (1974).

96. W. H. Fink, *J. Am. Chem. Soc.*, **94**, 1073 (1972); D. M. Hayes and K. Morukama, *Chem. Phys. Lett.*, **12**, 539 (1972).

97. T. Gillbro and F. Williams, *Chem. Phys. Lett.*, **20**, 436 (1973).

98. For general reviews, see reference 47; A. W. Adamson, W. L. Waltz, E. Zinato, D. W. Watts, P. D. Fleischauer, and R. D. Lindholm, *Chem. Rev.*, **68**, 541 (1968); P. C. Ford, J. D. Petersen, and R. E. Hintze, *Coord. Chem. Rev.*, **14**, 67 (1974).

99. W. Strohmeier, *Angew. Chem.*, **76**, 873 (1964).

100. A. D. Kirk, *J. Am. Chem. Soc.*, **93**, 283 (1971).

101. J. Zink, *J. Am. Chem. Soc.*, **94**, 8039 (1972); *Inorg. Chem.*, **12**, 1018, 1957 (1973); *J. Am. Chem. Soc.*, **96**, 4464 (1974); M. Wrighton, H. B. Gray, and G. S. Hammond, *Mol. Photochem.*, **5**, 165 (1973).

102. A. W. Adamson, *J. Phys. Chem.*, **71**, 798 (1967).

103. H. L. Schläfer, *J. Phys. Chem.*, **69**, 2201 (1965).

104. C. J. Ballhausen, *Introduction to Ligand Field Theory*, McGraw-Hill, New York, 1962.

105. S. C. Pyke and R. G. Linck, *J. Am. Chem. Soc.*, **93**, 5281 (1971).

106. M. Wrighton, *Inorg. Chem.*, **13**, 905 (1974).

107. R. G. Pearson, *Inorg. Chem.*, **12**, 712 (1973).

108. P. Natarajan and A. W. Adamson, *J. Am. Chem. Soc.*, **93**, 5599 (1971).

109. T. H. Whitesides, *J. Am. Chem. Soc.*, **91**, 2395 (1969).

110. M. Wrighton, G. S. Hammond, and H. B. Gray, *J. Am. Chem. Soc.*, **92**, 6068 (1970).

111. J. L. Reed, F. Wang, and F. Basolo, *J. Am. Chem. Soc.*, **94**, 7173 (1972); H. D. Gafney, J. L. Reed, and F. Basolo, ibid., **95**, 7998 (1973).

112. I. Burak and A. Treinin, *J. Am. Chem. Soc.*, **87**, 4031 (1965).

113. NBS Technical Note 270-3, National Bureau of Standards, Washington, D.C., 1968.

114. S. B. Jaffe and J. B. Anderson, *J. Chem. Phys.*, **51**, 1058 (1969).

115. J. B. Anderson, *J. Chem. Phys.*, **61**, 3390 (1974).

116. For a discussion of these methods and some of the results see articles by: (a) J. L. Kinsey, (b) T. Carrington and J. C. Polanyi, and (c) J. Dubrin and M. J. Henchman, *in Chemical Kinetics, MTP International Review of Science*, Butterworths, London, 1972, Vol. 9;

also articles by: (a) J. P. Toennies and (b) J. C. Polanyi and J. L. Schreiber *in Physical Chemistry, an Advanced Treatise*, Academic, New York, 1974, Vol. VIA.

117. D. H. Maylotte, J. C. Polanyi, and K. B. Woodall, *J. Chem. Phys.*, **57**, 1547 (1972).
118. (a) R. D. Levine and R. B. Bernstein, *Acc. Chem. Res.*, **7**, 393 (1974); (b) R. D. Levine, *Proc. Int. Symp. Chem. Biochem. Reacti.*, Academy of Science, Jerusalem, 1973, p. 317.
119. H. Eyring and M. Polanyi, *Z. phys. Chem.*, **12**, 279 (1931); M. G. Evans and M. Polanyi, *Trans. Faraday Soc.*, **35**, 178 (1939).
120. D. H. Williams and G. Hvistendahl, *J. Am. Chem. Soc.*, **96**, 6753, 6755 (1974).
121. S. H. Bauer, *J. Am. Chem. Soc.*, **91**, 3688 (1969).
122. For a theoretical discussion, see E. E. Nikitin, *in Physical Chemistry, an Advanced Treatise*, Academic, New York, 1974, Vol. VI A, p. 187; for a review see E. Weitz and G. Flynn, *Ann. Rev. Phys. Chem.*, **25**, 275 (1974).
123. See P. J. Robinson and K. A. Holbrook, *Unimolecular Reactions*, Wiley-Interscience, New York, 1972, for a discussion of the theory.
124. R. G. Cooks, J. H. Beynon, R. M. Caprioli, and G. R. Lester, *Metastable Ions*, Elsevier, Amsterdam, 1973.
125. J. H. Beynon, R. A. Saunders, and A. E. Williams, *Z. Naturforsch.*, **20a**, 180 (1965).
126. R. C. Dougherty, *J. Am. Chem. Soc.*, **90**, 5780, 5788 (1968).
127. See reference 124, p. 107.
128. K. C. Smyth and T. W. Shannon, *J. Chem. Phys.*, **51**, 4633 (1969).
129. (a) See reference 124, p. 190; (b) ibid., p. 199.
130. D. H. Williams and G. Hvistendahl, *J. Am. Chem. Soc.*, **96**, 6755 (1974).

# AUTHOR INDEX

# SUBJECT INDEX